D0777535

William Philpott

THREE ARMIES ON THE SOMME

William Philpott is Professor of the History of Warfare in the Department of War Studies at King's College, London. He is a specialist in the operations on the Western Front and has published extensively on the subject. He lives in London.

THREE ARMIES
ON THE SOMME

THREE ARMIES ON THE SOMME

The First Battle of the Twentieth Century

William Philpott

VINTAGE BOOKS
A DIVISION OF RANDOM HOUSE, INC.
NEW YORK

Dedicated to the memory of my grandfather
Sergeant James Erswell Philpott MM
and his daughters,
Joan Ella and Jean Gladys

Contents

Contents

Maps

Acknowledgements

I would like to thank the following individuals and institutions for permission to quote and cite material in their possession, or for which they hold the copyright: Her Majesty the Queen and the Controller of Her Majesty's Stationery Office; the Trustees of the National Library of Scotland; the Trustees of the Liddell Hart Centre for Military Archives, King's College London; the Trustees of the Imperial War Museum; the Master and Fellows of Churchill College, Cambridge; the Bodleian Library, Oxford; the Bonham Carter Trustees; the Councillors of the Army Records Society; the Cheshire Military Museum; the National Library of Australia, Canberra; the Australian War Memorial, Canberra; the State Library of New South Wales, Sydney; Archives and Special Collections, Queen Elizabeth II Library, Memorial University, Newfoundland; the Archives de l'armée de terre, Service historique de la défense, Vincennes; the Archives départementales de la Somme, Amiens; the Historial de la Grande Guerre, Péronne; Earl Haig; Lord Esher; Mr. M. A. F. Rawlinson. Finally I would like to thank Pen and Sword publishers for permission to quote from Jack Sheldon's excellent anthology of German combatants' writings, *The German Army on the Somme, 1914–1916*.

I also owe many personal and professional debts to those who have provided practical assistance and support during the writing of this book. Its preparation and publication have been greatly assisted by many individuals. Of these I would in particular like to thank Charlie Viney, my agent; Andrew Miller, my editor at Knopf; Maria Massey, production editor at Knopf; John Gilkes, who prepared the maps; and Philip Parr, who meticulously copyedited the manuscript.

Research for this book in France was supported by a small personal research grant from the British Academy. At the Archives de l'armée de terre, Vincennes, Colonel Frédéric Guelton and Lieutenant-Colonel Rémy Porte and their colleagues provided invaluable assistance and guidance. Laurent Henninger and his colleagues at the Centre d'études d'histoire de la défense offered a warm reception and valuable facilities while I was in Vincennes. Christopher Goscha provided a comfortable home away from home in Vincennes. The Australian government's Bicentennial Fellowships fund supported research in the Australian War Memorial, Canberra. Dr. Peter Stanley and his colleagues at the Memorial's research centre were welcoming and hospitable to the Pom who appeared in their midst. Professor Carl Bridge of the Menzies Centre for Australian Studies, King's College London, offered help and advice before, during and after that fellowship. The Arts and Humanities Research Council funded a term of research leave to enable me to finish the manuscript.

Over the years my research students, as well as undergraduate and postgraduate classes, in the Department of War Studies, King's College London, have been a constant source of stimulation and inspiration in understanding the complexities of the First World War. In particular the members of the "Coal Hole Club" military operations study group have provided a lively and sociable forum in which to refine my understanding of the dynamics of the Great War battlefield. They will see where their insights, sadly unattributable, have informed and improved this work. Similarly, my colleagues in the Department of War Studies have been helpful and supportive during the writing of this book. While it is invidious to single out individuals, nevertheless my successive Heads of Department, Professor Brian Holden Reid and Professor Mervyn Frost, must be thanked for supporting my periods of research leave without which this book might never have been finished. Similarly, Professor James Gow and Professor Theo Farrell gave valuable advice on my applications for funding which assisted the writing of this book. Over the years the Institute of Historical Research's Military History research seminar has been a lively forum for debate, and I thank Professor Brian Bond and its members for their friendship, and for hearing me out and offering professional insights on the occasions when I spoke on aspects of my ongoing research. The membership of the British Commission for Military History have also been stimulating, interested and patient in equal measure over the many years in which I have been obsessing about the Somme, to the point of letting me guide them over the less remembered corners of the battlefield. Professors Hew Strachan, Martin Alexander, David French and John

Gooch have always been wise friends, as well as strong supporters of this research project.

Family and friends too, if not possessing the expertise of colleagues and students, still demonstrated interest and patience while this lengthy project was completed. Many have been helpful or supportive along the way; a few deserve individual mention. Above all, my parents, my sisters and their families (who I will not name, since they know who they are); Professor Michael Neiberg, who read and provided insightful comments on the manuscript in draft; Sophy Kershaw, who read the manuscript in draft with a professional and personal eye, and was always supportive while the book rolled ever onwards (and remembering her grandfather Second-Lieutenant Raymond Kershaw MC, 2nd Battalion Australian Machine Gun Corps, wounded in the Battle of Hamel, 4 July 1918); Richard Kershaw and Jann Parry, who shared an author's ups and downs; the late Venetia Murray, who gave valuable advice on the ins and outs of the publishing trade when this project was in its infancy; Matthew and Carol Cragoe, my writing companions and hospitable friends; Rachel Mustalish, who looked after me in New York; Sian Evans, who kept me company in Paris; Jeremy Hughes, for his research assistance; Genevieve Ford-Saville, for our musings over one too many pints; Charlie, Sarah, Eleanor and Hebe Robinson, Kate and John Madin, welcoming friends always.

And finally, thank you to Elizabeth Greenhalgh: without her provocation, this book might never have been written.

THREE ARMIES
ON THE SOMME

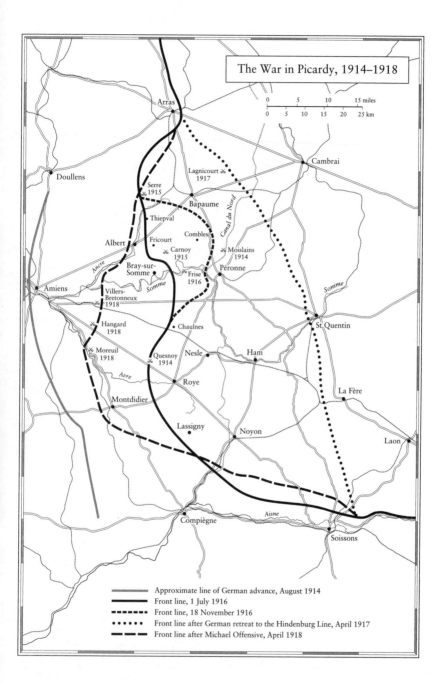

The War in Picardy, 1914–1918

Engagement

O N 1 July 1916, at precisely 07:28, a series of huge explosions tore open the ground around the small French hamlet of La Boisselle. A company of the 10th Battalion the Lincolnshire Regiment (the "Grimsby Chums") dashed from the British front line, outpacing men from the German 110 *Reserveinfanterieregiment* (RIR) to seize the newly blown Lochnagar mine crater. All along the front the deadly race was repeated, as thousands of British and French infantry rose in a long surging line, and German machine-gunners hastened to their weapons. The winners might live to see a different world.

"The Last Post" is sounded at Lochnagar Crater every 1 July, although nowadays there are no longer veterans among the mourners. Nevertheless they will not fade away. Dutifully each year their children, and after them their grandchildren and great-grandchildren, will gather at that sacred shell-hole to honour the memory of those who fought—and those who died—in Britain's greatest battle. Elsewhere small parties of French and Germans, Australians, Canadians and South Africans, pay their respects, as the Battles of the Somme pass from experience, through memory, into history.

The river Somme, which meanders across north-west France, bisecting the rolling uplands of Picardy, has lent its name to four battles. The first encounter was in September and October 1914. The second and greatest—the Battle of the Somme—lasted for four and a half months, from June to November 1916. The third took place in March 1918; the fourth in August 1918. Even today, the mention of this quiet river evokes deep atavistic sentiments: pride, anger, anguish, grief. The 1916 battle scarred the British national psyche in particular: up to that point Britain had had no Verdun, the "Meuse mill" which ground up the French army in early 1916, or Langemarck, the German army's *Kindermord* sacrifice of her eager young volunteers in autumn 1914. Yet the reasons for that scar, and the true nature of the historical phenomenon which inflicted it, are little understood. The 1916 Somme campaign was a struggle of three armies, and three empires. Each left its young men and their memorials in that patch of Picardy downland, now lush, quiet and mournful. On the hills opposite Albert, the British army made its longest and greatest sacrifice in the allied cause. Astride the river to the south their French allies were fighting their own gruelling battle, closely integrated with British operations but all but forgotten by posterity. To the east the German army fought its prolonged defensive Battle of the Somme, more bloody and morale-sapping than the Anglo-French offensive, yet rarely acknowledged or recounted.

In the summer of 1916 Walter Page, the American ambassador in London, noted that "war has come to be the normal state of life."[1] The world had paused as three great empires, championed by their armies, staked their futures in a single great battle. The scale of the Somme was immense, a global event impacting on the lives of everyone in Western Europe and resonating beyond European shores. Millions of Frenchmen, Germans and Britons, and many thousands of colonial volunteers and conscripts—Australians, New Zealanders, Canadians, South Africans, Moroccans, Algerians and West Africans—converged on that corner of a foreign field from twenty-five nations, all five continents.[2] It was no small corner: from north to south the battlefield stretches some forty miles, from east to west twenty: as large as London south of the Thames. Most went willingly, believing in the cause: the liberation of the invaded *patrie*; the defence or aggrandisement of a besieged empire; the crushing of rampant militarism; the forging of a national identity. Above all, honour and duty—loyalty to one's country, to one's army, to one's comrades—motivated the combatants. Three imperial superpowers found themselves enmeshed in a new kind of struggle, to the death: "a total war for the preservation of the nation" as General Ludendorff was to call it. The centrepiece of this industrial war was an industrial battle in which population, economy, industry and imagination all strained to sustain a battle longer and more intense than any fought before. Back at home, anxious families waited for news; a letter from a son, brother, husband or father, or the much-feared official telegram. Munitions workers toiled to feed the guns and farmers harvested to feed the men. Journalists reported, writers commented, poets rhymed and artists painted. Having committed their armies and nations to the field, political leaders of the three struggling empires were rapt, their credibility and authority on trial. Their rivals, of left and right, sought advantage from each turn of events. Allies watched their brothers-in-arms with interest and anxiety, their own nation's cause inextricably tied up with that being disputed in Picardy. Neutrals monitored world-changing events, judging and rejudging their position in the global struggle. Would the Battle of the Somme end the war? If so, what would be the peace? What if the war continued? Armageddon and Judgement Day seemed one.

As the summer drew on casualty lists lengthened; all, it appeared, for a few square miles of insignificant French countryside. The combatants knew that much more was at stake than mere ground and human life: honour; reputation; culture; tradition; the future. It must be victory, or extinction. Fighting for one meant change for all, as on the Somme the old familiar world perished and a new uncertain world emerged. As German veteran Ernst Jünger attested: "Chivalry here took a final farewell . . . The Europe of today appeared here for the first time on the field of battle."[3]

It is this 1916 battle which has left its mark on our collective psyche, but the Somme experience reaches much further. As well as the several million men who actually fought on the Somme—not just in 1916, but in 1914, 1915, 1917, and 1918—millions more then and later lived their experience vicariously, through newspapers and films, poetry and paintings, memoirs, plays, novels and histories. Many thousands took their experience with them to the neatly tended graves or unmarked plots they occupy to this day. Others survived yet left no record. Many, however, lived to tell their tales; soldiers' tales. In the war-of-words which follows any battle, official despatches from the commanders-in-chief sought to explain and justify their actions. Diaries, memoirs and letters of those who had fought proliferated. A few became celebrated classics: most are long-forgotten.

One memoir in particular constructed the Somme of popular memory. Winston Churchill, who lived by the pen as well as the sword, found the former an effective weapon in the so-called battle of the memoirs which followed the war. Although Churchill's career subsequently prospered, when writing *The World Crisis*, his memoir-history of the First World War, the former First Lord of the Admiralty was defending his own far from unblemished war record.[4] He shared responsibility for the ill-planned and badly executed Dardanelles campaign, which forced his resignation from the Cabinet in November

1915, after which he spent eighteen months out of office, in the trenches and on the backbenches. In 1916 he had been restless and critical, observing a war which apparently was not being won and looking for reasons and scapegoats, and his egocentric memoir revisited these wartime gripes. He had the unique perspective of someone who had both served at the front and moved in the highest political circles. Yet, like many of his contemporaries, he was too caught up in events to appreciate the bigger picture; too preoccupied with his own part in affairs to consider the roles of others. As he wrote to his wife while he worked on the first volume of his history: "It is a great chance to put my case in an agreeable form to an attentive audience."[5]

Published in 1927, Churchill's account of the Battle of the Somme in *The World Crisis* was to manufacture one of Britain's great historical myths. Ironically, the man whose account of the Somme has proved most influential had little to do with it. Churchill enjoyed the luxury of dissociation: he was not at the front and he had no ministerial responsibility. A troublemaking and under-employed backbench MP in 1916, he watched and judged. His critique of the genesis and conduct of the battle is founded on his own strongly expressed opinion, at the end of the battle's first month, that the British army's casualties were significantly higher than the enemy's.[6] Then his objection was ignored: the offensive was just getting into its stride. Yet Churchill remained too involved ever to rethink or revise this highly specific contemporary critique, which in *The World Crisis* is subsumed into a general narrative of what were to become familiar clichés, the product of a self-absorbed refusal to investigate the bigger picture: unimaginative and callous generals; ill-planned and futile offensive operations; high and unnecessary casualties; atrocious battlefield conditions; technophobic cavalrymen failing to appreciate the potential of new war-winning weapons, notably the tank. Churchill thus set the agenda for subsequent generations' perception of the Battle of the Somme, and the war of which it was the defining and pivotal event.

Written in Churchill's lucid and compelling prose, they proved popular accusations. The first printing of volume one of *The World Crisis* instantly sold ten thousand copies; the whole six-volume series (the third volume of which covered the Somme) sold more than a million copies in total.[7] Serialisation in *The Times* disseminated Churchill's message among the general public, and syndication in Europe and the empire reached the wider world. Cheaper, abridged, popular editions of *The World Crisis,* first published in the 1930s, reprinted in the 1960s on the fiftieth anniversary of the war, and most recently in 2007, kept Churchill's impression current. Here was a soldier, statesman and wordsmith who apparently could make sense of a conflict which defied understanding: who could explain the reasons why so many had died, and seemed to know who should be held responsible. Yet even on publication Churchill's words provoked concern, and they have stood up poorly to subsequent scholarly analysis. "The very attractiveness of Mr. Churchill's writing of itself constitutes a danger; for the layman may well be led to accept facile phrase and seductive argument for hard fact and sober reasoning," Lord Sydenham commented in introducing a collection of early criticisms of *The World Crisis,*[8] and in a later scholarly critique Robin Prior concluded that Churchill's best-selling account of the war was "to shape the popular historical memory of the recent past in its own image."[9] Indeed, Churchill's widely read nostrums and cavils have become the familiar staples of First World War literature, recycled again and again to a credulous public. Perhaps they reach their apogee in the influential work of A. J. P. Taylor, the pacifist historian whose scathing narrative of the war has dominated popular consciousness for over forty years.[10]

Churchill did not understand the First World War, or the central place of the Somme within it. As Gary Sheffield recently commented, Churchill's analysis of the Somme "combines a blithe disregard for what was possible in 1916 with an astonishing lack of understanding of the realities of combat on the Western Front."[11] Churchill showed no appreciation of the military skill, strength of purpose and moral courage required to fight and win such a battle and such a war: a battle and war of attrition. Although he could explain his own role and attitude well enough, as Sydenham recognised, "in accordance with the views he consistently holds," he was unable to admit that the Somme, a key element of a coherent whole, in fact came close to success.[12] More apposite was Lord Esher's post-war verdict, that "the battle of the Somme settled the inevitable issue of the War."[13] The British Empire committed its keen volunteer citizen armies to this three-nation encounter. France contributed the military expertise of her much-battered army. The German army offered resistance almost to the end of its resources and moral strength. Three empires engaged with and suffered through this blood-sacrifice, and its long, tragic denouement. It is wrong and morally reprehensible to dismiss this human phenomenon on the grandest scale as a futile engagement in a futile conflict, as so many do.

· · ·

To appreciate the true nature of the Somme the battle must be viewed through the eyes of those who fought and observed. Even before the guns had fallen silent Britain's official battle chronicler and future poet laureate John Masefield had begun work on *The Old Front Line,* the first of many guides to the battlefield "from which the driving back of the enemy began." As he surveyed the British front of 1 July 1916 he felt that the war and the world were turning:

> The old front line was the base from which the battle proceeded. It was the starting-place. The thing began there. It was the biggest battle in which our people were ever engaged, and so far it has led to bigger results than any battle of this war since the battle of the Marne. It caused a great falling back of the enemy armies. It freed a great tract of France, seventy miles long, from ten to twenty-five miles broad. It first gave the enemy the knowledge that he was beaten.[14]

Some sixty years later, as children and grandchildren rediscovered their forefathers' fire-trench or resting-place along the old front line, Colonel Howard Green made clear, in the introduction to the 1972 reprint of Masefield's work, how the world had indeed turned: "For the men who fought there, and the nations they fought for, it was never the same after the Somme. The songs, the writing, the poetry, became cynical, after the Somme."[15] Such war-induced cynicism has masked the true character of the Somme ever since.

As time moved on and events receded the weight of history accumulated. Historians assessed and judged these writings and recollections. Generals' reputations were made, lost and made again, and ordinary soldiers' stoic heroism gleamed ever more brightly from the greying mud. Military histories—official, scholarly, popular or sensational—sought to recount and critique the battle. One former subaltern in particular, Basil Liddell Hart, made his judgement on history and imposed his opinion on posterity. The most influential British military writer of the inter-war years had fought on the Somme before gas had cut short his front-line service. While recuperating he drafted a eulogy for the high command. Later, invalided out of the army and disillusioned with its superior officers, he turned his productive pen to a critique of the army's failings in the Great War, and of Britain's conti-

nental strategy in general, as a salutary example for the future. In particular, his history, *The Real War,* laid out many of the charges for the
trial-by-literature of the British commander-in-chief, Sir Douglas Haig,
and his subordinates, which continues to this day.[16] In this tone, Liddell
Hart set the agenda for subsequent scholarship. As well as such historical assessments, the Battle of the Somme left a rich artistic legacy. Serving poets, official war artists, demobbed novelists, unborn writers wove
a literary and cultural memory that obscured as much as illuminated the
battle and the war of which it was a part.[17] Gradually, inexorably, the
battle passed from personal memory to popular myth, both a paradigm
and a cliché of the Great War.

Why should this morass be churned up once again? Despite the
Somme's notoriety, we have only a partial picture of the battle, and a
rudimentary understanding of its true nature, impact and legacy. Taking
their cue from Churchill and Liddell Hart, English-language accounts
focus narrowly on the military operations themselves, and in particular
on key events and themes: the disastrous first day; the deficiencies of
high command; the bravery of the ordinary soldier; the use or misuse of
the tank. Fundamentally they dwell on the British Empire's experience,
and the role of ally and enemy alike receive scant recognition. In the
clutch of new (if hardly novel) battle narratives which appeared to mark
the ninetieth anniversary of the Somme, the prevailing paradigm went
unchallenged: even the most scrupulous accepted that "as a result of
France's grievous loss at Verdun . . . the predominant force undertaking
the offensive would now be the British army (with the French as a
lesser contributor), and the commander chiefly responsible for directing the operation would be not Joffre but Haig."[18] Nowadays Britain's
Somme is remembered as a national tragedy: "the glory and the graveyard of Kitchener's Army" as Liddell Hart dubbed his massacre of the
innocents.[19]

Yet the Somme can feign none of the heroic tragedy of Gallipoli or
Dunkirk, for the Somme was a victory, if an unappreciated one. As the
old battleground was trodden once again for its ninetieth anniversary,
the long-standing controversies showed no signs of abating: "Raw
emotive sentiments and folk myths [vie] with the academic assessments
of military historians," Peter Hart acknowledged. "There is no doubt
that the Somme *was* a tragedy and the massed slaughter and endless suffering it epitomises cannot simply be brushed aside by the justification
of cold blooded military necessity."[20] Is the root cause of this incon-
gruity that this victory was won at too high a cost? In an industrial war

just such an attritional campaign would have to be fought, and won, at some point and at some time. Casualties were expected to be high in the largest single battle fought by British arms in any war to date. The most quoted (and misquoted) statistic of British military history is that of the casualties of the first day of the Somme. There were 57,470 casualties of whom 19,240 were killed or died of wounds:[21] "the greatest loss and slaughter sustained in a single day in the whole history of the British Army" Churchill bluntly recorded.[22] They were excessively high on this particular day because an inexperienced and partially trained force went "over the top" with deficient artillery support and overambitious objectives. From this first-day setback the army learned a harsh but valuable lesson. Over the ensuing 140 days of battle tactical methods improved and losses diminished in proportion. In comparison with the ferocious single-day engagements of pre-industrial wars, where 30 per cent of an army might be left killed and maimed on the field, the Somme is unremarkable.[23] After the first day it was just another battle, albeit a continuous one: a campaign, in fact; the longest and the costliest of the war, with twelve major component engagements designated by Britain's official battlefield nomenclature committee, and innumerable smaller skirmishes. Moreover, as the French historian Pierre Miquel has acknowledged, this unprecedented loss of life was a "consensual sacrifice,"[24] something very hard to comprehend ninety years later in a casualty-averse age, in which individual battlefield losses are newsworthy events. In *Tender Is the Night* F. Scott Fitzgerald had his characters visit the site of their great battle, as so many others would do in the 1920s and 1930s. "See that little stream," one remarked:

> We could walk to it in two minutes. It took the British a month to walk to it—a whole empire walking very slowly, dying in front and pushing forward behind. And another empire walked very slowly backwards a few inches a day, leaving the dead like a million bloody rugs . . . This took religion and years of plenty and tremendous sureties and the exact relation that existed between the classes.[25]

In the second decade of the twentieth century nations accepted, understood and even welcomed such a blood sacrifice, for a cause and to a greater, noble, purpose.

The Somme was not simply a British campaign. Australia, New Zealand, South Africa, Canada and Newfoundland all left their young men and their memorials on the battlefield, stone testaments to imperial cohesion standing alongside English, Welsh, Scottish and Irish monuments. Committed after all hope of surprise had gone, the Newfoundland Regiment was cut to pieces at Beaumont Hamel on 1 July. The South African Brigade was destroyed in holding Delville Wood against repeated enemy counter-attacks. In six weeks at Pozières the Australian Imperial Force sustained 23,000 casualties, as many as in their eight-month stay at Gallipoli.[26] Such heroic sacrifices forged national myths.

Two other armies fought on the Somme. To understand the battle it is essential to consider the French and German roles and experiences alongside those of the British. The Somme was an Anglo-French offensive, carried out by French as well as British troops in roughly equal numbers, directed overall by Frenchmen, and one element of a broader allied grand strategy in which Russians and Italians also played their parts. This goes some way to explaining the location and duration of the battle. French casualties number around two hundred thousand: nearly half of the British figure, and deserving of more than a paragraph or two in the history books. Coming on top of those at Verdun, they were hard to accept with so little obvious reward, and consequently this further sacrifice is now all but forgotten in France, subsumed in the greater hecatomb of 1916. For France the Somme was not a novelty, merely a disappointment. The offensive was conceived as a deliverance: from the bloodletting that France had endured at Verdun since February; from nearly two years of German occupation of her northern *départements*. In planning and preparing the 1916 campaign General Joseph Joffre, France's commander-in-chief and de facto allied generalissimo, had envisaged a massive assault which would rupture the German trench line and drive the enemy from occupied France. Politicians and public were promised decisive victory after two years of sacrifice. It was a promise that Joffre could not keep. As the 1916 campaign developed, he came to realise that such a decision was impossible given the military circumstances of the Western Front in 1916. His field commanders, notably General Ferdinand Foch, who was charged with preparing the summer offensive, told him so. His manpower reserves, eroded by the battle at Verdun, were insufficient. *Matériel* and munitions of war were inadequate for an operation on such a scale. The inexperienced British army would have to be relied upon to bear the lion's share of the fighting. Yet national and international imperatives meant that the offensive must still go ahead. Verdun could be relieved, and enemy reserves could be kept from allied fronts. Above all, France's glorious army, tied down at Verdun, would once more be on the offensive.

When the French attacked to the south of the British front on 1 July 1916 they experienced a brief epiphany. With heavier artillery support and more sophisticated infantry tactics than their allies, they advanced two and a half miles and took six thousand prisoners in the first assault, suffering relatively light casualties. Yet this tactical fillip was far short of the strategic decision on which Joffre had staked his reputation. With British failure to the north to be redeemed, the French army's contribution to the offensive grew. Over ensuing months the pattern of piecemeal attacks on limited sectors of the front, occasionally punctuated by large-scale assaults, came to symbolise the Battle of the Somme. Although the French army was moving inexorably forward, literally grinding the German army into the mud, the snail's pace, and the lack of any obvious sign of victory—a town captured or a river crossed— sapped the morale of the troops and the will of the nation. The Somme seemed to be becoming another Verdun, which war-weary France could ill-afford. Redemption reverted to purgatory. Joffre and Foch were singled out for blame. Both lost their field commands at the end of the battle. France would win through the politico-military crisis which engulfed the nation in 1917 and, led by a rehabilitated Foch, drive the enemy from her soil the year after. Yet the fall-out from the Somme scarred the French psyche and divided the French public. The Third Republic lived on borrowed time until the next German invasion.

Ironically for France, the Somme had served its purpose, although this was not immediately evident. The German army's Somme, was a gruelling, morale-sapping experience, "the muddy grave of the German field army" in one participant's oft-quoted judgement.[27] At the end of 1915 the German commander-in-chief, General Erich von Falkenhayn, had hit upon pure attrition as the means to bring France and her allies to the negotiating table. The following summer on the Somme, Falkenhayn was to be hoist by his own strategic petard. Hundreds of thousands of men were to be lost defending this nondescript bend in the river. Added to the mounting losses at Verdun, by the end of the year the German army had suffered over one million irreplaceable casualties on the Western Front. For Germany, with her armies everywhere entrenched on enemy territory and making further conquests in the East in the summer of 1916, the Somme was a rude awakening. Modern industrial war was not about controlling territory, but involved deploying machines to kill men. Falkenhayn had expected to contain the allied offensive, as he had done so often in 1915. Indeed he did. But as Ernst Jünger later recorded, on the Somme, Germany's soldiers found themselves engaged in a new sort of combat: "What we had . . . been through had been the attempt to win the war by old-fashioned pitched battles, and the stalemating of the attempt in static warfare. What confronted us now was a war of *matériel* of the most gigantic proportions."[28] In this relentless *Materialschlacht* on the Somme, Germany's once proud and victorious divisions met their Nemesis—a tenacious, determined enemy who, despite no obvious strategic breakthrough, simply would not let go. By the end of the battle "the army had been fought to a standstill and utterly worn out."[29] Only the onset of winter saved it from collapse. Attacked on all fronts, Germany realised she was in a death struggle with an enemy with limitless resources and deeper pockets. Falkenhayn was relieved of his command and sent to redeem himself in Romania. The duumvirate who succeeded him, Paul von Hindenburg and Erich Ludendorff, had no illusions that Germany was fighting for survival and desperate measures were needed. Their military dictatorship, hardly obscured by a fig-leaf Kaiser, polarised German society and sowed the seeds of post-war political meltdown. One Somme survivor, Corporal Adolf Hitler, would be the ultimate beneficiary, and Hindenburg and Ludendorff would connive in his warped resurrection of German militarism.

IN HIS RESIGNATION speech to the Commons in December 1915, Churchill stated: "In this war the tendencies are far more important than the episodes. Without winning any sensational victories we may win this war. We may win it even during a continuance of extremely disappointing and vexatious events."[30] In this judgement he showed considerable foresight, even if his later memoir was infused with hindsight. The Somme, disappointing and vexatious though it may have been, unsensational in its immediate results, reinforced the tendency towards allied dominance of Germany. Recognition of that truth prompts many questions. Why was it fought? How was it fought? What did it achieve? What were its consequences? How is it remembered? Was it, as contemporaries suggested, the turning point of the war? Was it, moreover, a turning point in history?

The history of the First World War is overabundant with studies of what went wrong, what the consequences were for those trapped in the maelstrom, and how things should have been done better, a genre which Churchill's *The World Crisis* helped to start. Objective analysis of the reality of the war—the problems which it threw up for nations and their armies and the solutions that were found—is a more recent trend, with the old myths and clichés continually recycled alongside. Certainly the Battle of the Somme has much history. In particular, the nature of the Somme experience has been explained better than the battle's significance. If the pen has not proven mightier than the sword, it has nevertheless shaped a far longer struggle over our memory of the Somme, a struggle which still continues. From the propaganda-tinged reports produced by both sides even as the combat raged, through the diaries and memoirs published by its veterans, to the sensational or scholarly accounts of later generations, the Somme has never ceased to fascinate. *Three Armies on the Somme* inevitably offers much that is familiar, but also much that is new. As a military history its reinterpretation of the twentieth century's first truly modern battle adds breadth, depth and clarity to earlier narrowly focused histories. Yet it presents much more than a narrative of how the battle was fought. Over ninety years on there is also a need to consider why the Battle of the Somme was fought, and the profound impact that it had on those who fought it; to consider French and German experience alongside British; to observe the wider world as it watched this earth-shaking battle unfold; and to consider how these experiences shaped the memory and history of those who came after.

As memories fade and available records multiply it becomes possible to revisit "the old front line" once more, and to fashion a clearer understanding of monumental events than was feasible for those who experienced and were touched by them, and subsequently recorded their distorted impressions in partial or self-justifying memoirs. From this incomplete and tainted record many narratives have already been constructed. This one aims to explain for the first time why the armies of three great empires came to a trial of strength on the Somme three times, and why the pivotal clash in 1916 shaped our world and remains with us still. The experiences of three nations and their armies—shared, yet profoundly different—are at the heart of this history. The Battle of the Somme was as nothing yet experienced by Britain's volunteer "Tommies." Both a national glory and a tragedy for Britain and her empire, it was the beginning of the end for that empire and the culture of deference that sustained it. Conversely, their fight on the Somme was an all too familiar experience to French "*poilus*" and German "Fritze," an extension of the bloodletting of Verdun. Their death struggle had begun in August 1914, and two years on their very societies were at stake in the no-quarter battle of attrition raging along the Western Front. Both nations would shatter, rebuild and shatter again in the months and years to come.

Almost a century on, interest in the bloodletting of 1916 has not diminished. Why should it, for whose family history was not touched—and often scarred—by this notorious combat? Yet the battle's significance has been all but subsumed in the arcane minutiae of tactical analysis and the microcosm of trench experience, which fascinate but do not illuminate. While the men of the Somme are remembered, the conflict itself is misunderstood. Was it merely a futile battle in a futile war? Certainly it was nothing of the kind for the generation of 1916, those who fought and bore witness to the Somme. In the casualty-averse age in which we live now it is easier to respond to the state-managed mass death of the First World War with horror than to try to comprehend it. It was not possible to fight war in that era—an age of industrial weapons technology, entrenched ideology and mass mobilisation—without casualties, and the Western Front and the Somme especially are the supreme examples of the consummate killing power of the machine age. It will not do to shy away from this: the men of 1916 did not. A return to these men and their ideas, to the three armies of the Somme and the societies that spawned them, will help us, their descendants, to understand their struggle, rooted in their past, for their future. In doing so, the memory of those who fought, those who died, and those who lost on the Somme will be better served.

Three Armies

THE GERMANS RETURNED to Péronne on 27 August 1914. Rumours of their approach, of destruction and plunder, of casual atrocities, preceded them. A sense of fearful anticipation gripped the townspeople: "A desperate silence hangs over the town. Shutters and windows are tightly closed—but inside everyone is watching."[1] The young men were at the front, but the old men who watched silently from behind their windows had not forgotten the Prussians who marched down the rue St. Sauveur and into the market place in January 1871. The Germans remembered too: particularly, they feared the reappearance of *francs tireurs*, the civilian irregulars who had preyed on their advancing columns in the last Franco-Prussian War. Ruthless authority was their response. When the mayor refused to take down the tricolour flying over the *mairie* it was torn down by the furious enemy. In pique, and to deter resistance, the passing Prussians burned fifty-eight houses to the ground and looted the rest of the town, casually shooting passers-by who got in their way. At least it was not as bad as 1871, when nearly seven hundred houses had been damaged by the German bombardment.[2] For this small, sleepy market town on a bend of the meandering river Somme, which today houses France's museum of the Great War, it was the start of a four-year ordeal of war and terror.

The Somme *département* had known war since Julius Caesar's legions had marched through the region in pursuit of Gauls. Today it is still criss-crossed by the lattice of straight roads radiating from Amiens (Roman Samarobriva) along which later emperors' armies paced on their way to defend Rome's imperial frontiers. After the Romans, Flemings, Burgundians, English, Spanish and Prussians had invaded.[3] In 1414 King Charles VI of France repulsed the Burgundians on the heights south of Bapaume: today a small chapel commemorating that victory stands beside the Péronne–Albert road, reminding modern-day battlefield pilgrims of the region's bellicose history. A year later King Henry V's English army crossed the Somme near Péronne, on the way to its encounter with Charles's army at Agincourt. Péronne is still girded by the remains of its stout thirteenth-century castle and walls which discouraged Henry V's army from attacking it, reinforced by bastions erected in the sixteenth and seventeenth centuries. In 1536 the town resisted a lengthy siege by the invading Spanish. Until the seventeenth-century expansion of France's borders by Louis XIV, Picardy was frontier country, and Péronne, at a key crossing point, was "the key to the Somme."[4] Although war was to move away slowly northwards in the ensuing centuries, a British army passed by in 1815, pursuing the remnants of Napoleon's defeated forces after Wellington's victory at Waterloo: the British soldier and author Charles Carrington was surprised to find a discarded Brown Bess musket as he scoured the battlefield debris of his own fight around Péronne.[5] In the Franco-Prussian war, General Faidherbe's *Armée du Nord*—hastily raised by the new republic to continue the fight against the invader after Emperor Napoleon III's army had been trapped at Sedan and forced to surrender—drove off the enemy near Bapaume on 3 January 1871. This did not save Péronne, which capitulated a few days later,[6] and Faidherbe's army was finally overwhelmed on 19 January at St. Quentin.[7] The Somme, the focal point of the natural invasion route across northern France which Charles de Gaulle dubbed the "fatal avenue," was well acquainted with warfare long before the great attritional battle of 1916.

IN 1914 THE Somme was not ready for war. Before the outbreak of hostilities the methodical planners of the French General Staff had counted up the divisions on either side and concluded that there would not be enough men for the fighting to extend into the Nord region. When the Germans chose to march through Belgium and sweep round against the rear of the French armies concentrated on France's eastern frontier, there was little to oppose the invasion of north-west France. The surprise of those who opened their morning newspapers on 29 August 1914 to learn that the Germans were on the Somme was palpable.[8] Something had gone badly wrong with the French battle plan.

Both the French and the German armies, mass forces raised by widely applied conscription, fielded three sorts of formation on mobilisation. Their front-line divisions incorporated the currently serving annual contingents, topped up with reserves from the most recent contingents.* These divisions would lead the assault: the men were young, fit and up-to-date in their training, and their officers were skilled professionals. On paper, despite Germany's significantly greater population, the two countries' front-line resources seemed roughly equivalent in 1914. By virtue of a new three-year military service law introduced in 1913, France could field forty-six front-line infantry divisions in 1914, divided into twenty-one army corps.† Germany could field twenty-five army corps, although three of these were earmarked for the Russian front. Behind these in the second line were reserve divisions, with fewer professional officers and predominantly filled with reservists from earlier annual contingents, less well trained or up-to-date than the front-line divisions. France fielded twenty-five reserve divisions in all, organised in 1914 into groups that were distributed between her five front-line armies and the Paris garrison.‡ Germany organised her thirty-two reserve divisions into separate reserve army corps, supplementing her front-line army. Heavily outnumbered by France and her allies, Germany had to use her reserve divisions in the front line from the start while France held hers back for secondary operations. Germany had seventy-two divisions with which to attack France in August 1914. This was the basis of Germany's 1914 strategic gamble to knock out France with a rapid and overwhelming assault before the huge but slow Russian army could enter the fight. The supposedly unstoppable right hook which was the guiding principle of the infamous Schlieffen Plan brought the German armies through northern Belgium to the Somme. By leaving Germany's reserve divisions out of their calculations, French planners played into their enemy's hands. Both armies also fielded a third line, territorials in the French army and the *Landwehr* and *Landsturm* in the German. Composed of older reservists, such formations were used to hold defensive positions and carry out rear-area police duties. In August 1914 Germany deployed the equivalent of twenty-three divisions of third-line depot troops, and France twelve

---

*See Appendix for a note on the organisation of military formations during the war.
†France's initial Corps d'armée (CA) were numbered I to XVIII, XX, XXI and the Colonial Corps. The XIX CA was garrisoning Morocco. From October 1914 eight further CA, numbered XXXI to XXXVIII, were formed from reserve divisions and other formations.
‡These were roughly equivalent in size to army corps.

territorial divisions.* Finally, France deployed ten cavalry divisions and Germany eleven in August 1914.[9] The British Expeditionary Force (BEF) added a cavalry and six infantry divisions, and the Belgian army six further infantry divisions, a cavalry division and fortress troops to the allied order of battle for the first engagements in 1914.

By the end of the first three weeks of hostilities the German army's apparent advantage in numbers seemed to be producing the desired result. The French left wing was in precipitate retreat, taking with it the BEF which had deployed on its left flank. Between St. Quentin and the Channel coast France's northern *départements* lay undefended in the path of the sweeping German right wing; the BEF's open left flank and lines of communications through Amiens, the capital of the Somme *département*, were exposed. General Joseph Joffre, the French commander-in-chief, took the gamble of employing his own reserve divisions in the front line. General Ebener's 6th Group of Reserve Divisions, 61 and 62 *Divisions d'infanterie* (DI), which had formed the Paris garrison at the outbreak of the war, was transported by rail to Arras, with orders to move southwards to block the German advance on Bapaume and Péronne. Advancing down the road from Cambrai, Ebener's force drove back a thin German cavalry screen and marched triumphantly into Bapaume, past the monument marking their grandfathers' victory on the same field in 1871.[10] Theirs was to be an even briefer taste of victory.

At dawn on 28 August, Ebener's divisions resumed their march southwards, along the Bapaume–Péronne road, masked by the thick fog which frequently arose from the Somme valley early in the morning. Believing the enemy to have fled after their success of the previous evening, they were unprepared for action: their field ambulances were at the head of the transport column which proceeded down the main road. Few scouts had been sent out to front and flanks. Unknown to them, in the night General von Linsingen's II *Armeekorps* (AK) had reached the village of Moislains, a few miles north of Péronne, and was preparing to resume its march westwards. The leading elements of the

*From the age of 19, French conscripts served 28 years with the colours, 3 in the active army and 11 in the reserve, followed by 7 in the territorial army and 7 in the territorial reserve. German men were eligible for national service between the ages of 17 and 45. Active service varied between 1 and 3 years from the age of 20, depending on the arm of service, followed by service in the active reserve, *Landwehr* and *Landsturm*. With a much smaller population, France was obliged to conscript a greater percentage of her manpower to maintain some sort of parity with Germany.

French flanking regiments, 307 and 308 *Régiments d'infanterie* (RI) of 62 DI, passing in column-of-march from Sailly-Saillisel, skirting Bois St. Pierre Vaast, were surprised to hear firing to their front. French cavalry scouts had run into German sentries at Moislains. As the sun began to heat up the late summer morning the fog lifted suddenly, and German machine guns opened fire on the packed French infantry columns exposed on the gentle slope leading down to Moislains. Enveloped from a flank, and even fired on by their own comrades from behind, the two regiments broke and were massacred.[11] The French artillery belatedly came into action, and began to bombard Moislains, still filled with terrified French civilians. The village started to burn, adding to the confusion. German cavalry, sweeping in from the open left flank, soon silenced the French guns, cutting down the gunners at their posts. The shattered French divisions fell back in disorder, the remnants of 62 DI fleeing northwards back to Arras and 61 DI conducting an arduous fighting retreat westwards behind Amiens.[12] Further south, Péronne had fallen. The Prussians rounded up several thousand prisoners and hauled away the abandoned French guns before resuming their westward march. The sharp, brutal fight at Moislains was merely a foretaste of the full intensity and horror of industrial war, which for the next four years was to superimpose itself upon Picardy's pastoral calm.

THE GERMAN ARMY got the better of the first fight on the Somme, as it had done in most of the early battles on the French frontier. In this instance it was only to be expected because inept reservists had been pitted against front-line units. However, the French army gave a better account of itself in the early battles of August 1914 than historians, judging the overall result of the fighting on the frontiers, have given it credit for, even if it was not then a match for the more efficient Prussian-led military machine. Although a clichéd image of red-trousered infantry, chic white-gloved officers and cavalry with gleaming but useless breastplates is routinely rolled out to depict an army with outdated Napoleonic methods and misguided offensive doctrines, excepting its dress sense, the French army of 1914 was much like those of its enemy and its principal ally in the West. It was not so much that French military doctrine was fundamentally wrong in 1914; rather, when they went to war, the French commanders had not made up their minds on what that doctrine should be.[13] In the early years of the twentieth century all armies were wrestling with the same problems of tactical cooperation between the three main fighting arms (infantry, artillery and cavalry) on the industrial battlefield, the incorporation of new weapons and technologies into the army (not just machine guns and quick-firing artillery, but aeroplanes, motorised transport and telegraphy), and the direction of large units in the field.[14] The theorists of the offensive who stressed the élan and morale of the infantry—officers such as Colonel Louis de Grandmaison, who drew up the infamous infantry tactical instructions of 1913 which supposedly proved so costly in the first clash of arms and at whose feet much of the blame for the French army's disasters of August 1914 has been placed—were rivalled by the theorists of firepower, such as future generals Pétain, Fayolle and Debeney. But, as Grandmaison and others recognised, the real problem facing French military theorists was how to weld the two together in order to win on the battlefield: "The enemy must already be broken, and able to offer only a feeble resistance; the attack must have a marked moral and material superiority; and the distance [to attack] must be short. The principle does not change, that it is always fire that ensures movement."[15]

Resolving this problem on the industrial battlefield was to become the principal challenge facing the French army in the first year of war. Although important decisions had been made—to replace the nineteenth-century blue-and-red costume with a less conspicuous horizon-blue uniform, which was finally approved in July 1914, and to add heavier howitzers to the French army's firepower, initiated in April 1914[16]—the process of modernisation was not yet complete when war broke out, even if many of the possible tactical and operational responses had been identified and debated in the pre-war doctrinal ferment. Indeed, it was the very fact that methods and *matériel* were in transition that undermined the French army's military effectiveness. The reservists at Moislains would not have mastered Grandmaison's new theories—would probably not even have heard of them—and similarly their commanders would probably have forgotten the "Napoleonic" principles of manoeuvre adumbrated by Colonels Bonnal and Foch, professors of military history and tactics at the *École de Guerre* some twenty years previously; and, because of the French army's inadequate training facilities, the same might be said of many of the front-line divisions and corps.[17] The French army's weaknesses were apparent in the first encounters on the Somme and elsewhere: poor leadership, especially at the highest level; lack of coordination between artillery and infantry; weak reconnaissance and poor intelligence; lack of cooperation between higher formations; inadequate training in reserve formations. Since the German army had made better progress in assimilating these practical lessons, it had an edge in early encounters: it possessed heavier, if not technically superior, artillery; it utilised concentrated machine-gun fire; its reconnaissance was better; and there was effective cooperation between its infantry and artillery. But the German army itself was not without fault: it attacked in dense formations, which in turn rendered it vulnerable to concentrated French fire, especially from the expertly used French 75mm field guns, so-called 75s, the best artillery piece in use at the time. There are as many instances in early 1914 of heaped German bodies as of French, and at Moislains the French guns did good work against the enemy as well as their own civilians before they were overwhelmed. Therefore, while the French army was outgunned and outmanoeuvred in the first big battles, it was not outclassed. Although obliged to pull back when its exposed left flank was turned, it was far from beaten. There was still much fighting spirit in the men; and, after extensive sackings of elderly and out-of-date divisional, corps and even army commanders, its higher leadership was more capable of addressing the immediate task of halting and repelling the German invasion. At *Grand Quartier Général* (G.Q.G. — French General Headquarters) the unflappable General Joffre was already planning his counter-stroke.

Stout, gentle, self-deprecating and avuncular, the sixth-two-year-old generalissimo cut an unmilitary figure, his blue uniform stretched over the grand belly which attested to his love of the table: in looks alone he lived up to his nickname, "Papa." Yet his appearance was deceptive. Joffre's stolid peasant physique — his ancestors came from the Pyrenees — masked a sharp mind and firm will that had brought him to the top of the French army in a difficult period of civil–military tension and rapid change. Joffre was born in 1852, the son of a cooper. His intelligence allowed him to escape from the provinces, and he enrolled at the prestigious *École Polytechnique* in 1869, the youngest student of his year. The war with Prussia interrupted his studies, and Joffre was mobilised into the artillery, serving as a young officer in the defence of Paris. This formative and calamitous war overshadowed the rest of his military life: years later, planning for another war against Germany, he was determined not to make the same mistakes which his predecessors had made in 1870.[18] After graduation he joined the engineers, and as a young captain found himself supervising construction work in France's new fortress barrier, designed to keep the Germans out in future. Following the untimely death of his first wife, he volunteered for colonial service: for a while he would be married only to his profession. Success followed in Indo-China and West Africa, where, seconded from his prosaic task of railway engineer, he led one of the flying columns which captured Timbuktu, saving the day after the main column had been routed by Touareg raiders. This colonial triumph brought him to the attention of the government in Paris. He passed swiftly through a series of commands and staff appointments, before joining the Supreme Council for National Defence in 1909 as director of supply and communications. The professional understanding of movement and logistics he developed in this post would come into its own when war broke out. A solid republican in an era of acute political partisanship in France, in July 1911, Joffre found himself appointed, as the politically uncontroversial compromise candidate, to the post of Chief of Staff of the French army; commander-in-chief designate in the event of war.

It was a post which Joffre, with uncustomary modesty, felt he was poorly equipped to fill, a judgement which has been debated ever since.[19] If lacking the intellectual vigour to intervene actively in the ongoing doctrinal debates which animated the French army before 1914, he was no dunce: he had been Professor of Fortification at the artillery and engineering school at Fontainebleau earlier in his career. He was certainly more than the "solid shield behind which subtler brains could direct French military policy on the path to a crevasse which they had not perceived," as Liddell Hart dismissed him after the war.[20] Above all, he was a practical soldier who approached the job he was given with care and a technician's close eye for detail.

His principal task in his new post had been to prepare the army for the coming war with Germany. Joffre's solution to the problem of a German invasion of France—the infamous "Plan XVII," for a counter-offensive against the weak point of the advancing German armies— failed in the massive frontier battle of August 1914, principally because the numbers Germany deployed meant that there was no weak spot. However, the Battle of the Frontiers was a closer-run thing than is generally acknowledged, and the retreating French army, although shaken, remained an effective fighting force. It was now that Joffre showed his true character and worth, as well as his logistical skill. Imperturbable, phlegmatic, calculating, decisive, he pragmatically reorganised and redeployed his forces in anticipation of the moment when his army would strike back.

Joffre's plan was to shift troops quickly from his right wing to his left, forming a new "Army of the Somme," Sixth Army, under the command of General Michel-Joseph Maunoury, to strike at the exposed right flank of the German armies moving across northern France. Ebener's reserve divisions were moving southwards to join this force when ambushed at Moislains. Meanwhile, Maunoury's main force was advancing south of the river into the Santerre plain. It encountered General von Kluck's First Army around Proyart on 29 August, and gave a better account of itself than Ebener's force had done. Although heavily outnumbered, the regulars of 14 DI, redeployed from the east, blocked Linsingen's columns advancing along the left bank of the Somme, inflicting heavy casualties with concentrated artillery and rifle fire. However, Maunoury's army, composed as it was predominantly of reserve divisions and exhausted cavalry, was unable to check the German advance from Péronne for long, and with the routing of Ebener's divisions to the north Joffre was obliged to order a further retreat to prevent Sixth Army being enveloped in its turn. Although, in combination with Fifth Army's counter-stroke further east at Guise on the same day, Maunoury's attack succeeded in disengaging the French left wing and the BEF from the German pursuit, it could not save the Somme region from German occupation. The *département*'s capital, Amiens, was now an open city.

REMNANTS OF EBENER'S broken and exhausted regiments, who passed through Amiens in disarray on 29 and 30 August, confirmed the rumours that the Germans were advancing on the city. For some days the gunfire to the east had been audible in the city's streets. Crowds of Belgian refugees had passed through the city a few days earlier, bringing tales of death and destruction as the barbarians advanced.[21] Those who could packed up their possessions and took the roads eastwards, joining the throng of demoralised soldiers and dispossessed civilians. The garrison commander departed so precipitately that he nearly left behind his sword.[22] Those who could not flee awaited their fate. After the unusual traffic of the past week, the city became ominously silent, expectant. Just three weeks earlier the Amiénois had cheered off the city's departing regiments: "To Berlin: down with the Sauerkrauts."[23] Now some very sour Berliners were at the threshold.

The Germans marched in on 31 August. Monsieur de Langler, a clerk, encountered a German column going up the rue St. Fuscien as he headed into work at the town hall. He recalled the columns of blue-uniformed Prussians he had seen marching up the same street as a small boy in 1870. Arriving late at the town hall, he found the courtyard full of Germans, carrying off the weapons which had been collected there over the previous week: the municipal council had decided to disarm the city's citizens to prevent any *franc tireur* incidents after hearing of bloody German reprisals against civilian insurgents in Belgium. A joint proclamation was issued by the mayor of Amiens and the German commander, General von Stockhausen: the city would be bombarded if there were any acts of violence against German troops. The city was put under curfew and the procurator-general and twelve members of the municipal council were taken hostage as surety against the population's good behaviour and the payment of a 160,000-franc indemnity.

After the dense grey-uniformed columns had traversed the city on 31 August, Amiens became eerily empty. Only small German patrols could be seen, requisitioning the city's motor cars and plundering its shops and banks. Then, on 10 September, the Germans were back in force. All men of fighting age were ordered to present themselves at the Citadel. They were marched out of the city the next day. With them went the Germans. The prefect's secretary read out from the official communiqué that there had been a great Anglo-French victory over the German right wing. The triumph was confirmed when French cuirassiers entered Amiens on 12 September. The crowd were ecstatic: the assembled townspeople sang "La Marseillaise" as troopers from 9 *Régiment de cavalerie*, which had been the last to leave the city on 30 August, replaced the tricolour above the town hall. They had been delivered from the enemy, and the victory was France's. It was a heartfelt, but premature, celebration.[24] The city's young men had been transported to Germany, where they were to be interned for the duration of the war. Over those four years those who remained would listen daily to the guns devastating their *département*.

There had indeed been a great victory against the German right wing. In the second week of September, Maunoury's army finally delivered the counter-stroke which forced the overextended German forces to retreat. Taking its title from the river Marne, this huge battle east of Paris, the so-called Miracle of the Marne, which did so much to rescue France and enhance Joffre's military reputation, was only a temporary deliverance. It saved Paris, and pushed the enemy back fifty miles, but it did not break the German army. The simple truth was that great armies of millions of men could not be broken in a single battle, however vast and intense. The French themselves had already demonstrated this in

August. The same principle was reaffirmed time and again over the next six weeks in the so-called race to the sea, as the opponents jockeyed for advantage in Picardy and the Nord and the stagnating line of opposing armies extended itself inexorably to the Channel coast.

Liddell Hart subsequently asserted that once the German army's initial attack on France had failed Germany had lost the war.[25] But France and her allies still had to demonstrate that fact to the invader. It was, the historian and journalist Gabriel Hanotaux recognised, a fundamental fact that Germany, not France, was now besieged, and her circumvallation would have to be broken.[26] Few would then have predicted that it would take more than four bloody and catastrophic years to do so.

ALTHOUGH IN SEPTEMBER 1914 the German army pulled back hastily from the Marne and drew its extended tentacles in from Amiens, this was merely a tactical withdrawal. The advancing French and British armies were halted in their turn, along the river Aisne, and attention turned back to the open northern flank. Joffre and his opposite number, General Erich von Falkenhayn (appointed after his predecessor, General Helmuth von Moltke, had been removed after the repulse on the Marne), started to strip their stalled front lines of divisions, which would be moved to the open flank in an attempt to carry off the envelopment which so far both sides had failed to achieve.

In the last week of September the war returned to the Somme. South of the river, Joffre redeployed his Second Army under General Noël de Castelnau. Falkenhayn despatched Crown Prince Rupprecht of Bavaria, who had led the German Sixth Army which checked Castelnau's in Lorraine in August, to command the new forces he was gathering east of Péronne. Like the encounter at Moislains in August, the meeting between Castelnau's and Rupprecht's advancing armies was a confused and bloody affair. French reconnaissance was poor: Castelnau was reliant on G.Q.G.'s daily intelligence bulletin for information, which was necessarily superficial and generally out of date. Péronne was reoccupied by the enemy, while Castelnau's forces dug in on high ground to the north-west, between Maricourt and Hardecourt. French cavalry patrols reported that the town was now clear of the enemy. Indeed it was, because the Germans were crossing the river in force south of Péronne, moving south-westwards towards Chaulnes.[27] Castelnau's leading formations, IV and XIV *Corps d'armée* (CA), were advancing north-eastwards towards Péronne, driving in the German skirmishers. The Germans too were advancing blindly, probing for the enemy's left flank:

> It was a beautiful clear autumn morning. The flat land stretched away as far as the eye could see, field after field of sugar beet, punctuated by numerous villages and the high trees of the parks and châteaus. Everywhere hares raced away in front of the companies and coveys of partridges rose out of every field ... We expected to come under artillery fire at every moment, but all was calm.[28]

On 24 September, when the main bodies of the advancing armies crashed into each other and locked, this bucolic vista was obliterated.[29] The German guns set to work. Soon a long band of flames, the farming villages of Maucourt, Chilly, Lihons, Vermandovillers, Chaulnes, Herleville and Foucaucourt, illuminated the Santerre plain.[30] The church bells in Lihons rang out defiantly until a shell toppled its steeple. No victory peals were to be heard in the Somme until 1918.

The fighting south of the Somme on 24 September was intense and confused. The artillery of both sides did great execution. This put an end to both armies' flagging belief that battle could still be conducted in Napoleonic style, bivouacking by night and advancing in close formation during the day, infantry at the double and artillery at the gallop. Bivouacs in the towns and villages—Chaulnes, Chilly, Maucourt, Fouquescourt—had to be fought for, often hand-to-hand, and movement in the open was suicidal if field guns were deployed. At Chilly one German battery sergeant noted the destruction of a French infantry company advancing in close formation 1800 metres away with a few well-placed rounds.[31] In this confused, ill-coordinated fight German firepower and weight of numbers again prevailed, and the French infantry were pressed slowly backwards.[32] However, it was now abundantly clear to the Germans that the French were fighting much more effectively than they had only a month before. "The battles in Lorraine had given our troops a feeling of superiority," Colonel Schulz of 22 Bavarian *infanterieregiment* (IR) recollected:

> They received an extremely rude shock when they had to accept ... that the French, even though they were from the same corps that we had brushed aside in Lorraine, had changed. They

were tough, daring and self-confident. The Miracle of the Marne had so raised the morale of the French that ... ill-prepared, uncoordinated and over-hasty minor attacks were dashed to pieces around Maricourt.[33]

By the end of the day the adversaries had fought each other to a standstill. The survivors dug in where they were. (They would still be on roughly the same lines when the "Big Push" was launched two years later.)

Little forward progress was possible on the Santerre plain, although the reciprocal fusillade and violent attacks and counter-attacks with the bayonet continued over the following days.[34] The enemy might still be outflanked to the north. Castelnau had one corps at his disposal, General Balfourier's elite XX CA, the so-called Iron Corps. Its two Divisions, 11 DI, the "Iron" Division (*division de fer*), and 39 DI which aspired to be the "Steel" Division, were the best in France, having been maintained at full strength in peacetime and trained to cover the deployment of the rest of the army in the event of war.[35] Balfourier's divisions were advancing astride the Somme from Amiens, seeking the open German flank. Thirty-nine DI, moving south of the river, turned southwards towards Chaulnes. Eleven DI moved along both banks of the river, hoping to reach Péronne and then turn southwards against the enemy's rear. There was "nothing in front of them," Second Army's two-day-old orders optimistically stated.[36] They pushed a little further forward than the troops to the south, creating a pronounced salient on either side of the river, before stalling in their turn in the face of concentrated artillery fire. In the "race to the sea," one French general remarked, they were always twenty-four hours and one army corps behind the Germans.[37] This lack of the initiative was critical: instead of falling on a vulnerable flank, the French were usually trying to cover their own, with only exhausted cavalry and territorial divisions. Thus 11 DI, advancing on Combles and Péronne under heavy artillery and machine-gun fire, found itself outflanked from the north on 26 September, after territorials had abandoned Bapaume. Its commander, General Ferry, diverted his reserves to covering the bridgeheads to the north of the Somme, between Mametz and Maricourt, and blocking the enemy's advance on Albert. On his left, XI CA, the first formation of yet another new French army (the Tenth), hastily deployed to the exposed northern flank, had mistakenly abandoned Fricourt village. They could not recapture it.[38] The 11 DI's hurriedly prepared attempt to infiltrate around either side of the village on the morning of 29 September made little progress.[39] The attack was renewed in the afternoon, and 26 RI fought its way into the village. Illuminated by a searchlight brought up for the purpose, hand-to-hand combat continued through the night and into the next morning. Only half the village was held, and over the next two days further companies were sent in to push back the enemy, house by house, from the higher ground on the northern side of the main street. Pinned down by accurate sniper fire from the northern half of the village — 26 RI lost its third commander since the outbreak of war, and its second in a week — the French could not advance.[40] They were eventually driven out of Fricourt, but the Germans in their turn could go no further westwards. A surprise night attack north-west of Fricourt on 7/8 October, intended to break through to Albert, was a complete failure; four hundred prisoners and as many dead and wounded were lost.[41] So Fricourt village would mark the far point of the German advance north of the Somme, the heavily fortified apex of a sharp salient which the British army would have to assault on 1 July 1916.

It was apparent that the front was stalemating, as G.Q.G. acknowledged in a tactical note that was unfortunately leaked to the press. Modern war was turning out to be butchery, Hanotaux commented, and means would have to be found to lessen its impact.[42] Even at this early stage of the war, the value of the spade and the gun on the modern battlefield was obvious to the more perceptive soldiers.[43] For now the French army was ordered to dig in deeply where it stood: new tools were issued for the job (although the *poilus'* inherent reluctance to dig proved awkward). In future, it was indicated, operations would be slow and methodical, progress less than a kilometre in each advance.[44] This directive was the first acknowledgement by G.Q.G. that they were engaged in a new sort of *matériel* war, and that the nature of operations had to be rethought. In future, attacks would be well prepared by the artillery, with a new purpose, "to wear down the enemy, to sustain our moral ascendancy, and geared towards creating a favourable moment for breaking the stalemate."[45] It would not now be a quick, easy war. Old methods of fighting, in particular the belief in the supreme value of the offensive, would have to be reconsidered in the light of experience.

A war of flesh was going to become a war of steel, of weaponry and machinery, science and technology. It was an immediate and obvious lesson, but one which could not be implemented overnight: the home front would have to be mobilised to sustain the voracious appetite of modern industrialised war, and that would take time. Over the next three years the apparently static battlefield became, paradoxically, a vibrant and dynamic military school, as tactical innovation and technical novelty tripped over each other, as offence and defence battled for superiority. "This very cycle of action and reaction, designed to break the deadlock . . . confirmed it," Hew Strachan has perceptively identified.[46] The future fighting on the Somme was to be a microcosm of this lively martial academy, which was to revolutionise the nature of warfare, even if it could never master its fundamental horror and brutality. This was to give another novel and paradoxical dimension to the war, for despite the massive material effort of the belligerents, human lives still lay at the root of strategy. Hanotaux hoped that France's generals would be careful with their men's lives, would not throw them against the enemy's guns without proper support, for this was now a war of attrition. It would be a "long, bloody, painful" war for France, in which "the whole world must suffer." "Not giving an inch, holding on till the last minute, will bring victory" was his sombre counsel.[47]

It was well to be gloomy and fatalistic, to rally France for a gruelling ordeal, for the sacrifice of her young men and the devastation of her gentle countryside, for a prolonged and exhausting war of liberation. Even as the epicentre of the battle moved northwards, towards Arras and then Ypres, the fighting in Picardy assumed characteristics that would subsequently become notorious. In late September and early October the German artillery pounded the improvised French trench lines, forcing out the defenders, if not through shelling alone then at the point of the bayonet. In their turn the weary *poilus* counter-attacked, seeking to retake lost ground or improve their own tactical position, trench by trench.[48] It was small-scale fighting, under a constant rain of shells and bullets, in which the opposing line would move back and forward a few hundred metres at a time, the casualty count mounting all the while. "The 124th has fought non-stop for a fortnight! Fifteen days of very hard battle!" noted Sergeant Alfred Joubaire:

> The men have no strength left. The regiment is in a terrible state; a shadow of its former self, a skeleton. No officers; a battalion commander has replaced the colonel, the companies are commanded by their adjutants or even sergeant-majors. Most of my comrades are dead, wounded or missing. We are at less than half strength. We crave rest; but we must hold on, no one will retreat.[49]

Attrition did not start on the Somme in 1916, even if there and then it was to reach its apogee. By late 1914 the thinning French ranks were hard pressed, but held on. Where flesh and steel were in short supply, willpower would have to suffice. Joffre sent General Ferdinand Foch, recently appointed his deputy, to coordinate operations on the northern flank. The gloomy Castelnau wanted to pull his army back behind the river Somme, but was shamed by the fighting spirit of Foch, who a few weeks earlier had been his subordinate: "Retreat is impossible. You must hold on at all costs and die rather than give up an inch of ground. I won't be argued with in this situation."[50] Little did Foch realise that he was choosing the ground for his own battle of wills two years later.

The final French attack south of the Somme, to take the village of Quesnoy-en-Santerre, showed signs of a new approach to operations. Although the army was not yet provided with the vast supply of munitions which set-piece attacks would require — indeed, one of the factors which curtailed operations for the winter was a shortage of shells on both sides — the attack by General Baret's XIV CA had all the hallmarks of the planned, "scientific" battle that was to be adopted as the French army's standard procedure by 1916. All the available heavy artillery of the corps and supporting formations was assembled to batter the enemy's trenches and defence works, and to counter the supporting German batteries. On the eve of the attack Castelnau ordered a reduction in the infantry detailed to assault, to be held back as a reserve to exploit any success, rather than thrown *en masse* against the enemy's trenches. While this might slow down the operation, it would save lives. At 12:30* on 29 October the heavy artillery opened up on the village and the German guns, and the field artillery on the enemy's trenches. It was not the prolonged softening-up that was to come; the infantry went forward at the same moment. Initial progress, as so often, was swift, with the infantry of 117 RI advancing through the German defences to the outskirts of the village, and on to "Hill 98," the dominant high ground to the south. However, as the infantry tired and suffered casualties, the attack lost its momentum, and supporting infantry companies were sent forward to reinforce the assault. It finally stalled a few hundred metres from the village, and Baret ordered his men to dig in on the captured ground. Field artillery was brought forward to bombard the village at close range, before the infantry assault was renewed at first light on 30 October. Two enemy counter-attacks prevented this, but at 13:00 the next day

---

*All times are given in the twenty-four-hour clock.

the French infantry went over the top again. The Germans counter-attacked, and hand-to-hand fighting ensued in the village. A final French charge, with cries of "*Vive la France!*," cleared the village in the evening. Two cannons, four machine guns and four hundred prisoners were the spoils. Four days later Hill 98 finally fell. President Raymond Poincaré, observing the spectacle, noted that off to one side a peasant ploughed a field while his wife gathered sugar beet.[51] Despite war coming to their fields, for civilians daily life must go on as best it could.

Such small-scale attacks—on a village, a hill or a strongpoint, to straighten the line or harass the enemy—which gained a few hundred metres and a couple of lines of shattered trenches at a time were to be repeated thousands of times over the following years, contributing in aggregate to the attrition of the enemy, tiny steps on the path to victory. If well managed and appropriately resourced, they could succeed with an acceptable level of casualties. If hasty and inadequately provisioned, they might prove disastrous. They represented the fundamental problems facing First World War commanders: how to combine the different fighting arms to overcome the enemy's defences economically, and how to merge small-scale operations into a grander, war-winning strategy. The French army's commanders were engaged with these problems from the start, although universal, effective solutions were not immediately apparent, not least because the enemy faced the same problems and was innovating in turn. But one overriding truth, pithily summed up in General Mangin's famous aphorism, prevailed: "Whatever we do, there will be casualties." But in the choice between manpower and *matériel* as the key to unlocking the stalemated battlefield, the keener minds were already aware that guns and munitions were the answer, not men's lives.

MUTUAL EXHAUSTION, AND an increasing shortage of ammunition, finally put an end to the tooth-and-nail struggle in Picardy in late October (although it continued further north until late November, principally around Ypres). With thousands of men burrowed beneath the earth, the Santerre plain became strangely silent and forlorn. "What wide loneliness!" the journalist Maurice Barrès reported on a visit to a forward command post. "One can see nothing and it is as flat as a billiard table. And how quiet! I wanted to walk straight ahead just to find something interesting."[52] After the battle for Quesnoy the troops engaged with a new adversary, albeit one familiar to armies down the ages: winter on the Somme, with its associated cold, wet and clinging mud. Of the journey to the front line Barrès noted: "The rain was torrential; the mud slippery. To get along I had to pull myself through the sticky yellowish mud with my elbows. One moved very slowly."[53] The predominant element of war, long periods of boredom—as great a strain on military morale as the enemy's aggression—set in. "A month staring at this flat horizon, with hardly a tree, a month wallowing in the mud, sleeping on the ground, through cold nights," scribbled Eugène Pic, a disgruntled *poilu* confined to "Jena Avenue," a fortified section of a sunken farm track outside Lihons. "It's tough and monotonous, atrocious. Who would dare to think that at war one would get bored?" It was time to dig deeper and prepare to hibernate, "in that little corner of a Picardy road that would be our world and our life for long months."[54] Over the winter the elaborate systems of opposing trenches, named after the nation's heroes and victories, lost comrades and fondly remembered homelands, took solid, permanent shape. Dugouts, for protection as much from the elements as from the enemy's shells, were burrowed into the chalk. Barbed wire was stretched between the lines, and machine-gun posts excavated to cover no man's land. Mortar pits and artillery emplacements were built for the guns, and support and reserve trench lines, communication trenches, first-aid and command posts—the "Lihu Imperial Palace," the damp *poilus* of Jena Avenue dubbed that of their well-sheltered officers—deepened the defensive zone and kept the men warm and occupied. The Germans were digging in for a long stay: "Now we were making the trenches into real fortresses. They became very deep and communication trenches were dug between them. Comfortable shelters were excavated for the officers and soldiers. They had floors, beds, chairs and mattresses. At battalion headquarters there was even a piano. In December electricity was installed."[55] Further back, bivouacs, field hospitals, bath houses, supply depots and *estaminets* were organised to house and comfort the men when out of the line. The linear city, stretching from Nieuport on the Belgian coast to the Swiss frontier, the "Western Front" as the Germans dubbed it, was assuming its final shape. As one French officer noted, a new, limitless siege of Sebastopol was starting on French soil.[56] The *poilus* were becoming resigned to a long war of attrition with determination if not enthusiasm. "We must be patient and wear out the enemy . . . We will not give a metre of ground," Commandant Bénard wrote home from his muddy billet near Albert. "We are certain we will win. From the general down to the humblest soldier all know this and the conversation is all about how we will break German power for ever." But army life was not all bad in the "phoney war" if basic human needs were met. Bénard had his own cow, found wandering abandoned on the battlefield, and he was getting fat on this ready supply of fresh milk. Eels from the rivers and rabbits from the fields supplemented the monotonous army diet of corned beef and rice pudding. In idle moments he played cards and could think about more homely matters, such as the design of his garden, which he planned with his bugler, who before the war had been a seed merchant.[57]

THE FRENCH ARMY prepared one last hurrah before the year's end. In mid-December Joffre ordered an attack all along the front, in the hope of destabilising the enemy's trench line: it was time, he informed the army, "for the final liberation of the invaded national soil ... for your heart, your energy, your willpower, to conquer at any price."[58] His ebullient yet quaintly anachronistic rallying-cry was well received, if hopelessly premature.[59] Atrocious weather conditions and a lack of artillery ammunition doomed the assault to failure. On the Somme front 53 DI tried to break through in front of Albert in a small-scale rehearsal for the British attack on 1 July 1916. But with virtually no artillery preparation, inevitably there were heavy losses for little gain. Here and there a few trenches were taken, written up in the French official communiqué as the capture of a heavily fortified defence work: for the troops, momentarily shaken out of their numbing trench routine, it was only "an advance across a couple of rows of sugar beets,"[60] and a shocking forewarning of things to come. "I've just passed the most tragic hours of my life ... terrible days. Fights, nights in the trenches, muddy night marches under shell and machine-gun fire ... I don't know how I am still in this world ... I saw death up close," Bénard wrote home after the ordeal. Two whole companies had been killed or captured; storming the enemy trenches with cries of "*En avant la France*," they were simply swallowed up. Only a few wounded stragglers returned. Amid the carnage, the first cases of shell-shock were appearing.[61] For the Germans, it was a victory to crow about: 1200 prisoners and many enemy dead.[62]

THE LINE IN the Somme sector had settled down into the configuration that it would retain until July 1916. It was a well-sited defensive position. The Germans held a series of key villages on the forward slope of the dominant Thiepval Ridge. Their front line stretched north to south in front of Gommecourt, Serre and Beaumont Hamel to Thiepval itself atop a spur jutting out defiantly towards the French positions which clung to the slopes on the other side of the Ancre valley. South of Thiepval the line veered south-eastwards, towards Fricourt, which was to be turned into a heavily fortified strongpoint, a second key bastion at the apex of a pronounced salient. At Fricourt the line turned sharply eastwards, past Mametz and Montauban, turning southward again in front of Maricourt, joining the river near Curlu. South of the impenetrable wetlands which choked the valley of the slow-moving river below Péronne the line resumed its southwards course, across the gently undulating Santerre plateau, and on into the Oise. More fortified villages delineated the German front line: Frise, Dompierre, Fay, Soyécourt, Vermandovillers, Chilly.

As the days shortened, an uncertain pastoral calm returned. Those farmers who had not fled to Amiens resumed their timeless routine as if in a Millet painting, gathering what remained of their devastated crops, and sowing what they could for the next year's harvest amid the intermittent shelling. Their huge, heavy-wheeled farm carts competed for space with the military convoys which now clogged the narrow country lanes.[63] War or no war, the Picards needed to make a living and invaded France needed their produce. An uneasy juxtaposition of static war and normal economic life developed. After all, the soldiers' incessant quest for food and drink represented a lucrative new sideline for the impecunious peasantry.

The regime was more oppressive, and less profitable, behind the German lines. The town of St. Quentin became the headquarters of the German Second Army. The principal municipal buildings were taken over for barracks, hospitals and offices. A levy of 220,000 francs was imposed on the town. With the onset of winter, wine and coal were requisitioned by the occupying forces, while the German soldiery pilfered foodstuffs. People as well as property were at the mercy of the invaders. Stragglers from the retreat, hidden by the townspeople, were hunted down—or on occasion sold out by the townspeople—and shot as spies. The same fate awaited those found concealing weapons or suspected of spying. The civilian population was exploited as a readily available labour force. Men of fighting age were required to register with the German authorities. Often, as in Amiens, they were rounded up and sent to Germany. Others were put to work bringing in the harvest or clearing the rubble of shattered buildings. The women of Roye were organised to wash the Germans' clothing and to fill sand bags. They were paid in war notes; the contributions extracted from the occupied towns were in gold or silver. Meanwhile, the women of Péronne were conscripted for more basic tasks, as Henri Douchet recorded: "The Germans think of everything! They have just organised prostitution . . . a first list of 35 women . . . who have been required to undergo medical inspection."[64] In spring the French civilians were put to work in the fields again; but at the same time as they produced food for the occupiers, their own rations were cut. Any "useless mouths" were forced out, often via Switzerland back to unoccupied France.[65] Such a regime of terror and depredation may have kept the French population docile, but it was more grist to the allied propaganda mill which thrived on tales of horror and atrocity from behind the German lines.

Christmas came and went on the Somme. Although the ground was white with snow, the mood was black: the famous Christmas 1914 truce was only nominally observed in this sector. The autumn's fighting had been hard, and the adversaries had no great sympathy for each other. Men of 205 RI holding the front line in the Fricourt–Mametz sector briefly responded to the call of the Germans opposite to fraternise. But their colonel soon ordered them back to their trenches.[66] Elsewhere, in response to the German carols— *"Stille Nacht, heilige Nacht"*—the French sent back bullets and shells.[67] The fusillade which accompanied the New Year was more appropriate.[68] What would 1915 bring, Pic mused: victory and peace perhaps?[69] Across no man's land, the Germans wished for the same.[70] Such fraternisation as there was took place for practical purposes. After the costly December fight a party of Frenchmen approached the enemy's line under a white flag to arrange for the retrieval of the numerous blue-and-red-clad corpses which still lay in no man's land. A temporary truce ensued: but the German officer commanding made good use of the opportunity to reconnoitre the French trenches.[71] After heavy rain, bailing out the flooded trenches became a common priority and firing ceased temporarily.[72] Under the daily trial of cold, rain and sticky yellow mud, morale was weakening. A few disgruntled, hungry deserters crossed over from the German lines. Such small signs were encouraging for the French; but there was still much work to be done, and the cost would be heavy. It would take several years: "Germany will give way in the long term," Bénard wrote confidently, "but she will ruin Europe in men and money first."[73] Nevertheless, the *poilus* remained determined and morale was high: "We will get these Boches out," a cart full of wounded men had assured Poincaré when he passed them on his way to the front.[74]

The home front was also solid. In August 1914 French politics, customarily ideologically driven and divisive, had been transformed in response to national crisis. The last act of political partisanship, the assassination of the supposedly anti-war socialist leader Jean Jaurès by a right-wing extremist on the eve of war, had been unnecessary, as had the government's plan to round up and imprison left-wing pacifists and agitators on the outbreak of hostilities. Jaurès, a Frenchman and patriot as much as a socialist, had called on his followers to rally to the cause of national defence. Premier René Viviani reshuffled his Cabinet at the end of August to accommodate socialist deputies in a government of national defence, and a sacred union (*union sacrée*) between France's diverse political parties and interest groups was established in the face of foreign invasion. The union seemed to be holding. In January 1916 the leading socialist newspaper, *La Guerre Sociale*, rechristened itself *La Victoire*. France was united and determined to drive out the invader.

IN 1915 THE fighting on the Western Front went elsewhere. France and her British ally delivered a succession of increasingly large attacks against the flanks of the large German salient that jutted into France's occupied northern provinces, in Artois to the north and Champagne to the east. The results were meagre. A limited amount of shell-smashed territory was liberated, at great cost in human life. There was no sign that the German defensive line could be broken, or that the Germans were likely to be forced to retreat by such casualty-intensive methods.

Although the Somme front was not chosen for a large-scale offensive in 1915, it was not the "quiet sector" which has often been suggested in accounts of the lead-up to the 1916 battle, at least not before the arrival of the British army. Subterranean warfare commenced in the spring.[75] In March a large mine was blown under the French front line at Carnoy, and fifty of Commandant Bénard's men were lost. In the skirmish for the new crater which ensued Bénard's men understandably gave no quarter.[76]

In early summer, as a decoy from a larger attack planned in Artois, Second Army, now commanded by General Philippe Pétain, attacked the German salient at Serre, across the shallow valley which separated the opposing lines north of the river Ancre. The sector was heavily fortified, but a thorough preliminary bombardment enabled 21 DI to seize more than a kilometre of the enemy's front on 7 June. But the French infantry could not fight their way up the slope to Serre, which remained in German hands. Four further attacks over the next few days were driven off with heavy losses. Eventually the French dug in at the bottom of the valley.[77]

The fight for Serre demonstrated the tactical problems of the offensive in trench warfare. With sufficient and prolonged preliminary bombardment, the enemy's front line could be captured with relative ease and acceptable casualties. However, the follow-up attacks against reinforced German rear positions with weaker, less precise bombardment generally resulted in heavy casualties without appreciable progress. From such attacks the French army was learning that the offensive needed to be careful and methodical, with individual attacks well supported by the artillery, which was emerging as the decisive arm on the modern battlefield. Conversely, the Germans were learning that strong, in-depth fortifications and reserves close to hand could blunt an enemy attack. Chastened by this surprise, the defenders of Serre set about strengthening their positions, turning the village into a mini-fortress.[78] A year later a new army would have to attack it again.

ON 12 AUGUST 1914, COLONEL Charles à Court Repington, the military correspondent of *The Times,* had encapsulated the clash of empires which was about to begin on the Western Front. It would be "the most frightfully destructive collision of modern history. Prussian military despotism on the one hand, and the fate of fair France and little Belgium on the other, are the stakes in this giant contest."[79] It seemed to be a fight of right against might. France was invaded and, Viviani proclaimed, was "fighting for the liberties of Europe which France, her Allies and her friends are proud to defend . . . for the honour of the flag and the soil of the country." The Entente would battle for "liberty, justice and reason," Poincaré declared.[80] For "the final fight which shall settle for ever our great position in the world," German Chancellor Theobald von Bethmann Hollweg propounded a rival trinity, as *The Times* reported:

> When the German sword again glides into its scabbard everything that we hope and wish will be consummated. We shall stand before the world as its mightiest nation, which will then at last be in a position, with its moderation and forbearance, to give the world forever those things for which it has never ceased to strive—peace, enlightenment and prosperity.[81]

Two armed rival alliances, with shared European heritage but fundamentally opposed political and cultural systems, were ranging to settle their differences once and for all on the battlefield. This contest between the ideals of parliamentary democracy (whether French republican or British monarchist) and authoritarian Germanic *Kultur*—both mediated by their common imperialistic capitalism and militarism—was to dominate and devastate the continent for the next thirty years. Although the stakes were wider, and the First World War more far-reaching, than such matters of Western European power politics and nationalistic hubris, these remained fundamental truths for Germany, France and Great Britain as their armies marched towards the Somme; towards the cataclysm which was to accentuate the disparity between such straightforward, deeply held and comforting patriotic aims and the incomprehensibility and vast horror of modern industrialised war. For, as Repington would report humbly after his first visit to the front, the war "transcends all limits of thought, imagination and reason . . . This war, for once, is bigger than anybody. No one dominates it. No one understands it. Nobody can."[82]

Perhaps one man *did* understand the war (even though he could not dominate it): Field Marshal Lord Kitchener, Britain's newly appointed Secretary of State for War. He had told Repington, in his first and only press interview, that the war would be long, and that his purpose was "to wage war on a great scale." Germany, with a population of seventy million, would, he anticipated, "fight to the last breath and the last horse." France, heavily outnumbered, could sustain the principal burden of the war for only so long, so it was up to her allies, Russia and later Britain, to assist her in defeating the Central Powers. To this end, Kitchener immediately began to mobilise the empire's military resources. The Field Marshal's duty in a long war was "to prepare our land forces so that they may not only second the efforts of our friends by weak contingents, but may by their steadily expanding numbers and their constantly increasing efficiency enable us to play a part worthy of England in the war and at the peace impose terms most in consonance with our interests." Here now was a new Chatham or Pitt, a man who might fulfil Britain's destiny in enhanced post-war imperium: "at last—since the race of war is not only to the swift but to the pertinacious—we may figure in arms in a manner befitting the wealth and spirit of our Empire and the legacy of a great and honourable past."[83]

An arch-imperialist, unhappy with Britain's pre-war reconciliation with France—he himself had blocked French pretensions to expansion in Central Africa at Fashoda in 1898 — Kitchener was also a global strategist (possibly the only one produced on either side during the war). He appreciated the extent and nature of the fight to which Britain was committed. Britain would have to win the war before she could win any peace, and this was the first task of the mass army Kitchener started to raise. It would be a hard and unprecedented effort, he told the House of Lords in his maiden speech on 25 August, a challenge to the fabric of British government and society: "exertions and sacrifices beyond any which have been demanded will be required from the whole nation and Empire, and when they are required we are sure they will not be denied to the extreme needs of the state by Parliament and people."[84]

Given the wildfire spread of hostilities across the world's most developed and prosperous continent, and the immediate deployment of over ten million men in arms, Kitchener was right to expect a long and disruptive war of attrition, although in a nation gripped by war fever such balanced and far-sighted perception was likely to be overlooked. In Britain attention was focused on the small area of Europe directly across the Channel. The cause of brave little Belgium, invaded and pillaged by the Hun, rallied public support at the start of what was essentially a traditional *realpolitik* imperialist war, of the sort which had made Britain great over previous centuries. Into the war alongside Britain came her empire: the patriotic zeal of her white-settler colonies, the Dominions of Canada, Australia, New Zealand and South Africa; the vast but ill-organised resources of India; and small but heartfelt contributions from many of her lesser territories which dotted the globe. Kitchener was the man with the task, and vision, to organise these men for war. Hailed posthumously as "the architect of victory," in August 1914 his foresight was less trusted, and his methods viewed as frighteningly radical. Ignoring those around him who naïvely expected a short war, from the start he anticipated a three-year conflict: a war of attrition in which allies and enemy alike would fight themselves to a standstill. It would be the resources of the British Empire, military, industrial and financial, which would sustain the Entente into the final battles that would secure victory. The linchpin of this commitment was the army he raised in 1914 and 1915, which would be equipped and trained in time to go to the aid of Britain's hard-pressed allies in the third year of the conflict. In the meantime, sustaining the Entente became a cornerstone of Kitchener's strategy: men for France; *matériel* for Russia; money for both.

Kitchener was the last of those popular, heroic Victorian soldiers which the British Empire produced in the same way modern societies produce sporting heroes. Even today he remains famous for his heavy moustache and penetrating blue eyes, staring out from the iconic recruiting poster whose image has been constantly recycled and reinvented over the years. Yet Kitchener of Khartoum — "K of K" or "King of Kings" to the less respectful — was more than "a great poster," as the Prime Minister Herbert Asquith's wife once memorably dismissed him. Herbert Horatio Kitchener had a lifetime of imperial soldiering and administration behind him when, at the age of sixty-four, he reluctantly joined the Cabinet. Against his better judgement, he felt he could not refuse Asquith's summons in his country's hour of need. Kitchener knew he was no politician: he accepted the new appointment on the express condition that it was apolitical and only "for the duration."[85] He was a man of strong will and fixed habits, used to the firm discipline of the army and the quasi-dictatorial authority of senior rank and high imperial office. Kitchener's appointment was, the Prime Minister himself admitted, "a hazardous experiment." The genteel ways of London society were alien to him, and his new job and responsibilities were unfamiliar. He had always been his own chief of staff, and the administrative structure of the War Office was as uncongenial to him as the political practices of Whitehall. The disruptions occasioned by his arrival at the Horse Guards earned him a further epithet, "K of Chaos." Yet, for all his eccentric habits, Kitchener possessed a shrewd mind that grasped the nature of modern mass warfare and the fundamental truth of this war right at the start. "There is no army!" he declaimed on arrival at the War Office.[86] In a fight to the death Britain would need an army, and that army would be instrumental in an allied victory. And he had the confidence of the people, which in a mass war was essential. When he issued his enduring summons to the fight — "Your king and country need you!" — there was no shortage of volunteers.

Britain was above all an imperial and naval power, and until very recently her army's role had been confined to home and imperial defence. Yet, although controlling the greatest empire the world had yet seen, in the years before 1914 Britain had perceived that her greatness was resting on increasingly unstable foundations. Her defeat of Napoleon in 1815 and the control of the world's oceans that it secured had allowed Britain to remain more or less aloof from continental quarrels for the best part of a century, during which time she had developed her commercial and industrial hegemony to the point that Queen Victoria's empire was the world's first superpower. But by the end of Victoria's long reign, other states, the so-called Great Powers—France, Germany, Russia and the rising United States in particular—were closing the gap on Britain. She was, Repington and other imperial commentators opined, overstretched: her inability to meet her imperial commitments had been exposed only too clearly in the three-year struggle against a couple of thinly populated Boer republics at the turn of the century. There followed a diplomatic and military revolution. Since such an extended empire had too many vulnerable points, in the new century Britain's statesmen decided that they had to reduce the number of her enemies, and work more closely with her friends. The results, an alliance with Japan in 1902 and ententes with France and Russia to settle outstanding colonial differences in 1904 and 1907, brought Britain back to the centre of the European Great Power diplomatic system from which she had all but abdicated after 1815. In harness with her diplomatic realignment there developed a strong movement for national efficiency and imperial organisation, which would allow her weight to count in an increasingly precarious European balance of power.

The flipside of this reorientation was that Britain made new enemies. Germany in particular became the object of British animosity, both official and popular. Kaiser William II, one of Queen Victoria's many grandsons who occupied the thrones of Europe, had embarked on an active foreign policy in pursuit of colonial territory and world power after he came to the throne in 1890. Elements of this so-called *Weltpolitik,* especially the building of a High Seas Fleet to challenge British maritime supremacy, pushed the two empires on to a collision course. Britain entered the camp of Germany's traditional rivals, the Dual Alliance of France and Russia, who sat menacingly on her western and eastern frontiers. By the second decade of the twentieth century, Germany was diplomatically isolated, domestically divided, and facing the threat of a war on two fronts. This climate of fear, when catalysed by the volatile Balkan politics in which Germany's only reliable ally, the declining Austro-Hungarian Empire, was mixed up, would eventually push Europe into a continent-wide war.

The natural concomitant of Britain's developing hostility to Germany was a friendlier relationship with France. By 1914, this amounted to a tacit agreement to support France militarily in the event of a Franco-German war provoked by Germany. There was active speculation as to how many bayonets the British fleet was worth to France in the event of war—anything from zero to 500,000—but whatever its value, it was acknowledged by all that it could do nothing in the first clash of arms between France and Germany. What was needed was an army, and this military commitment was the cornerstone of Britain's continental diplomacy before 1914. Although the issue of military intervention on the continent was too politically sensitive for Asquith's Liberal government to give France a definite promise of military support, Britain's military leaders actively prepared for war alongside France, and her Francophile politicians did their best to encourage such preparation, and to furnish an effective military force.

Unlike the mass conscript armies of France and Germany, the pre-1914 British army was a small, professional, volunteer force, dwarfed by those of the continental powers. However, Richard Haldane, Secretary of State for War from 1905 to 1912, did his best to organise what Britain had into an efficient military system after the shock of the South African War exposed the deficiencies in Queen Victoria's red-coated imperial police force. Whatever the value of her fleet, by 1914, Britain definitely had the 160,000 bayonets of her Expeditionary Force, now clothed in khaki, the centrepiece of Haldane's military system. The home-based regiments and battalions of the regular army were organised into one cavalry and six infantry divisions, and equipped on a continental scale, ready for immediate despatch overseas. The mobilisation and transportation plans worked out between the British and French General Staffs before 1914 would take them to the French left wing, to a concentration area around the fortress of Maubeuge, where they would arrive in time to take part in the first clash of battle. Careful nurturing and mentoring of this small force by the French had created "six divisions . . . of extremely well-led battalions, constituting a valuable reinforcement whose intervention in the European theatre would be a real factor in success."[87] Indeed, the British official historian, Sir James Edmonds's, verdict that the BEF was "incomparably the best trained, best organized and best equipped British Army which ever went to war" has not been disputed.[88] However, in the words of one less than satisfied War Office official, the modest BEF was "a pill to stop an earthquake."[89] Moreover, what stood behind it was less reliable. A Special Reserve existed to keep the BEF's battalions up to strength (but calculated on the scale of imperial operations rather than the high rates of attrition of continental warfare). Several extra divisions of regulars might be scraped together from imperial garrisons in time. In the second line stood the Territorial Force, fourteen divisions of semi-trained local volunteers recruited for defence of the British Isles when the BEF was overseas. These were an element in Haldane's grand design of an imperial army, some forty-six divisions in all, designed to supplement the troops raised at home with Indian army units and volunteer forces raised in the white-settler colonies.[90] This elaborate scheme was far from complete when war broke out, although the Dominions possessed their own local recruiting organisations which could be used to raise reinforcements and expand colonial forces. It was this multifaceted and diverse system which Kitchener took over in August 1914.

Clearly, the British army was very different from those of its continental friends and adversaries, and its arrival at the front in 1914 was something of a wild card. Because of the British government's refusal to enter into a formal military alliance with France before 1914, it was unclear whether the BEF would even come to France's aid; and if so where it would go, and how much of it would be deployed (in the event, only four of the BEF's six infantry divisions were sent out in the first instance, and it was initially intended that they be deployed in reserve at Amiens). However, thanks to carefully worked-out transportation timetables, the deployment of the BEF to fight on the left wing of the French battle line went smoothly, and it arrived to engage the advancing enemy's right wing at Mons on 23 August. Once there the British troops gave a good account of themselves: the fire-discipline of the regular battalions, trained to fire fifteen aimed shots a minute, was devastating against the close-formed ranks of the advancing German columns. But while the BEF was certainly the best-trained and most effective force to take the field in 1914, there were hardly enough of them. The Kaiser had famously instructed his generals to brush aside Britain's "contemptible little army." The "Old Contemptibles," as the survivors of Britain's original Expeditionary Force proudly dubbed themselves, were not brushed aside, but they were forced back and badly battered during the allied retreat, and further decimated in the intense defensive fighting south of Ypres in October and November. By the end of the year the BEF was almost spent. It had fought hard and well for four months, sustaining the highest rate of casualties among British forces in any phase of the war. At least the resources of the British Empire were being mustered to come to its support.

Kitchener's aim was to raise an army of seventy divisions (with more coming from the colonies) by the third year of the war.* Between

---

*In all, between 1914 and 1918 the British Empire raised 104 infantry divisions: 12 regular, 1 Royal Naval, 32 Territorial Force (including 8 home defence divisions), 30 New Army, 18 Indian and 11 imperial.

1914 and 1918, Britain put 5,704,416 men into khaki, an eightfold expansion on her pre-war military strength.[91] Her empire added some 2,000,000 more. As well as forming six new regular army divisions and expanding the Territorial Force, creating home and overseas service divisions, Kitchener's best-known achievement is raising the thirty divisions of the so-called New (or Kitchener) Armies—five New Armies of six divisions each in all, alongside divisions raised by Canada, Australia, New Zealand and South Africa. These formations effectively comprised four separate forces, each with its own particular terms of service and *esprit de corps*. Men serving in the Territorial Force, for example, were liable only for home service, unless they took an "imperial service" oath. Many did in 1914, but many others did not, hence the splitting of the Territorial Force into home and foreign service units. Kitchener's New Armies were recruited to serve for three years, or the duration of the war.

These formations were raised on the voluntary principle. The rush of young men to the colours in 1914 and 1915 is a unique phenomenon: public schoolboys instilled with a sense of adventure and Homeric destiny; clerks and factory workers wishing to escape their numbing daily routine for the excitement of war; agricultural labourers seeking to escape the poverty of the countryside; Irishmen redeploying from their own, undeclared civil war to fight in a real war. Overseas, close ties with the motherland drew those who had recently emigrated, or the sons and grandsons of older settlers, to fight for the imperial cause, to join the great foreign adventure before it was all over. Although the newspapers were full of "brave little Belgium" and "the rights of small nations," this was merely the pretext for deep-seated patriotism (which often escalated into jingoism) and "an innate sense of obligation to King and Country" to take hold.[92] The widespread experience of collective belonging—to public schools, Boy Scouts troops, sports and fitness clubs or rifle associations—leavened by pre-war imperial propaganda, had bred in Britons a strong sense of national pride and a latent militarism, less obvious than that on the continent, but equally potent. Now that the predicted showdown with Germany was upon them, an international family of militant Britons was spoiling for the fight.

Preparing such huge numbers for war was a challenge. Britain had neither experience of, nor the infrastructure for, raising a mass army, and Kitchener's decision to bypass the existing organisational structure of the War Office only added to the mess. Army recruiting stations could not cope with the first mad rush to join up. Local corporations and aristocratic patrons stepped in to fill the gaps in the recruiting organisation in towns and counties. Under this new impetus, the ranks swelled. The *Accrington Observer and Times* of 8 August 1914 announced a "Great Rally to the Flag": the mayor, John Harwood, himself a retired captain, called on the town to raise 1100 men, its own battalion, for the conflict. British society was still sufficiently imbued with the Victorian ethos of established social hierarchy and deference that young men were ready and willing to answer the summons of their elders or betters to war. The Accrington battalion was duly raised within ten days.[93] Kitchener's "first 100,000" had reached 761,000 by the end of September 1914.[94] Factories, offices, sports associations and other local organisations started to enlist *en masse*, forming the many locally raised "pals" units from which the Kitchener Armies drew much of their *esprit de corps*. They were, according to their historian Peter Simkins, the sacrificial warriors of a dying era:

> among the last manifestations of late Victorian and Edwardian Liberalism and of a unique combination of political, social, economic and military factors which disappeared in 1916 in the face of conscription and the unrelenting demands of modern war. Above all they were the product of the British voluntary tradition and of that peculiar mixture of patriotism, parochialism, accelerating urbanization, and civic pride which characterized early-twentieth-century Britain.[95]

Men came from all walks of life. Young officers, already committed to the military life, found the opportunity to practise their profession. Bernard Montgomery, a platoon commander in the 1st Battalion Royal Warwickshire Regiment, proceeded to France with the 4th Division and received his baptism of fire in the Battle of Le Cateau on 26 August.[96] Lieutenant Alan Brooke arrived some weeks later in Flanders with a battery of horse artillery attached to the Indian Cavalry Corps, in time to fire a few salvos in the First Battle of Ypres.[97] But most of the warriors were amateurs. Liddell Hart's pondering whether he might take a military commission on leaving Cambridge was decided by Kitchener's call. He was commissioned into the King's Own Yorkshire Light Infantry, and a year later found himself deployed with his New Army division to the Somme front.[98] Educated aesthetes who would never have thought of putting on a uniform in peacetime were both sympathetic to the national cause and enthused by the prospect of

Byronic adventure. Their emotive poems and memoirs, penned after any sense of excitement and purpose had receded, would become central texts in a canon of Great War literature of deep emotional resonance and enduring fascination. Even before Kitchener's call to arms went out, Siegfried Sassoon had abandoned the sporting life to join the territorial cavalry. A commission in the Royal Welch Fusiliers followed. Half-German Robert Graves, who had given up officer training at school in favour of poetry-writing and preparing for his application to Oxford, renounced his place at St. John's College and was commissioned into the same regiment when war broke out.[99] A later volunteer, Wilfred Owen, was to reach the front in December 1916, a replacement subaltern in the Manchester Regiment. But such literati were not typical of the crowds of young men who flocked to the colours. Four Etonian contemporaries, Harold Macmillan, Oliver Lyttelton, Bobbety Cranborne and Harry Crookshank, left university to join up. But having gained their commissions in ordinary line regiments, they exploited their family connections to transfer to the Grenadier Guards. The sons of members of the present Cabinet were obliged, and willing, to set an example. The Prime Minister's eldest son, Raymond Asquith, too mature at thirty-one for the first mad dash to enlist, left his legal practice, wife and young family in the New Year to join Macmillan and his friends in the Grenadier Guards.[100] His younger brothers were all already in uniform: Arthur had been commissioned in the Royal Marines at the start of the war, Herbert was in the artillery, and Cyril, too sickly for foreign service, found a place in the Territorial Force.[101] The Chancellor of the Exchequer, David Lloyd George, in a final flourish of his pre-war Liberal pacifist credentials, hoped to dissuade his eldest son, serving in the Territorial Force, from volunteering for service abroad: "Beat the German Junker but no war on the German people and I am not going to sacrifice my nice boy for that purpose," he protested to his wife. But Gwilym and his younger brother Dick did go, although their concerned father secured them staff appointments in his "Welsh Army," the 38th (Welsh) Division, which he had been instrumental in raising in the principality.[102]

As well as these future leaders and opinion-formers, many thousands of ordinary men swelled the ranks. Colonists in Canada, Australia, New Zealand, South Africa and Newfoundland joined their own queues to take the King's shilling. For some, it was not just the cause but also a free trip home which appealed. James Bamford had emigrated from a respectable, stultifying Victorian household in Wigan in 1912 at the age of twenty: he left an Englishman and returned an Australian.[103] For others, patriotic duty appealed. Owen and James Steele, the sons of a Manchester-born St. John's tableware merchant, answered the New-foundland Patriotic Association's first call for volunteers. Owen was among the first contingent of recruits to the Newfoundland Regiment, the five hundred "blue puttees," so-called because of their improvised military attire. His younger brother joined the second contingent.[104]

Volunteering was a collective as well as an individual enterprise. The larger British cities were able to raise small armies of their own. London's regiment, the Royal Fusiliers, raised 59 battalions during the course of the war, some 235,476 men, with "Stockbrokers," "Public Schools," "Kensington," "Frontiersmen" and "Jewish" battalions among them.[105] Smaller communities mustered their own more modest but equally significant contingents, so-called pals battalions which proudly if unofficially bore their local identities: the 10th Battalion the Lincolnshire Regiment, the "Grimsby Chums"; the 11th Battalion the East Lancashire Regiment, the "Accrington Pals"; the two Barnsley battalions, the 13th and 14th the York and Lancaster Regiment. Public and private associations, professions and trades followed the lead, rais-ing battalions with unfamiliar but distinctive epithets: the "Post Office Rifles"; the "Hull Commercials"; the two "Sportsman's" battalions, 23rd and 24th the Royal Fusiliers. It was a national army whose spirit was founded on deep-seated regional and local affiliations, and personal loyalties.

There were insufficient uniforms, equipment or weapons for these hordes, and no barracks to house them. The unlucky ones spent their first winter in uniform drilling in hastily improvised tented camps in windswept fields (appropriate acclimatisation for service in the trenches); the more fortunate were sent home to await their turn. Those in training camps were clothed temporarily in quickly manufactured blue serge uniforms—"Kitchener blue." Similarly, there were few trained officers or non-commissioned officers (NCOs) to lead them. Retired officers and old soldiers were "dug out" to command and train the new formations, while junior officers and NCOs were speedily promoted from the ranks: students and schoolboys from Officer Train-ing Corps, chief clerks or factory foremen found themselves in posi-tions of command and responsibility.

This system produced battalions commanded by old soldiers with antediluvian attitudes and trained by old men in outdated methods, and companies and platoons led by keen youngsters with only superficial military experience. Lieutenant Winston Churchill, who had left the army in 1899, found himself a temporary lieutenant-colonel, in command of a New Army battalion at the front, after he left politics at the end of 1915. His attempt to drill his infantrymen with "cavalry commands, and fairly archaic ones at that," attests to the rudimentary skill of such dug-out and overpromoted officers.[106] Weapons were in short supply, and the few that the British army possessed were prioritised for the front. Rifles, machine guns and artillery would be issued to the new formations only shortly before embarkation for France, and consequently training in their use left a lot to be desired. It was obvious that it would be some time before the New Army could take the field with any hope of success. But the way the war was shaping up, there was plenty of time.

During 1915, Britain's expanding forces deployed to France and the Near East, making their first tentative forays into the trenches with mixed success. Although Great Britain had put more men into uniform than she had since her protracted struggle with Napoleon a century before, the horrendous casualty rates of the Western Front, as well as the need to sustain her armies for the prolonged attritional conflict that was developing, soon indicated that mere popular enthusiasm was not enough to see the nation through to victory. By the second half of the year, the lines at the recruiting offices had shortened, and the ranks in France were already starting to thin. The only solution, conscription of men of service age, although controversial and bitterly contested, would finally be adopted in January 1916. Voluntarism, the liberal principle on which Victorian and Edwardian Britain prided herself, was outdated. The crux of an attritional conflict was that the nation which could put the last man and the last gun into the field would win; to do that meant rethinking old ways and exercising new powers. So Britain ultimately followed the continental lead: the citizen became a soldier, the servant of, and cannon-fodder for, the expanding martial state. Her first conscripts would be ready to fill the gaps in the ranks that would inevitably result from the "Big Push."

TOWARDS THE END of July 1915, German headquarters had begun to receive reports of novel activity on the front north of the Somme. Shells fired at the German lines were of a new type, and observers reported seeing men clad in khaki among the blue-uniformed Frenchmen in the trenches opposite. Raiders were despatched to investigate, and returned with English-speaking prisoners.[107] The British army was taking over the line north of the river. After fighting alongside them for a year, the French had concluded that the inexperienced British troops were better suited to defending the line than attacking the enemy's. The British were on the Somme to train and acclimatise.

The sector which General Sir Charles Monro's newly formed Third Army took over was relatively quiet. Liddell Hart remembered that "it was possible to drill a battalion in full view of the German lines without a shot being fired."[108] That was just as well, since there was still much drilling to be done in the partially trained British divisions. Two British army corps, some six divisions, were deployed on a fifteen-mile front between the river Somme and Hébuterne.[109] For the Tommies, it was a welcome change of scenery and lifestyle after the flat monotony of Flanders. "Here the country is a rolling one," noted the 5th Seaforth Highlanders' war diarist:

> Trees are everywhere, and there are large woods . . . our men are now in the ruined village of Authuille, nicely ensconced in the beautiful valley of the Ancre . . . Down this valley flows a nice little river where the water is clear and pure and the current can actually be seen, so different from the sluggish rivers of muddy, dirty water further north.

Not surprisingly, bathing in the pretty stream was a popular distraction from military routine. The war, if quiescent, was not far away. The severed railway line from Albert to Bapaume was barricaded against the Boche, some seventy yards away; on either side stretched the trenches, "like so many zig-zag, ugly white scars on the beautiful scene."[110] Mine warfare had not abated, and small-scale "camouflets" were exploded almost daily.[111]

A steady routine of trench life commenced, "under the shadow of the mighty German fortress of Thiepval, which frowns down upon our happy valley and lets us know of its predominating position by its frequent outbursts of trench mortar and artillery fire."[112] The cycle of a few days in the front line, a few in reserve and a few at rest was ritually followed, as the onset of the second winter of the war brought the inevitable battle with the elements, the mud, rats and lice. Simply getting back and forward through the mud-choked trenches, and shoring up their collapsing sides, became the soldiers' main preoccupations.[113] Although shells were exchanged on Christmas Day when the Germans came out of their trenches to fraternise[114] — the troops were less benevolent after a year of reciprocal killing — it was to be another quiet winter, with no portent of the cataclysm to come.

By the time it reached the Somme, the British army had been fighting hard for a year alongside its French ally, and its leaders and rank and file had started to learn some important lessons about the new style of warfare. But their education was far from complete. When an army expands tenfold it necessarily becomes temporarily de-skilled, and has to relearn its trade. When that trade is also changing, such a process of adjustment is doubly difficult; especially when, after the heavy casualties amongst its pre-war cadres, the bulk of its junior field commands, senior ranks and staff appointments were filled by officers promoted rapidly and with limited experience of the responsibilities of command. In recent years, historians of Britain's First World War military effort have started to talk of a "learning curve" along which the army progressed from 1915 to 1918. In practice, this process of improvement was far from the smooth parabola suggested by this term, and some lessons were ill-absorbed, even if the British army by 1918 was as good as the others on the Western Front. Subsequent, overtly nationalistic, claims that by then it was the best army in the field probably need to be taken with a pinch of salt, since each army had strengths and weaknesses. However, that was yet to come. Certainly, Britain approached the Battle of the Somme with a naïve, partly prepared army, "New" in name and limited in competence, knowledge and experience. Military necessity dictated that its school would not be the parade ground and training camp, but the battlefield itself.

IN CONTRAST TO the hurriedly raised and equipped British army, at the beginning of 1916 the German army was at the height of its morale and physical effectiveness. The allies would have to grind down both of these before victory could be achieved. While civilians in Germany were tightening their belts, hardship had yet to bite at the front. The soldiers were living off the fat of the enemy's land; and in France, with its crop fields, orchards and well-stocked cellars, the land was particularly fat. Still buoyed with the success of its 1914 advance, and battle-hardened by its resolute defence throughout 1915, the German army remained confident of ultimate victory. In the field it was aggressive and tactically skilled. In late January, a *coup de main* was launched by 11 *Infanteriedivision* (ID) against the village of Frise, a French salient on the southern bank of the Somme, "to show the enemy once more that the old Prussian offensive spirit was alive and well in our troops and that the enemy press was lying when it maintained that our troops had lost their spirit." After eight hours of heavy bombardment the German infantry assaulted. At a cost of fewer than 100 German casualties, over 1300 French prisoners and a large quantity of *matériel* were captured. The corps commander congratulated his soldiers for their courage, devotion to duty and sacrifice; they were serving their emperor and fatherland well.[115]

The Kaiser's army was not one but many. A sub-clause in the 1871 German unification settlement had reserved to the larger monarchies and principalities obliged to join with Prussia the right to raise their own troops under the imperial Hohenzollern banner: the Kingdom of Bavaria even retained a separate army. As well as three Bavarian army corps, alongside the Prussians marched distinct regiments, divisions and corps of Saxons, Württembergers, Hessians, Hanoverians, Brunswickers and many other lesser regions.[116] However, not all shared the military spirit of the Prussians, the mainstay of the imperial army: Saxons, in particular, were notorious for fraternising with the British troops across the line, united in their loathing for all things Prussian.

Notwithstanding the local variations and antagonisms which characterised the imperial army, its martial spirit was strong and its social base solid. In pre-war Germany a man could first put on a uniform at the age of seventeen, and he would not finally be freed from martial obligations until he was forty-five. This national service was seen as an honourable duty, not an unwelcome imposition by the state. The army was the pre-eminent institution in the Kaiser's empire, and also its cement, with a heritage and ethos that could be traced back to the foundation myth of the Prussian nation. Germany's population was large enough that a certain degree of selectivity was possible in recruitment. The army favoured rural conscripts, steeped in German traditions and values, over the sons of the industrial towns and cities, who were potentially infected with the virus of socialism. Those rejected for regimental service would not, however, escape military duty entirely: their names would be put down for the *Ersatz* (supplementary) reserve, around one million semi-trained men liable for call-up in the event of war. Conscripts would strive to serve in the better regiments—the guards, cavalry or horse artillery—while a commission in the reserves brought social cachet. The higher command remained the preserve of the nobility (except in the technical branches of the armed forces), but the junior officer corps was a proving ground for the sons of the middle classes, who might be inducted into the ways of the officer and gentleman, and the martial values of the regime. Many chose the reserve officer's commission as a way of shortening their military service while enhancing their social status. Wearing the uniform of their regiment would instil pride and earn respect, as well as the admiring glances of pretty girls![117]

In addition to its pervasive social influence, the army demonstrated professionalism and competence from top to bottom. At its head sat the Great General Staff, Prussia's supreme military institution, and one which had been copied (with mixed success) across Europe after Prussia's victories against Austria and France in the 1860s and 1870s. The General Staff was a small, self-selecting elite of highly trained and influential officers, a veritable institutional brain for the army. Its various sub-sections—for intelligence, mobilisation and transportation, military education and training, and history—overseen by the Chief of the General Staff, planned operations and prepared the nation for war. It also developed a common doctrine by which the army could execute the directives of the higher command with speed and consistency.[118] General Staff officers were the demi-gods of the army, with the authority to give orders to senior commanders in the name of the commander-in-chief. In its lower echelons, the German army relied on a dedicated, highly trained and professional body of NCOs to run its administration and smaller commands. NCOs were either educated in specialist secondary schools from the age of fourteen or selected from the annual intake of recruits. A man signing on for a twelve-year engagement would be guaranteed a job in the imperial civil service after his tour of duty (provided his service record was good, which encouraged high standards). The active German officer corps, while equally professional, was small and socially exclusive, hence greater responsibility devolved upon NCOs than in the other continental armies. Even allowing for the large pool of reserve officers who would be mobilised in wartime, the army's reluctance to dilute the aristocratic officer corps meant that senior NCOs would often be found in command of platoons and companies on the battlefield.[119]

The imperial army was an institution moulded by, and ideally suited to, the Kaiser's belligerent regime. Militarism was deeply rooted in Germany's history and traditions. Historians have long argued over whether Germany started the war in 1914; whether she had premeditated aggressive intent or was merely lashing out against neighbouring states which she felt were increasingly hemming her in. Whatever Germany's motives, she had certainly prepared the means for an attack on France via Belgium. One inquisitive retired British intelligence officer made a tour of the German border with Luxembourg in June 1914, inspecting the massive railway sidings which were being completed in the barren countryside around Trèves, a small, sleepy frontier town: "elaborate detraining stations were passed every few miles . . . on one stretch were to be seen eight lines running parallel to each other."[120] The completion of her railway works coincided fortuitously with the assassination of the Austrian Archduke Franz Ferdinand by a Bosnian nationalist, furnishing the opportunity for some sabre-rattling. Even if she did not want, or anticipate, a general European war, there are strong grounds for believing that Germany had no intention of backing down if her Austro-Hungarian ally's actions precipitated one.

The plain fact was that Germany had a penchant for, as well as skill at, war. Prussia—or the more grandiose German Empire (which was merely a euphemistic fig-leaf for the imposition of Prussian control over the smaller Germanic principalities of Central Europe after 1871)—had always been belligerent. The expansionist policies of Frederick William of Hohenzollern, the Great Elector and founder of modern Prussia, in the seventeenth century had brought Prussia more land, and his son a royal title. His great-grandson Frederick the Great followed his example in the eighteenth century; and in the nineteenth victories over two Napoleons, uncle and nephew, brought Prussia to her zenith as the greatest, but also most insecure, power on the continent. It was said of Frederick the Great's Prussia that it was "not a state with an army, but an army with a state," an aphorism which still had much validity at the turn of the twentieth century.

If one thing united the allies above all, it was their determination to combat what they designated Prussian militarism, which had brought German armies into their own lands. It apparently went hand-in-hand with arrogance and barbarism. Atrocities against civilians and allied soldiers, destruction and plunder, large and small acts of vandalism worthy of the Germans' barbarian heritage—the library of Belgium's leading university, Louvain, was burned to the ground, as was Hanotaux's personal library, as the invaders marauded across France[121]—sent the allied propaganda machine into overdrive. German *Kultur*, spreading at bayonet-point across Belgium and northern France, was another thing against which the allies could unite. This helped to mobilise public opinion behind the war effort. Soldiers at the front needed no such persuasion: they could see at first hand the impact of war, of German aggression, on the people and property of northern France. It was a grim reality, a daily reminder of why they had taken up arms, of what they were fighting for and against.

The German army facing the British and French in 1916 was confident in itself, well led and disciplined, well positioned and experienced, with the knowledge that it had contained all previous attempts to drive it from its defences. Its soldiers knew that their skill and tenacity had given Germany the advantage, and that if they could hold on to the foreign soil they had seized at the start the final peace would vindicate their efforts. Germany would be freed from the crushing embrace of her rivals, which had forced her into war; then she would step up to her rightful place in the world. Edmonds recognised that "final victory could not be expected until there was a breakdown in German morale, and the Germans of 1916 were such doughty opponents that nothing but continuous hard pounding could bring them to this state."[122] The allies were planning just such a pounding, unlike anything the German army had endured to date.

The Strategic Labyrinth

JUST AFTER BREAKFAST-TIME on 6 December 1915, the usually monastic calm of Chantilly, a leafy town twenty miles north of Paris, was disturbed. A cavalcade of staff cars passed down the town's central avenue and turned into the courtyard of the Hotel du Grand Condé, the imposing luxury hotel in which G.Q.G. had installed its offices. The hard-working French staff officers briefly left their desks to peer at the generals and staff officers in colourful, unfamiliar uniforms, delegates of all the allied armies: British, Russian, Italian, Serbian and Belgian. The visitors paused momentarily to take the salute from the guard of honour, before being ushered inside to take their places at the conference table. This meeting meant much to Joffre, who had struggled for some months to arrange it. In the lull of winter, well away from the sound of the guns—and the meddling of politicians—the allies' military leaders would finally sit down together for the first time and work out a common plan. Above all, this was to be Joffre's personal achievement, for as well as being the host Joffre was now at last de facto generalissimo of the alliance.

The year to that point had been difficult for Joffre, and for the allies generally. As worrying as the enemy to his front were the enemy behind him: disgruntled, carping politicians, dissatisfied with their commander-in-chief's conduct of the war, who plotted for his removal. The German army still occupied France's nine northern *départements,* and Joffre's increasingly intensive frontal assaults on its well-placed entrenchments had achieved little other than lengthening casualty lists. While his British allies were keen, they were still few in number and poorly trained and equipped for modern warfare. In 1915 they had given little material assistance to France's war of liberation; and their political masters in London seemed more preoccupied with imperial adventures away from the Western Front. Meanwhile, Russia's ill-armed hordes—France's great hope before 1914—had spent 1915 in hurried retreat across Poland. Fresh to the fight, Italy's overconfident army had rushed headlong into the craggy hills beyond the river Isonzo in May, where it had sat ever since. Most recently, Serbia had been occupied and her army evacuated to Corfu. The inert Belgians—now an army rather than a country—sat on the end of the allied line, waiting for the inevitable peace conference. They preferred to let their allies do any fighting in France, rather than risk the devastation of Belgium.[1]

Joffre had so far made little progress in managing this disparate coalition, with the allies conducting the war "each upon his own front, and each according to his own ideas ... constantly obliged to bow to the will of its adversary." As the year drew to a close the enemy appeared to be on top everywhere. There was but one small crumb of comfort: although far from defeated, the German army was starting to show the strain of its continued battering in 1915, and it seemed that its commanders were equally unable to find an answer to the strategic stalemate in the West. Proper coordination, Joffre believed, might now wrest the initiative back from the enemy, and he felt he would achieve this at the conference.[2] He hoped that over the next three days the allies' military leaders would decide how to bring the full weight of allied force to bear on the German army. Then the enemy would be crushed, and France liberated.

Joffre approached the Chantilly conference with renewed vigour. At last his domestic troubles were beginning to abate. On 2 December the new government in Paris, headed by a firm, pragmatic lawyer, Aristide Briand, had nipped parliamentary plots against the general in the bud by appointing him as France's supreme commander, with responsibility for operations in all theatres.[3] However, the appointment as War Minister of General Joseph-Simon Gallieni, his rival for the laurels of the Battle of the Marne, meant that a close eye would still be kept on Joffre. Nevertheless, "poor Joffre," as he was accustomed to refer to himself, fully expected to prosper in 1916. The new year would be different: for the first time, the allies would have a leader and a plan.

Joffre's greatest asset was authority, which he was determined to put to good use. If one positive thing had come out of the chaos attending Serbia's defeat, it was the recognition that the alliance needed better direction. The "solution" (in reality a poorly thought-out compromise, as was much alliance policy during the war) was for France, personified in her commander-in-chief, to coordinate allied strategy, subject to the approval of all the allied governments, which would meet periodically in conference.[4] The meeting of representatives of the allied armies at Chantilly was the first step towards this coordinated strategy. Joffre had a plan, and he was determined to secure its adoption for next year's campaign.

At 09:15 Joffre entered the conference room with his entourage, and sat down next to General Sir Archibald Murray, Britain's Chief of the Imperial General Staff (C.I.G.S.), who had been seated in the place of honour on the commander-in-chief's right-hand side. The representatives of the other allies, arranged down the table in order of importance, and the aides and translators who crowded the chairs which lined the walls of the conference room, fell silent. After a few brief words of welcome from his chief, General Pellé, Joffre's chief of staff, delivered a lengthy summary of Joffre's plan.[5] The way to win the war was through better coordination and concentration on a single objective: "the destruction of the German and Austrian armies." Although Germany had won the early battles and entrenched her armies on foreign soil, the allies had an overwhelming superiority in manpower and *matériel*. Therefore, Germany's plan to outlast the allies would not work, and her defeat was only a matter of time. So far, procrastination and lack of coordination had handed the initiative to the enemy, but if the allies were agreed and ready by the spring they could finally take the war to the enemy. "The essential aim of all our efforts," Joffre proposed, should be a "simultaneous and combined offensive" by the four principal allied armies, "with the maximum of troops possible on their respective fronts, whenever they are in a position to undertake it, and when circumstances appear favourable." Although vague—the details were yet to be agreed—this common undertaking was at least a first step towards genuine coordination, essential if the allies were to secure a military decision in 1916. Meanwhile, attrition, an important element in the overall strategy to defeat the German army, should continue on all allied fronts, to wear down the enemy pending the general offensive. In particular, the British, Italians and Russians should intensify their efforts, while "France will assist in this wasting of the foe as far as her resources in men will permit."[6] Thus the outline for the 1916 campaign was presented: coordination of allied action, the attrition of the enemy, followed by a decisive blow.

What really mattered at this conference was to secure agreement on these principles. With it would come de facto acceptance of both French leadership and Joffre's personal power. Worryingly, an alternative plan had been put forward at the last minute by the Russian army's chief of staff, General Alexieff. Having worked hard to bring about this first allied strategic conference, Joffre was in no mood to have his authority challenged. Alexieff's plan—for a concerted allied effort in the Balkans to knock out Austria-Hungary—would undoubtedly have won favour with allied politicians who feared the lengthening casualty lists. Yet, even if the Russian army was up to it, which Joffre doubted, this strategy would not liberate France; nor add fresh lustre to Joffre's tarnished reputation; nor increase pressure on the main enemy, Germany. Alexieff himself could not cross an embattled continent to put his case in person, and his representative General Gilinski, head of the Russian mission at Joffre's headquarters—"a very old man; he had been chief of staff in Russia but had been removed on account of his incapacity," Douglas Haig later noted dismissively[7]—lacked the authority to stand up to Joffre and the assembled allied generals. As a result, the Russian plan was quickly thrown out.[8]

The British were less likely to object. Their representatives were last year's men. Field Marshal Sir John French, the commander-in-chief in France, had tendered his resignation the day before, after conducting a succession of indifferent offensives culminating in the costly Battle of Loos. Demoralised and powerless, he had little to say at the conference. Murray had the authority to speak for Britain, but although he was a consummate staff officer, he was no diplomat; and he was well aware that he was only a stop-gap C.I.G.S., pending reorganisation of the British War Office. Both French and Murray stressed that they could not commit Britain to any plan without first consulting their government—although it was clear that they preferred Joffre's proposed battle with the German army to Alexieff's Balkan sideshow. In the event it was left to their successors, General Sir Douglas Haig and General Sir William Robertson, respectively, to persuade the sceptical politicians in London.

When Joffre bid farewell to the allied delegates two days later he was in good heart because the conference had endorsed his plan: the "charter of the coalition," as he was to call it.[9] Yet there was still much to be done to bring his plan to fruition. The practicalities of when, where and how to attack would have to be worked out. In the conference's conclusions all of these details were, perhaps inevitably, imprecise. The allies were simply to make every effort to maximise men and *matériel*; to coordinate their efforts to the extent that their offensives would begin simultaneously, "or at a date sufficiently near to prevent the enemy bringing his reserves from one front to another"; to be ready "as soon as possible"; and to carry on the war of attrition "with the greatest intensity" in the interim. Provisionally, they were to attack at the beginning of March, with the precise date depending on the weather, the enemy and the political situation.[10]

The Somme campaign was to be the Anglo-French element of this allied plan—"the General Allied Offensive," as the press would call it. Although that was by no means certain as the delegates left the calm of Chantilly to return to the noisy reality of the war. Planning a First World War battle was a multi-stage and multi-layered process. At the highest strategic level military operations had to fit the broader political and military objectives of the alliance and its individual members: crucially, as French and Murray had intimated, the lead taken by the military chiefs at Chantilly still had to be endorsed by the allied governments. Next the time, place and nature of the offensive had to be agreed by the British and French commanders-in-chief. Then operational and tactical objectives on the battlefield itself had to be identified, and a plan of assault drawn up by the planning staffs. Troops had to be trained and deployed for the coming attack. Weapons and munitions had to be provided. The home front had to be prepared to support the fighting front. And throughout this lengthy process the plan had to be flexible, responsive to changes in the strategic, political, tactical and economic situations. Planning started in the quieter months of winter and would last until the eve of battle itself, nearly seven months in all. Inevitably, the battle that began on 1 July 1916 was to be very different from that contemplated in December 1915.

Even when these meticulous preparations were complete, it remained unclear when, or if, the enemy would break. The wearing out of the German army—attrition—was a central element of the 1916 plan, and on the basis of all known statistics that would take time. Of course, the enemy would not remain passive while the blow against them was prepared. So the allies had to anticipate what Germany would do in 1916, and how that would affect their plans. Joffre had warned his guests not to be over-optimistic: all the French General Staff's calculations indicated "that the decisive moment of the war may still be far off." But there were some reasons for hope. The Austro-Hungarian army was all but exhausted, and Germany was having to stretch her resources to reinforce her main ally. Strain was now showing on the German home front too.[11] However, the battle-hardened and confident German army was still entrenched on French soil, a fact that would dominate French thinking as detailed planning progressed.

. . .

AFTER THE CHANTILLY conference, Joffre's task was twofold: to prepare the French army to attack and to arrange effective coordination with the British. The first of these tasks would be relatively straightforward, and Joffre assigned it to his experienced and trusted subordinate Foch, commander of the French *Groupe des Armées du Nord* (GAN, or Northern Army Group).* Foch is perhaps the only First World War general who deserves recognition as one of the great captains of history. Before the war his British friend Brigadier-General Sir Henry Wilson had prophesied that Foch would lead the allied armies to victory in the coming war.[12] However, when he received Joffre's order to plan the coming offensive, his glory days were yet to come.

Like Joffre, Foch came from the Pyrenees region. He was born in Tarbes on 2 October 1851, the son of a civil servant. But there was martial blood in his veins: his maternal grandfather had risen from the ranks to be one of the first Napoleon's generals. It was a conservative, Roman Catholic area and Foch was brought up in, and continued to practise, the faith of his forebears. This proved to be a brake on his military career in a strongly anti-clerical republic at the turn of the century, until more practical considerations intervened. Foch had joined up to fight the Germans in 1870—he did not see action, but did see a shambolic French army with poor leadership and organisation—before studying, like Joffre, at the prestigious *École Polytechnique*. In 1873 he joined the artillery, then passed through the French army staff college in the 1880s. He finally became a general in 1907.

He was small and grey-haired, and thickly moustached in the style of so many French generals, and all who met Foch noted his peculiar mannerisms, which were mimicked in the lighter moments in the officers' mess at Chantilly.[13] He jabbered incessantly in short, incomplete sentences; he waved his arms and punched his fists; he twitched his head and shouted; he puffed constantly on a cigar. But behind this idiosyncratic exterior there lurked a supremely active and gifted mind, steeped in the principles and practice of war. Interspersed with his peacetime regimental duties, Foch had taught at the French staff college, and been appointed its head in 1907.

When war broke out Foch was commanding XX CA, part of the French army's covering force. It was to be badly handled at Morhange

---

*Foch's planning is examined in Chapter 3.

in August 1914, after Foch had disobeyed his army commander Castelnau's order and sent it forward unsupported against superior German
forces. Foch's star was momentarily occluded. At the same moment he
had to endure personal tragedy: both his only son and his son-in-law
fell in the first week of battle. But Joffre retained faith in his long-time
comrade. As senior French commanders were purged left, right and
centre during the retreat, Foch was promoted to command the newly
formed Ninth Army, and given a chance to redeem himself. His memorable report to Joffre as he fought to hold the Germans in the vital centre of the French line on the Marne — "My right is driven in — my centre
is giving way, the situation is excellent, I attack"[14] — epitomises this
tough, indomitable fighter. His determined defence and counter-attack
during the Battle of the Marne cemented his favour at G.Q.G.

Here Foch was joined by Colonel Maxime Weygand, appointed as
Ninth Army's chief of staff. In the gritty defence of the crumbling
French centre was forged one of the most successful and enduring military partnerships of the war. A cavalryman and trained staff officer,
calm, thoughtful and highly intelligent, Weygand proved a foil to
Foch's irrepressible live-wire. He possessed the knack of understanding
and translating his chief's half-formed and carelessly expressed ideas
into effective military plans. Weygand remained at Foch's side as he
rose through the higher echelons to the post of allied supreme commander in 1918. "*Weygand, c'est moi,*" Foch was fond of exclaiming.
He was to follow in his chief's footsteps, rising to head the French army
between the wars.

By the end of 1914, Joffre had promoted Foch to command the
GAN, with special responsibility for coordinating the three allied
armies on the northern end of the front, experience which would prove
invaluable in managing a joint offensive. Of all the French senior commanders, Foch had the most experience of working with the British. It
had begun before 1914, when he had visited London regularly to monitor the British army's preparations for war. On one memorable occasion, when asked by Wilson how many troops Britain should send to
France in the event of war with Germany, he quipped: "One single private soldier; and we would take good care that he was killed."[15] Foch
certainly expected that the latent military potential of the British
Empire would be a valuable support for France in a war with Germany,
although like other Frenchmen he realised that harnessing it would be
difficult. A blood-sacrifice, however small and symbolic, was the best
way to secure wholehearted British commitment to the joint cause.

Under Foch's management, during the First Battle of Ypres in autumn 1914, and again in the Anglo-French attacks in Artois in 1915, that sacrifice grew.

In these battles Foch worked closely with Haig and General Sir Henry Rawlinson, the British principals who were to direct the 1916 campaign. This was just as well, because Joffre's second task, coordinating with the British army, would prove difficult. First and foremost, the British and French high commands had to agree on the place, time and method of attack. Some two months of consultation would eventually produce agreement that the British and French armies should prepare to assault at their point of junction astride the river Somme on 1 July 1916. But as important as where to attack was how to attack. In 1915 attempts to break through the German defensive line had failed. The logical conclusion that such a breakthrough was impossible had yet to be drawn, although those who had already tried and failed (Foch, Fayolle and Rawlinson, the three field commanders on the Somme, among them) were starting to think this way. In 1916 two options existed for improving the chances of ending the trench stalemate: to wear down the enemy's power of resistance until such a breakthrough could be forced ("wearing-out" or "*usure,*" as British and French staffs termed it, respectively) or to attack on such a scale that the breakthrough could not be contained. Both methods had their advocates; and both involved engaging the German army in a lengthy and well-planned battle on a much greater scale than any 1915 attack. The Somme was to be an uncertain compromise between the two, as plans evolved with changing circumstances.

Joffre wanted three things of the British. First, he requested that they take over the French Tenth Army's line in front of Arras, to free French divisions for the coming offensive. Second, he wanted the fresh British army to launch preliminary "wearing-out" operations on its front, to draw in the enemy's strategic reserves before the French army delivered the decisive blow. Third, he expected that the British army would be ready to support the French army to the utmost of its power when the main offensive was launched. The man he would have to persuade on all of these issues was General Sir Douglas Haig, the incoming British commander-in-chief.

WHEN HAIG ASSUMED command of the British armies in France and Flanders on 19 December 1915 it was not expected that six months later he would preside over Britain's most disastrous day of battle. As a consequence, history and historians have largely been unkind to him, refusing to accept the magnitude of his task and the personal qualities which he brought to it. The story of the Somme is central to the story of this intelligent, purposeful and determined Scotsman. The popular image of Haig is something of a cliché: the "strong silent man, blue eyes, white moustache," an image which contemporary journalists were wont to recycle unimaginatively,[16] stares out from Haig's official portraits and stiff, posed photographs, an all too easy representation of gruff, old-fashioned stolidity for anti-Haig polemicists to exploit. The real Haig is a more complex character, a man of surprises and contradictions, quiet and private on the one hand, sharp and strong-willed on the other. He was a cavalryman, which alone is enough to damn him in the eyes of many of his critics, although before 1914 he had metamorphosed into a consummate staff officer. As his most respected biographer John Terraine recognised, he was an educated soldier.[17] Born in 1861, he was educated at Clifton College in Bristol and Brasenose College, Oxford. He was a conscientious student but, owing to illness, did not take his degree. His passion was riding—he represented England at polo—and this prepared him well for his chosen career, the army. He entered Sandhurst in 1884, passing out top of his class and winning the sword of honour. Reputedly his instructors tipped him to go right to the top of his profession.[18] At the Royal Military Academy he exhibited the hard work, rigid self-discipline and aloofness that were to be his defining characteristics; traits which earned him the respect of other men, but rarely their affection. In fact Haig's behaviour masked a deep shyness, manifested in a marked inability to communicate with his fellow officers. Ambitious and self-assured, intelligent and diligent, professional and a martinet on the parade ground, yet also reserved and aloof, Haig was always more at home in the office than the mess, and hence the trajectory of his career was set.[19]

His other defining characteristic was a deep Christian faith, which he had inherited from his strict Presbyterian mother.[20] As the burden of command came to weigh more heavily upon him, he increasingly placed his trust in God, as well as in his own judgement. His chaplain, the Reverend George Duncan, who became his spiritual comforter, noted in a post-war memoir that as the war progressed Haig began "to view in a more definitely religious light both the issues at stake . . . and the part which he himself was being called to play."[21]

After Sandhurst Haig joined the 7th (Queen's Own) Hussars, and set about the routine of regimental soldiering in India. Unlike most of his fellow officers, for whom sport and the mess were the main attractions, the punctilious Lieutenant Haig showed a real interest in the profession of arms. More remarkable still for a late-Victorian colonial soldier, he was genuinely curious about European armies. While on leave in the 1890s he attended the manoeuvres of both the French and German armies, on the latter occasion meeting Britain's future Nemesis, Kaiser William II. Haig, it was later recalled by his then chief Field Marshal Sir Evelyn Wood, "knows more about the German Army than any officer in England."[22] The knowledgeable and ambitious young officer was a natural choice for nomination to the army staff college at Camberley, which he entered in 1896. In an intake that was to produce three field marshals, his instructors again predicted that Haig would go right to the top.[23]

Now a fully qualified staff officer, Haig soon gained active service experience, first in the Sudan under Kitchener, then in South Africa, where he served as chief of staff in Sir John French's Cavalry Division, becoming a protégé of the rising star of the Edwardian army. Back in London, he was appointed aide-de-camp to King Edward VII. By now a major-general, in possession of a modest fortune, Haig was in need of a wife, if only to manage the domestic responsibilities that went with his new rank. More Collins than Darcy, he set to his uxorial quest with customary decisiveness, proposing to Dorothy Vivian, one of Queen Alexandra's maids of honour, within days of their first meeting, and marrying her within the month. Despite the haste, the match was to prove a good one: as well as proving an excellent hostess and providing Haig with four children, Dorothy became a solid rock on whom he could depend for comfort and support. He would write daily to her from the Western Front, and Lady Haig in turn would devote herself to her husband's interests at home. Moreover, in making such a union, Haig adroitly added royal favour to his existing high-level patronage in the army.[24]

In peacetime Haig held staff appointments in India and London: Inspector General of Cavalry in India; Director of Military Training at the War Office (where he helped to get the BEF ready for war); and Chief of Staff of the Indian Army. After the 1904 Entente with France, as the General Staff began to prepare for continental war, he resumed his study of the associated military problems. In particular he recognised the potential problems of coalition warfare: "Military history teaches us that the whole question of cooperation with an ally is fraught with difficulties and danger. When the theatre of operation lies in the country of the ally, these difficulties increase, for war can rarely benefit the inhabitants on the spot, and ill feeling is certain to arise."[25] Later, no doubt, Haig was to reflect frequently on the wisdom of this judgement. For their part, the French, who closely monitored the British army's preparations for war, recorded a favourable assessment of the future British commander-in-chief. Haig "is considered as one of the best British generals. He has a good eye and wide command experience. He speaks French well enough, and willingly."[26]

After peacetime command of the prestigious Aldershot Army Corps, he took them to war as I Army Corps in 1914. In January 1915, when subsidiary armies were formed in the expanding BEF, he assumed command of First Army, which he directed in the principal British offensives of 1915: Neuve Chapelle, Aubers Ridge and Loos. After early success at Loos turned to bloody massacre, Haig decided to act. Taking advantage of his private links with the King, he made sure that Sir John French, who had held back the reserves for the battle, took the blame for failure. When French was recalled, Haig was the obvious successor. To justify his earlier duplicity, Haig wrote sanctimoniously to his wife of his reception at British headquarters: "All seem to expect success as the result of my arrival, and somehow give me the idea that they think I am 'meant to win' by some superior Power."[27]

Haig was fifty-four years old and at the peak of his profession when he took command in France. He was among the most educated, professional, experienced and resolute senior commanders in the British army at a time when such qualities were sorely needed. He possessed supreme self-confidence and a sense of divine mission. So was he really out of his depth on the Western Front or merely, like so many others, obliged to re-learn his profession from first principles?

HAIG'S TASK IN 1916 was to plan the largest military operation yet undertaken by a British army. That meant determining the time, place and nature of the British attack, and coordinating it with that of the French. He was as fitted for this task as anyone, yet it was a challenging and invidious one. In practice his role in the wider offensive scheme was closely circumscribed. British historians invariably attribute too much power to Douglas Haig. The Somme, it is routinely asserted, was his battle, and in histories of the campaign the French generals with whom he worked so closely are relegated to walk-on parts: vexatious foreigners interfering with the commander-in-chief's carefully laid plans. Such Anglo-centric history has misconstrued Haig's position in the complicated allied chain of command. When he assumed command he directed a group of three British armies, soon to expand to four with the creation of Fourth Army in March. As such, his position was comparable to that of the three French army group commanders, to one of whom, Foch, Joffre had devolved the direction of the 1916 offensive. At the same time, as the commander of a major allied force, Haig was primarily responsible to his own government, not to the French command hierarchy, and retained a right of appeal against decisions taken by Joffre. Moreover, he had to clear any important agreements made with Joffre with his political masters in London. All this was symptomatic of the confusion in allied direction on the Western Front until the formal appointment of Foch as allied generalissimo in 1918.

Nearly two years of practical experience had shown up the difficulties in effective cooperation in the field: one of the factors behind Sir John French's recall had been his often fractious relations with the French high command. The instructions Kitchener issued to Haig reflected the ambiguity of the situation. Officially Haig was

> to support and cooperate with the French and Belgian armies . . . to assist the French and Belgian Governments in driving the German Armies from French and Belgian territory . . .
>
> The defeat of the enemy by the combined Allied Armies must always be regarded as the primary object for which the British troops were originally sent to France, and to achieve that end the closest cooperation of French and British as a united Army must be the governing policy; but I wish you distinctly to understand that your command is an independent one, and that you will in no case come under the orders of any Allied General further than the necessary co-operation with our Allies above referred to.[28]

Before issuing this official directive Kitchener had cautioned Haig privately "to keep friendly with the French. General Joffre should be looked upon as the Commander-in-Chief in France, where he knew the country and general situation well . . . we must do all we can to meet the French Commander-in-Chief's wishes whatever may be our personal feelings about the French army and its commanders."[29]

In principle Haig had no objections to collaborating with his allies, although he was well aware of the practical difficulties after eighteen months of close cooperation when a corps and army commander. Worryingly, however, Haig harboured a deep-seated grudge after French forces had seemingly abandoned him during the retreat in August 1914. He remained intensely suspicious and mistrustful of French intentions and promises, while believing the French to be too demanding as well as duplicitous; such "ungentlemanly" behaviour undermined his own willingness to act as a loyal comrade-in-arms. When Joffre demanded British labourers in "compensation" for the loan of French guns, for example, Haig was infuriated. "The truth is there are not many officers on the French staff with gentlemanly ideas," he noted. "They are out to get as much from the British as they possibly can."[30] Although this wariness would influence his operational judgement, often for the worse, Haig's political and strategic judgement was more astute. Right from the start he had been aware that victory depended to a great extent on the French, whose army represented in the early years of the war the main obstacle to a German conquest of Western Europe, and that this implied a certain deference to France. By 1916, he was well aware that France and its army, which had already lost nearly one million men, were beginning to tire from the exertions of the first two years of war, and that it was time Britain and her army assumed a heavier share of the burden of defeating the German army.

Individual Frenchmen, more than France, were the source of Haig's trials.[31] Joffre, whose moment of glory had long passed, was in Haig's view past his best, "the old man" when he provoked Haig's displeasure,[32] too beset by political critics, ill-advised by a shifting entourage of self-seeking staff officers, and lacking the mental acuity required for the new style of warfare.[33] Nevertheless, he recognised Joffre's right to direct, if not command, his ally, and professed himself "very anxious to do everything possible to meet the Generalissimo's wishes."[34] But a willingness to cooperate did not translate into a real affinity with or affection for his opposite number, and, given Haig's natural reserve, their relationship remained coldly formal. As Foch purportedly told his friend Wilson, "Haig and Joffre exchange letters every 15 days, and as neither says anything they are in 'perfect' agreement."[35] Foch, with whom Haig was to work closely in the preparation and conduct of the offensive, had earned greater respect from the Scotsman after their mutual efforts in the battles of 1914–15, although his manners left something to be desired.[36] He judged Foch to be intelligent, reliable and competent, despite his quirks. Importantly, he also knew Foch was able to conduct relations with his allies with great tact, a quality which Joffre sadly lacked. However, despite working side-by-side for four years, Haig and Foch never became close: their characters contrasted too much for that. So Haig's view of France and its army and his relationships with its leaders were complex and volatile. On assuming command he found himself committed to a coalition battle. He was himself committed to the coalition. Realising that commitment in practice was, nonetheless, to be fraught with difficulties and would lead to moments of acute allied disagreement.

This was only to be expected in such a close military alliance, which entailed cooperation on a daily basis. It did not help that Joffre was overbearing, and had pretensions to command the British army.[37] It was the task of the liaison officers between the two allied headquarters to manage the day-to-day business of the two armies. Colonel Sidney Clive headed the British liaison mission at G.Q.G., while Colonel Pierre des Vallières represented Joffre at British General Headquarters (G.H.Q.). Haig warmed to des Vallières immediately, not least because he was half-Irish, and having grown up there demonstrated the unostentatious, gentlemanly manners which in Haig's eyes so many Frenchmen seemed to lack. Haig naïvely took des Vallières into his confidence, inviting him to attend his daily staff briefings.[38] Yet des Vallières, for all his apparently British (even Haig-like) qualities, was deeply loyal to his nation and his chief, as acknowledged in Joffre's understated posthumous tribute: "a most useful instrument in effecting the collaboration of the French and British on the Western Front."[39] He would act as a spy in the allied camp; and with direct access to the British commander-in-chief would keep Joffre abreast of the real opinions at G.H.Q.

AN ATTACK HAD three elements: time, place and method.* At Chantilly the allies had agreed to be ready to attack in March, and had committed themselves to the principle of "simultaneous combined" offensives. How simultaneous and combined they would be in practice remained to be seen. Although it soon became apparent that the Russians would not be ready until the summer, the Anglo-French forces nevertheless looked to be prepared by the spring, not least to forestall or counter a renewed German attack on the Russians. Potentially the assault could be delivered anywhere along the Western Front. On the

---

*The first two were matters of strategy; the third of operations, discussed more fully in Chapter 3.

British front there were two realistic options: an attack on the northern flank, from Second Army's position in the Ypres salient against the Belgian coast; or an attack on Third Army's front, north of the river Somme against the Thiepval–Martinpuich–Combles range of hills. On the French front GAN could attempt to force its way across the river Somme, or Central Army Group could assault the German positions on the Aisne heights and in Champagne. As for method, on this the two headquarters were in agreement. Given the need to draw in and use up the enemy's strategic reserves behind the Western Front before any decision could be achieved, both had concluded that the main offensive should be preceded by preliminary wearing-out operations: a subsidiary attack or attacks on such a scale that Germany's reserve divisions would be pulled into the battle before what Joffre termed the "rupture" of the enemy's defensive front was attempted elsewhere along the line. Both agreed too that the main offensive had to be on as wide a front as possible to ensure that the attempt to break through could not be held up by pressure from the flanks: twenty-five miles in Haig's opinion, forty kilometres in Joffre's—the same length, but by different national measures.* However, this consensus in principle masked differences in practice, and it was to be almost two months before all these matters were resolved.

With these thoughts in mind Haig and Joffre met to consider plans for the first time at Chantilly on 23 December 1915. There was none of the pomp and ceremony of earlier in the month. Haig and Lieutenant-General Sir Launcelot Kiggell, his new chief of staff, were shown along a short, draughty corridor to the spartan office which Joffre retained in the Villa Poiret, the small house in the Boulevard d'Aumale where he lodged to avoid the daily bustle of headquarters. After brief introductions, they got down to business, setting out their respective points of view on the coming year. "Altogether it was a very satisfactory interview," Haig noted in his diary, although a final decision was reserved pending examination of each other's proposals.[40] Joffre asked Haig to consider relieving Tenth Army in front of Arras, which Haig agreed to in principle, even though he believed that the relief should be delayed until after the offensive, otherwise it would use up divisions which needed to be trained for the attack. Haig resented the implication that British troops were second-class, line-holding troops, but he saw the

---

*The same distinction between miles and kilometres in British and French operations has been maintained in this book.

relief as a useful bargaining counter in the coming negotiations with the French. As a goodwill gesture he agreed soon after to take over part of Tenth Army's front in early January. The rest would await the final plan.[41]

In fact Haig had already decided how he wanted to proceed. Ever since the front had stabilised at the end of 1914 Britain's leaders had had one eye on the Belgian coast, where the enemy had based destroyers and submarines in order to challenge British control of the Strait of Dover. The Royal Navy was severely overstretched and, since maritime security was paramount to Britain, an offensive along the Belgian coast to capture the ports of Ostend and Zeebrugge was the one way in which the army could assist the senior service in its task. Having already convinced Murray of the importance of a coastal offensive, Admiral Sir Reginald Bacon, commander of the Royal Navy's Dover Patrol, was knocking on Haig's door almost as soon as he arrived at G.H.Q. Haig was responsive to Bacon's initiative. He instructed his staff to work out plans, with the qualification that "the time of execution must depend on General Joffre's plan for the general offensive in the spring."[42] This was where Haig wanted the British army to launch its decisive attack, and he indicated as much to Joffre when they met again on 29 December.[43] Although as yet undecided, Joffre's thoughts inclined to an attack on the Somme front, where the two armies could attack on a longer combined front; the terrain was suitable, and surprise might be achieved in a previously quiet sector.[44] For Haig, the Somme front was an option for a spring attack—the waterlogged ground in Flanders would be unsuitable for an offensive before May—but otherwise he preferred this sector for any wearing-out operation before the main attack.[45]

Differences were acute over the wearing-out phase of the offensive. France was running out of manpower, and it was judged that her army could sustain only one major offensive in 1916. Therefore, Joffre had argued at Chantilly that the fresh British army should make the preliminary attack—the donkey work, although he did not say as much—leaving the more experienced French to lead the main attack and deliver the *coup de grâce* to the exhausted enemy. Murray had raised no objection, but Haig was less accommodating, reacting with surprise and concern to Joffre's proposal that his army make a large-scale attack in April or May, and then join the French in the main combined attack in July. The new British commander had a very different conception of the "wearing-out fight," as G.H.Q. called it. He thought that both armies should play a part—the British making one or two attacks and the French one—and that these should immediately precede the main attack.[46] They might do this side-by-side astride the Somme.[47]

TO UNDERSTAND HAIG'S reasoning it is necessary to step back from the Western Front, and to examine the broader political context in which the offensive was prepared. Well aware of the British government's aversion to large-scale offensive action after the indecisive attacks of 1915, on his return from the Chantilly conference Murray outlined in detail the rationale for the combined offensive, and the case for concentrating Britain's effort on defeating the German army on the Western Front.[48] His strongly made and convincing case was the climax of a year of civil–military struggle over control of British strategy, and prevarication over Britain's military commitment to France and the Western Front.

From the start of the war Britain's strategy had been ill-managed. Cabinet ministers set to distributing troops and redrawing lines on maps with gusto: "[We were] more like a gang of Elizabethan buccaneers than a meek collection of black-coated Liberal Ministers," Asquith wrote to a confidante of the first wartime Cabinet meeting.[49] Such amateur strategy, redolent of the age of Pitt and ill-suited to modern industrial war, was to be the bane of Britain's military professionals. It resulted in two particularly inept campaigns in 1915, at the Dardanelles and in Mesopotamia. The novel experiment of appointing Kitchener as Secretary of State for War in August 1914 further complicated the situation. In this new appointment, Kitchener was the proverbial square peg in a round hole: "a man of strong will and excessively stubborn," the French military attaché presciently warned his superiors.[50] After decades of semi-autocratic command in the colonies, and used to the centralised control of military and imperial administration, Kitchener was unfamiliar with the niceties of Cabinet government and notoriously intolerant of verbose, dithering civilians. The forceful, despotic field marshal rode roughshod over the War Office and the government's statutory military advisers, the Imperial General Staff (an institution for which he held no affection), while at the same time bullying his Cabinet colleagues mercilessly. Kitchener began the war as a military colossus, but as things went wrong for British arms his reputation plummeted: unfairly perhaps, for he understood the nature of the war and the complexities of strategy better than any of the amateurs with whom he sparred on a daily basis. Since he could not be sacked without provoking a public outcry, his exasperated Cabinet colleagues sought to marginalise him. Towards the end of 1915 the General Staff was reconstituted at the War Office, first under Murray and then under the intelligent, strong-willed and pragmatic Robertson, from no-nonsense Lincolnshire farming stock and the only man to rise through the ranks of the British army from private to field marshal. Robertson had returned from France, where he had been French's chief of staff. He knew the terrain, the army, the French (a "peculiar lot,"[51] in his bluff judgement) and the enemy; and he forcefully took up Murray's call for an offensive in the West as the best way to beat Germany.

Kitchener probably possessed the best strategic vision on either side in the war. Believing that the Western Front was stalemated, he was convinced that in 1915 the western allies should adopt a holding strategy until his armies were ready, and that available British resources could be more profitably employed elsewhere, which inevitably caused trouble with the French, who were fighting an intensive war of liberation on their own soil and saw Britain's growing strength as the means of their salvation. Kitchener played a cat-and-mouse game with Joffre for the first six months of 1915.[52] After the failure of the French commander's spring 1915 offensive in the Champagne region, Kitchener diverted New Army divisions to Gallipoli, placing great strain on the Anglo-French alliance. Eventually, however, it was the British War Secretary who capitulated to French pressure. At the first Anglo-French political summit which met in Calais on 6 July, Kitchener had agreed to provide a timetable for despatching British forces to France, on condition that Joffre would not mount any more large-scale offensives in 1915.[53] In coalition war, "We must wage war as we must, and not as we would like to," he cautioned his sceptical Cabinet colleagues,[54] and by the end of the summer Kitchener felt obliged to consent to an autumn offensive "to relieve pressure on Russia and keep the French army and people steady," even though he expected no significant results.[55] His apprehension proved fully justified: for the British army, the Loos offensive proved a small-scale rehearsal for the first day of the Somme.

Kitchener's actions reflected the importance of sustaining the coalition with France. Without the French army, Germany would control Europe; and the French army and nation were wearying. But so were the Germans, while the British were reaching their peak. Kitchener had always intended that Britain's first mass army would go to France when it was ready; the failure of diversions elsewhere only strengthened his resolve. By the end of 1915, he was committed to the Western Front and Joffre's offensive. His strategy of attrition might pay off; and even if it did not, it was time for the British to pull their weight.*

However, since the results of Britain's 1915 campaigns, both on the Western Front and outside Europe, had been disappointing, strategic policy was being pulled in two directions. Haig's plan would therefore have to be formulated in the context of an acute political and strategic row at home.

Put simply, the soldiers argued for a more complete commitment to the Western Front: this would sustain and appease the French, and increase the likelihood of a successful offensive. Conversely, certain politicians called for a limit to be placed on the British war effort: this would reduce the strain on the Treasury and save British lives. So the atmosphere in the Cabinet Room in 10 Downing Street was tense when the War Committee (the Cabinet committee charged with directing the war) met on 28 December. Kitchener was glad to have Robertson back from the front to second him,[56] for some of his Cabinet colleagues, with whom he had previously crossed swords over the Dardanelles, were eager for another fight. After the heavy losses at Loos, both Arthur Balfour, First Lord of the Admiralty, and David Lloyd George, Minister of Munitions, questioned the utility of an offensive in the West. Reginald McKenna, Chancellor of the Exchequer, opposed the expense of a mass army, which might bankrupt Britain before a victory was won. In a long, heated row, Robertson went head to head with Balfour and Lloyd George, the first clash of what was to prove a fraught relationship with the future Prime Minister. Lloyd George argued against becoming a French pawn: "The French General Staff was . . . unavoidably biased, for it was quite a natural desire of the French nation to drive the Germans out of France." Balfour was more anxious about the method than the principle of a Western Front offensive, fearing further costly and indecisive attacks. Backed up by Kitchener, Robertson stood his ground. Britain could not stand out alone against their allies; concentrating at the decisive point was sound strategy; and, learning from the mistakes of 1915, the allied armies would be better prepared and equipped. Faced with "a rare and remarkable unanimity of military opinion," the Cabinet Secretary Maurice Hankey noted, "whatever

---

*Fate dictated that Kitchener would never see his strategy put into full effect on the Somme: he drowned when HMS *Hampshire* sank en route to Russia in June 1916, just as the armies which bore his name were preparing to start the offensive.

misgivings they might feel—and some of them felt them acutely—the civilian members of the War Committee had no alternative but to give their consent." The army's opponents reluctantly agreed that effort should be concentrated on the Western Front in 1916, honouring the agreement with Britain's allies. There was no other way effectively to hit at Germany.[57] But it was a highly qualified assent: they were authorised only "to prepare" for a spring offensive, with a further proviso stipulating, "although it must not be assumed that such offensive operations are finally decided upon."[58] So the die was cast, albeit with Balfour's and Lloyd George's anxieties far from assuaged. The dispute rumbled on outside the Cabinet Room as they started to promote the alternative of an offensive in the Balkans, which would be more sparing of British lives. Only after further assurance that the army did not intend to try to "break through" the enemy's lines—notwithstanding the rather indiscriminate use of this expression at G.H.Q.—but rather to launch a systematic series of operations to exhaust the enemy's reserves before a decisive blow, was Balfour finally appeased.[59]

Robin Prior and Trevor Wilson have concluded that such civil–military struggles over strategy were no more than

> play acting, springing from the reluctance of sensitive men to acknowledge military necessity. As far as Britain was concerned, the war would be fought on the Western Front or it would be abandoned. None of the political masters who decried the actions of the generals wished to call off the war. Hence they could do little but wring their hands and dispatch additional men and weapons to the cauldron.[60]

Whatever the caveats of their post-war memoirs, right from the start ministers were given a clear and detailed assessment by the General Staff of the purpose and nature of the coming offensive, which was to be reinforced over the following months. They would not call it off: fundamentally, alliance considerations meant that they could do no such thing. But their anxieties were real, and their exchanges with the military were heartfelt. They might ultimately be empty rhetoric, but they could not be ignored.

HAIG WAS WELL aware of these domestic rumblings as he planned the offensive with Joffre, for Robertson was not slow to pour out his troubles to the commander-in-chief. "It is deplorable the way these politicians fight and intrigue against each other," he grumbled in early January. "They are my great difficulty here. They have no idea how war must be conducted in order to be given a reasonable chance of success, and they will not allow professionals a free hand . . . It is all very unsettling and takes up much of my time explaining what these people cannot understand and do not want to understand."[61] Fractious civil–military relations were to weaken the war effort on both sides of the Channel. In his dealings with Joffre, Haig had to take care that politicians—French as well as British—would not renege on any joint undertaking. Lloyd George was still arguing for a delay until the summer, when the guns and munitions which his ministry was responsible for providing would be ready. He was off to see his opposite number in France shortly, to find out how much ammunition was needed and how long it would take to manufacture it: "the thin end of the wedge for deferring matters," Robertson wrote tartly.[62] Lloyd George was one to be watched, Haig thought: "Astute and cunning, with much energy and push but I should think shifty and unreliable," he noted presciently when the minister called in at G.H.Q. to discuss munitions matters on the way back from this meeting.[63] At least Lloyd George appeared charmed and mollified. In thanking Haig for his hospitality, he remarked: "The visit . . . left on my mind a great impression of things being *gripped* in that sphere of operations; and whether we win through or whether we fail, I have a feeling that everything which the assiduity, the care and the trained thought of a great soldier can accomplish, is being done."[64] It was a noteworthy expression of confidence in a man Lloyd George was later to condemn roundly.

Since Haig's political masters favoured a summer offensive, when all the allies would be ready,[65] he would have to proceed carefully in his negotiations with Joffre. Haig was reluctant to commit the British army to undertaking any preparatory wearing-out attacks, fearing this would cause the politicians to get involved. Moreover, an attack in Flanders, which Murray had pointed up as the most important military operation from a British point of view, and one likely to secure a clear strategic prize, would naturally find greater favour with anxious politicians than an attack on the Somme, which aimed at no obvious objective. Politicians thought rather too simplistically, measuring success in terms of trenches, towns and territory captured, and had difficulty grasping the abstract strategic concepts of attrition and moral dominance. Deferring the attack also seemed to make sense. A premature offensive would be folly.[66] Foch, who dined with Haig on 11 January, thought that the allies would not all be ready before May, and Haig accepted that they should wait till then.[67]

On 20 January, Joffre returned Haig's visit. On arrival at G.H.Q. in St. Omer he presented Haig and Rawlinson with French decorations—the customary French embrace took the British officers somewhat by surprise! Haig and Joffre then adjourned for private discussions. In the interim Joffre's staff had considered Haig's proposals for two consecutive attacks, which des Vallières had reported with some alarm.[68] Since Haig had indicated his willingness to be ready by mid-April to make a preparatory attack on the Somme, which he would follow up with his main attack in Flanders, coincident with the other allied armies' attacks, the French felt this would be a good compromise.[69] It was not right that the British, who were to play the main attacking role in 1916, should be told what to do; moreover, this solution would alleviate the difficulties of trying to coordinate a combined offensive.[70] This formed the basis of the agreement reached between Joffre and Haig on 20 January. Joffre got the British to agree to a wearing-out offensive in the spring. The French would give limited support (in the form of artillery[71]) to a British attack on the Somme in mid-April, "to get enemy's 1st line trenches." In return, Joffre assented to an offensive against the Belgian coast, which, if the allies were not attacked first, would coincide with a French offensive in June.[72]

Far from satisfying both parties, as might have been expected, both left the meeting with grave concerns about the compromise. Haig experienced his first doubts about the French army's ability to sustain another year of heavy fighting: "General Joffre told me privately last Thursday that although his companies are quite up to strength he has no longer large reserves behind them," he reported to the Prime Minister. "The French army is thus only capable of undertaking *one* big offensive effort."[73] This worry was to dominate Haig's thoughts about his ally over the ensuing months. Conversely, atavistic fears that Britain was trying to fight the war by proxy, economising her own manpower, resurfaced among the French, and were to persist over the following months, sowing suspicion in the allied camp.[74] Joffre thought that more was needed, so a few days later he requested a second British wearing-out attack in May, should the offensive be postponed to the summer to accommodate the Russians. This came as something of a shock to Haig, who noted dryly in his diary on 28 January:

such attacks as he proposes will be of the nature of attacks in detail, which will wear us out almost more than the enemy, will appear to the enemy, to our troops and to neutrals, like the French and our own previous attacks which could not break through and be classed as "failures," thereby affecting our "credit" in the world which is of vital importance as money in England is becoming scarcer. In my opinion when we start the preliminary attacks we must be able to go right through with the campaign to a decisive issue.[75]

Moreover, Robertson had just warned him that the politicians "quite definitely opposed . . . an attack on our part which is independent of a general allied offensive." Haig was caught in a cleft stick, as Robertson pointedly reminded him: "If we cry off the preliminary wearing-down attacks the French will probably consider that we are not playing the game and are trying to avoid the losses involved. On the other hand, if Joffre tries to force on our War Committee his proposals for a preliminary offensive by us it may well lead to trouble with our government." The volatile political and military situation necessitated a further round of negotiations: nothing short of a second Chantilly conference, Robertson believed.[76]

Haig duly drove the long road back to French headquarters on 14 February for a second confrontation. Earlier he had made strong representations to Joffre on the question of the wearing-out phase; if carried out in the proposed disjointed manner, he argued, it would wear out the allied forces as much as those of the enemy. "In my opinion it is most important that once we start to attack the operations should be carried on simultaneously and without any break and as quickly as possible *by all the allies* to a decisive issue," he told Sidney Clive on 31 January.[77] The next day Clive delivered a seven-page letter which set out Haig's views in detail, because, Haig curtly informed the French commander-in-chief, "it seems necessary that there should be no misunderstanding as to what I am able to undertake and what the British Government is likely to approve of my undertaking."[78] Haig's volte-face, des Vallières reported with some alarm, had resulted from political pressure.[79]

Joffre had his own, rather different, political anxieties. Volatile French parliamentary politics always exercised a pernicious effect upon military policy, as Haig noted disdainfully as the next conference approached: "The French General Staff is much influenced by politics and do not always give a sound *military* opinion." Although Briand's government backed his plan wholeheartedly, Joffre feared that a military failure could topple the government and plunge France into domestic crisis.[80] Hence a new plan had landed on Haig's desk. If the British—government and army—were acting up, it threatened Joffre's whole plan, which had taken such effort to arrange. Since a change of command had not brought a change of attitude, and magnanimity had not secured loyal cooperation, the time had come for Joffre to take his recalcitrant allies in hand, and to revise plans accordingly. The British must take over more of the French line, and be led into the attack, side-by-side on the Somme. The preparatory attacks would be much reduced, to maximise the weight of the main attack.[81]

Haig welcomed the change, and another serious quarrel was avoided. To the French he presented a theatrical show of comradeship. When des Vallières appealed to him personally for a greater British contribution, in an uncharacteristic and almost Gallic show of emotion Haig seized him by the arms and exclaimed, "Never fear! We will do what we must."[82] But in reality there were insurmountable practical obstacles in the way of his own plan for an attack on the northern flank. On 7 February Haig had visited King Albert of the Belgians at his modest seaside villa at La Panne, in the small corner of unoccupied Belgium where he sat out the war, to discuss the coming offensive. He had returned annoyed: "Does the King express himself as badly in French as in English? I couldn't understand anything he said," he remarked disdainfully to des Vallières of his interview.[83] Desperate to save his kingdom from the ravages of war, the notoriously obstinate King had refused to cooperate with any offensive into Belgium, preferring that the allies attack somewhere in France. The sovereign's "purely selfish view of the case," annoying as it was, could not be challenged, and it obliged Haig to rethink.[84] Joffre's new proposal at least allowed him to change tack without losing face. An attack with the French on the Somme offered certain advantages: the front of attack would be longer; there would be fewer vulnerable flanks; it could be made earlier in the year, preventing the enemy seizing the initiative; and, as Joffre indicated, it would replace the piecemeal wearing-out attacks to which Haig had objected.[85]

It was with these converging thoughts that Joffre and Haig approached their Valentine's Day summit. If Cupid's arrows did not exactly strike—"I had an anxious and difficult struggle. I had to be firm without being rude," Haig noted preciously—both men came away reasonably happy. Joffre immediately gave up the idea of a separate wearing-out offensive. Any such operations would now be in the fortnight immediately preceding the main attack. It was settled that the two armies should be ready to attack astride the river Somme on 1 July. The British would have twenty-five divisions with which to seize the commanding heights north of the river; the French would have forty to break through to the south once this had been done. In total more than one million men were to be committed in the greatest battle since the huge encounters of August and September 1914. Joffre had his offensive, and his allies would play their part. However, the matter of taking over Tenth Army's front was not resolved. Haig refused to consider relieving it before the next winter—his divisions needed time to prepare and train for battle, and should not be used up holding the line—so this remained a point of difference after the conference.[86]

The compromise of 20 January had been a military one. That of 14 February was far more political, the product of a lengthy and difficult slalom through the labyrinth of national and international politics and strategic planning. This journey was not yet over, and more compromises were to follow.

IT IS AN old military adage (one that Haig was fond of quoting)[87] that no plan survives the first contact with the enemy. That first contact in 1916 came on 21 February, when the German army launched its own attritional offensive at Verdun. This took nobody by surprise. Even as the Valentine's Day summit met, the massive German artillery concentration was firing its ranging shots against the apprehensive French defenders, and Haig was well aware of French fears that a strong German attack was imminent.[88] The details of the newly minted agreement were rapidly thrown into question. In panic, Joffre called for an immediate British offensive to distract the Germans, and the relief of Tenth Army to free French reserves for Verdun. Haig was reluctant to do either—he feared the Verdun attack was only a preliminary to an attack on the British front—and was unable to do both. Des Vallières was despatched by an angry and anxious Joffre to London, where he hurried to put the French point of view directly to Robertson and Kitchener at the War Office. He was received with cold formality. The Germans, Kitchener informed him, in terms that the British commander-in-chief would echo over the coming months, "were playing our game by offering us the battle of attrition we want at Verdun." Nevertheless, briefed beforehand by the French military attaché on the large number of troops the British had available at home and in Egypt, des Vallières was able to argue a forceful case. Robertson eventually agreed to send six divisions from Egypt and one from home immediately to France, and to make up deficiencies in the ranks of the army in France. Kitchener grudgingly consented. Waiting overnight at the Ritz for a written note of the agreement, des Vallières was struck by the bright lights and gaiety of London society, partying while his countrymen were slaughtered in the allied cause.[89] French patience was failing, and their increasingly clamorous demands for British action were to trouble alliance relations in the ensuing months.

After the War Office's intervention, Haig agreed to relieve Tenth Army immediately, even though this reduced the number of divisions he would have for the summer offensive from twenty-five to fifteen. At least this was preferable to a British counter-attack. The British army, under-strength and inadequately trained, was nowhere near ready to attack: such an ill-prepared and premature engagement to relieve the French would have been just what the politicians had decried. However, just such a possibility became an ongoing theme of the ensuing months' discussions, as Joffre fluctuated between panic and calm over the situation at Verdun. Every fresh German assault brought urgent appeals for the British to do something, which Haig did his best to appease, if not to satisfy. Although it was not his intention to attack before his army was fully ready "except in an emergency to save the French from disaster,"[90] the worsening position of the French army became a factor—increasingly the predominant one—in his thinking, and dominated his dealings with the French. Since the French were discontented, panicky and importuning, and the British anxious yet unready, Anglo-French relations in the lead-up to battle were to remain volatile.

Verdun had two pernicious effects. First, it used up the French army, whose fresh reserves shrank inexorably over the following months. It remained uncertain how many divisions they would be able to commit to the offensive, or even if they would have any at all. Second, it worried the French government, and provoked renewed crisis throughout the country. Although Joffre's initial assessment that the French army's reserves would be used up in three months proved overly pessimistic, the forty fresh divisions in hand in February had been reduced to twenty-two by May, and Haig remained concerned whether his allies would attack at all, particularly as there were rumours circulating that Joffre was to be replaced and the French government was about to change the strategy.[91] Haig could comfort himself that the wearing-out fight was being fought by the French, even if this meant that his army would now have to take the lead on the Somme. Thus, paradoxically, the decline of the French army strengthened his commitment to the offensive while inexorably forcing him to rethink it. Doubts started to surface, at both headquarters, as to whether the decisive blow could be delivered in 1916, and consequently what purpose the offensive would serve. Moreover, petty-minded suspicions of France's real intentions started to develop. After the latest allied strategic conference, at which the troublesome issue of the allied campaign at Salonika in the Balkans once again dominated and soured proceedings, Kitchener dined at G.H.Q. on 29 March. Haig noted:

> He thinks the French are aiming at a development of their dominions in the Eastern Mediterranean, and will not now fight actively to beat the Germans in France. They mean to economise men, consequently it is possible that the War will not end this year. Lord K. wished me for that reason to beware of the French, and to husband the strength of the British army in France.[92]

Robertson too felt that Joffre had "no idea of ever taking the offensive if he can get other people to take it for him."[93]

Against this uncertain background Haig sought political sanction to attack: he was preparing as authorised, and the changed situation in France now required a decision by the British government, whose stance remained ambiguous and dilatory. Here the French inadvertently came to Haig's assistance. Passing through Paris on the way back from a trip to the Italian front at the beginning of April, Asquith was ambushed by Briand who demanded that Haig take the offensive to relieve pressure on Verdun.[94] Asquith took the matter to the War Committee, who agreed on 7 April to let the offensive go ahead. Hankey noted privately that their decision "was so worded as not to give the impression that the War Committee liked it,"[95] but alliance obligations and military developments had left the politicians no alternative.[96] Haig confirmed this when he appeared before the War Committee at the end of May. The offensive's purpose had by now subtly shifted: it was intended to "*dégager*"(relieve) the French army.[97] Joffre seemed to be abandoning hope of a decisive victory, accepting that the adversaries were engaged in a protracted battle of wills which the side whose morale could stand up the longest would win.[98]

It was clear from the heated discussion at the next allied conference (which again focused on the thorny subject of the allied expedition to Salonika) that Briand's government was shaky, and that the French were wavering in their commitment to an offensive on the Western Front. Haig urged the British government to remain firm: "sound policy . . . required that all our resources in men and ammunition should be sent to the decisive point, viz., France, and not wasted against the Bulgars in the Balkans, or on any other secondary objective." To clarify any lingering uncertainty as to whether an attritional or breakthrough battle was intended, he told the Prime Minister privately that a *"durée prolongée"*(drawn-out) offensive, as Foch consistently described it, was now contemplated.[99]

In truth, Haig had been through a difficult period of change and indecision, to such an extent that he no longer knew for certain what the French intended, or even if they meant to attack at all. His lack of trust in Britain's principal ally resurfaced, and he resurrected his scheme for an attack in Flanders, which could either relieve the hard-pressed French or deliver the decisive blow, as a contingency plan should the French pull out.[100] The Somme might be a preparatory blow — the two-week wearing-out attack that had originally been contemplated — before British efforts shifted to a more promising front.[101] But such an attack remained contingent on the French, so by May, Haig was very anxious to know their intentions. Robertson had earlier cautioned him: "In every way we possibly can we must take the lead, or at any rate refuse to be led against our own judgement."[102] For Haig, this meant waiting as long as possible, so that his new divisions were trained and ready for what seemed likely to be a predominantly British battle.

Haig was right to be concerned, for the French were divided among themselves. From his experiences in 1915, Foch had concluded that a decisive victory would not be possible with the means available in 1916. He argued that the offensive must be postponed until the following year, when France would have all the guns and munitions she needed. Pétain, now directing the defence of Verdun, believed that all French effort should be concentrated at that vital point, where any counter-attack should be made. As the bloody stand-off around Verdun continued, and the German army consistently failed to break through to the city after its initial success, even Joffre started to doubt that final victory was possible with the forces available, although he refused to give up hope entirely. France might not win the war by her own efforts, but the attack must still take place; the army and France's allies needed an offensive. Above all, the British must play their part: Joffre would not countenance any change of plan to allow them to go it alone in the north.[103] With Foch absent and unable to put his case for postponement (he was recovering from a car crash), France's other leaders met on 17 May to resolve their differences. It was agreed that the British should be asked to attack alone elsewhere on the front while the French concentrated their effort at Verdun.[104]

Aware that his government favoured postponement, Haig urged Joffre to delay the attack until mid-August, when he expected his men to be fully trained. Joffre demurred; he thought that any increase in British strength would be offset by a decline in French manpower. More worryingly, the French political situation was growing increasingly unstable, and by the middle of May there were growing rumours that Briand or Poincaré would call off the offensive.[105] It was time, Haig informed Robertson, "to march to the support of the French,"[106] even though he remained unsure that the French themselves were going to march. He went right to the top, making a personal representation to Poincaré in order to secure a decision.[107] On 31 May a council of war, to which Haig was invited, assembled in the dining salon of the presidential train, which had conveyed a delegation of ministers to Dury, where Foch had his headquarters. Haig had not been invited to argue with the massed French ministers and generals, merely to assent to their plans, but at least it seemed they were determined to act. Joffre's confidence had returned: he wished to attack on the Somme—"an early attack to extricate the French at Verdun," he explained privately to Haig[108]—and he hoped for significant success. Moreover, it was necessary for the French army to resume the offensive to raise morale, both military and civilian, rather than leaving it up to their ally. It would be better to use its last reserves for a counter-attack than to feed them into the Verdun mincing machine, which would have to happen if the attack were delayed until August, as Haig still wanted. Eventually he backed down, admitting that he could be ready to attack by the beginning of July, as originally agreed. The waverers—Pétain and Foch—were put firmly in their place by Briand, much to Haig's satisfaction: Foch's "excuses seemed very lame, he ate humble pie and I thought he looked untrustworthy and a schemer."[109] In the final scheme the British army would take the lead in a combined offensive on the Somme by 1 July, and the French army would support them to the best of their ability.[110] Now that matters were finally settled there was a sober recognition of the true purpose and nature of the coming attack. The battle would "help our Allies" by taking pressure off Verdun, Haig acknowledged. But he did not "think that we can for a certainty destroy the power of Germany this year. So . . . we must also aim at improving our positions with a view to making sure of the result of campaign next year."[111]

THE OFFENSIVE'S SHAPE and nature altered drastically during the months of negotiation which followed Joffre and Haig's Valentine's Day agreement. France's leaders were increasingly panicked, and Britain's increasingly constrained, by the ongoing events at Verdun. By the end of May there had been both a change of purpose and a role reversal. The contemplated attack was no longer the ambitious spring scheme of a decisive offensive which would break the static front and win the war for the allies. Foch made it explicit at the 31 May conference that he had no intention of trying to break through.[112] The "Big Push" had shrunk to another limited and potentially indecisive attack, no larger than that delivered by the Anglo-French armies in Artois in 1915, although the respective contributions had been reversed. While it was still to be coordinated with attacks on the other allied fronts—the Russians attacked on 4 June and the Italians on 15 June—its strategic rationale had changed under the cosh of Verdun. Its immediate purpose was now to relieve pressure on the French army, and its method was to be a long-drawn-out battle of attrition. Given time and good fortune, such a battle might sap the will of the enemy and ultimately lead to victory; but in the era of industrial war this was by no means assured. Nevertheless, the offensive had to go ahead, because promises had been made to allies back in December 1915, and anxious politicians had to be appeased with a success. Moreover, France and her hard-pressed army needed a fillip, and it was one of Joffre's cardinal principles to seize the initiative from the enemy whenever possible. France would therefore resume the offensive again, before it was too late. Haig, if he was to maintain his air of a loyal comrade-in-arms, could not hold back, even though he viewed July on the Somme as neither the time nor the place for an attack: his army was insufficiently trained and he would have few tangible prizes to show for any advance. Many other senior commanders shared his scepticism: the Somme field commanders in particular—Foch, Fayolle and Rawlinson—knew from prior experience that method and *matériel* were deficient for quick victory. On their plans and preparations, and those of the enemy, the fate of empires now rested. Unfortunately, in the change of emphasis lay the seeds of confusion and potential disaster.

Planning the Attritional Battle

A T DAWN ON 21 February 1916, 1220 German guns unleashed a new sort of devastation on the French positions in front of Verdun: "A storm, a hurricane, a tempest growing ever stronger, where it was raining nothing but paving stones," one survivor remembered, "around us everything is shaking, everything is breaking apart, everything is about to capsize, and it's as if we are in a ship scraping its bottom on a reef in a sea of mud."[1] Not for the German gunners a slow, methodical bombardment, such as the French had employed the previous autumn. Instead an intensive, continuous hurricane of fire swept across the French trenches till dusk, when the first assault troops advanced to probe what remained of their defences. The formula was repeated the next day, and by evening the French front had all but ceased to exist. Over the next three days German infantry pushed steadily forward, and on 25 February the weakly garrisoned Fort Douaumont, key to the Verdun perimeter, fell without a fight. It appeared that Verdun was open for the taking. But France threw her best troops into the fight, Balfourier's XX CA, who shored up the line behind Douaumont in the last days of February. The energy of the initial German assault had been absorbed, and under Pétain's steady direction the German advance was slowed and the line stabilised. Verdun would be defended—it *had* to be defended—with the blood of Frenchmen. The eyes and hopes of the nation turned towards this venerable frontier fortress, where a new and horrendous form of war was being trialled.

BY THE END of 1915, German strategy had arrived (not for the first or last time) at a crossroads. Before 1914, the empire's military thinkers and planners had wrestled with an unpalatable truth: that a future war might be long and draw on all society's resources. A true *Volkskrieg* ("people's war"), with all its concomitants of political change and social upheaval, such as the French had initiated after their battlefield defeats in 1870, would not be good for the fragile political consensus which papered over growing social divisions in the German Reich. Before the war the fractious debate on the empire's predicament spilled over into the groves of academe, where Hans Delbrück, Germany's leading military historian, had challenged the Great General Staff's contention that the solution to their strategic dilemmas—of a two-front war and potential social upheaval in a long struggle—was a short, sharp military one: a "strategy of annihilation" that would sweep the armies of Germany's enemies from the battlefield in quick time and allow her to impose a "Carthaginian peace" on her defeated rivals before the home front imploded. It was a Bismarckian solution, one which had, after fifty years of economic development and social and political modernisation, become increasingly detached from the realities of modern war and the workings of industrial mass society.[2]

Left-wing analysts such as Karl Marx and Friedrich Engels, who monitored the tensions and traumas of developing industrial societies as they fragmented along class lines, had earlier posited a far more frightening prospect in the event of war:

> No war is any longer possible for Prussia-Germany except a world war and a world war indeed of an extension and violence hitherto undreamt of. Eight to ten millions of soldiers will mutually massacre one another and in doing so devour the whole of Europe ... The devastations of the Thirty Years' War will be compressed into three or four years, and spread over the whole Continent. We will see famine, pestilence, general demoralisation both of the armies and ... the people; hopeless confusion of our artificial machinery in trade, industry and credit, ending in general bankruptcy; collapse of the old states ... to such an extent that crowns will roll by the dozens on the pavement and there will be nobody to pick them up ... only one result is absolutely certain: general exhaustion and the establishment of the conditions for the ultimate victory of the working class.[3]

Although the officers of the Great General Staff were probably not great readers of the works of Marx and Engels, they were acutely aware that the leaders of Germany's powerful socialist movement were, and they feared such a future. Germany's rulers were desperate to avoid this nightmare of social dislocation, and the political revolution which would inevitably follow; even if they grossly overestimated the likelihood of such upheaval before August 1914. By then, control of strategy and policy was firmly in the hands of the military: the first the preserve of the Great General Staff, the second that of the capricious Kaiser and his Military Cabinet and Prussia's War Ministry. Together (or more accurately separately, since the three bodies tended to work in isolation, and often at cross purposes), these unaccountable military cliques prepared Germany for combat and control once war broke out.[4] The consequences—a misguided belief in war as a solution to Germany's domestic and international problems, mass mobilisation to swell the ranks of the front-line army for the first, potentially decisive military encounter, and a battle plan which staked all on annihilating one enemy quickly in order to turn all Germany's forces eastwards against the slower but mightier Russian behemoth—meant that German strategy was deeply flawed, if entirely understandable in such unnerving circumstances. An increasingly isolated Germany—hugely powerful yet dangerously insecure (if countries can be such, independent of their leaders and citizens, who perhaps embody such powerful insecurities), armed, frightened, bombastic—was prepared to take the gamble, fully aware of just how great a gamble it was. Bethmann Hollweg dubbed it "a leap in the dark," and even the men charged with putting it into practice were dubious about its chances of success. As Helmuth von Moltke (the nephew of Bismarck's famous war manager, Helmuth von Moltke "the older"), who inherited Alfred von Schlieffen's infamous idea for a knockout blow against France, cautioned:

> How and whether it will be possible to command mass armies, as we have formed them, cannot be known to any man, in my opinion . . . [A future war] will become a *Volkskrieg* that will not be settled by means of one decisive battle, but rather it will be a long, difficult struggle with a nation that will not give in before the entire strength of its people is broken and until our nation, even if we are victorious, is almost completely exhausted.[5]

When Germany's huge armies did not respond to his control, Moltke's nerve failed. So it was left to another, General Erich von Falkenhayn, to manage Germany's long-anticipated nightmare. Moltke's strategy of annihilation had not won the war. But Germany had not lost it either. Or at least she did not know yet that she had lost it.

Despite early German victories against the French and Russian armies, they were far from annihilated. Mass armies were vital, resilient things, able to absorb a great deal of punishment before collapse, as Germany's was to demonstrate over the next four years. But the societies which sustained them were more fragile, and although all classes and creeds had rallied to the colours in August 1914, the political truce which prevailed in Germany would need to be sustained during a prolonged struggle. For Germany, the war was to be a life-or-death conflict, in which her future would be assured only by imposing her authority on Europe and consolidating her power in the world. Since her menacing neighbours had forced her to defend herself, Germany would fight on until, as Bethmann Hollweg put it in his notorious programme of war aims drawn up in September 1914, she had won "security for the German Reich in west and east for all imaginable time."[6]

Germany's war aims were couched in the typical imperialistic, nationalistic rhetoric of early-twentieth-century international diplomacy. Such bombastic sabre-rattling had promoted war in the first place, and now it would sustain and expand it. Bethmann Hollweg's extensive inventory of territorial annexations, economic supremacy and colonial expansion set the agenda for a protracted conflict. Germany's hegemonic imperial ambitions could be added—apparently with no sense of irony—to the sins of barbarism and militarism which steeled her adversaries against her. It would therefore be difficult to force Germany's enemies to the peace table. Indeed, even as Bethmann Hollweg was penning his hubristic memorandum, British, French and Russian representatives were meeting in London to sign a convention to the effect that none of them would make a separate peace until Germany had been defeated. Therefore, after barely one month of war, compromise and negotiation were off the agenda: it would be a fight to the finish, even though the strategy of annihilation (to which the Entente had subscribed as well) had failed.

When the series of bloody yet indecisive fights of autumn 1914 only served to underline this truth, Germany had to find another strategy in order to achieve her hegemonic ambition. The alternative—that advocated by Delbrück—was a strategy of attrition. Although Germany was not strong enough to take on the rest of Europe and win, Falkenhayn recognised that she was also not easy to beat, and that the cost of defeating her might be more than the allies were prepared to pay. Such a strategy had served Prussia's most illustrious military hero, Frederick the Great, well in the Seven Years' War (1756–63), in which he had kept the combined might of France, Austria and Russia at bay until they were forced to accept a negotiated peace. Now, if Germany could hold her defensive lines long enough, and inflict painful defeats on her surrounding oppressors, one or more might then be forced to negotiate, making Germany's final victory easier.[7] Although, of course, such a strategy overlooked the fundamental truth that the aims of war had shifted since Frederick the Great's more rational age. Since the French Revolution, national interest, popular will and ultimately national survival had become the touchstones of war. Germany's strategic policy had not yet caught up with the realities of the modern world.

Falkenhayn's appointment to the *Oberste Heeresleitung* (Supreme Army Headquarters, or OHL) on 14 September 1914 followed Moltke's nervous collapse.[8] He inherited a military situation which was not of his making, and a political climate which was not to his liking. His predecessor's plan to defeat France quickly was in tatters, while in the East two of his subordinates, Eighth Army commander General Paul von Hindenburg and his capable chief of staff Major-General Erich Ludendorff, were winning plaudits for the series of victories that had saved East Prussia from the invading Russians. In formulating his strategy Falkenhayn had to balance the demands of Eastern and Western Fronts. By choosing defence in the West and offence in the East in 1915, he handed Hindenburg and Ludendorff more laurels but did not bring final victory appreciably nearer. In the circles of intrigue that animated the higher echelons of the German politico-military establishment, there were mutterings that Falkenhayn had failed, that he had no coherent plan for winning the war, and that he should be ousted in favour of the victorious Eastern Front duo.

Falkenhayn's problem was not one of ability, but one of affinity. He was born in 1861, into a Prussian landowning family, as were so many of Germany's senior generals. His family's genteel poverty obliged him to follow the traditional noble profession of arms. He passed the *Kriegsakademie* exams in 1893 and served for three years with the Great General Staff in Berlin, but then his path diverged from that of most of his fellow officers. Between 1896 and 1902 he served in China, found himself caught up in the Boxer Rebellion, and therefore gained the active service experience that most German officers lacked. He did not rejoin the Berlin staff after returning from the Far East, instead taking up a series of provincial staff appointments. But his achievements in China had not gone unnoticed at court. In 1913 he was promoted lieutenant-general and appointed Prussian Minister of War. This was both a surprise and a provocation to Moltke and the officers of the General Staff.[9] When elevated to the supreme command Falkenhayn was young—at fifty-three, he was younger than all Germany's corps and army commanders—talented and confident: his official portraits show a man with upright, military bearing and a steely gaze. However, among the army's leadership he was seen as a political general who had no great understanding of, or liking for, the Great General Staff's methods and pretension to direct matters; and they, in their turn, did not agree with his strategic reorientation. For now he enjoyed the emperor's confidence: the Kaiser's firm support protected him when the pro-Hindenburg cabal first schemed for his dismissal in January 1915.[10] But since the monarch was to become increasingly marginalised as the war went on, Falkenhayn's position was always insecure.

When Moltke's bold gamble to smash France quickly before turning Germany's might against Russia did not come off, Falkenhayn found himself directing the two-front war that German military leaders had hoped to avoid. His first task was to reorganise for a defensive struggle against a coalition with more men, more money, more industrial capacity and, despite many differences, a clear common purpose. Over the winter of 1914–15, existing German divisions were restructured, giving the high command more operational flexibility. Brigades were suppressed, to free up officers and men for new formations.* In future, each division would have only three infantry regiments, but more machine guns and supporting arms, a clear shift towards a war of *matériel*. By this process, and by mobilising more men to create additional divisions, Germany began a large-scale expansion of her army, which by January 1916 would have 171 infantry divisions in the field, with deep manpower reserves for an attritional struggle.

At the end of 1915, with victory no nearer than at the year's start, Falkenhayn took the strategic gamble which took the war into a more intense phase. At the Battle of Rossbach in 1757 King Frederick had inflicted such a defeat on the French army that it had been hamstrung for the next six years of the Seven Years' War. It was felt that the same could be done to France in 1916, and the best place to do it was the vital frontier fortress of Verdun. As the Prussian demagogue of war, Carl

*See Appendix for a note on the organisation of military units.

von Clausewitz, had decreed, the strategic objective of war should be to break the enemy's capacity to resist by destroying its army and its will to fight. At Verdun the French army would have to fight or risk the collapse of national morale, and there its units could be systematically bled to death. It was the ultimate strategic logic of mass industrial war, in which the huge, resilient war-making capacity of the capitalist state and empire had to be ground down. Attrition would force France to the peace table, sundering the Entente and allowing Germany to defeat Britain in her turn. By taking the offensive at Verdun, Falkenhayn was initiating the war of attrition in earnest, and inviting the Entente to retaliate in kind: while Joffre planned to squeeze Germany to death in 1916, Falkenhayn intended to grind France into submission.

ATTRITIONAL BATTLES NOW provide the prevailing images of the Great War, and 1916 is remembered as the conflict's crowning folly. In fact, casualty rates on the Western Front were higher in the mobile operations of 1914 and 1918, partly due to the scale and intensity of the fighting but also because the shelter of well-made trenches was missing. Accepting this paradox—that on a battlefield dominated by artillery, machine guns and other industrial weapons trench warfare was the safer option for the vulnerable infantryman—the fighting in 1916 was certainly horrific and prolonged. Verdun, the first truly industrial battle, was a step up from those of 1915, which had themselves been unprecedented in their destruction and terror. At Verdun the artillery smashed all before it, and the infantry cowered in hastily scraped holes, isolated and fearful of the indiscriminate death falling from the sky. Even when the battle was in a relatively calm period an intermittent drizzle of metal persisted. When it flared up a steel storm lashed everything in its path. The earth itself was smashed and reshaped. Trees and buildings were obliterated, and whole villages were reduced to brick-dust smudges in an otherwise brown morass. For the first time the now familiar moonscape of the shell-cleared battlefield, which appeared to airmen flying overhead like "the humid skin of a monstrous toad," was seen.[11] Here the infantry had little role but to hold on, to survive, to betoken the respective front lines at which the opposing artilleries would aim: as one survivor put it, "to act as standard bearers marking the zone of superiority established by the artillery."[12] This was only the prototype for the Somme. Yet there was an emerging method in this madness, a reconfiguration of battle in response to the weight of metal which swathed and eviscerated the landscape: as Pétain and his counter-parts were fond of pronouncing, "the artillery conquers, the infantry occupies." At Verdun and on the Somme the common soldier would have to learn new ways of warfare which redefined his role on this machine-age battlefield.

From its start Verdun formed a grotesque, co-dependent, conjoined twin with the Somme. The battles were not separate engagements but the two components of a sustained campaign of attrition in 1916. Everything that happened in one battle had consequences for the other. It has already been seen how the German attack at Verdun warped and undermined the Anglo-French contribution to the General Allied Offensive, such that by June 1916 the Somme had become the Verdun counter-offensive. Similarly, in its planning, preparation and execution the Battle of the Somme was always conditioned and compromised by the French army's continued engagement on the Meuse. It was some compensation for the erosion of French resources for the "Big Push" that military method was reforged and refined in the hellfire: in Joffre's estimation "if the hardest, at the same time the best practical school of application the French Army ever had."[13] If a quick, massive break-through and a decisive victory over the German army were still contemplated, if only by the naïve or over-optimistic, at the turn of the year, Verdun affirmed what the realists had come to surmise: that indus-trial battle, and modern mass war, was intensive, slow, grinding and attritional by nature. Such a battle, as Verdun was demonstrating, involved taking on the enemy at his strongest point and beating him. It would be the army which brought the most to the battle and suffered the least, the nation which produced the most and endured the longest, which would win in the end. And come the Somme, Verdun would appear merely a rehearsal for Armageddon. This reality, the cost in lives which it entailed, would inevitably scar the nations that fought it and the soldiers who fought for them. But in a war in which fundamental principles, national values, the empire's very existence were at stake, giving everything to the fight was the only way to win. Thus the horror and killing accelerated inexorably.

ALTHOUGH IT WAS Falkenhayn who began the battle of attrition in spring 1916, the same strategic conception had not been overlooked by Joffre and the other allied commanders. Joffre's Chantilly memoran-dum had made it clear that the strategic purpose of the 1916 campaign was "the destruction of the German and Austrian armies." Kitchener had raised his New Armies with such a strategy in mind, and over the course of 1916 they were to deploy to France to increase the allies' numerical superiority over the Central Powers. Joffre and Haig had argued over the timing and nature of the attritional phase of their offensive before agreeing that their contribution to the allied general offensive would be a joint attack north and south of the Somme bend at Péronne, to be launched around 1 July.

Theoretical analysis of the strategic and operational problems and possibilities of the Western Front now had to be translated into practicable attack plans. By early 1916, the allied general staffs were part-way along a continuum that led from the bloody and stalemated small-scale battles of early 1915, such as Notre Dame de Lorette and Aubers Ridge, to the large-scale mobile offensives of 1918. Now the lessons of 1915 and Verdun had to be assimilated, certain problems addressed, and a more effective offensive doctrine promulgated.

The British and French high commands were broadly in agreement on the nature of the operational and tactical problems which they faced after their failure to break through the German defences in 1915. It was a matter of scale, timing, method and coordination: what emerged was the idea of a carefully planned and managed "scientific" battle. The idea that a quick breakthrough could be achieved was losing favour. Instead, the collapse of the German defensive front would be brought about by sustained pressure. The offensive would therefore have two phases: the first attritional, to draw in and defeat the enemy's strategic reserves, which hitherto had been able to move forward quickly enough to plug any breach made in the German defensive system; the second decisive (the "decisive attack," or *bataille de rupture*), in which a massive blow would smash into the thinned enemy front at another point, break it and roll it up, thereby restoring mobile warfare. It was not a subtle or speedy method, and Joffre's stark post-war summary acknowledges the toll it would take on both winner and loser:

> The object we had in view was victory, and since the Germans had undertaken a war of attrition, it was for us on our side to conduct it with due economy, and turn the situation to our advantage by making attacks only when we could profitably assume the role of attackers. In this way we would cause a melting of the enemy's reserves and reduce his front to nothing more than a thin barrier which, if we economised our own reserves,

could be broken down by the combined blows of the Allies, and permit the passage of our victorious battalions. In this game it was certain that we would also suffer wastage, but the enemy would be worn down; the whole question lay in conducting our affairs with such wisdom as to enable us to last longer than he did. In war it is the final battalions that bring victory.[14]

However, in a war of big battalions, attrition—somewhere, somehow—was an essential precursor to victory.

This strategic conundrum engendered a tactical parallel: how were the infantry to take ground, and hold it against the enemy's inevitable counter-attacks, with acceptable losses to themselves; and how were they to follow up early success to sustain the pressure which was required to draw in and defeat the enemy's strategic reserves, which was the real objective in the attritional phase of an offensive? By autumn 1915 it had become clear that with sufficient artillery German front-line defences could be captured relatively easily; and if that lesson had not yet been fully absorbed, the German attack on Verdun sharply reinforced it. In September the British army had taken a large section of the German front line at Loos; and in Champagne the French army had stormed most of the forward defences which had been holding it up since the previous winter. However, both attacks had then stalled in front of the German army's second defensive position, where hastily organised follow-up attacks had been checked. Counter-attacks by fresh reserves had then recaptured much of the ground lost to the first assaults. From this experience the allies developed a further tactical concept, "bite and hold," for seizing sections of the German defensive system and breaking up the inevitable counter-attacks with concentrated artillery and machine-gun fire. Only when the captured ground had been defended and consolidated could the advance be resumed, under an overwhelming curtain of artillery fire.

Foch, who would have to translate this theory into practice, was, like his fellow generals, struggling with the problems thrown up by static siege warfare. The man chosen to plan the Somme offensive was the most experienced French senior commander, as well as the most dynamic. His indomitable spirit and boundless energy were legendary, and he drove his subordinates hard to get things right. His eccentricities have been used by his detractors as evidence for a lack of clarity, narrow-mindedness and absence of vision.[15] In fact, this irrepressible personality belied Foch's shrewd brain, which was forever reflecting on the principles and practice of war. Before the war, he had combined the professions of academic and soldier. In the 1890s he lectured at the *École Supérieure de Guerre,* the French staff college, and these lectures were subsequently published.[16] His ideas were therefore widely disseminated in the higher echelons of the French army in the two decades before the war. In 1907 he was selected to head the *École Supérieure de Guerre* by Premier Georges Clemenceau, who decided that Foch's patriotism and clear-thinking mind were more important qualities than his politically suspect Catholicism and conservatism.[17] By 1916, his professor's mind was starting to understand the complex military machine that was evolving from the primordial trench mud, and to conceptualise the proper way of using it on the battlefield. Although his contemporaries were often nonplussed by his apparently simplistic formula—"Attack! Attack!"[18]—this basic principle remained the key to battlefield victory. Certainly Foch always believed in the moral ascendancy of the offensive; armies could not win wars without it, and for France before 1914 it was a valuable counterpoint to the German army's greater size, skill and material advantage. But this quasi-Napoleonic theory did not translate in practice into the tactical offensive under all circumstances, with little thought given to the safety of the men. Like others who witnessed industrial battle at close quarters, Foch was immediately aware of its true nature, apparent from the tactics he prescribed in the Marne counter-attack: "Infantry was to be economised, artillery freely used and every foot of ground gained was to be at once organised for defence."[19] This was the cardinal tenet of the new style of battle—*matériel* and defence works were to be used in increasing amounts to preserve flesh—even if its implementation was as yet rudimentary.[20] "Lots of artillery, few infantry," and the proper integration of the two was Foch's formula for tactical success on the modern battlefield.[21] The artillery would prepare the way for the infantry's advance, a few kilometres at a time, but their élan would get them across the battle zone to engage the enemy at close quarters. "This concept of the battle of the future suggests large, repetitious offensives, with plenty of guns, and lots of munitions," he concluded. Foch recognised too that the days of linear operations, seeking to turn an exposed flank, were over. Battle was now to be developed in depth rather than breadth, hacking through the enemy's defences by successive forward strokes, and would continue for an indeterminate period of time.[22] For Foch it was the "break-in," not the "breakthrough"—a subtle but vital difference—which determined the outcome of the battle.[23]

By the start of 1916, with a year of unsuccessful attacks behind him, Foch was already sceptical of Joffre's ambitious "breakthrough" strategy, which was unrealistic, given current weapons and tactics. "Lack of suitable *matériel* had . . . halted our offensive in the war of manoeuvre; the want of it was even more strongly felt in the war of position," he recalled in his memoirs.[24] These doubts would become more pronounced as the French army was ground down at Verdun. The war, Foch felt, would last another eighteen months, and it would be another year before the allies had sufficient *matériel,* in particular heavy artillery and munitions, for the final, decisive battle.[25] But even though he made no secret of the fact that he thought the 1916 plan was overambitious,[26] Foch remained committed to an offensive strategy, considering it both politically and morally essential; and as a loyal subordinate he was determined to carry out his chief's instructions, his own better developed theories of war notwithstanding.[27] He would therefore have a difficult and unenviable task in 1916, applying evolving doctrine to Joffre's overambitious, but necessary, offensive. The battle which he would direct from June to November 1916 would be on a smaller scale than that contemplated in February, but it would broadly follow the scheme established the previous winter. Attrition of the German army would be sustained, and perhaps its resistance would break on the Somme front itself.

WHILE JOFFRE AND Haig had been negotiating the place and time of the joint offensive, Foch had been working on the operational method. He would employ his tactical formula to make a steady systematic advance, in what he dubbed *"un effort de longue durée."* Taking one German defensive position at a time by "a sustained action, involving a series of stages as close together as possible," would draw the enemy's reserve division into the fight. After that a decisive victory might be sought elsewhere on the front. The keys were to attack on a sufficiently long front—Foch suggested between 100 and 200 kilometres—and to possess adequate *matériel.*[28] Joffre's initial conception, however, was more ambitious: with the British now committed, Foch's suggested wearing-out battle between the Somme and Lassigny was to become a decisive breakthrough battle, with the wearing-out taking place elsewhere.[29] For this, Joffre planned to place three French armies at Foch's disposal. Sixth Army currently held the front south of the Somme. Pétain's Second Army was in reserve, and it was Joffre's intention that it would form the training organisation and exploitation force for the offensive. Finally, Tenth Army, currently interposed between two British armies in the Arras sector, would be relieved by the British and so be free to take over part of Sixth Army's front. The provisional date for the assault was 1 July, when the Russians should also be ready, but Foch was warned that everything needed to be ready for April, in case the Anglo-French forces had to respond to a German offensive in the East.[30]

However, before Foch had even drawn up a preliminary offensive scheme the plan started changing. When the Battle of Verdun began, Joffre had thirty-one fresh divisions in reserve. Over the ensuing weeks and months these were to be fed into "Meuse mill." Pétain's Second Army was called immediately to the defence of Verdun, and by April thirty-nine divisions had passed through its hands.[31] Joffre had managed to secure further reserves as a result of the British relief of Tenth Army in March, but this was double-edged as it reduced the number of divisions that Britain could commit to the offensive. Against this background of ongoing attrition and diminishing resources, Foch had to formulate a plan of attack, and integrate the efforts of the two allied armies.

THE SOMME SECTOR, where the allied fronts joined, was fair, if not ideal, fighting country. The landscape of the battlefield, approximately delineated by a quadrilateral of small market towns, Bapaume, Péronne, Chaulnes and Albert, has changed little in over ninety years. A large, square, metallic aircraft factory sits just outside Albert, and the parallel lines of *autoroute* and TGV railway track cut across the rear of the combat zone. But otherwise the Somme remains a rural backwater, specked with woods, scattered with small villages among its wheat, potato and sugarbeet fields. The main roads still trace their long-established pattern, although most of the early-twentieth-century railways are now gone. The road between Bapaume and Péronne to the east and the two straight Roman highways linking those towns to Amiens triangulate the battlefield. Through its middle the river meanders, across a wide, marshy flood plain, flowing south to north, then turning westwards at Péronne and looping to Amiens and the sea, bisecting another outcrop of the chalky downs on which the Germans chose to site their defences in 1914. Chalk hills gave the defender several advantages. Most obviously, high ground gave observation over the enemy's lines. Chalk drained well, and was also easily mined. Dry, deep, solid entrenchments could be engineered. South of the river lies the Santerre, another chalky upland, but flatter and more open, gently undulating in contrast to the rolling landscape to the north. Here the riverine bluffs rise more gently to a low, flat plateau tucked in the Somme bend opposite Péronne to which the village of Flaucourt, in a dip at its western end, gives its name. It was a commanding position, dotted with artillery batteries, from which fire could be concentrated on either river bank. Four kilometres west of Flaucourt lay the German front line south of the river, with the French front around five hundred metres beyond. West and south of the Flaucourt plateau the ground is relatively level and featureless, dotted with small villages and woods, incorporated as strongpoints into the German defences. Along the river Frise, Feuillères and Biaches villages hug the southern bank; Curlu, Hem and Cléry nestle between successive spurs of the chalk downland which rises more sharply to the north. Here there are more villages and larger woods, either in the dips or on the crests of the succession of parallel folds which rolled out behind the German front line. Together they form a high ridge, named after the village of Thiepval which surmounts its westernmost promontory, a highly fortified bastion from which German sentinels looked out over the British front, stretching away north to Beaumont Hamel and south to Fricourt, the apex where the parallel fronts turned eastwards. It was a strong natural position, made stronger by the ceaseless work of the German army's field engineers. Haig's despatch on the battle reported a well-sited line, which the Germans had spent two years making impregnable:

> The first and second systems each consisted of several lines of deep trenches, well provided with bomb-proof shelters and with numerous communication trenches connecting them. The front of the trenches in each system was protected by wire entanglements, many of them in two belts forty yards broad, built of iron stakes interlaced with barbed wire, often almost as thick as a man's finger.
>
> The numerous woods and villages in and between these systems of defence had been turned into veritable fortresses . . . The salients in the enemy's line, from which he could bring enfilade fire across his front, were made into self-contained forts, and often protected by mine fields; while strong redoubts and concrete machine gun emplacements had been constructed in posi-

tions from which he could sweep his own trenches should these be taken. The ground lent itself to good artillery observation on the enemy's part, and he had skilfully arranged for cross fire by his guns.[32]

Haig's decision to attack such a formidable position has subsequently been questioned; but he had no choice. Moreover, after two years of fortification most of the rest of the enemy's front line was much the same.

FOCH'S FIRST PLAN, submitted to Joffre on 16 March, was for an attack by the three French armies on a forty-five-kilometre front between the river Somme and Lassigny.[33] The centre army would break through at Roye, pushing forward to cross the Somme between Cizancourt and Offoy. The others would cover its flanks, the southern one seizing the high ground north of Lassigny, and the northern one the Flaucourt plateau. The British would extend this front a further twenty-five kilometres north of the river, where they would assault the Thiepval heights and push eastwards, initially to cover the flank of the French advance, and subsequently to advance on their southern flank between Maurepas and Cléry to turn the bend of the river at Péronne. On this seventy-kilometre front the allies would employ sixty-seven divisions—forty-two French and twenty-five British. Thus the original British operational objective was to outflank the Somme bend to facilitate the French advance. While the main British objective on the ground, the Thiepval–Combles massif, would not change over the ensuing months, the significance of their attack and its method would develop as the French army's role in the battle was reshaped because of events at Verdun.

The first impact of Verdun was to use up the French army's reserves. Less obvious, but more significant, Joffre's attention turned eastwards, and Foch found his operation was losing its momentum even while the plans were still being drawn up. Although Joffre approved Foch's plan,[34] in late April the latter learned that the offensive was to be scaled down.[35] Foch's reservations about the offensive were heightened. In a self-justificatory note which he prepared at the end of the battle he emphasised the prophetic warning he had sent to G.Q.G. in early May: "In any case, and whatever method you decide to adopt, after some tactical successes at the start of the action we must anticipate that we will

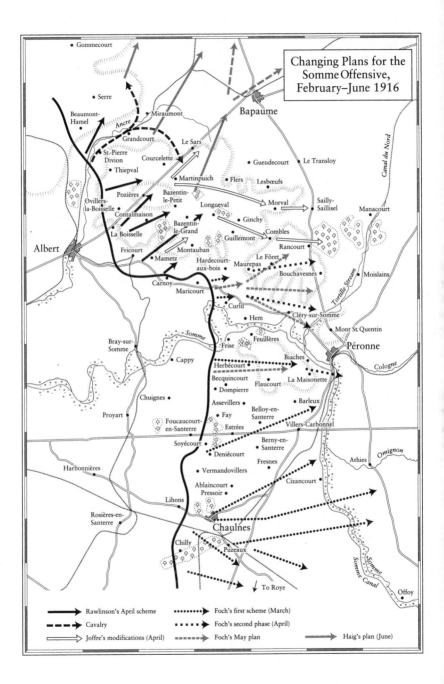

Changing Plans for the
Somme Offensive,
February–June 1916

Gommecourt
Serre
Beaumont-Hamel
Miraumont
Bapaume
Ancre
Grandcourt
Le Sars
St-Pierre Division
Courcelette
Thiepval
Gueudecourt
Le Transloy
Martinpuich
Flers
Lesbœufs
Pozières
Bazentin-le-Petit
Longueval
Morval
Sailly-Saillisel
Ovillers-la-Boisselle
Contalmaison
Manacourt
La Boisselle
Bazentin-le-Grand
Ginchy
Combles
Fricourt
Montauban
Guillemont
Rancourt
Mametz
Hardecourt-aux-bois
Le Fôret
Albert
Carnoy
Maurepas
Bouchavesnes
Moislains
Maricourt
Curlu
Hem
Cléry-sur-Somme
Mont St. Quentin
Bray-sur-Somme
Somme
Frise
Feuillères
Péronne
Cappy
Herbécourt
Biaches
Cologne
Becquincourt
Dompierre
Flaucourt
La Maisonette
Chuignes
Asservillers
Belloy-en-Santerre
Barleux
Proyart
Fay
Foucaucourt-en-Santerre
Estrées
Villers-Carbonnel
Soyécourt
Berny-en-Santerre
Fresnes
Harbonnières
Deniécourt
Athies
Omignon
Vermandovillers
Cizancourt
Ablaincourt
Pressoir
Lihons
Rosières-en-Santerre
Chaulnes
Chilly
Puzeaux
Somme Canal
Offoy
To Roye
Canal du Nord
Tortille Stream

→ Rawlinson's April scheme
┅┅ Cavalry
▷ Joffre's modifications (April)
••••• Foch's first scheme (March)
▪▪▪▪ Foch's second phase (April)
═══ Foch's May plan
➤ Haig's plan (June)

soon be checked; that is to say there will be losses and sacrifices for no obvious gain."[36] There could be no doubt that the battle would now be "hard and long, whose outcome would be the attrition of the means which the enemy engages in the theatre."[37] Furthermore, the British would now be making the main attack, with the French playing a supporting role.

WHILE FOCH AND Joffre had been sparring over the size and nature of the offensive, the British had not been idle. Lieutenant-General Sir Henry Rawlinson, commander of the newly formed British Fourth Army, was responsible for planning the assault on the Thiepval Ridge. "Rawly"(to his friends) was an experienced, intelligent and efficient general, but not the most popular man in the army. From Sandhurst days he had carried the unfortunate nickname "the Cad," a reflection of his ambition and sense of self-importance, which have made him fair game for those historians who tar all Britain's First World War generals with the same dismissive brush. He certainly looked every bit the stereotypical British general: tall, lanky, balding and moustached. Born in 1864, he was educated at Eton and Sandhurst, before being commissioned into the Rifle Brigade, transferring later to the Coldstream Guards. An early posting to India brought him to the notice of a powerful patron, Lord Roberts, and he gained his first campaign experience against the Burmese Dacoits in 1889. He attended the staff college in the 1890s, before serving on the staff of his second powerful patron, Kitchener, in the 1898 campaign in the Sudan. Further staff service with Roberts and Kitchener in the South African War was followed by his first independent field command, of a flying column pursuing Boer commandos across the veldt. Back in London, Rawlinson joined the War Office staff, before becoming commandant of the staff college between 1904 and 1907. After that he held a number of home commands, and when war broke out had just stepped down from commanding 3rd Division on Salisbury plain. In the small, professional Edwardian army, the support of high-ranking patrons had done much to further Rawlinson's career, but his own talents and intelligence had also contributed to his rise. He studied his profession thoroughly, and in particular thought carefully about the changing nature of war at the turn of the century, and the new military technologies which the army was starting to use. As the most thorough study of his First World War career concluded:

It is fair to say that when war came to Europe in 1914 Rawlinson was well equipped to exercise command in battle. He had directed troops under fire, and shown capacity as a staff officer ... He showed commendable concern for that facet of war which was to loom so large in the endeavours of the British Army during the Great War: the marriage of infantry with ample fire power—in particular the fire power of machine-guns and artillery.[38]

When war broke out Rawlinson was on half pay, but his old patron Kitchener appointed him Director of Recruiting at the War Office. He set the process of raising the New Armies in motion, before being hurried out to France to take over from the wounded commander of 4th Division. In early October, Kitchener called him back to command IV Army Corps which he had mustered to raise the siege of Antwerp. Rawlinson found himself next in Belgium, his corps linking up with the BEF in the hard-fought defence of Ypres. After his corps was broken up he returned briefly to London, but in January 1915 he was back in command of the re-established IV Corps, where he found a new patron, the commander of First Army, Sir Douglas Haig.

IV Corps took part in all three large-scale British offensives in 1915. In the first, at Neuve Chapelle in March, in what was little more than a large raid, a section of the enemy's front line was seized after a sudden, surprise, hurricane bombardment. Then, in a foretaste of the future, the attack quickly bogged down with heavy losses. In the post-battle inquisition, Rawlinson was nearly sacked. Haig's advocacy saved him.[39] The relative success at Neuve Chapelle contrasted with IV Corps's failure in its next attack on 9 May in the Battle of Aubers Ridge. When a lighter but more methodical bombardment failed to destroy the enemy's defences 8th Division's infantry advanced towards undamaged trenches manned by machine-gunners.[40] The resulting carnage was a small-scale rehearsal for 1 July 1916.

The Battle of Loos, in September, was to be Haig and Rawlinson's greatest test so far. Again the first day went well. If not a success along the whole front, First Army's assault seized its objectives along much of the enemy's first line. IV Corps, the most experienced attacking unit, broke through on the right and pushed its leading battalions up to the German second line, where they halted to await support. But on day two the battle started to go badly wrong: inexperienced reserves, thrown against the intact German second position with inadequate artillery support, were cut to pieces. On this occasion Haig—ably seconded by Rawlinson, who wrote to Kitchener and other political contacts in London[41]—was able to shift the blame from his shoulders to those of French, who had held the reserves back under his control. When Haig replaced French a few months later, his loyal and experienced subordinate naturally benefited too, being appointed to command Fourth Army in the Somme sector in February 1916.

Thus, when they started to plan the Somme offensive, Haig and Rawlinson had a long-established and effective working relationship. Haig trusted Rawlinson, and Rawlinson deferred to Haig, if not always without an argument: they had often disagreed on the detail of operations in 1915, but Haig's opinion had generally prevailed. Now each assumed a new and higher level of responsibility. Rawlinson had never planned a battle for a whole army; Haig had, and he was determined to impose his authority on the battle that would make or break his reputation. In this lack of clear command responsibility lay the potential for disaster. The hybrid and inadequate attack plan that resulted presents a sorry tale of confusion and compromise.[42] Not only the micromanagement of the commander-in-chief but the changing situations of ally and enemy would leave Rawlinson with a plan that was poorly suited to the circumstances he faced on the ground.

But this was all in the future when Rawlinson took up his army command, and he initially addressed the task with relish: "It will be a great battle, and I am delighted at having the duty of thinking and working it out," he scribbled in his diary after discussing the operation with Haig.[43] "Capital country in which to undertake an offensive," he informed the King's aide-de-camp after his first tour of the front lines.[44] Privately he noted optimistically, "one gets an excellent view of the enemy's positions which could I think be captured without much trouble."[45] The Fourth Army's operational objective was quite straightforward: to seize the Thiepval Ridge and push on to the Bapaume–Péronne road. This would involve taking three successive German defensive positions. Rawlinson had learned from his 1915 battles "that a line of trenches can be broken with suitable artillery preparation combined with secrecy,"[46] and that for the infantry to make progress with acceptable casualties it was necessary for the artillery to concentrate its support.[47] With a greater concentration of guns than ever before, this was not thought to be problematic. He had also identified the need to prepare carefully but promptly the assault on the enemy's second position: both Neuve Chapelle and Loos had shown that pressing an attack without proper artillery preparation once the enemy's defence had stiffened was a recipe for chaos and heavy casualties.[48] His basic principle of "bombard and storm" held good, and his recognition that the infantry's task was subordinate to that of the artillery suggests that Rawlinson had learned the fundamental principle of offensive warfare on the industrial battlefield.[49] Perfecting this in practice, however, was still a matter for reflection. The nature of the bombardment and speed of the advance were moot: intensive and quick to retain the element of surprise and destabilise the defence; or slow and methodical to ensure the proper destruction of the enemy's defensive works and minimise casualties? Haig had learned a different lesson from Loos—the need for speed and to push the reserves through quickly to support early success—which was to underpin his own approach to the offensive. This key issue was Rawlinson and Haig's major point of disagreement when planning the first assault.

Rawlinson's first plan, submitted to G.H.Q. on 3 April, was unacceptable. He proposed a slow, methodical advance, assaulting and capturing one German defensive position at a time before moving up the guns to attack the next (the method which the more experienced French army advocated). His four army corps would take the front between Serre and Mametz (the "first objective" or "blue line," as the staff defined it), some twenty thousand yards long, establishing defensive flanks to north and south. The German second position (the "green line") would then be assaulted north of Pozières, where it was more easily targetable by the British guns. This, Rawlinson suggested, "would take a fortnight or more."[50] This was a prime example of the "bite and hold" method, a term Rawlinson himself had coined after Neuve Chapelle.[51] In it the infantry's objectives would be determined by the support the artillery could provide, which itself depended upon having sufficient heavy howitzers and good observation over the enemy's line.[52] Initially only the German front position could be observed from the forward slopes of the shallow valley across which the adversaries eyed each other. The second line was four thousand yards away, on a reverse slope, and a third line was under construction in front of the Bapaume–Péronne road a couple of miles behind that. Haig's objection was not to Rawlinson's tactical method, but to his operational conception. Haig's diary comment of 5 April has often been cited: "[Rawlinson's] intention is merely to take the Enemy's first and second system of trenches and 'kill Germans.' . . . I think we can do better than this by aiming at getting a large combined force of French and British across the Somme and fighting the enemy in the open!"[53] This was not, as the editors of Haig's diary noted, a neat summary of "the difference between Rawlinson's 'bite and hold' philosophy and Haig's desire for mobile warfare."[54] Rather, it was a comment on the unsuitability of Rawlinson's limited operation for the bigger offensive scheme. In the less well known marginal notes Haig appended to Rawlinson's plan, his superior acknowledged that a single rush against the enemy's second position would be "impracticable," and that his tactical objective—to "kill Germans" rather than to seize ground—was "correct."[55] However, he was concerned that Rawlinson, despite earlier discussions with Foch and Fayolle, had "not realised the nature of the manoeuvre which devolves on his command."[56] The Somme offensive was no longer simply a wearing-out battle, and Rawlinson's first plan did not acknowledge that "the objective of the operation is to gain the ridge north of Péronne with the object of helping the French army on the right to cross the Somme."[57]

Although the basic principle behind the offensive had been accepted by the British commander-in-chief, the fine detail of how the British and French would operate together on the Somme still had to be resolved. Just how coordinated the two attacks were to be, and what role each army would play, remained uncertain. In Foch's first plan the roles of the two allied armies were quite distinct. The French would make the main assault—which made sense because they were both the larger and the more experienced force—and the British would support them north of the river by taking the Thiepval heights and advancing eastwards. This objective was subsidiary, but vital to the success of the whole operation, for unless the British reached the Bapaume–Péronne road, which would outflank the German defences along the river south of Péronne, the main French assault was likely to stall when it came up against this natural obstacle. Consequently, Foch's initial proposal was that the British attack should precede the French offensive by several days.[58] Thus it was a preliminary operation, designed both to draw in the enemy's reserves and improve the tactical position for the main French operation. Unfortunately, this seemed to be just the sort of separate, small-scale operation to which Haig had consistently objected.* Moreover, with the situation at Verdun delicate, Rawlinson was concerned that he would be hurried into an attack before he was properly ready.[59] Joffre certainly wished to keep his options open, and did not rule out the possibility of the British attacking alone.[60]

*See Chapter 2.

In a series of letters exchanged in late March and early April, Haig and Joffre therefore painfully thrashed out such differences. It was not the most efficient way of working but it did at least facilitate progress, if not clarity, in their joint venture. There were three principal points of contention: coordination, particularly in the joint attack north of the river; the nature and objectives of the first assault; and the subsequent development of operations. The convoluted discussions which ensued, and the compromises which resulted, were indicative of the complex and sensitive nature of joint planning.[61]

In its early operational scheme Fourth Army was to establish its left flank on the spur north of the river Ancre, between Serre and Miraumont, to cover the main advance on either side of the Fricourt salient against the Pozières–Bazentin-le-Petit–Longueval crest line in the centre. The right flank would be refused, with only the enemy's front-line trenches being captured between Fricourt and Montauban. At the same time, French troops in the Maricourt sector immediately north of the river would seize the German first line. This, however, left a widening gap between the British right and the French left, which the enemy could exploit from the valleys between.[62] Moreover, Foch wanted the British to seize the ridge before the main French attack south of the river. Joffre disagreed, proposing a truly combined attack, in both time and place. Consequently, he determined on the seizure of the ridge line on the southern sector, which stretched from Mametz, through Montauban, to La Briqueterie, both to cover the British flank and to facilitate the French advance to the south. After securing this first objective, subsequent attacks would push the line eastwards to the Bapaume–Péronne road.[63] This represented a necessary clarification of the strategic focus of the offensive; indeed, by this point Haig was so unclear on the purpose and objective of the allied offensive once the first two enemy positions were taken that he had to request a directive from Joffre. As the British attack was conceived, its axis of advance was northeastwards towards Bapaume, divergent from the main French assault. An advance towards Morval, Combles and Rancourt on the Bapaume–Péronne road would reorient the British advance eastwards, and coordinate better with a French advance along the north bank of the river.[64] The French had no complaint about this. The key to exploiting the first assault was to get French troops in large numbers across the Somme south of Péronne, and this would be much easier if the river line had been turned north of the Péronne bend. The ultimate goal was to establish the Anglo-French armies in open country on the line Bapaume–Rancourt–Péronne–Ham.[65]

This negotiation over objectives, while convoluted, is significant. As well as the specific difficulties of coordination at the junction of two allied armies, general problems of perception and focus were emerging. Moreover, such issues diverted the French high command's attention from the main endeavour south of the river to the subsidiary but more problematic operation to the north, and fuelled suspicions that the British were not prepared to pull their weight, despite the trial their ally was enduring at Verdun. The British still had one eye on an attack further north in Flanders, so Joffre felt obliged to cancel any preliminary operations to compel Haig to focus on the Somme. Furthermore, Haig was indicating that he would now have fewer divisions available for an attack; and that the French would need to provide the reserves to exploit the first success.[66] These were worrying developments. Joffre felt obliged to give Haig a remedial lesson in conducting an offensive: he should concentrate all his forces for a concerted effort; this should proceed by successive stages, determined by the coverage of the artillery barrage; the cadence of the operation would be dictated by the time it took to redeploy the artillery for each stage; and, somewhat patronisingly, "the point of our manoeuvre should be to beat the enemy."[67] This last point, the French were coming to suspect, did not seem to be uppermost in British minds. In his own memorandum on the strategic development of the offensive, Foch noted that the British might not be able to reach Rancourt, and that significant French forces acting north of the river would be needed to seize Bouchavesnes to turn the river line.[68]

THIS TASK WOULD fall to General Marie-Émile Fayolle, who took command of Sixth Army, holding the front south of the Somme, on 29 February 1916. Fayolle was among the best field commanders that the Western Front produced. In spring 1916 he was untried as an army commander (his appointment was provisional), although vastly experienced in trench operations. Like Foch, he had studied at the elite French technical school, the *École Polytechnique*; they had been in the same class. Fayolle had also made his career in the artillery, retiring with the rank of *général de brigade* just before war broke out. As a practising Catholic in the years when French politics had been gripped by fervent anti-clericalism, in peacetime Fayolle had been denied promotion to the senior ranks of the army which his talent and conscientiousness merited. But war was to give him the chance to demonstrate his skill as a field commander.

Fayolle came from central France: his father was a lace merchant in Le Puy in the Massif Central. He cut a large, somewhat stolid figure with a broad face, topped with the "high bald head . . . of a savant." His upper lip was adorned with the regulation bushy moustache that seemed to be standard issue for so many French generals, with a pair of bright eyes glinting above it. He was a man of few words, with a rich but abrupt turn of phrase and noted idiosyncrasies that were mocked affectionately by his staff.[69] Fayolle was an educated and thoughtful soldier: he had taught at the *École de Guerre* at the same time as Foch and Pétain, as professor of artillery tactics, and had been an active participant in the doctrinal debates which had animated the French army before 1914. In these, Joffre recollected, he had shown himself to be "a man of sound common sense, always giving preference to simple solutions."[70] Fayolle's artillery theories, stressing *matériel* and the importance of concentrated artillery fire, may have offended the prophets of élan, but they were to prove prophetic and highly appropriate for the war that was to come.

In May 1914 Fayolle had reached retirement age. But he had been tending his rose garden at Clermont, his other passion, for only a few weeks when war resuscitated his career. On mobilisation he was given a reserve brigade, and almost immediately promoted to command 70 DI when its commander broke down. This reserve division, which ever after bore the epithet "division Fayolle," did better than most in the confused, bloody encounters at the end of August 1914. Having the well-practised batteries of the artillery training camp at Mailly assigned to it undoubtedly did its gunner chief a special favour. Fayolle was soon established as one of France's most competent field commanders while others around him were losing their heads and commands.[71] His biographer notes that he was a calm, no-nonsense man who got things done through a combination of common sense, clear thinking, moral and intellectual integrity, a capacity for hard work and a firm grasp of what was possible.[72] Such practical, down-to-earth qualities were essential for engaging with the military problems thrown up by the First World War battlefield: his trained mind saw war "as a problem in which all the facts must be gathered before seeking a solution."[73] As an artillerist he was quick to grasp the nature of the new style of warfare, and as early as August 1914 established the basic operational principle that would serve him (and later the whole French army) well: "I attack with the greatest care, with all the artillery and the fewest infantry necessary."[74]

Before 1916 Fayolle had a steady stream of successes, on which he built his rise to army command. In August 1914 he saved Foch's XX

CA from being outflanked by boldly deploying his guns forward (not for the last time) to halt the German advance. His theories, well executed by competent gunners, immediately proved their worth. Later, in the "race to the sea," his heavily outnumbered division blocked the German advance on Arras. Fayolle's division joined Pétain's XXXIII CA on the Artois front in the autumn, where it was engaged in the small-scale but bloody trench fighting around Notre Dame de Lorette. In the May 1915 Artois offensive XXXIII CA performed heroically, gaining a reputation as one of the elite corps of the French army. When Pétain was promoted to command the Second Army in July 1915, Fayolle's reward was command of XXXIII CA. This corps, blessed with its successive leaders, a fortuitous duo of infantry and artillery professors (Pétain had been Fayolle's pupil before becoming his colleague), became a paragon of the new defensive and offensive methods appropriate to trench warfare: steady, limited attacks, well supported by the artillery, not wild rushes against unbroken enemy defences. When Fayolle received his own army these methods could now be applied more generally.

As well as being a skilled field commander, Fayolle was what would be dubbed a "soldiers' general." Not for him the comfortable life of headquarters. He earned the sobriquet *"Général Caillebotis"* — literally "General Duckboard" — from his habit of walking through the front-line trenches when a divisional commander. He knew what it was like to be shelled, and on several occasions came close to death himself.[75] His heart was with his men. He did not see his wife and daughter for the first twenty months of the war: "They have not changed," he noted tersely when he did.[76] Relegation to a château behind the lines did not suit his temperament.[77] He was a self-important and irascible man, morose and critical. His overwhelming trust in God contrasted sharply with his contempt for his fellow man. He did not tolerate fools or dissenters: both superiors and subordinates regularly came in for scorn in the candid diary which he kept during his command. But he had many redeeming qualities. His young British liaison, Captain Edward Spiers, remembered this white-haired gentleman fondly after the war as "one of the very sweetest and wisest old men I have ever met." And Fayolle's "scholarly, gentle mentality" was complemented by the energy of his younger chief of staff, Colonel Duval, "one of the finest soldiers the war produced," in Spiers's judgement. He was in the habit of meeting the chiefs of staff of Sixth Army's corps nightly to review developments and future plans. Spiers was "certain that these meetings under Colonel

Duval provided the most efficient means of command I saw during the war."[78]

Fayolle held no illusions about the full grim reality of mass industrial war, "not merely between two armies, but between two nations. It will continue as long as they have resources."[79] Agonised by the heavy losses suffered by his men—he had burst into tears on first witnessing the carnage of the industrial battlefield[80]—Fayolle was disillusioned with operational methods which cost the French army so much for so little, and appalled by the apparent idiocy of those around him.[81] Like Foch, his agile mind was grasping for a solution based on these material resources, but putting theory into practice remained problematic. If the enemy's artillery could not be neutralised, Fayolle concluded, an attack would never make great progress. In time more munitions, particularly gas shells, and greater use of aerial bombardment might prove effective. But in early 1916 the requisite *matériel* was clearly lacking. Despite this, the high command expected an attack later in the year.[82] Inevitably, as Verdun was demonstrating, that meant attrition, for all its faults: "That is where civilisation has got to! And it is by no means over. The next battle will cost 200,000 men, and one must question whether in that case it should take place. Can one expect a decisive success from attrition? I don't think so, it is still too soon."[83] But since responsibility for that battle had fallen on him, Fayolle would try his best to ensure that his men were not exposed to a repeat of the bloodletting of 1914 and 1915.

Nevertheless, he doubted the rationale for the battle from the start: "Do they hope to break through? I do not think so. Then what is the point of this battle? Attrition they say. Fighting to wear down the enemy. Hum! That is hardly enough."[84] He therefore chose to proceed methodically, to husband his soldiers' lives, and not to attempt more than his artillery could comfortably support. This would impose a certain rhythm on the offensive. It was quite clear to him that with enough artillery and skilled infantry the enemy's defensive positions could be captured relatively easily. His men had achieved as much on a number of occasions in their 1915 attacks. And if Fayolle needed a final lesson in the principle that the artillery conquered and the infantry occupied, it was taught to him starkly just before he gave up his corps command. In a feint attack to accompany the opening assault on Verdun, the enemy stormed a section of his front:

The Boches have captured the front-line trench and the support trench. How do they do it: all their attacks succeed . . . they knock over everything with a horrifying bombardment after concentrating superior means. Thereby they suppress the trenches, the supporting defences and the machine guns. But how do they cross the barrage? Probably their infantry infiltrate, and since there is no one left in the fire trenches they get in, and when they are there to get them out we need to have the same artillery superiority.[85]

At Verdun the Germans demonstrated the efficacy of this method on a grand scale. But this, Fayolle recognised, was only half the solution, for such methodical progression nullified any element of surprise and potentially dictated that any offensive would grind to a halt, as happened soon enough for the Germans at Verdun.[86]

Sixth Army's attack plan was therefore focused on the need to maintain enough momentum to carry the infantry forward over successive German defensive positions into the open country beyond. "It will be a battle of a month or more which must be continued without a break, the artillery always firing, the infantry always moving forwards," Fayolle anticipated. There were certain practical issues which also needed to be addressed: "It would seem that the most serious problem will arise from confusion and the closeness of the opposing lines. Therefore, the army corps should delineate successive lines which need to be attained." Above all, and in this Foch concurred, the enemy's second position had to be taken quickly.[87]

Moreover, as the planned operation took shape, Sixth Army's role was growing in importance. In preliminary discussions with Foch, Fayolle stamped his mark on the plan in its early stages, pointing out the difficulties of attacking south of the river: too many villages, and a river obstacle behind. The north bank was the decisive point, so forces should be concentrated there.[88] Foch, who shared Fayolle's doubts about breaking through,[89] reoriented his plan accordingly in the detailed instructions issued to his three army commanders on 14 April. Each was to have sixteen divisions, and discrete, but integrated, missions. General Micheler's Tenth Army was to set up a defensive flank to the south; the centre army (as yet undesignated) was to push to and across the Somme; and Sixth Army was to support this breakthrough to the north. Fayolle was set three tasks. He was to capture the German defensive system between the Maricourt salient north of the Somme and Lihons, twelve kilometres south of the river: in the initial assault the first position was to be captured, with a foothold secured in the second position at Assevillers, which was subsequently to be occupied down to Puzeaux. Then the Flaucourt plateau was to be taken, to deny the enemy observation over an advance north of the river. Finally, in the northern sector, "the left of the army will assume a leading role; in cooperation with the British army . . . it will push towards the Bapaume–Péronne road to open the way for our armies held up by the river above Péronne."[90] For that task, the attacking force in the Maricourt sector would expand from one division to several army corps.[91]

The first task was fairly straightforward, but would have to be carefully prepared. The key to a successful assault was an effective preparatory artillery bombardment, followed by a swift exploitation by the infantry. Ever the pessimist, in Fayolle's judgement his infantry were not a match for the enemy's: they were poorly led and did not know how to manoeuvre. His generals and artillery commanders were equally mediocre.[92] But, provided there was sufficient artillery support and his offensive principles were adhered to closely, they could do the job. His staff—"slow but methodical, too slow nevertheless . . . ridiculous in wartime"—drew up an attack plan that would take them beyond the second German position. Subsequent stages would depend on the progress of operations elsewhere.

What happened over the next six weeks, however, shifted the orientation of the French offensive from south to north. By May, as guns and divisions were diverted to defend Verdun, Foch had reduced his plan to a single-army attack astride the Somme, on a shorter front, determined by the number of guns he now had available for support. At the end of May, Fayolle was told that he was now supporting the British attack to the north, even though the tactical objectives of Sixth Army's offensive remained as set out in April. A breakout south of Péronne was now no longer part of the plan. Instead, the attack north of the river became the main axis of advance, "which would develop depending on the rate of English progress, and would be reinforced in order to push as far forward as possible in coordination with them."[93] With the scaling down of Foch's original three-army battle, there was no operational rationale for a French attack astride the Somme except to keep their allies marching beside them, an objective which had been growing in importance as the negotiations continued.

For Fayolle, such a reorientation was immaterial, if not insignificant. His task was simply to get his army forward as speedily as the need to minimise casualties would allow. In practice this meant that north of the river the initial French objectives would be limited to the enemy's front position, between the British right and Curlu village on the river. The line in front of Maricourt formed an advanced salient, so the French attacking force, now increased to a whole army corps, would be starting from a point in front of the British attack. The division on the left was charged with covering the right of the British advance by capturing Bois Favière while the British took Montauban. The right-hand division faced the German strongpoint at Bois Y: this would be stormed quickly, and the advance would then be pushed on to the plateau above Curlu village, which would secure good observation for an attack on the German second position.

South of the river a second army corps would seize the German front position between Frise and Dompierre, between 1200 and 1500 metres deep, and push on through the German second position between Herbécourt and Assevillers one kilometre beyond. A third army corps would establish a protective flank between Fay and Estrées. The final objective of the southern attack remained the high ground behind Flaucourt. Once this plateau was in French hands an advance along the northern bank could be supported from the south by French artillery, while denying the Germans that key observation point.[94] For this plan to succeed, the German second line would have to be stormed quickly in the southern sector. Fayolle's assault plan therefore envisaged a rapid conquest of the German first position and an advance immediately to the second. If this was weakly defended, it would be assaulted in short order; there would be artillery close up with the forward infantry battalions to offer support. If not, then provision was made to deploy the artillery rapidly forward to pre-prepared emplacements to allow an assault on the second position the next day, before the defence had consolidated.[95] In this way, Fayolle planned to reach open ground south of the Somme before the enemy had time to bring up reserves. The general principles by which Fayolle would conduct his attack were clear: "move from objective to objective, without compromising the élan of the men, at the same time leaving nothing to chance. Each attack will therefore be against a fixed objective, of limited width and depth, always preceded by an artillery preparation and reconnaissance of the enemy's wire."[96] If the artillery did the hard work, and the infantry suffered few casualties and remained fresh and organised after the first assault, then forward exploitation would be possible. In this way, Foch's deep battle, through successive lines of German defences, could be fought with a cadence that allowed relatively rapid progress.

· · ·

THE BRITISH ARMY faced similar problems when selecting objectives on their sector of the front, but their planning was neither as methodical nor as effective as that of the French, who by this time had learned to calculate their front and depth of attack in proportion to the available artillery and munitions.[97] (They had therefore adjusted ambitions downwards when resources were diverted to Verdun.) British tactical planning was determined by intention, and resources would have to be deployed accordingly. Unfortunately, between February and June 1916, intention was not fixed, and resources, although much better than in 1915, remained inadequate. Haig and Rawlinson's debates over the objectives and methods of the first assault have been pored over and critiqued often and at length, mainly because of their consequences. Much of the detail is abstruse, but the principles of the disagreement are clear, and parallel to some extent Foch and Joffre's differences over the French attack. Was the operation to be slow, methodical, *matériel* intensive and attritional, designed to grind down the enemy's power of resistance until it collapsed? Or was it to be a sudden, powerful, disruptive thrust aimed at doing as much as possible in the first assault and rapidly exploiting the resulting confusion in the enemy's defence, as Haig intimated in his reaction to Rawlinson's first proposal?[98] Rawlinson is generally credited with planning the former, although he is usually damned for giving in to Haig, who pressed for the latter. In fact, both men wavered from one to the other as resources, French intentions and enemy action fluctuated from week to week. They were not alone in their uncertainty, and the contradictory and shifting guidance received from Joffre did nothing to help them draw up a consistent and viable plan of attack. The fact that Joffre's final briefing to Foch and Haig, three days before the bombardment opened, still indicated two possible courses for the impending offensive—a quick, sudden collapse of the enemy's resistance or a long, hard, attritional battle[99]—is indicative of the indeterminate nature of the battle to which all were committed, come what may.

After digesting Haig's objections to his first scheme—which called for refocusing his axis of advance westwards towards Combles to assist the French, and accelerating his operation to exploit the initial destabilisation of the enemy's defence[100]—on 19 April Rawlinson submitted a second, more ambitious plan to extend the attack front to Montauban, and capture the green line in one go. Nevertheless, while admitting that the circumstances might necessitate such a risk being taken, he repeated his earlier arguments against such a course, and cautioned that if the green line were not taken in the first rush, "the whole operation may be retarded to a greater extent than would occur should the attack be made in two phases as I originally proposed."[101] But Rawlinson's revisions indicated capitulation to Haig on the fundamental hypothesis of the operation: it was not to be bite and hold, but rush and hope. Unfortunately, it seemed that Haig had absorbed only the final point of Joffre's remedial lesson. Rawlinson's suggestion that the operation would be "sustained over a considerable period of time" provoked a marginal scribble: "The enemy must be beaten!" As for "the necessity of maintaining a sustained offensive for a considerable period of time," Haig commented, "There is no such 'necessity.' The object is to capture and hold the Pozières–Combles ridges: and these should be seized as rapidly as possible once the attack starts."[102] For Haig, speed was of the essence. After coming to grief in front of the German second line at Loos, he believed that the attack had to break through the defended zone before the enemy had time to deploy his strategic reserves against the breach. This was a matter of a day or so, no longer, a very short window of opportunity.

Haig's intervention possibly threw away any chance of genuine success. Rawlinson's initial plan had suggested that his staff had absorbed the principles by which the French intended to maintain the momentum of an assault against successive German positions, with the attack on the second position being pre-prepared so that it could follow rapidly on from the capture of the first, before the enemy brought up reserves.[103] If this principle had been adhered to, there would have been a realistic chance of combining surprise with speed to destabilise the defence.

Two other elements of the operation were moot in this exchange of views: the nature of the bombardment and the employment of cavalry. Most problematic was the bombardment. With many more heavy guns available to the British army than in 1915, it was judged that the German defences could be crushed by weight of artillery fire, even if in practice artillery techniques were not yet sophisticated enough to neutralise the enemy's resistance completely. As far as the bombardment was concerned, the choice lay between a short, intensive artillery preparation and a less intense, more methodical barrage fired over a number of days. Haig favoured the former, as it would produce surprise and greater disorganisation in the defence, and supposedly facilitate a quicker, deeper advance. Rawlinson argued repeatedly and ultimately successfully for the latter (not least because that was the method chosen by the French), which would allow the destruction of the enemy's strongpoints and wire-cutting to be properly monitored. Unfortunately, the commander-in-chief's insistence on an increase in the front and depth of attack spread the artillery fire available to Fourth Army too thinly, and would therefore require more from the untested infantry, two factors which had caused Rawlinson to err on the side of caution in his first proposal. The change in the plan meant there would be fewer guns per yard of front attacked on the Somme than at Neuve Chapelle or Loos, and this could be only partially offset by prolonging the bombardment.[104]

Second, Haig wished to find a role for his own arm of service, the cavalry, in the offensive. Of the many sticks which have been used to beat him, this is probably the most frequent and resilient. Yet Haig has been unfairly treated in this respect. Despite inherent obsolescence in the face of barbed wire and the machine gun—the image of vulnerable horsemen falling before the enemy's un-silenced emplacements is one of the enduring clichés of the Western Front—the cavalry retained a role for much of the First World War. Before the advent of fast tanks in 1918, it was the most mobile arm on the battlefield. Even if it could not function amid the trench lines, it remained the arm of exploitation, should such an opportunity arise. Before the war there had been a lengthy and acrimonious debate in the British army on the role of cavalry on the modern battlefield. Were they still a shock force, able to disrupt and demoralise the enemy? Or were they now demoted to the role of mounted infantry, effectively horsed riflemen, more mobile but otherwise the same as their marching comrades? Haig, who had drafted the 1904 cavalry training regulations at the height of the so-called cavalry controversy, appears rather equivocal on the subject (believing the cavalryman should become a hybrid who could both charge and dismount and shoot), perhaps in order not to offend powerful patrons on either side of the argument.[105] The Western Front finally settled that debate. Before the Somme the French army reorganised its cavalry divisions into prototype "mobile divisions," including more infantry, armoured cars and greater firepower.[106] Although British cavalry divisions retained their old model, their *raison d'être* similarly changed. Haig had no illusions about this, even though in his thinking there remained a certain ambiguity regarding their use. In 1916 the cavalry's role was to seize and hold ground quickly, in cooperation with the other arms, to take full advantage of the enemy's temporary disorganisation. Alongside Rawlinson's casual suggestion that he might "push the cavalry through" if the first attack demoralised the enemy, Haig noted, "this seems to indicate the intention to use the cavalry as *one* unit. This is not my view of its employment *during* the fight."[107] Fourth Army's cavalry, distributed by regiment among its army corps, were to cooperate closely in small units with the other arms, in a mobile exploitation role, pushing up the Ancre valley and fanning out behind the German second position between Miraumont and Grandcourt. It was not to be employed in sabre-wielding charges against disorganised infantry and gunners; although, if such targets of opportunity presented themselves, they were to be exploited. At the same time, massed cavalry might still have a role later in the battle if the line were broken: a Reserve Army commanded by Lieutenant-General Sir Hubert Gough was being created, with two cavalry divisions at its disposal for operating in the open behind the German defensive front.[108]

The British plan was set. The idea of a manoeuvre battle had been formulated,[109] to help the French attack by seizing the German defences atop the Thiepval Ridge and pushing westwards along it in support of the French attack to the south.

THE ANGLO-FRENCH OFFENSIVE was assuming a definitive shape by May 1916. Then, barely a month before the start, everything changed. Ironically, just when the French were coming to accept the slow, attritional, indecisive nature of the offensive, they gave Haig just enough slack to allow him to review the battle one more time. As the struggle at Verdun ground on, Joffre was urging the British to hasten their preparations; Foch was recasting the joint plan in the light of reductions in the French army's contribution and resources; and Haig took this as a cue to expand the scope of the British offensive, despite his professed unpreparedness. This late, hasty and hopelessly over-optimistic revision would seriously compromise the careful and thorough planning effort that had gone before.

On 31 May, at the final Anglo-French plenary conference, it had been decided that the offensive would take place side-by-side on 1 July, although the armies' roles would be reversed. Haig was now off the French leash as far as his Somme attack was concerned, even if his hopes for a delay or a change in the front of attack had been thwarted.* Three factors contributed to his final, imprudent revision: the changed French role and its impact on the strategic purpose of the offensive;

*See Chapter 2.

continued confusion over the nature and aims of the battle; and the developing intelligence picture of the enemy's situation. Given the reduction in the French army's contribution, and the reversal of roles, Haig sought a final clarification of the purpose and objectives of the offensive, now that his operations were not designed to support a French crossing of the Somme.[110] While Haig remained determined to conduct his offensive within the parameters of the generalissimo's broader strategy, in the changed circumstances he was, des Vallières warned Joffre, anxious to re-orientate his axis of advance northwards in the direction of Bapaume. Des Vallières suggested that "it was necessary to insist on the idea of a sustained battle (still inadequately understood by the British staff), and so any directive should insist on *distant* objectives, to be certain that they would engage *all their reserves* to achieve them."[111] For the French, distant objectives were useful to ensure the British pulled their weight in the offensive. To Haig, they seemed to offer the change to win a decisive victory. Joffre, in reinforcing the methodical nature of the offensive, unfortunately and rather ambiguously had hinted that this was within his grasp. Given Verdun and the recent Russian offensive:

> In our joint offensive on the Somme . . . *we can envisage knocking out the German army on the Western Front, or at least an important part of their forces.* The success of our offensive is initially and fundamentally linked to successively taking and holding all German defensive positions on the attack front; in continuing the battle without rest until this first result is achieved.
>
> But our experience of previous attacks indicates that, to drive the enemy from his prepared positions, we have to conduct a long drawn out battle, the final form of which it is impossible to predict.
>
> Consequently it does not seem opportune to me to fix, at the moment, distant objectives for our armies beyond the German third position. The plan for exploiting our success is essentially a concomitant of the level of attrition of the enemy and their dispositions, as well as our own reserves when we attain the last German position.[112]

No wonder Haig remained confused! Unfortunately he took this as a cue to go all out for the third position and beyond.

Haig had formed his conception of military operations at the staff college more than twenty years earlier—"the late-nineteenth-century ideal of war," as Tim Travers, one of Haig's more trenchant critics, has dismissed it—and this still underpinned his intentions on the Somme. He saw defeating the enemy's army as the principal objective in war. To do so, he envisaged battle developing through four successive stages: manoeuvre for position, wearing out, a decisive blow to break the enemy's resistance, and exploitation.[113] The manoeuvre had taken place in 1914; the wearing out had continued ever since, especially at Verdun. Haig now resurrected the idea of a decisive blow. In the second week of June he went to London for an allied conference and discussed the situation with Robertson and the Cabinet. Then he reminded the Prime Minister that it would be a sustained, attritional offensive.[114] However, his own thoughts were tending elsewhere. On his return to France he issued a final, ambitious agenda to Fourth Army's confused commander, just over a fortnight before the guns were due to start firing. Joffre had hinted that the German army had been significantly worn down at Verdun. Haig's own intelligence confirmed this, and this was the final, decisive factor in his decision-making process.

Haig's intelligence chief, Brigadier-General John Charteris, has often been made a scapegoat for the overambition and fanciful operational planning which characterised G.H.Q.'s mind-set during the attritional campaigns.[115] Only recently has the role of intelligence in G.H.Q.'s planning process been properly analysed.[116] Military intelligence is all about perception. Unfortunately, Haig tended to read too much into the reports of the enemy's military position which issued from Charteris's office, seeking confirmation for his own preconceptions, rather than information on which to base his judgements. Hence, having watched the French army engage and wear down the German army at Verdun for four months, he made the plausible but erroneous assumption that the wearing-out phase of the campaign had been progressing smoothly. The most recent intelligence estimates of German forces and reserves opposite the British front reinforced this perception: there were only thirty-two German battalions on Fourth Army's front, with sixty-five more in reserve which could be drawn into the fight in the first week. The British had more than two hundred in hand for the battle. A high proportion of the German reserve divisions thought to be available (five out of seven) had already been through the mill at Verdun. There were also signs that the Germans were transporting reserve divisions to the East in response to the successful Russian offensive. Moreover, German ranks were thinning, with men of the

1916 class appearing among the prisoners.[117] Attrition was all about using up the enemy's strategic reserves, and in June, Haig looked at the evidence and concluded that this had been accomplished to such an extent that there was now a realistic chance of breaking the German front and defeating the German army in the open field.

New orders were issued accordingly in the third week of June. The rapid capture of Bapaume and exploitation northwards to destabilise the front opposite Arras replaced the advance eastwards to outflank the river line at Péronne as the operational objective of the offensive. "With the object of relieving the pressure on Verdun and inflicting loss on the enemy," Fourth Army was to seize the green line between Serre and Montauban as its first objective. Then, if the enemy's resistance broke down, Rawlinson was to "advance . . . eastwards far enough to enable our cavalry to push through into the open country beyond the enemy's prepared lines of defence." After which the axis of advance would shift northwards, with a defensive flank being established between Bapaume (captured by the cavalry) and Monchy-le-Preux. This new attack plan set distant objectives, changed the direction of attack so it diverged from the French advance, and emphasised a breakthrough and exploitation with cavalry.[118] At the same time, Haig was advising Rawlinson to reduce his bombardment.[119] Not surprisingly, Rawlinson was nonplussed: "As the day of battle approaches so is one worried by alterations and modifications." His chief of staff retired to bed with a headache![120]

To be fair to Haig, a breakthrough was by no means assumed to be a certainty and provision was made should Rawlinson's attack stall. Since that provision was to shift the focus of the British attack elsewhere, however, it rarely features in an analysis of Somme planning. If the Somme attack did not break the enemy's resistance, Haig contemplated renewing the offensive at Ypres while Fourth Army kept the enemy "fully employed": the Somme might yet prove to be the preliminary holding attack.[121]

Haig strove to get the British army's plan for the Somme right. But ultimately he got it very wrong. Was that predictable, and could it have been avoided? Historians have debated this point ever since. It was a concomitant of coalition planning as well as military *mentalité*, which differed depending on which side of the river you occupied. Throughout the planning process Haig had taken his cues from Joffre, something that has been inadequately acknowledged, but these prompts were shifting and contradictory, and then further warped by Haig's own military logic. The rigidity and obtuseness traditionally attributed to the British high command might have served better here than the improvisations of a fertile mind. Having clashed with Haig over such matters already,[122] Rawlinson for one maintained a pragmatic detachment from his superior's hazy vision, telling his final corps commanders' conference: "Before I read [Haig's orders] I had better make it quite clear that it may not be possible to break the enemy's line and put the cavalry through at the first rush . . . it is impossible to predict at what moment we shall be able to undertake this, and that decision will rest in my hand to say when it can be carried out."[123] Right at the start of the planning process Haig had instructed General Sir Edmund Allenby, then commanding on the Somme, that for the "preliminary wearing out offensive to get the first line of trenches he will need 15 divisions and 3 in reserve. Under certain circumstances it might be necessary to increase the strength of this attack into a decisive one."[124] This is roughly what Rawlinson had available on 1 July 1916, and Haig had clearly forgotten his earlier qualification. For the same British force, as yet amateurish and partly trained, to smash the enemy's resistance the Germany army would have to have been very badly mauled at Verdun.

Nevertheless, in mid-June Haig resurrected the original plan for a decisive breakthrough, even though the French had by then abandoned it. Following the wearing out done by the French at Verdun, Haig's army was now to break the German front and exploit the disorganised troops behind it. While this was not entirely unreasoned, it was highly speculative, just like much of the planning which had gone before. Foch was certainly unhappy, but Haig was by now going his own way. After receiving Foch's protestations in writing, on 28 June, Haig met Joffre and Foch in a bid to iron out their last differences. In his diary he was dismissive of the Frenchmen:

> The old man [Joffre] looked tired and rubbed his head. He evidently saw the force of my argument but Foch had got at him. Foch seems anxious that the British should do all the fighting required to get the French on to the open ground between Bapaume and Péronne, and he ignores the danger on our left flank which is very real if we do not enlarge the gap northwards as soon as possible.[125]

After six months of careful, deliberate negotiations, the allies would still be starting their joint offensive at cross purposes. Haig had the impression that the French expected his army to do most of the fighting; while conversely, Foch thought that French participation was the only way to get the British army to fight.[126]

Haig was aware that with no increase in resources this might not be the decisive battle that would end the war. Like Kitchener, who when asked when he thought the war would end had refused to make a prediction, merely intimating that it would start in earnest in April 1916, Haig knew the allies still had much fighting to do. In London in June he had been non-committal when the King had speculated that the war might last until October 1917: "I said that we should be prepared to carry on the war into next year, but I said that signs were not wanting to show that the Germans might bargain for peace before the coming winter."[127] But that would not happen without a battle of "*durée pro-longée,*"[128] of which Rawlinson's attack would be only the first phase. In June 1916 no one could determine how the defeat of the German army would be accomplished or how long it would take. It would be to Haig's continual discredit that he was too sanguine. Rawlinson was more realistic. From the start he acknowledged that it would be a slow, drawn-out affair. But as early as March 1916 he had been convinced that it would happen:

> During the summer they will have to meet a strong offensive on all fronts simultaneously and those who think must realise that they cannot count on ultimate success on all fronts. The moment they begin to get pressed back then our people will know that the game is up and we shall then have to see how they will fight a losing game. I am inclined to think they will stick it out manfully though the internal troubles they will have to meet will probably be more damaging at first to them than the external pressure ... After that it is only a question of time before we enter Germany from east and west.[129]

The length of time was the only uncertain variable. Rawlinson himself judged: "The Bosches are not as short of men as some people make out. They are feeling the pinch of the blockade and will feel it worse as time goes on but they are quite capable of continuing the war into 1917 without bringing on a revolution."[130] Nobody, including Rawlinson, expected Germany to hold out for almost two and a half more years.

On the day, given sufficient élan in the attack, Haig hoped the German army might suffer a crushing defeat on the battlefield. In his final briefing, after the bombardment had started, he cautioned the methodical Fourth Army commander not to halt his advance for an hour on the green line, but to push advanced guards forward as soon as the second objective was taken, before the enemy could deploy reserves to fill the breach. "In my opinion it is better to prepare to advance beyond the enemy's last line of trenches," Haig recorded, "because we are then in a position to take advantages of any breakdown in the Enemy's defence." He did not want a repeat of the initial success and second-day failure against reinforced defences he had experienced twice already at Neuve Chapelle and Loos. But he was realistic enough to acknowledge that "if there is a stubborn resistance put up, the matter settles itself! On the other hand, if no preparations for an advance are made till next morning, we might lose a golden opportunity."[131] Moreover, these final orders only reflected the ambiguity of Joffre's last instructions, which Haig had received a few days before.

GARY SHEFFIELD'S ACCOUNT of the offensive noted "the ambiguities inherent in the British concept for the Somme." Often the complex and convoluted planning for the biggest coalition offensive yet seen has been oversimplified into a debate between two British generals over the practicalities of an artillery bombardment and the use of cavalry in the field. The real issues and disagreements went much further and deeper. How could attrition be conducted effectively in the field? In what way should the battle develop and how long should it last? What were its local objectives and its wider strategic purpose?[132] The British plan was the product of two minds, certainly with different conceptions of the nature of industrial battle, often competing over details, one thinking strategically, the other tactically. As yet the link between these two levels of war—what modern armed forces define as the "operational level of war"—was embryonic. How were the initial attack executed by Rawlinson's corps and the as-yet-sketchy operations which were to follow it up to contribute to the strategy formulated six months previously at Chantilly for a General Allied Offensive which was to bring the German army to its knees? Moreover, these ambiguities were not only British, for Joffre, Foch and Fayolle wrestled with the same conceptual difficulties inherent in linear siege warfare and attritional battle, as well as the practicalities of tactical success against strong field fortifications backed by massed artillery. Joffre certainly still retained an increasingly misguided confidence that, if enough pressure were applied, the Western Front would break (although he thought it might shatter rather than be pierced). Haig too had not abandoned such a conviction, although both were unsure just how much pressure was necessary. In both men's final briefings to their subordinates these ambiguities are obvious. A shared and apparently naïve hope for a quick, violent success, in keeping with traditional conceptions of decisive battle, for which both have been routinely castigated, juxtaposes sharply with an acknowledgement—admitted, perhaps feared, and probably expected—that such a victory would not be forthcoming; that instead there would be a drawn-out, costly and indecisive fight which would wear out the attacker as much as the defender. This was an acknowledgement that military operations between well-armed, equally matched, entrenched forces were inherently attritional in nature, and that the side which inflicted the most damage, which endured the longest, would emerge the final, punch-drunk victor.

Only Foch seemed to possess the intellectual apparatus to bridge the gulf between tactics and strategy. His planning of the offensive and even his reservations suggest that he was conceptualising industrial war not as a single, decisive clash, but as a succession of operations geared to destroying the fighting capacity of the enemy's army. He stressed on 31 May that he would not try to break through.[133] His battle would be one of *durée prolongée,* and such pragmatic thinking wormed its way into Joffre's and Haig's conceptions in time, even if it did not entirely oust more grandiose Napoleonic conceptions of decisive battle. For Foch, 1 July 1916 would not be the end of the fight, but the start of a sustained process of beating the German army. Successive French and British attacks, coordinated but not necessarily coincident, would grind down the enemy to a point where Joffre's "rupture" might ensue. This was the purpose of the Somme, and through progress and setback Foch was to stick to this goal. The Somme might not finish it, but, as Masefield acknowledged, "It was the starting place. The thing began there."[134] It would be Germany's Nemesis, not that of the allies.

Over ninety years on we should be able to get a clearer picture of the nature and practice of attrition than those who were improvising it in 1916, even though most analysis to date has been based on the false premise of evaluating strategic attrition in terms of defective operational planning and execution: "the tactics of Malplaquet and Borodino . . . combined with the killing power of modern technology" in one quotable summation.[135] Those who claim that attrition replaced breakthrough as the objective of the Somme offensive after the disappointment of the first day are wrong. Such misperception arises from a narrow fixation on the minutiae of the British attack plan. Strategic attrition was a key element of the plan for 1916 from the first conference at Chantilly; it was a necessary precursor to any decisive defeat of the German army. Its tactical realisation remained problematic. As the scale of the "Big Push" shrank, so expectation of victory did too, and attrition as an end in itself assumed greater importance over particular territorial or operational prizes. Charteris, the man responsible for assessing the wearing out of the German army, confided to his wife on the eve of battle:

> We do not expect any great advance, or any great place of arms to fall to us now. We are fighting primarily to wear down the German armies and the German nation, to interfere with their plans, gain some valuable position and generally to prepare for the great decisive offensive which must come sooner or later, if not this year or even next year. The casualty list will be long. Wars cannot be won without casualties. I hope people at home realize this. We are *winning,* even if we do little more than we are doing this time. But it will be slow and costly. If we face losses bravely we shall win quicker and it will be a final win.[136]

On the same day Haig was maintaining his customary optimism. He entrusted his thoughts to his wife, and his fortunes (and errors) to God: "I feel that everything possible for us to achieve success has been done. But whether or not we are successful lies in the Power above. But *I do feel* that in my plans I have been helped by a Power that is not my own. So I am easy in my mind and ready to do my best whatever happens tomorrow."[137]

Joffre, less trusting of divine providence, was also quietly confident. "He hopes for—even if he dares not predict—a strategic success," President Poincaré had noted the previous month.[138] Foch retained his energy and enthusiasm, while Fayolle was his usual mordant self: "acrobatic demonstration of the battle by Foch. What a strategy!"[139]

It was a strategy of attrition, which had already been practised for four and a half months. If nothing more, it would take pressure off the French army, restore initiative to the allies and bleed the German army in its turn. The idea of attrition has since become strategic anathema: Jack Sheldon criticised "the moral bankruptcy of attrition theory" in his study of the German army on the Somme.[140] But it made sense in the deadlocked circumstances of 1916. For the allies, strategic attrition meant overstretching and grinding down Germany's powers of resistance to such a point that she would break; operational attrition meant using up the army's manpower reserves to break the Western Front stalemate, restore mobile warfare and force the German army back to the Rhine; and tactical attrition meant killing Germans on the Somme, with inevitable loss to the attackers at the same time.

No one was sure what would happen after 1 July: Joffre was undecided; Haig overconfident; Foch realistic; Rawlinson confused; and Fayolle pessimistic. Early on the latter had spotted the inherent paradox of the scientific battle:

> We have understood that we cannot run around like madmen in the successive enemy positions. Doctrine is taking shape. If there are so many defensive positions, there will need to be as many battles, succeeding each other as rapidly as possible. Each one needs to be organised anew, with a new artillery preparation. If one goes too quickly, one risks a check. If one goes too slowly, then the enemy has time to construct successive defensive lines. That is the problem, and it is extremely difficult.[141]

Although a definitive solution to this problem was not to be found on the Somme in 1916—it was not then possible, as Fayolle wished, to organise a succession of attacks at several points along the front[142]— what he reasoned, did and learned in 1916 formed the basis of the offensive doctrine which would eventually carry the allied armies forward to victory in 1918. All the commanders knew they had a long, tough fight to come, because for six months these principals had been scripting the approaching military spectacular. Meanwhile, a supporting cast of millions, the soldiers and workers of three great empires, had been preparing to play their roles.

Preparing the Big Push

PRING CAME LATE to France in 1916. But when it came it was glorious, even against the backdrop of war. For Australians who had wintered in Egypt, "that perpetual land of sameness," the leisurely train journey through France was enchanting.[1] Bands welcomed them on the quayside at Marseille, and French civilians crowded to see the fine, strong men in their distinctive slouch hats who had come to fight for them. The new arrivals also saw their first Germans, prisoners-of-war working on the dockside.[2] The troop trains took a slow, circuitous route, up the Rhône valley past vineyards and orchards, dawdling across Burgundy and the wheat fields of the Île-de-France, and skirting round Paris (much to the disappointment of the soldiery, which had been boxed up for two days by that point), and on to the base camps behind the British front. The French civilians were welcoming throughout the journey, pressing food and drink on the occupants of the stuffy carriages at every halt. But the troops could not fail to notice the sombre reality behind the welcome: "Everywhere are women and girls in black . . . looking beneath the sorrow of every face, I can see the quiet, calm resignation that France is not making all these sacrifices in vain, and that the end of it all must culminate in victory."[3] The people clearly hoped that these men might be their deliverers.

Nine British and imperial divisions were transported from Egypt to France in the first half of 1916. The men knew that they were heading for the real war and something important: "Constant rumours are afloat that the big offensive is shortly to commence. We all hope it will for we are eager to see this thing finished."[4] It was an open secret that there would be a major allied offensive in 1916. It was the natural thing to do to win the war. The concentration of soldiers, the stockpiling of *matériel*, the factories working at full capacity, all pointed to one thing, a major offensive. The British press and public began to speculate about the "Big Push," as they called it. It was only a matter of time: not if, but when and where?[5] Some tried to see for themselves. One intrepid (and obviously charming) female journalist managed to secure a pass from a town marshal, a uniform from a gullible private and a ride from a provost marshal, then spent a week living among the men of the 51st (Highland) Division, until an apoplectic general cursed her on her way.[6] Even without the inquisitive tourist or spy, media speculation and public gossip, there was no disguising what was taking place on the Somme front. "Here I am on the right flank of the British army, next to the French," noted Captain Harry Bursey on 21 June:

> The sights one sees here are just wonderful. Guns, guns and more guns. From a machine-gun to a 17-inch. What are they here for? ... They are here for "the Day." The day we have looked forward to for many weary months, the thoughts of which have helped to cheer us when we have been saddened by rain and our bodies numbed by the bitter cold. And it is coming soon.[7]

M*ATÉRIEL*-INTENSIVE INDUSTRIAL BATTLE depended on those who produced arms, as well as those who fought. The warring empires were in the process of making machine-age armies, appropriately equipped for trench warfare. Many more of the existing weapons were needed quickly, particularly heavy artillery and machine guns. Large numbers of close-combat weapons also had to be manufactured, particularly light machine guns and hand-grenades. Furthermore, many new weapons had to be developed from scratch in response to the particular needs of the front line. The First World War was a war of invention, where scientific-industrial complexes competed with each other to produce more efficient or deadlier weapons for the fight, increasing the killing potential, if making little impact on the stalemate. Gas, flame-throwers, grenade-launchers, sub-machine guns, trench mortars and cannon, fighter and bomber aircraft, tanks and self-propelled artillery all made their battlefield debuts between 1914 and 1918, in many forms and variations. The amount and types of ammunition to feed this voracious arsenal also expanded rapidly. Finally, the paraphernalia of static siege warfare had to be manufactured in ever-increasing quantities: military engineering and hard labour were both integral to trench combat. Foch's initial shopping-list for the Somme offensive included 1069 pieces of heavy artillery and 1908 field guns; over 5,000,000 shells of all calibres; 2,500,000 grenades; 330 tonnes of barbed wire; 70,000 square metres of canvas; 1,000,000 logs; 2,200,000 sandbags; 110,000 picks and shovels with 22,000 spare handles; 310 barrack-huts for hospitals and administrative offices, plus 6000 more for quartering troops; 3200 telephone sets with 13,200 kilometres of telephone cable and 4000 telegraph poles; 700 water bowsers and 15 ground-water pumps; 500 kilometres of railway track; 32,000 cubic metres of aggregates, and 3 steamrollers.[8] Ships and barges, railway locomotives and trucks, caterpillar tractors, lorries, cars, motorcycles, carts and limbers had to be provided in equally large numbers to move everything.

This dependency on industrial manufacturing linked the home front and the army with iron shackles. One could not do its job if the other did not, and the "second front"—which possibly should be termed the "first front," for without the guns there was no chance of success on the battlefield[9]—had to be as well organised and managed as the armies in the field. Understanding the phenomenon of mobilisation, and the nature of mass society at war, is therefore key to understanding the First World War. Battle could not be joined half-heartedly; workers and civilians had to be attuned to the war effort for industrial warfare on the scale of the Somme to be feasible. This had been realised by 1915, but only during the course of 1916 did its full import sink in. The vast scale and commitment of this collective endeavour must be recognised. Society was being dismantled and reformed, the values of citizens and the principles of their leaders challenged, all in the cause of victory.

When Kitchener told his colleagues at his first Cabinet meeting, "We must be prepared to put armies of millions in the field and to maintain them for several years,"[10] he was initiating a process that would permanently alter British society. There were some, such as the Home Secretary and future Chancellor of the Exchequer, Reginald McKenna, who thought that the world's leading commercial and financial power could carry on world war with little real change to her social fabric. But this policy of "business as usual," which impeded full British mobilisation for the next eighteen months, could be pursued only for so long and so far. By 1916, the fatuousness of such reasoning was evident.

To mobilise the nation and empire, four resources had to be itemised and utilised. Fundamental to everything was manpower. As well as men to fill the ranks, workers—both male and female—had to be organised to extract raw materials, labour on the production lines and drive transport. In practice this could not be done without winning over the workers' official representatives, trade unions and socialist Members of Parliament. The industrial process, from the extraction of raw materials and production of foodstuffs to the invention, manufacture and maintenance of weaponry, also had to be overseen at each stage, to ensure quantity, quality and cost control. This meant engaging with or taking state control of private industry: each of these alternative solutions attracted adherents, depending on political ideology. The distribution of goods had to be ensured. Britain's insularity made this more of a problem, because everything deployed for war would have to be shipped overseas, and much of the wherewithal to produce it had already come in the same way. At least Britain could make use of the world's greatest commercial infrastructure and its largest trading fleet. Finally, the so-called fourth arm of defence, banking and finance, had to be regulated to cover the cost. Again Britain had the benefit of the world's richest empire and most solvent banking system. But even this was to creak under the strains of financing extended and extensive global war.

The process of mobilisation entailed a decisive shift in the relationship between the state, private capitalist enterprise and the individual citizen. The solutions adopted took slightly varying forms in different states, but all shared a common premise, which economists classify as corporatist. The state mediated the economic relationship between private producers and the workforce, while directing both in the pursuit of the national goal, victory. Corporatism was supposedly a short-term and temporary wartime solution, effective if not entirely efficient. However, the unprecedented efforts of 1916 were to consolidate this process, endorsing an innovation which over the longer term was to reshape domestic politics. The principles at stake, the rights of private capital versus the needs of state enterprise, determined the battle lines for the socio-political "civil war" that came to define the twentieth century.

When asked in late 1915 when the war would end, Kitchener had been non-committal. But he had confirmed that it would start in April 1916.[11] However, while progressing at an unprecedented rate, Britain's war mobilisation was only partially complete by then (if it ever could be truly "complete"), because the process was piecemeal, and at times chaotic. Nothing like this had been done before, so much was improvised as obstacles were encountered and demands accelerated. What progress Britain had made is normally attributed to one man in particular, by himself certainly. David Lloyd George, Chancellor of the

Exchequer in August 1914, became Minister of Munitions in May 1915 after a "shells scandal" rocked Asquith's government and forced him to reconstitute his ministry on an all-party basis. The War Office could not manage the rapid expansion of Britain's armed forces and the industrial restructuring which it entailed (although it was making a better job of it than Lloyd George's war memoirs later suggested).[12] What was needed were "men of push and go," of whom Lloyd George was one. The "Welsh Wizard," a former lawyer and Member of Parliament for Caernarfon Boroughs, had made his reputation as a firebrand Liberal-Radical in the ideological and constitutional disputes which characterised turn-of-the-century British politics. Self-important and determined in his opinions (which all too often combined a poor grasp of the facts with a lawyer's firm conviction), changeable and scheming (for his own self-interest as much as the particular cause he was pursuing), loquacious and argumentative (too much so for the more measured and thoughtful senior British commanders), Lloyd George aggravated his political colleagues and the generals with whom he would increasingly come into contact on a regular basis. Short, trim, with bright blue eyes and flowing white hair and moustache, charming and charismatic, Lloyd George's attractiveness to and passion for women are attested by his other nickname, "the Goat." Although a pre-war pacifist, he had rapidly transformed into a warmonger and within months of the start of the war was spoiling for a "fight to the finish" with Germany.[13]

By 1916, the British war effort was accelerating. As well as providing battalions for Kitchener's New Armies, the industrial towns of England—and America, where large orders for field guns and ammunition were placed—were making the *matériel* with which they would take on the Kaiser's host. This had profound consequences for the structure of British society and industrial relations. The July 1915 Munitions of War Act and its January 1916 amendment gave the new munitions ministry unprecedented control over British industry and labour. Factory workers christened it the "Slavery Act," for it removed their rights to move freely from jobs in "controlled establishments" producing munitions or jobs otherwise vital to the war effort, or to strike, and put an end to trade unions' restrictive practices in the workplace. On the other hand, the same acts outlawed excess profits and employers' lock-outs were also to be restricted.[14] Such quick, radical change could not produce immediate results, but at least it promised that in time Britain's citizen army would be appropriately equipped for its task.

When this would be achieved was uncertain, however, and this was another element in fraught civil–military and inter-allied relations. "The amount of ammunition will be *very large indeed*," Haig anticipated when the planning process began: one of the factors behind his pleas to Joffre to postpone the offensive until August was that by then he anticipated having more heavy guns and munitions.[15] Robertson thought the "wobbling" politicians would use insufficiency of munitions as an excuse to delay the offensive and look elsewhere. Lloyd George's insistence that the British army be as well provisioned as the French, and his plan for an Anglo-French armaments conference in the spring to "discuss with them how much ammunition they think they will want before they can be ready and how long it will take to get it . . . [are] the thin end of the wedge for deferring matters," Robertson moaned to Haig.[16] The latter, as usual, was more sanguine, noting after Lloyd George visited him at the end of January: "The conclusion is that the amount of ammunition available at the end of April seems to be ample and that the amount coming in from May onwards should easily meet all our demands."[17] As it turned out, Lloyd George's prudence was sensible and far-sighted. The hasty expansion of the munitions industry had parallels with the rapid expansion of the army: shortages of plant and raw materials, de-skilling of the workforce, bottlenecks in manufacturing, and inadequate quality control all delayed and compromised the production process. Although the acute shortages of 1915 were easing, in practice, as Haig acknowledged at the end of the war, "It was not until the mid-summer of 1916 that the artillery situation became even approximately adequate to the conduct of major operations, throughout the Somme battle the expenditure of artillery ammunition had to be watched with the greatest care."[18] Of the 7908 guns ordered for delivery in the first nine months of 1916, only 4314 had reached the front by the end of the year. Shell manufacture was taking off, testimony to Lloyd George's dynamism and ability to improvise: ammunition production rose nearly fivefold during 1916, and over fifty million shells of all calibres were delivered by the end of the year. However, both guns and munitions proved faulty: the official history of the war lists in detail the technical reasons for "prematures," "duds," "blinds" and inaccurate shooting which characterised the June 1916 bombardment.[19] It meant that the British offensive would not enjoy the luxury of unlimited munitions.

Britain's financial and credit system had to be reconfigured to pay for her and her allies' mobilisation. The import of key raw materials, foodstuffs and manufactures, and the liquidation of assets to pay for them, was an expanding responsibility for the Chancellor of the Exchequer and the many new ministers whose portfolios—munitions, labour, shipping control, food control, supply and research, reconstruction— corresponded to elements of the state-managed war effort. Luckily for Britain, the war broke out on a bank holiday, and Lloyd George kept the banks' doors closed for three more days while he introduced emergency legislation and printed cheap Treasury banknotes, averting an immediate financial crisis.[20] Nevertheless, by the end of 1916 Britain's war effort was in hock to the United States; although, through the system of inter-allied war credits set up in February 1915, this was more than offset by the credit which the City offered to France, Russia, Italy and Britain's lesser allies. Financial expedients, such as the weakening of the gold standard, expansion of paper money in circulation and the raising of war loans on the home market, kept the war economy afloat, although not without harmful inflationary consequences. But, as with the industrial economy, after 1914 the domestic and international financial systems experienced profound changes that would undermine post-war stability. Although the City's pre-eminence shielded Britain from the worst of it, debt spirals, bank failures, currency inflation and collapse, public sector retrenchment and mass unemployment were to scourge Europe through the 1920s and 1930s.

In July, when British effort commenced in earnest, Lloyd George had just moved to the War Office, following Kitchener's untimely death. He was going to oversee the application of the military-industrial machine that he had played a central role in creating, and he would be a pivotal figure in the battle he had helped to prepare, the battle that would ease his rise to the top of the greasy pole.

FRANCE'S FRACTIOUS REPUBLICAN political system adapted less easily to the requirements of total war mobilisation, despite the show of national solidarity evinced by 1914's *union sacrée*. There was a steady turnover of ministries in Paris—France was on her second premier and third ministry by 1916—and a forceful hand at the helm was lacking until the last year of the war. In the interim, parliamentary factions jockeyed for power and influence, political parties strove to promote their political and social agendas, a forceful president tried to keep everything under control, and all used the high command as a scapegoat for the failure to evict the enemy from French soil. For his part, Joffre took on the politicians at their own game, which too often distracted his attention from more pressing matters of strategy. Thus civil–military relations became a source of regular friction in the affairs of state, and gossip in wider society, where a radical press fuelled the rivalries between front and rear: *l'Homme Enchaînée*,* the mouthpiece of the Radical Party journalist and former premier Georges Clemenceau, France's elder statesman and deepest patriot, took the lead. At the root of this political turmoil was the French system of national service. Like everyone else, parliamentary Deputies were mobilised to their war stations in 1914, facilitating a direct conduit from the front to Paris, the party machines and the press, free of the censorship restrictions imposed on the forward zone by martial law. They could undermine the war effort from the inside, rather than snipe from the outside. Most famously, Colonel Émile Driant, patriotic right-wing Deputy for Nancy, sparked a long-running argument over the state of Verdun's defences before his martyr's death in the front lines there.[21] This culminated in secret parliamentary sessions to investigate the conduct of the high command which were sitting when the guns began firing on the Somme.[22]

While France's army had been about as well prepared and equipped as might be expected given fractious civil–military relations and government parsimony before 1914, it was nowhere near ready for the war it was going to have to fight. Military mobilisation, which disrupted France's war economy in its early months by putting industrial workers in uniform and getting them killed, and Germany's occupation of the Longwy–Briey basin, France's principal coal- and steel-producing area, dislocated things further. Fortunately, France found her own Lloyd George, the socialist Deputy Albert Thomas, who entered René Viviani's coalition government as Under-Secretary for Artillery and Munitions in May 1915. (His appointment was as much the result of a parliamentary intrigue against Joffre's guardian angel, Minister of War Alexander Millerand, as a calculated effort to improve the management of the war effort.[23]) Thomas, "so covered with hair and spectacles" that he looked older than his thirty-nine years, was "a fine energetic man," in Haig's summary judgement,[24] and such men were sorely needed. Industrialists' heads needed to be banged together, to refocus them from their own commercial interests to those of the nation. But the process of full mobilisation was greatly complicated in France by the

*Literally, "the chained man," the rechristened *l'Homme Libre* ("the free man"), a direct comment on France's wartime press censorship laws.

many interest groups and points of view which had to be taken into account. Being highly ideological, legislation on state intervention in industrial management, taxation of excess war profits and regulation of industrial manpower, all contemplated in 1915, was not to be finally agreed and passed until 1917.[25] In the meantime, wage and price inflations increased the social gap between the men on the front line and those retained on the home front.

As in Britain, mobilisation was not quick or smooth, but matters were progressing in the right direction. Joffre had put in a huge order for modern heavy artillery and munitions when the war stalemated in 1914. Disruptions in the labour market, the problems of negotiating contracts with manufacturers and issues of quality control in the rapidly expanding armaments industry all delayed completion of the army's first supply programme; and the confused situation was only exacerbated by further large orders for guns and shells and new types of weaponry placed during 1915. While French shortages and bottlenecks were not as acute as those that had provoked the "shells scandal" in Britain, her war effort was only starting to approach full capacity in 1916. As we have seen, Foch recognised the insufficiency of armaments in his repeated calls to postpone the offensive until 1917. However, politically this was impossible; and by June 1916, there were signs of improvement. The thousands of obsolescent fortress guns that had been pressed into front-line service to meet the army's heavy artillery requirements in 1915 were being replaced in significant numbers by the modern, quick-firing pieces ordered by Joffre in 1914. Sixth Army would have more and better guns to support its offensive than had been available in the spring. Shell production was also increasing. If the supply was not yet unlimited, with resources having to be carefully monitored and distributed between the different army fronts, Fayolle's army would have priority, and there would be more than enough shells for his first attack.

Full mobilisation was still ongoing when the 1916 campaign was planned, and productive capacity was still expanding: munitions programmes ordered in 1915 would come on stream in the summer and autumn to sustain the mass armies in the field. To that extent, Kitchener's prediction that the real war would begin in 1916 was correct. But prolonged industrial battle would expose the inadequacy of the early effort: much more would be required to prosecute mass war to its denouement.

. . .

NAPOLEON FAMOUSLY REMARKED that armies march on their stomachs. By 1916, paper was the nourishment of the military: the unsung machines of the First World War were the typewriter, the Roneo duplicating machine and the printing press, which enabled large numbers of documents to be produced and distributed in a relatively short time. Appropriately for the "scientific" battle, every operation was planned down to the finest detail: directives were issued from top to bottom, and reports and minutes circulated to check that all went according to plan, or to determine improvements if it did not. All this bureaucracy has left a wealth of source material for the historian, although at the time it could have done little to promote efficiency or initiative. It reflected the methodical way in which armies approached the novel challenges of industrial war. The staff were acutely aware of the problems that the modern battlefield presented, and of the need to adapt pre-war practices. Fourth Army's May 1916 tactical notes signed off with the disclaimer:

> The principles laid down in Training Manuals in peace have stood the test of war, but various causes, such as modern inventions and the numbers engaged on each side, have very materially changed, and will continue to change, the form that modern warfare takes.
>
> This being so we must profit by our experience to adapt the principles of war to the new conditions as they develop.[26]

Here was an acknowledgement that the industrial battlefield was an unfamiliar, complex, dynamic environment. Proper preparation could do much to offset its challenges, but there was a vast gulf between paper prescriptions and battlefield practice, and it was never going to be bridged easily.

A First World War battle was a multifaceted event, which would pass through several stages and entail the coordination of diverse elements. By 1916, quasi-Napoleonic forms of battle were being supplanted by a new style of warfare, which Foch dubbed "*une action profonde,*"[27] and Jonathan Bailey has subsequently defined as "modern," three-dimensional, deep battle—"large-scale high-intensity conflict" founded on the precepts of indirect artillery fire.[28] If not yet a thing of the past, linear engagements, in which both sides could see each other and strong frontal assaults or the turning of a flank would be decisive, were impractical on the congested battlefront in northern France. On the Western Front most of the enemy's force was behind the front lines, and attacking required a tactical method that allowed engagement with all his defences, two or more successive positions several miles deep. An offensive entailed going through or over them, not round them, which meant the artillery had to hit targets that could not be seen with indirect fire. The allies faced this type of battle in 1916, so they had to develop appropriate forces and tactics for it.

The first task was to observe and gain an understanding of the enemy. Second, the commanders and staff had to plan, supply and control the battle. Third, the soldiery had to fight: to hit targets with firepower and occupy and hold on to ground with manpower. Finally, they had to communicate what had happened to the rear, in the short term in order to respond to events on the ground, in the medium term to reassess the situation before the next fight and later to explain the bigger picture to those at home. After all that had been achieved, the process could begin again.

Assessing the enemy, his dispositions and defences, and more generally his reserves and intentions, was the task of military intelligence. At the simplest level, one needed to see what the enemy was up to: to this end, static observation posts were established by battalions, brigades and divisions.[29] Observation was the key to mastering the entrenched battlefield: overlooking the enemy's positions facilitated mapping of where he was, reporting of what he was doing and planning of an attack on his lines. Artillery fire could also be observed, and the progress of a bombardment monitored. In this respect, the Somme front was suitable for an attack in successive stages as Rawlinson originally proposed. The German first position was sited on the forward slope of the Thiepval Ridge, so it could be watched from observatories in the British front line across the Ancre valley. A detailed picture of the enemy's defences could therefore be made: photographic panoramas of his front, and neat, accurate trench maps.[30] The problem on the British front was that observation did not extend to the enemy's second position, situated on the reverse slope of the ridge, dominated by the Pozières heights, the key to observation over the whole area. Rawlinson pointed this out to Haig in his first attack plan, noting that "parts of it, though not actually out of range of our guns, would be difficult to deal with, as they are only observable from the air."[31]

In the French sector, good observation existed over the German front position north of the river from the observatory at Vaux, on a promontory jutting out above the Somme valley. Colonel Mangin of 79 RI took his junior officers, NCOs and men up there to observe the shells falling on the German lines. Seeing that that the artillery was doing its job would boost their confidence for the attack.[32] South of the river the ground was flatter, better for movement, but with poor observation except from the Flaucourt plateau, the main French objective in this sector.[33]

Where there was no direct line of sight, aerial observation could be used to learn what was going on in the enemy's rear areas. In 1914, when air observation was in its infancy, the retreating Germans sensibly sited their forward defences on high ground, giving them a clear view of the enemy: on the Somme, for example, the whole British front line to north and south could be observed from the Thiepval spur "and [the enemy] had skilfully arranged for crossfire by his guns."[34] By 1916, developments in aircraft and photographic technology and observation techniques were starting to offset this early advantage. The modern concept of "air superiority" was just beginning to assume military significance. In addition to tethered observation balloons—one German infantryman noted twelve flying behind the allied lines when the bombardment began[35]—each French corps was assigned one squadron and each heavy artillery regiment half a squadron of aircraft for observation of fire.[36] Seeing behind the enemy's lines gave a significant advantage in the planning and conduct of operations. British air reconnaissance established a reasonably accurate map of the German second position and gun emplacements, and learned that the third position was little more than a single shallow trench. But air observation still had its limitations. Despite rapid progress in aerial photography and skilled photographic interpretation, it was not completely accurate; and camouflage techniques were developing to hide guns and positions from eyes in the sky. Moreover, observation of artillery fire from the air was intermittent, partly because clear, settled weather was necessary for effective aerial observation and photography, and partly because the technology did not yet exist for rapid correction of fire. One-way transmitting wirelesses were starting to be fitted to observing aircraft, speeding up the process, but messages still had to be relayed through ground stations to artillery batteries by telephone. More usually, an aerial reconnaissance would be carried out after a barrage, with photos taken, developed and passed to the gunners, after which a further bombardment might be fired, depending on the success or failure of the first. This was one of the factors that forced the armies to opt for slow, steady fire plans before 1 July 1916. In a bombardment lasting several days, most of the time was spent measuring, checking and re-firing.

Aircraft in 1916 were slow-moving and vulnerable, to both enemy aircraft and ground fire, particularly the two-seater observation aircraft such as the British BE2c and French Caudron G4. Despite the glamorisation of First World War fighter "aces," their job was quite prosaic: simply to protect the slower observation aircraft. The tethered balloons which looked permanently over the enemy's positions were even more vulnerable, fairground targets for the fighter pilots. The allies organised a concerted attack on the German "Drachens" on 25 June, to deny the enemy the opportunity to fire back at their massed batteries: "I went to the grandstand and watched [as] airmen attacked all three Bosch balloons that were up at 4 and brought them down in flames," Rawlinson recorded with satisfaction.[37]

Seeing where Fritz was was one thing; knowing what he was doing was quite another. Listening posts monitored telephone calls on the other side of no man's land.[38] German telephone-intercept techniques were more advanced than British. G.H.Q. standardised such practices only after the battle had started, realising from captured documents that German interception of front-line telephone calls before 1 July had contributed significantly to their effective response to the first assault.[39] The British had formed a new Intelligence Corps of linguists in 1914 to question enemy prisoners (often captured in trench raids mounted for the purpose of gathering intelligence) and deserters, and to collate the morsels of information collected in order to build up a clearer picture of the enemy's strength, morale and activities.[40] It was French practice for each unit to report daily on enemy activity, casualties, prisoners taken and shells fired and received, listed by time, number and calibre: 76 on 79 RI's front on 14 June; only 30 on the seventeenth; and 152 150mm shells in a concentrated bombardment of the Suzanne–Maricourt road early on the nineteenth.[41] Such seemingly trivial details helped to build up a complete picture of the enemy's dispositions, fighting spirit and intentions. All quiet on the Western Front it may have appeared, but there was a daily routine of small-scale interaction, listening, watching, shelling and raiding, as each side prepared for the coming fight.

What was going on far behind the enemy's lines was more difficult to assess, but equally vital to the intelligence picture. An accurate estimate of the enemy's strength and deployment allowed strategic calculations of the broader military position, as well as more precise operational planning. The enemy's front-line order of battle could be compiled through the regular capture of prisoners. A network of agents behind the lines supplemented this by reporting on reserve formations and troop movements. By 1916, G.H.Q. had an extensive network of agents in northern France and Belgium monitoring the German rear areas, and the War Office operated a wider Secret Service network reporting on Germany from neutral countries (principally the Netherlands). A parallel French train-monitoring network covered north-east France and the Rhineland.[42] In June 1916 the spies' reports were encouraging. After the opening of Russia's offensive, troop trains were moving eastwards, reducing Germany's reserves in the West. In the final intelligence summary before the attack, the Second Bureau* calculated that there were 121 German infantry divisions opposed by 103 French and 54 British. It reckoned that nineteen German reserve divisions were immediately available for the fight, seven behind the Somme front (five facing the British, and two the French) and twelve that could be redeployed from other sectors of the Western Front.[43] This was an overestimate—the German army actually had ten divisions in reserve in the West in late June, with six behind the Somme front—and would have suggested to Joffre that the wearing-out battle was not yet over. Conversely, G.H.Q.'s assessment concluded that there were only three divisions of "reduced quality" behind the Somme front, and six in reserve in the West,[44] a severe underestimate that impacted unfortunately upon Haig's operational planning.†

All this intelligence fed into the detailed planning of the battle, from the strategic level down to the preparations to capture sectors of the enemy's front. Tactical guidance issued to French units in 1916 was extensive and precise. In April GAN published an eighty-two-page briefing document, the essence of which was distilled down to battalion commanders, carefully delineating the stages and processes of the coming offensive. The attritional nature of the coming battle was pointed out on the first page:

> The offensive battle involves the conquest of a succession of enemy positions, well organised and arranged in depth . . . The battle once begun is therefore a drawn-out affair, which must be conducted methodically until the enemy's power of resistance is broken by moral, material and physical degradation, but without exhausting our own offensive capacity.

---

*The French General Staff had four bureaus. The second was responsible for intelligence.
†See Chapter 3.

The necessary preparations—planning, preliminary works and tactical training—and how to mount the successive attacks on the first and second German positions were set out in minute detail. The key principle was the effective coordination of the two main arms. If the artillery cleared the ground, the infantry could progress with relative ease and fewer casualties. Therefore, "the artillery preparations determined what the infantry could achieve," and the depth of the infantry assault should be determined by the range of the guns.[45] Moreover, the French had learned that a successful infantry assault depended on reconnoitring the enemy's defences effectively, giving each attacking unit a single, clear objective coordinated with the units on the flanks, and maintaining the cohesion of the assaulting units. This required foresight and practice; nothing should be improvised on the day. On this theoretical basis Sixth Army and its subordinate formations would draw up their plans. Nevertheless, as General Amédée Thierry, 2 *Division d'infanterie coloniale's* (DIC) chief of staff, recollected, it was one thing to outline these novel methods on paper, quite another to provide the wherewithal and military skill required to execute them effectively.[46] Moreover, prescription and rigidity were to have their own adverse consequences. Hidden in a footnote, GAN pointed out that between the general attacks on successive enemy positions, "all the enclaves which survived the general attacks would need to be reduced."[47] In this instruction— that a jumping-off line needed to be secured for each large-scale assault on a trench position—lay the seeds of the small-scale attritional operations—*grignotage* (literally, "nibbling"), as Joffre defined them— that would frequently cost the attackers more heavily than the large set-piece attacks that they prepared.

The most important element of the assault plan was the artillery bombardment, which had to be coordinated with the infantry advance.[48] Here French methods showed greater sophistication, although there is evidence that British gunners were taking an interest in French practice. Fourth Army's tactical notes certainly acknowledged the French principle that the attack should not be pressed beyond the range of the artillery, even if in practice it was to spread its bombardment too thinly.[49] If not yet as complex or heavy as those which were to be fired in the last two years of the war, by 1916, a preparatory bombardment was taking on a distinct form. The artillery had various tasks, and different calibres of guns were assigned to each. The guiding principle was the destruction of the enemy's fixed defences and the killing of his defenders, to allow the infantry to progress. The artillery was in essence a large broom sweeping the ground before the advancing foot-soldiers:

"the artillery devastates, the infantry overwhelms" was G.Q.G.'s January 1916 formulation of this cardinal principle.[50] At the same time neutralisation and interdiction—temporarily incapacitating the defence, and isolating the battlefield to prevent reinforcements reaching it—were beginning to be developed alongside destruction as elements of a comprehensive fire plan.

The first problem was how to clear the way to the enemy's defences. This meant cutting the barbed wire, "the infantry's worst enemy," which protected the successive lines of trenches: Thierry counted twelve successive belts of wire, including some left over from the time the French held the village, on the front facing 2 DIC at Frise.[51] The best tools for this were high-explosive shells (which exploded into a number of sharp-edged fragments) fired at relatively short range by the field artillery, British 18-pounders and 4.5-inch howitzers and French 75s. By 1916, French factories were producing high-explosive shells in huge amounts and so wire-cutting on the French front would prove relatively unproblematic. Even so, to ensure success, each French regiment was assigned special wire-cutting sections that would go forward on the night before the attack to clear any surviving entanglements.[52] British high-explosive shell production was less advanced. Although such shells started to be delivered in significant numbers during the spring, the British artillery still had large stocks of the older shrapnel shells (anti-personnel shells that exploded into a hail of small, round bullets) that had to be employed for wire-cutting, making the process far less efficient on the British front.[53]

The field artillery's other job was to fire the barrage which supported the infantry as they advanced. The first supporting barrages had been linear, lifting from fixed line to fixed line (often the enemy's successive trench lines) as the infantry moved forward. A more sophisticated model, the creeping or rolling barrage, had not yet been universally adopted. Major Alan Brooke, who prepared 18th Division's fire plan for 1 July, remembered that he had learned the principles of the creeping barrage from a French artilleryman, Colonel Herring, that March.[54] Fourth Army was less enthusiastic about the principle, preferring a barrage which lifted from one fixed line to the next. Its tactical notes cautioned:

> The ideal is for the artillery to keep their fire immediately in front of the infantry as the latter advances, battering down all opposition with a hurricane of projectiles.
>
> The difficulties of observation, especially in view of dust and

smoke, the varying rates of advance of the infantry, the varieties of obstacles and resistance to be overcome, the probable interruption of telephone communications between infantry and artillery and between the artillery observers and their guns, renders this ideal very difficult to obtain.[55]

GAN's tactical notes also advocated a linear supporting barrage, which would concentrate fire on enemy trenches or the edges of woods and villages where the defence was likely to be concentrated as the infantry went forward. Batteries of 75s (which, despite the trench stalemate, had not lost their old battlefield manoeuvrability) would also accompany the infantry advance to bring close fire support against pockets of resistance.[56]

Destroying the enemy's fixed defences was seen as the key to the infantry's success. Howitzers—British 4.5-, 6- and 8-inch guns and French 120, 155 and 220mm weapons—and trench mortars, which both had a short, high trajectory, enabling shells to be plunged into the enemy's lines, were the most effective weapons for smashing trenches and strongpoints. By 1916, the French army had worked out the best weapon for each job: the howitzers destroyed the trenches themselves, as well as machine-gun and observation posts; light mortars were also good for this job, since they fired over a short range and could be effectively targeted on key points, particularly in the enemy's front line; more powerful mortars, 270mm calibre or larger, were used against the stronger defences, fortified villages or concrete strongpoints. Enemy dugouts, if they could not be destroyed, were to have their entrances blocked in this way, or would be neutralised with gas projectiles.[57] Underground mines were also to be used against particular strongpoints in the front line, either to destroy them directly or to mask them with a heap of raised earth. Nineteen mines were dug on the British front, eleven small ones and eight large.[58] The Hawthorn Ridge mine near Beaumont Hamel and two near La Boisselle have left their permanent marks on the Somme landscape.

As well as clearing the way for the infantry, the artillery was tasked with silencing the enemy's artillery so that it could not interfere with the ground attack. This counter-battery fire, recognised by G.H.Q. as "the most important, as it is the most difficult task of the artillery under present conditions,"[59] was the weak element of the bombardment in 1916, relying as it did on being able to locate and destroy enemy batteries. Trench artillery could be spotted across no man's land and engaged with trench mortars or heavier artillery. But field and medium guns, emplaced between the first and second positions, and sometimes behind the latter, were harder to spot. Most would be in reverse-slope positions, so could be observed only from the air. And even if camouflaged enemy batteries could be located—usually by photographic reconnaissance, since flash-spotting and sound-ranging techniques were in their infancy in 1916—they were hard to hit as they were small and well dug in. The French were starting to realise that temporarily incapacitating the crews with an intensive high-explosive or gas-shell barrage that would force them into cover was a more effective way of silencing guns, at least while the attack was under way, than trying to destroy them.[60]

Finally, the artillery had fire missions behind the front defences. Super-heavy, long-range guns, usually large-calibre ex-naval pieces mounted on railway carriages, would fire on targets in the enemy's rear areas. As well as eliminating artillery beyond the range of the other guns, interdiction fire, as it was called, was intended to disrupt the enemy's communications and slow down his response to an attack. Key communications centres—crossroads, railway sidings and junctions, field headquarters—would be targeted, as would troop concentrations and ammunition dumps. A new element of warfare, aerial bombing, was to be incorporated into this long-range barrage. On 1 July squadrons would attack railway lines bringing reinforcements to the battle, and ammunition dumps as far away as Mons, Namur and Lille.[61] Most of these techniques, which would be commonplace in later battles and wars, were in their infancy in 1916, and they would not be perfected without extensive experience.

No bombardment, however heavy, could eliminate every last defender concealed over a large area. But a prolonged bombardment would exhaust, disrupt and demoralise the defence, making it easier for the infantry to subdue any survivors. If the artillery had done its job properly, the infantry's role would be practicable. However, while the bombardment was the vital element, the infantry assault itself, the leap "over the top" into the unknown, remained the main and defining event of battle. Infantry tactics were adapting to the reality of this frantic rush across no man's land, preparing the troops for the confused, vicious, close-quarter combat which would be met on the other side; for sweeping up scattered survivors and overwhelming pockets of resistance that stood between them and their final objectives. By 1916, the French army's tactical preparations were of a high order: they had been testing and refining their methods for more than a year already. Rawlinson knew this. He had sent staff officers to study French practices, and his army's tactical notes acknowledged allied influence.[62] Germany is traditionally credited with the initiative when it comes to the development of appropriate infantry tactics for the industrial battlefield, but all armies were engaged with the same problems, and worked out similar solutions. "Storm-troops," infiltration tactics, combined-arms tactics, heavy support weapons, infantry specialists and small-group formations are all identified in the paeans to German military skill. But all were employed by the French army on the Somme in 1916. G.Q.G. revised training manuals based on their battlefield experience in 1915 and at Verdun. Indeed, they concluded that German offensive methods there vindicated the tactical principles which they had formulated over the previous winter.[63] Costly and hard-to-control "human wave" tactics were abandoned. Instead, the French infantryman was to become a specialist in one of the arms of trench fighting—the rifle and bayonet, hand-grenade, rifle-grenade or light machine gun—and assault tactics were to become more dynamic. Assault waves (which should not be simplistically equated with "straight lines," since their layout varied between lines, groups and single-file columns, as required) were to be thinned out, and infantry companies trained to manoeuvre in small, mutually supporting groups. Fire-and-movement tactics would allow the infantry to overcome points of enemy resistance that had not been neutralised by the artillery. Experimental at the start of the year, such tactics proved their worth and would become standard by the later months of 1916.[64]

Some French tactical principles had clearly been absorbed by Fourth Army, such as the need to equip the assault troops with appropriate weaponry, grenades and automatic grenade-launchers, light machine guns and light artillery weapons for taking on surviving enemy machine-gun nests and defending conquered positions. In this the British were some way ahead of the French. The Lewis gun, first issued to troops in 1915, had proved its worth at Loos and was now the standard battalion light machine gun. One key difference between British and French tactics was in the emphasis placed on "mopping up" the enemy's trenches once they had been entered. *Nettoyeurs* (literally "cleaners"), armed with grenades and revolvers, were designated to follow each assault wave in every attacking French company, tasked with clearing specific dugouts, neutralising isolated pockets of enemy resistance, and rounding up prisoners.[65] This would maintain the momentum of the assault, since the leading waves could press forward confident in the knowledge that their rear was protected and they would not be cut off. In comparison, although Fourth Army's tactical notes suggested employing "clean-up parties" to guard the flanks, they gave no guidance on how exactly to clear ground that had been crossed.[66] Those soldiers who were detailed to mop up received only superficial training, as Corporal Harry Shaw remembered:

> They said to us, "You lot are moppers up, that's what you've got to do, follow in after the first wave and mop up." But they never told us what mopping up was, and we only had a vague idea. No training as such, except that we were supposed to chuck bombs at these flags that were supposed to be dugouts.[67]

Under this regimen, the British assault troops were liable to lose their momentum and become bogged down fighting in the first position, rather than pushing on to their further objectives, or to be cut off and counter-attacked if they got beyond it.

This deficiency notwithstanding, British assault tactics were more flexible than posterity acknowledged. It is one of the enduring Churchillian clichés of 1 July 1916 that the British infantry advanced in long, straight lines directly into the enemy's machine-gun fire.[68] Rawlinson's apparent lack of confidence in the inexperienced New Army troops, especially the prescription that they "push forward at a steady pace in successive lines," has been regularly deployed against him by critics of British generalship.[69] In fact, Fourth Army was promoting the deep formations needed to sustain the momentum of the attack in a deep battle. Its tactical notes suggested four assault waves were needed to secure a position, "each line adding fresh impetus to the preceding line . . . and carrying on the forward movement," not prescribing any particular assault formation.[70] Robin Prior and Trevor Wilson's detailed study of the various corps attack plans for 1 July has demonstrated that there was a high degree of flexibility (and a certain amount of ambiguity) written into these notes,[71] and considerable variation in the tactics adopted along the British front: local ground and objectives, as well as the mind-set of senior commanders, prompted adaptation. Some units stuck to steady straight lines, with mixed success: Lieutenant-General Pulteney's III Corps, attacking Ovillers and La Boisselle in the centre (Churchill's main focus in his account[72]), did so. Most other divisions did not.[73] The staff may have issued training pamphlets, tactical notes and operational orders, but it remained the responsibility of the commander on the spot to execute them in the most appropriate way.

French formations also advanced in successive (non-linear) waves: 2 DIC's plan for 1 July varied the number of assault waves in relation to the depth of the objective.[74] Experience in the field suggested that the first two waves should be composed of skirmishers, with light machine guns for close fire support. Supporting waves were to advance in single file.[75]

Differences between the allied armies are understandable when their previous experience is taken into account. It is overly simplistic to judge that the British army was too rigid or conservative in its tactics and command. It was keen to learn, engaging with its task thoughtfully and professionally; but to date its commanders, officers and men had relatively little experience of large-scale offensive operations.

Haig had informed Clemenceau in May: "my divisions . . . want much careful training before we could attack with hope of success."[76] However, a great deal of "training" was merely military routine: long route marches along tree-lined country roads (the fields with their sprouting crops were out of bounds); rifle and bayonet practice, for the moment when they would close with the enemy; grenade and machine-gun practice for offensive and defensive fights; gas drills in which men would be plunged into chlorine-filled trenches in their primitive gas masks.[77] All this "bull" evoked a certain degree of disdain in the keen, overconfident but unmilitary soldiery. "Tomorrow we are being inspected by all sorts of Generals and people with red things on their hats," Bernard Ayre wrote home from his Fourth Army training school. "Of course there will be a tremendous amount of eyewash etc."[78]

Effective field training for the fight to come was more problematic. This was where the British army lagged behind its French ally, having neither a pre-war training infrastructure to hand, nor appropriate experience on which to base that training. GAN's instructions prescribed that formation attacks were to be practised behind the lines, on elaborate mock-ups of the ground to be attacked. Fourth Army followed the same principle: lines of mounted men or marshals with flags indicated the progress of the artillery bombardment, while enemy trench lines were delineated with tape or shallow ditches.[79] "War under these conditions certainly was very enjoyable," noted one subaltern after a day manoeuvring in the sunny countryside.[80] They were hardly realistic, but such staged rehearsals were perhaps the best that could be done with inexperienced troops. Preparations in 90th Brigade suggest that familiarisation and formulaic repetition were the order of the day:

> There is a large area here, mapped out with trenches, representing no man's land, the Boche trenches beyond, and a plan of the village of Montauban, which we are to capture and hold, all to scale as obtained by aerial photographs. Trenches, streets etc. are marked with names that we shall christen them when we get across there, and here the Brigade goes en bloc every day to practise step by step (in accordance with the timetable fixed for the actual day) our part in the big offensive.[81]

If there had been no enemy left in the opposing trenches to contest the passage it would have been a perfect preparation. But at least such training built confidence (or perhaps overconfidence). "The men are in splendid spirits. Several have said that they have never before been so instructed and informed of the nature of the operations before them," Haig noted the day before the assault. Maybe it simply provided reassurance, not least to the commander-in-chief himself, who had worried constantly about the state of readiness of his civilians in uniform: "With God's help, I feel hopeful," he signed off.[82]

If the commander-in-chief was hoping for the best, others were preparing properly for the test which they faced. Some New Army formations, under the careful tutelage of professional regular officers, were learning the soldier's trade. General Walter Congreve's XIII Corps, which was to attack on the right of Fourth Army's front, and which had 90th Brigade as its reserve, would perform best on 1 July. In the corps as a whole, training and preparation were of a high order. Major-General Ivor Maxse, the commander of Congreve's left-hand division, the 18th, was one of the best British divisional generals of the war, an intelligent and forthright Guardsman renowned for strict military discipline and the thoroughness of his training regime (so much so that in 1918 Haig would appoint him to command the British army's new training establishment). Maxse grasped both the theory and the practice of battle. One of his subordinates recollected that his conferences were "like a university course on how to make a fine fighting division."[83] "Training is everything," he was fond of pronouncing to anyone who would listen: at the end of July he went so far as to complain to Fourth Army's staff that excessive fatigue had interfered with preparation and training for the attack.[84] Maxse's artillery was controlled by Brigadier-General

Casimir van Straubenzee, with the gunnery staff work in the hands of Major Alan Brooke, who would rise to be C.I.G.S. in the next war. Their innovative barrage, modelled on French practices, would use trench mortars for close-range wire-cutting (rather than shrapnel fired by field guns), would include timed artillery "lifts" sweeping from trench to enemy trench (thirty-five in all at a measured pace of fifty yards every ninety seconds so the infantry could keep up) and would be plotted carefully on accurate maps: all of these methods would become standard as the battle progressed.[85]

This combination of artillery bombardment and modern infantry tactics was designed to carry the assault on to its objectives. That was only the first stage of battle, however, and experience had taught the allies that holding their objectives in the face of the inevitable German counter-attacks was much more problematic. Less haste, better leadership and greater strength were seen as the solutions. GAN prescribed attacking with three carefully controlled waves: the first to rush the enemy's line and push on to within striking distance of the next objective; the second to reinforce the first; and the third acting as a reserve.[86]

While British generals certainly understood the principles of this tactical method, putting them into practice with limited experience and raw troops would be a challenge. Haig's note of a conference with Hunter-Weston, whose VIII Corps had come from Gallipoli, is revealing:

> I impressed on him that there must be no halting attacks at each trench in succession for rear lines to pass through! The objective must be as far as our guns can prepare the Enemy's positions for attack—and when the attack starts it must be pushed through to the final objective with as little delay as possible. His experiences at Gallipoli were under very different conditions... now his troops can be forward in succession of lines in great depth, and all can start at the same moment![87]

The fact that the commander-in-chief was giving tactical revision classes to a senior field commander six weeks before the attack is indicative of the deficiencies of the rapidly expanded, inadequately trained and led British army as it prepared for its first real test; and it shows that neither had grasped the importance of calm, steady progress and maintaining order and control, as the French had.

In contrast to the British assaulting divisions, which were a mixture of regular, New Army and territorial divisions,* the French had selected their best formations for the assault, and trained them thoroughly. North of the river the shock troops of XX CA, "considered, in peace and war, the best corps in the French Army,"[88] were back, rested after their latest feat of arms, blunting the German strikes north and south of Verdun in February and March. Its two original divisions from France's standing peacetime covering force, 11 ("Iron") and 39 ("Steel") DIs, were to attack on 1 July, supported by two recently created "assault" divisions: 153 DI (with one brigade of Moroccan troops and another of elite *chasseurs à pied*) and 47 DI (composed entirely of *chasseur* battalions). South of the river, I *Corps d'armée colonial* (CAC), with four divisions (1, 2 and 16 DICs and the Moroccan Division) came from France's pre-war Colonial Army, a standing force with considerable combat experience overseas. Although greatly expanded after the outbreak of war—16 DIC was a new formation raised from conscripts—the Colonials maintained an *esprit de corps* and tough military discipline worthy of their pre-war professional roots.

Once its objectives were seized, the infantry would need sufficient firepower and cohesion to keep the enemy's counter-attacks at bay until reinforced. The artillery could assist by firing an isolating barrage in front of newly won trenches to keep back hostile infantry; but in practice the infantry had to be able to defend themselves until the new position could be consolidated. The keys to holding on to captured ground remained reorganisation, reinforcement and resupply, digging in and wiring. Supporting waves would therefore carry picks and shovels, sandbags, wire bales and other engineering stores, as well as extra ammunition, food and water for the assault troops. In effect the whole line would lurch a few hundred metres forward, stop, consolidate, report back, rethink and then start all over again.[89] Such was the pace and nature of trench warfare. Whatever the hopes of over-optimistic military planners (and subsequent criticisms of armchair generals), it would not be hurried.

THIS SLOW, STEADY pace was dictated as much by the relatively primitive nature of communications on the First World War battlefield,

---

*Although, in practice, regular and New Army brigades had been swapped between divisions when they reached France, so for many this distinction was not clear-cut.

both for the passage of information and for the movement of *matériel*, as by tactical and operational factors. Often overlooked because they are seen as prosaic and routine compared with the excitement of the fight, logistics are nonetheless central to success and failure in the field—and in 1916 supply and control were integral components of the ongoing industrialisation of war. As Martin van Creveld has identified, the Western Front was the first campaign in history which required more manufactured goods—bullets, grenades and shells, machines and replacement parts, explosives, preserved food, fossil fuels—than locally sourced raw materials (predominantly food and fodder) for its conduct, and a modern transport infrastructure to deploy it all.[90] Unfortunately, in 1916 the capacity to send messages and move men and munitions lagged behind the ambitions of the generals. Once battle was joined commanders' ability to communicate with and control their formations was limited. Similarly, supplying and reinforcing troops in the field was a slow, labour-intensive and confused process, necessitating careful management and much hard work behind the lines. The demands of the Somme campaign were to stretch such a logistic infrastructure to breaking point.

Military communications were on the same technological cusp as weaponry in 1916. Wireless communication, which was to revolutionise battlefield communication between forward and rear formations as the twentieth century wore on, was in its infancy. Too large to be carried into battle, wireless sets, which at this time transmitted only telegraphic messages in Morse code, could at least be mounted in aircraft, which could observe the progress of operations and report back to headquarters. Therefore, troops on the ground had to be identifiable from the air. The best means available were metal or coloured-cloth badges sewn onto their backs, or coloured groundsheets that were spread out to indicate a unit's position. The customary method of field communication was by telephone; artillery and infantry could be directed fairly speedily by a call from headquarters, provided the telephone wires had survived the bombardment (they were usually buried six feet deep to protect them from all but the heaviest shells). But the elaborate network of lines and field telephone exchanges that linked the defended zone stopped at the front line. So when troops went forward, parties of signallers would trail telephone wires across no man's land to the new front line. Of course, these vulnerable wires laid in the open (or the signallers themselves) would often not survive the fight. As a result, more traditional methods were the mainstay of communication from front to rear once battle was joined. Despatch runners (brave men who suffered heavy casualties), carrier pigeons, message-carrying dogs, flares and other visual signals were all vital, if slow and uncertain, means of passing information back from the battlefield. Information that reached the colonels' dugouts or the generals' châteaux would often be inaccurate or out-of-date, if it reached them at all. Therefore, since orders could not be altered quickly once an attack had gone in, higher commanders and staff were relegated to a spectator's role; hence the elaborate preparations to ensure that the battle went precisely to plan. If anyone could exercise initiative, it would be the junior officers closer to the action. Paradoxically, however, the elaborate and detailed instructions as to timing, method and objectives issued from headquarters limited the scope to do so.

The movement of hundreds of thousands of men and thousands of tons of *matériel* into and around the battle zone was an even greater challenge, testing the organisational skill of the army staff and the civilian civic and transport authorities on whom they relied. The paraphernalia of a modern industrial army was immense and complicated. Between March and June an enormous shanty-town grew up behind the forward lines: huts, tents and dugouts to house troops; artillery parks and ordnance workshops for the guns; stables and wagon lines for horses and vehicles; dumps for munitions; reservoirs and water and sanitation pipelines; aerodromes; hospitals; fuel depots; electricity generators and sub-stations; training grounds; light railways, roads and telephone networks to connect them together; all behind miles of newly excavated trenches and barbed wire. Altogether a huge swath of France was earmarked for combat, with a spider's web of railways, roads and canals linking this teeming hub to the wider world; through Amiens to interior France and the world's sea lanes; through Péronne, Bapaume and St. Quentin to Germany's industrial heartland in the Ruhr.

British lines of communication had been developing along with the expansion of the army. Fourth Army's supply lines stretched back from the forward railhead at St. Roch, just south of Amiens, through the British advanced base at Rouen, to the principal base port at Le Havre. Along this railway would come everything needed for the battle which could not be sourced locally; and back would go wounded men destined for the base hospitals around Rouen and Le Havre, or at home. By 1916, the British army's consumption was prodigious: over two million gallons of petrol per month for the motorised transport that was becoming a mainstay of the supply chain, 148,000 tons of munitions for the first month of battle, and a million new steel helmets issued to the front-line troops between January and June all had to be shipped from Britain, distributed and accounted for.

Preparation was very labour-intensive. When 11 DI entered the line at the beginning of June its regiments were instructed to itemise the preparatory labour required. On 37 RI's sector, immediately north of the river, 91,420 man-hours of work were listed: the majority, 72,000 hours, for digging six kilometres of new trenches; the rest for excavating jumping-off points, command posts, dugouts, machine-gun emplacements and munitions stores, wiring and ongoing repairs. It was estimated the work would take two months to complete.[91] Whether all would be ready in time was a constant worry for the high command. British divisional engineer companies and pioneer battalions were insufficient for the vast programme of field works, so the ordinary soldiers bore the main burden of this preparatory labour. Fatigues—digging, shifting, building or maintaining the wherewithal of war—were a constant drain on the soldiers who had come to fight, a regular source of complaint as well as an ongoing diversion from battle training.

The transport infrastructure immediately behind the battlefield presented particular problems. Industrial armies could not function without good railways. An infantry division consumed between one and a quarter and four trainloads of supplies per day, depending on whether it was resting or fighting: Fourth Army would require between thirty-one and seventy trains per day once battle commenced. The main double-track railway line between Amiens and Bapaume had been cut by the trench lines just north of Albert. This town's railway sidings were too exposed to be used in the preparatory phase of the assault, so the main British railheads were located at stations between there and Amiens: Corbie, Heilly, Méricourt and Buire. A light railway connected Albert with Doullens, but a new line had to be laid to a railhead at Acheux, which would supply the northern sector of the front. Another single-track spur branched off from Albert, past Fricourt to Bray, just north of the river; but since part of this now crossed no man's land a second new line had to be built from Dernancourt to link the southern sector of the front into the rail network. Despite this expansion, railway capacity remained inadequate. However, the decision was taken not to use up valuable capacity moving road-stone (it was not anticipated that the army would be using the same roads for the rest of the year), which, given the reliance on road transport to supplement the railway network, would prove a mistake.[92]

It was a modern, mechanised transportation system that gave way the nearer it came to the front. The British army calculated that supplies could be moved up to thirty-two miles from the railhead to the front-line troops—the first twenty-five by Army Service Corps motor transport to divisional refilling points, the final seven by divisional horse-drawn transport and pedestrian carrying parties up the line.[93] Each link in the logistical chain was slower, more difficult and more dangerous than the last: what began as bulk goods in railway trucks would end up as individual man- or mule-loads inched slowly forward across shattered terrain under enemy fire. Road transport was the central link in this supply chain, so efficient traffic control and road maintenance were vital. The fan of main roads linking Amiens to Albert, Bapaume, Péronne and Roye were the only major communication arteries; the side roads linking the scattered communities and isolated farms were little more than tracks, often sunk between steep earth banks and deeply rutted from the passage of heavy farm carts. These Roman-straight highways were not the main roads of modern times: their thinly metalled surfaces, resting on a ballast of chalk aggregate, could not take the constant passage of heavy military vehicles. They quickly broke up into chalky rubble—dusty and rutted in dry weather, "a mass of liquid mud without bottom" when it was wet. Maintenance was carried out by British pioneers and French territorials during the brief interruptions and hold-ups in the traffic. But these were hasty, temporary repairs, filling in pot-holes and improving drainage: the lack of road-stone—and a shortage of steamrollers—made proper maintenance impossible.[94]

On Sixth Army's front, logistics would need to be carefully organised. Only one railway line ran from Amiens to the front, with a railhead at Bray, north of the river. It was to be supplemented by a network of light narrow-gauge railways to move munitions and other supplies to the front. North of the river, the French sector was hemmed in by the British lines of communication to the north and by the Somme itself to the south. As the French advance fanned out, so the bottleneck through which it had to be supplied became more congested. The Somme canal, which followed the line of the river from Amiens to Péronne, could be pressed into service to move heavy stores and casualties to and from the battle. But again, particularly south of the river where there were no railways, roads were the mainstay of the French lines of communication. The "*voie sacrée,*" along which lorries passed continuously to sustain the defence of Verdun, is celebrated as a model of military improvisation.[95] Another road-train, less sacred or celebrated but equally vital, would sustain the French army's offensive south of the Somme, along the road from Amiens to Foucaucourt.

Amiens was the focal point of this logistics network; not just its railway lines and sidings but its roads and parks became choked with the paraphernalia of war. The *poilus* rechristened it "*camionville*" — literally "truck town" — on account of the thousands of military vehicles that lined its boulevards and clogged its narrow medieval streets.[96] This was only one of the inconveniences which Picardy's inhabitants were obliged to suffer as the great battle took shape. Still, as had always been the case when armies took the field, there was much profit to be made from the thousands of men in blue and khaki who crowded their streets and camped in their fields. One passing brigadier-general recorded, "All was bustle . . . Amiens was like nothing we had yet seen in France — civil life continued, all the shops were doing a roaring trade, and all sorts of vegetables and fruit found its way into our various messes."[97] Baser needs, for drink and female company, generated profitable sidelines, from unofficial village *estaminets* to the brothels that burgeoned in Amiens. For the less prurient, such as Lieutenant Gerald Brenan, there were more conventional attractions. "We would arrive in time to eat a splendid lunch at the big hotel in the square," he recollected, "after which Ralph would go off to meet the little tart with whom he corresponded while I, to his jeers and disgust, would visit the cathedral."[98] The British army had its Claims Commission to pay for local supplies and requisitions, and rent for buildings and fields occupied by the army's many offices and camps (although not rent for the front-line trenches, as the myth would have it). It received claims for damage to property and persons occasioned in the day-to-day business of running an army (although this did not cover the inevitable impact of "*faits de guerre,*" today's "collateral damage").[99]

So far, the engrained rhythms of rural life, the cycle of cultivation, had seemed to defy any short-term martial disruption. Yet gossip and rumour about the war, the trial at Verdun and the invasion and occupation of their own fields merged with the farmers' usual seasonal talk of the spring sowing. By the end of May, the assault troops were starting to concentrate, and they became more certain, more apprehensive. Then those who still cultivated their fields within range of the guns were given forty-eight hours' grace to leave: otherwise they would have to stay put for ten days. No civilian traffic would be allowed when the attack got under way.[100] Driving their carts westwards to an uncertain future as refugees, the peasants could but ponder and marvel at the mass of humanity, the guns and the vehicles streaming in the other direction; and at the sudden eruption of heavy gunfire.

The gathering warriors were equally pensive. All who marched behind the British front were struck by the leaning Virgin of Albert, gleaming atop the town's basilica and visible for miles around: the statue of the Madonna and Child had been struck by a German shell in 1914 and now listed precariously. The soldiers' chit-chat contended that if Our Lady fell to earth, the war would end; and rumour had it that French engineers had wired the icon in place to prevent that. But now, they expected, that moment had come. "We know what we have come for and we know that it is right," one Australian clergyman in uniform preached to the men as they trans-shipped from Egypt:

> We have all read of the things that happened in France. We know that the Germans invaded a peaceful country and brought these horrors to it . . . Here we are on that great enterprise and with no thought of gain or conquest, but to help to right a great wrong . . . With our dear ones behind, and God above, and our friends on each side, and only the enemy in front—what more do we wish than that?[101]

OF COURSE, THE significance of all this activity was not lost on those watching, waiting and worrying across no man's land. Captain Harry Bursey noted the enemy's gallows humour the day before the bombardment started: "The Huns put a board up yesterday in their front line trenches and on it was pinned a paper with the following. 'We know you are going to attack. Kitchener is done. Asquith is done. You are done. We are done. In fact we are all done.' "[102]

Whatever the *Frontschweins'* gloomy expectations, Germany's commanders were not yet aware that she was done (although Falkenhayn had suspected it from the start). There were signs that the war was intensifying, and conditions on the home front were worsening, but war news remained upbeat and few doubted that victory would come eventually, even if not very soon. Christmas and New Year in Germany had lacked the usual festivities: meat was rationed, traditional Christmas cakes were hard to find, Christmas trees were expensive—although they were still distributed along the front lines to cheer the soldiery—and there were no fireworks on New Year's Eve as all explosives were reserved for the front. Newspapers reported victories in foreign lands, but some were starting to call for peace before victory. Karl Lieb-knecht's left-wing socialist faction withdrew their support for further war credits in the Reichstag. Formed into an Independent Social Democratic Party, they turned their vitriol against both the warmon-gers, with their imperialistic, bloodthirsty agenda, and their more mod-erate Social Democratic brethren who colluded in the capitalist conflict.[103]

Nevertheless, Germany's factories were still working at full capac-ity. The War Ministry had taken quick steps to put the economy on a war footing, establishing a War Raw Materials Office, directed by the industrialist Walter Rathenau, within a fortnight of battle commencing in 1914. As well as domestic coal, in which Germany was self-sufficient, key imports such as Scandinavian metals and Romanian oil would be managed centrally to keep Herr Krupp's vast armaments business going in the face of the allied blockade. The rich resources of occupied Belgium and France would also be exploited for the Reich's war effort, while her world-leading chemical industry would be put to work creat-ing new weapons of war (poison gas being the most notorious), and artificial "*ersatz*" commodities to replace those denied by the blockade, particularly synthetic nitrates for explosives and synthetic rubber derived from coal. Key strategic metals—copper, chromium and nickel—had been stockpiled in advance of war. The output of weapons increased nearly fivefold between August 1914 and December 1915, and munitions production sevenfold by the end of 1914. Thus, in the early phase of the war Germany escaped the "munitions crises" of the allies, even if her potential for mobilisation, greater in August 1914, was starting to stall by early 1916.[104] It had not been too difficult to keep Germany's armies supplied for two years, blockade notwithstanding; but they still had to deliver the victory her leaders had promised. Now an allied offensive threatened to expose the hollowness of those promises.

Positive news from Verdun in the spring kept civilian spirits up, but as the casualty lists lengthened and wounded and shell-shocked soldiers returned to Germany's towns and villages a note of pessimism and war-weariness started to emerge: the war would go on indefinitely, there would be no victory, the Western Front was deadlocked and young men would continue to die. The military euphemism "human material" was banned by the press censor in a vain attempt to hide the mechanical reality of attritional war. At home, food was becoming scarcer—tea and coffee disappeared from the shops, bread was rationed, and meat became an occasional treat, at least in the towns—and goods more expensive as the blockade began to bite. In May 1916 the War Food Office had been opened, to oversee the extension of the rationing system. The capture of Fort Vaux at Verdun and the reported "victory" over the British fleet at Jutland in early June momentarily perked up flagging domestic spirits.[105] At the front, though, there were no grounds for optimism: "The signs of a British offensive in our area increased day by day and the first storm signals became ever clearer. That they would come was certain, only the extent of the operation was unclear."[106]

Falkenhayn had factored an allied counter-attack into his attritional strategy from the beginning. Once the French were ground down at Verdun, and any British counter-offensive repulsed, the German army would be able to deliver the final blow that could bring the Entente to the peace table. But when and where would this counter-attack come? By early summer, the OHL staff were coming to realise that they had not broken the French army at Verdun. If their own reserves were diminishing, they could comfort themselves that the French had fed so many divisions through the Meuse mill that they would be incapable of any effective counter-stroke. So the offensive would be left to their inexperienced British allies, and therefore would be easily contained. Yet the prospect of a final victorious German offensive declined as spring turned into summer. The Russian and Italian armies were starting to fight back. Divisions were needed to shore up the fractured Austro-Hungarian front,[107] while the remaining reserves in the West would be committed to another push at Verdun, to keep the French tied down. Fort Vaux fell in early June, and Thiaumont and Fleury were within striking distance.[108] Falkenhayn did not know it, but the successful assault on the latter would be the high point of Germany's war. The day after Fleury was taken the allied guns opened fire on the Somme.

In Picardy the impending cataclysm was more obvious. Since 1914, the Somme sector had remained the domain of the German Second Army, led from 1915 onwards by General Fritz von Below. From a Prussian military family (his cousin Otto also commanded an army), Below had been appointed after a successful campaign as a corps commander on the Eastern Front. Although he has not left much of a mark on history (he did not live to write his memoirs and is only a shadowy presence in those of others), he was a competent, resourceful field commander who found himself facing an unprecedented challenge. He was "one of our best army commanders," remembered one of his staff officers, the future Field Marshal Erich von Manstein.[109]

By late May, the signs that an offensive was being prepared were unmistakable, and Below anxiously informed OHL that his front needed reinforcement. Falkenhayn was not totally unreceptive to this plea, as some have suggested,[110] but having just resumed the offensive at Verdun he did not have much to spare. By mid-June, German military intelligence had a fairly accurate picture of the impending allied attack: it calculated that the British had twenty-two and the French nineteen divisions available. Falkenhayn moved four of his reserve divisions so that they could intervene in Second Army's sector if necessary, although he believed that the main battle would be further north, on the Arras front, where Third Army was making similar preparations for the later stages of the offensive. He trusted that Second Army's front would hold.[111]

If men could not be spared, strong field fortifications were the next-best thing. Since the French had surprised them at Serre in July 1915, Second Army's field engineers had been working to make their defences impregnable. The front system, which comprised three interlocking lines of trenches—forward, support and reserve—was dug deeper and wired more extensively, after which similar attention was given to the second position. Villages along the line were fortified for all-round defence, and redoubts (such as Hawthorn, Leipzig and Bois Y) were built where the German line jutted into no man's land, from which advancing troops could be enfiladed with machine guns. An intermediate line of strongpoints sited (Schwaben, Stuff, Pommiers and other redoubts) for all-round defence was also constructed to hold close reserves who could reinforce the front lines or counter-attack rapidly, and to protect vital points in the defensive zone.[112] A third position had been mapped out three to five kilometres behind the second over the winter: OHL's defensive memoranda now prescribed at least three positions after the enemy had reached the second system in autumn 1915.[113] This line hugged the forward slope of the low ridge across the valley behind the main Thiepval–Combles heights, covering Bapaume and the road that linked it to Péronne. Below the river bend, the third line covered bridgeheads on the west bank. Digging and wiring of this new line commenced in earnest in May and June. Defending was as labour-intensive as attacking. In 26 *Reserveinfanteriedivision* (RID)'s sector north of Thiepval there was

> feverish development of the positions, including the intermediate and the Second and Third positions, especially the Grallsburg and the Schwaben Redoubt. The Ancre Valley obstacle was strengthened. Stop lines were constructed, as were additional communications trenches. Numerous new battery positions were constructed, ready to accommodate reinforcing batteries. Dugouts were improved, deepened to at least seven metres and equipped with two or three exits. Specially organised concrete squads built sector observation posts. The wire obstacles were strengthened and mine galleries were extended, as was the telephone network ... Civilians were moved to the rear and large quantities of ammunition, including hand grenades, were placed in shell-proof shelters, as far forward as the frontline itself.[114]

It was, as Haig subsequently described it in his despatch, "one composite system of enormous depth and strength."[115]

A defensive position was, however, only as good as the troops who manned it. The regiments holding the line in June 1916 were experienced, of high morale, and familiar with the ground they had occupied for many months. They were finished products of the German military system, hardened by the outdoor life and honed by combat. They could be expected to fight well when the allies attacked, and the defensive tactics which they employed were sound. The bases of the German army's defensive doctrine in 1916 were using pre-registered artillery to isolate ground and the fixed machine gun to control it, elastic defence and the rapid counter-attack. Linear defence, holding the forward trenches against all comers, had been abandoned after the 1915 battles had demonstrated that it was a costly, inflexible method. At the same time, the principle that the defensive zone was an integral system, the loss of any part of which would compromise the whole, still held. Nevertheless, the first lines would not be contested tooth and nail. Instead, they would be weakly garrisoned, with a machine-gun-based defence absorbing the shock of the enemy's charge. "Correct use of the machine guns is of fundamental importance," one late 1915 tactical document pointed out. "If they are properly placed off to the flank, preferably not in the front-line trench, but in an elevated position and properly handled, their effect can be practically devastating."[116] German machine-gunners were specialist troops, better paid than their riflemen colleagues. They would have the key role of breaking the cohesion of the allied assault. After the attackers had been sucked into the battle zone behind the front trenches, where artillery and machine-gun fire would inflict further casualties and sow confusion, the defence would bounce back. Specially trained companies would counter-attack the exhausted and fragmented enemy, eliminating them from the German lines and restoring the integrity of the position.[117] This was German defensive doctrine, and it was well understood by the troops. The worry was that there were not enough of them. Once the forward position was garrisoned there were few formations to reinforce the defence. Expecting that the main attack would come north of the river, Below had to strip his defences in front of Péronne, moving 10 Bavarian ID northwards to furnish a reserve. This left the units holding the front south of the river stretched dangerously thin.[118] So perhaps Haig's optimism was not entirely without foundation: the German defences, while formidable, were certainly not impregnable. Rearward works were still ongoing, sections of the defences were weakly held and local reserves were in short supply. If the tough forward crust was broken, there was little behind it to resist an immediate British surge.

By June, the war was hotting up along with the weather. The newspapers that reached billets and trenches made great play of the Russian offensive in Galicia, where General Brusilov's armies had smashed in nearly fifty miles of the Austro-Hungarian front. In Italy an Austro-Hungarian surprise attack in the Tyrol had been repulsed, and the Italian army was once again on the offensive on the Isonzo front. The General Allied Offensive was clearly under way, and it was time for the Anglo-French contribution. "My world is full of joy; it's vengeance for Verdun," wrote Mangin as the last preparations were completed. To his regiment had fallen the honour of beginning the liberation of *la patrie,* of taking the fight once more to the enemy, and of showing the watching world what the French army, when well led and properly equipped, could do.[119]

By the last week of June, the build-up was complete. Rawlinson had eighteen infantry divisions at his disposal, eleven in the first line and four in close reserve, plus two further infantry and one cavalry division in his army reserve (one division was holding the quiet sector of the front between Fourth Army's left wing and Third Army's subsidiary two-division attack on the Gommecourt salient). Gough's Reserve

Army had three infantry and three cavalry divisions in hand. The French deployed eighteen divisions for the battle,[120] roughly equivalent to the British strength: those in the front line were reinforced with extra infantry battalions and field artillery such that they resembled small army corps in themselves.[121] Although they were now supporting the British attack, the French were not bringing significantly fewer divisions to the battle. From the way they were deployed it is clear that they planned to feed them one-by-one into a long battle of attrition, rather than hazard all in the first rush. Only five French divisions were in the front line for the first assault. Fayolle had six more in reserve behind Sixth Army's front and three more holding the line south of his attack, with the other four in GAN and G.Q.G. reserve. Foch also had four cavalry divisions, but they were still in their cantonments. They were not, as Churchill falsely surmised, waiting to go forward in a rush with their British counterparts.[122] There was no role for cavalry in Foch's offensive until the enemy had been pounded into submission and his fixed defences captured. That was one key principle of the scientific battle. The other was that the artillery would do the hard work.

The preparatory bombardment commenced on 24 June: around 3000 field and heavy guns and more than 1400 trench mortars were mustered for the work of obliterating the enemy's defences. They would fire over two and a half million shells into the German lines, ranging from 58mm mortar bombs up to huge projectiles from 12-inch naval guns and 15-inch howitzers.[123] It was scheduled to last five days, but was extended by forty-eight hours because low cloud, intermittent drizzle and fog reduced visibility and prevented flying. This meant (by coincidence rather than design) that the attack would be launched on 1 July, the provisional date agreed by Joffre and Haig back on 14 February. The bombardment was a spectacle to behold, a vast firework display created by different-coloured shells.[124] "Armageddon started today and we are right in the thick of it," wrote Captain Cuthbert Lawson, a forward observation officer in the 29th Division.

> I get a wonderful view from my observing station and in front of me and right and left there is nothing but bursting shells. It's a weird sight, not a living soul or beast, but countless puffs of smoke, from the white fleecy ball of the field-gun shrapnel, to the dense greasy pall of the heavy howitzer high-explosive.[125]

Mangin's infantry "climbed on to the parapet and cheered every shell": they were now giving back to the Germans what the invaders had dealt out at Verdun. But it was not simply a *son et lumière*. Its purpose was to cut the wire, destroy the German defences and demoralise or eliminate the defenders.

First, the enemy was blinded, to prevent any effective response. After the German observation balloons were shot down, other prominent observatories were targeted: south of the river the bell tower at Assevillers, the factory chimneys at Flaucourt and the lookouts at Dompierre and Becquincourt were toppled. Infantry patrols went out daily to monitor progress, reporting back on uncut wire and the state of the German front line. When possible, aircraft flew over to observe the results of the bombardment further back. For the first couple of days its impact was limited. The enemy's artillery fired back, blindly but not without effect—one shell hit 3 DIC's headquarters, seriously wounding its commander, General Gadel—and infantry patrols were chased off with rifle and machine-gun fire. In the French sector destruction of the fixed defences was incomplete and the breaches in the wire insufficient. If enemy counter-fire indicated that their batteries had not been silenced, this revealed their positions to the aerial observers, so they could be neutralised over the next few days: hence the postponement of the assault. Moreover, the logistics were already under strain. The new light-gauge railway was unable to deliver shells at the required rate, and men had to carry the trench-mortar bombs forward, delaying their use. But by 30 June the results were deemed satisfactory.

> The enemy's artillery . . . responded much less than yesterday . . . In front of the whole enemy first line the wire had been breached; in many places the wire was totally destroyed; the front-line trenches were blown in; the enemy's infantry were inactive . . . our patrols went out in broad daylight as far as the enemy's line without being disturbed and completed the clearing of the wire with wire-cutters.[126]

As intended, the French artillery had conquered, so there should be little contest when the infantry went forward to occupy the enemy's lines the next day.

The situation on the longer British front was rather less promising. The French had concentrated their firepower, so that their bombardment was denser than those fired in 1915. In contrast the British deployed fewer guns per yard of trench to be destroyed than in their previous offensives,[127] and a smaller proportion of heavy guns and high-explosive shells than the French.[128] Moreover, much of the hastily manufactured ammunition—between a quarter and a third—did not explode. Therefore, while the barrage was longer and heavier than any British gunners had fired before, it was inadequate for such an ambitious operation. It was most effective on the right wing, where fire from French 75s to the south added weight to XIII Corps's fire plan. There close-range fire with trench mortars and high-explosive shells destroyed the German wire. Elsewhere on the front, particularly north of Thiepval, the artillery had only cut lanes in the wire, but these merely provided a focus for concentrated defensive fire.[129] Counter-battery fire was also deficient. Not enough guns and shells had been allocated to this task; moreover, many German guns remained concealed and silent until the day of the attack, and new batteries had been positioned in response to the allied build-up. Destruction was also incomplete. Patrols reported that some sections of the front line had been "flattened," but the section opposite VIII Corps in the north remained reasonably intact. Because the barrage was too thin, the destructive fire could barely cover all the German first position; the second position went largely untouched. Prisoners brought in mixed reports. Some dugouts had been destroyed and their occupants buried. Others remained undamaged.[130] Fourth Army headquarters chose to interpret this intelligence (which came mainly from the southern sector, where the bombardment had been more effective) in a positive light.[131] The show must go on, even if there were considerable uncertainties about the size and nature of the audience in the auditorium across no man's land.

On the eve of battle Charteris was sent to one unnamed corps headquarters, delegated to cancel its attack if preparations were deemed inadequate. "It had little chance of complete success and there was a certainty of many casualties," he noted, but since corps and divisional commanders were convinced that a great victory awaited them, Charteris demurred. Still, he "came back feeling very miserable."[132] Not for the last time, enthusiasm had got the better of sensible military judgement.

If the variation in the British and French bombardments was to prove significant when the two allies attacked on 1 July, for those underneath the *Trommelfeuer* (literally "drumfire") there was little to distinguish the British "toffee apples" from the French "*marmites*" raining down on their shelters. Under such prolonged, intense fire these would have seemed perilously thin, but several metres down in the chalk their garrisons were impervious to all but a direct hit from the heaviest shells. Except for the unfortunate sentries who had to man the tops of the dugout steps and watch for the enemy's assault, flesh was relatively safe. But seven days of shelling was more than nerves could take. Many men lost their minds and had to be forcibly restrained by their comrades. This strain on German nerves was all to the good for the allies, of course, especially given the uncertainties about wire-cutting and destruction. "The prolongation of the bombardment will increase the demoralisation of the enemy who according to prisoners are starving in the front trench having had no rations for 48 hours," Rawlinson assumed.

> The pipe lines are also cut and they get no water. If this is true and I see no reason to disbelieve it, we ought not to have much serious trouble in gaining the green line. This constant holding them in suspense as to when the attack will come is very nerve shaking for them and I hope we may be able so to reduce their moral[e] that we shall have a fairly easy job at any rate at first.[133]

Exhausted and nervous the defenders might have been after a week of earth-shaking torment. But there would still be plenty of them when the allied armies went over the top on 1 July.

THE 1915 WESTERN Front offensives had produced meagre results. Two main factors were to blame: the allies' home fronts had yet to pro-vide the weapons and munitions of modern war in sufficient quantities to sustain their armies' offensive operations over long periods while economising on soldiers' lives; and Anglo-French efforts had been poorly coordinated. After six months of careful planning and prepara-tion, the 1916 offensive promised to be different. Huge armies had been organised, trained and armed with the wherewithal to launch a sus-tained offensive on a wide front. British and French operations would also be better coordinated, and integrated with operations on other fronts. Perhaps the allied generals had cause to be optimistic.

Between March and June 1916, Foch prepared the first machine-age battle, to be fought with machines in unprecedented numbers and con-ceptualised as a smooth, multi-stage industrial process, in which every stage followed on naturally from the last, and in which each part worked in harmony with the others, under the direction of the high command, the engineers of modern war exercising "a perfect intellectual and moral discipline." The troops were merely parts in this war machine. The durable artillery could be kept in place and would function at full capacity throughout the process. The infantry, which wore out more rapidly, would need to be regularly replaced. Preparation, foresight, command and discipline managed the process, armaments were the tools, and "the energy of the combatants" provided the fuel that would keep the wheels of battle turning.[134]

As far as Foch was concerned, "Attacking a defensive position is like a climb, in which the details have to be studied and carefully anticipated and the condition of success is timely and precise execution."[135] Sixth Army had prepared accordingly. Although Fourth Army had tried to absorb the basic principles of the scientific battle, with limited experience some elements were deficient. In the absence of military professionalism the inexperienced British volunteers at least brought enthusiasm: "The junior ranks of both armies are keen to get to work as soon as possible. The spirit of the men is excellent and they will I feel sure render a good account of themselves."[136] However, as Captain Bursey mused: "Many a bright-eyed boy laughing on this 21st day of June, will ere many days are past be laid beneath the sod. That is what 'the day' means." But that was only the half of it. "It means something else too. It means the liberation of Europe from a mad tyrant, William II of Germany. God grant that our arms may be successful."[137]

"Varying Fortune": 1 July 1916

AT DAWN ON 1 July 1916 the first aeroplanes took off into a pale blue, cloudless midsummer sky. They buzzed along the frayed, meandering line that separated friend from foe, making one final check on the destruction wrought on the German army's defences. The ground, swathed in early morning mist, presented an eerie panorama: "It was like looking at a bank of low cloud, but one could see ripples on the cloud from the terrific bombardment that was taking place below. It looked like a large lake of mist, with thousands of stones being thrown into it," remembered Lieutenant Chetwynd-Stapleton.[1]

Below, in the trenches, thousands of dry-mouthed and silent men with pounding hearts checked their equipment once more, a routine repeated many times already to fill the tense hours of waiting. A double rum ration, issued to the nervous men, momentarily relieved anxious minds, giving brief, inadequate Dutch courage: "You don't get enough to make you feel like telling anyone the history of your past life, let alone getting to the stage of being drunk."[2] Veterans, who knew what awaited them over the parapet, stoically endured the deafening row. Eager volunteers and new recruits, for whom the tension preceding zero-hour was a novel and frightening experience, struggled to stay calm, unwilling to show themselves up or let down their mates. Those few who could snatched some fitful sleep in a dugout or on the duckboards of the trench floor.[3] The many who could not were pensive, their thoughts drifting as far away from the heaving field as possible: to families, friends, loved ones. Hasty, perhaps last, letters were penned, mixing hope, sentiment and a longing for normality. "The Germans are frightened to death," wrote Private Laidlaw of the Glasgow Stock Exchange Company to his wife:

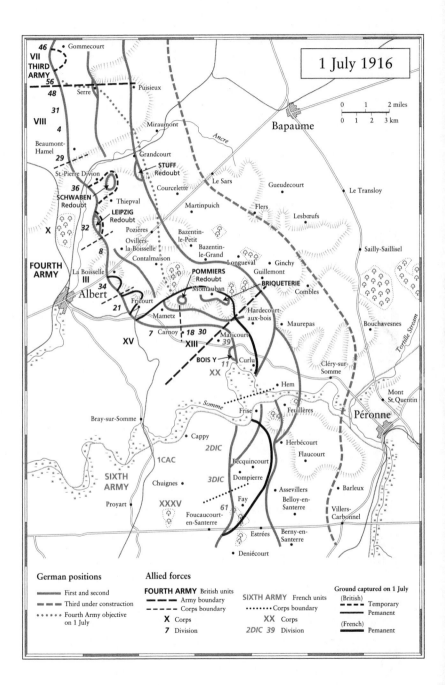

1 July 1916

| 0 | 1 | 2 miles |
| 0 | 1 | 2 | 3 km |

German positions

—— First and second
– – – Third under construction
•••••• Fourth Army objective on 1 July

Allied forces

FOURTH ARMY British units
– · – Army boundary
– – – Corps boundary
X Corps
7 Division

SIXTH ARMY French units
•••••• Corps boundary
XX Corps
2DIC 39 Division

Ground captured on 1 July
(British)
– – – Temporary
—— Pemanent
(French)
—— Pemanent

And I don't wonder as our own guns make my heart jump sometimes. What a scene it is . . . nothing but smoke and fire, hell let
loose, and yet our men go into this hell singing and danc-
ing . . . The French troops are just wonderful too. Oh there is no
doubt about it Germany is finished. Another thing which
affected me tonight are the pipers. I could just shut my eyes and
think of home. Oh my Darling I hope I am home soon.[4]

Whatever their personal apprehensions, the troops were united and resolute in their purpose. As Brigadier-General Rees of 94th Brigade
exhorted, they were "about to fight in one of the greatest battles in the
world, and in the most just cause . . . Remember that the British Empire
will anxiously watch your every move . . . Keep your heads, do your
duty, and you will utterly defeat the enemy."[5]

At 06:25 the gunfire rose to a crescendo, beginning the final intensive hour which heralded the assault. At 07:22 the trench mortars added
the last note to the roaring hurricane of fire, raining thirty bombs a
minute onto the enemy's already pulverised front line. Lieutenant
E. Russell-Jones, commanding a trench-mortar battery in 30th Division, hurriedly scribbled, "We're within a few minutes of what is to be
the beginning of the end of German culture."[6]

Across no man's land, other men, dazed, exhausted and hungry,
remained determined to defend that culture: as they struggled to calm
their nerves they thought of Prussia's neat medieval towns and
Bavaria's forested mountains. For months they had been waiting,
preparing, training to do their bit, and now the time had come.
Unteroffizier Friedrich Hinkel was ready to take his revenge:

Artillery fire! Seven long days there was ceaseless artillery fire,
which rose ever more frequently to the intensity of drum fire.
Then on the 27th and the 28th there were gas attacks on our
trenches. The torture and the fatigue, not to mention the strain
on the nerves, was indescribable. There was just one single heartfelt prayer on our lips: "O God free us from this ordeal: give
us release through battle, grant us victory; Lord God! Just
let them come!" And this determination increased with the fall
of each shell. You made a good job of it, you British! Seven
days and nights you rapped and hammered on our door! Now
your reception was going to match your turbulent longing to
enter![7]

At 07:28 the underground mines were blown, and the men of the first wave tensed for the off. "I felt jovial, hilarious and an absorbing excitement made me almost long for Z hour to be up and over one of a victorious sweeping army," remembered Lieutenant Will Mulholland.[8] The eyes of the world were upon them, and now they would fight for King and country, regiment and comrades, family and friends, God and a better world tomorrow, should they live to see it. Then, at precisely 07:30, the world fell strangely, momentarily silent, as the British guns lifted off the German front trench. Birdsong was heard before the shrill of whistles — the signal to assault — drowned out that final pastoral note. On a sixteen-mile front between Gommecourt and Maricourt, 73 infantry battalions, some 55,000 British and French soldiers, left their trenches and swept towards the German front line; "lined up as if on parade, they set off for Berlin," Russell-Jones noted.[9] Behind them another 100,000, supporting waves and reserve battalions, awaited their turn to follow up, push on and, if luck was with them, break through.

It was not the first, last or biggest push, yet the events of 1 July 1916 were to make it the most notorious. By the end of the day, some 62,000 allied troops had become casualties, the majority of them on the British front. It was "the greatest effort and the greatest loss that the British army has ever crammed into one day";[10] the French army's best battlefield performance to date; a shock for the German army, yet only the beginning of their slow, grinding attrition.

WITHIN THE HOUR, two men were shaking hands, congratulating themselves on a job well done. At zero-hour Colonel Fairfax, commanding the 17th Battalion King's Liverpool Regiment (1st Liverpool Pals), and Commandant Le Petit of 153 RI had set off together, reportedly arm-in-arm,[11] tasked with coordinating the British and French troops attacking shoulder-to-shoulder at the junction of the allied armies. Not only had they stayed together, but they had advanced with speed and determination, their fired-up men sweeping the remnants of the battered enemy before them. "It was a magnificent spectacle," recounted one French signaller attached to the British headquarters. "Ahead of me in the distance was the plain stretching to Bois Favières. The English went across it at a brisk pace, in well-formed lines, with fixed bayonets. I had a magnificent view of the battlefield, just as if I were Napoleon."[12] Soon the entirety of the enemy's first defensive position was in allied hands; they sealed their achievement at Dublin

Redoubt, a strongpoint in the German third line. Thirtieth Division and 39 DI had taken their first objective with light casualties. In front lay the villages of Montauban and Hardecourt, and beyond them the weakly garrisoned German second position, "a single shallow trench."[13] Thereafter, nothing but open country.

Balfourier's XX CA had attacked from the familiar ground north of the river which it had defended so gallantly in autumn 1914.[14] Carefully husbanded to spearhead the attack, well rested and reinforced after the ordeal of Verdun, the men of 39 and 11 DIs had been fresh and in high spirits. "Go, it's the end of the war," General Vuillemot of 11 DI had told his officers.[15]

Seventy-nine RI's carefully prepared attack on Bois Y went like clockwork. This forward-thrusting protective bastion that covered the position immediately north of the river was packed with machine-gun posts from which attacks on the line to north and south could be enfiladed. A thickly wired forward position, Menuisiers Trench, lay in front of the dip in the ground that concealed the wood, linked by communication trenches to the main German line on a crest two hundred metres behind. Behind that the position dropped steeply into a valley, where the enemy's guns were located. The regiment's final objective was fifteen hundred metres away. As planned, the bombardment had done the real work before the infantry left their trenches. By 1 July, the ground was cleared; the enemy's trenches were flattened and his wire obliterated, the soil churned up. The only lingering worry was that it would be difficult to cross such ground, but Mangin's men remained "fully confident of success." At 05:00 the colonel and his officers held a short mass in their command post before beginning the final preparations for the assault. At 07:30 four waves of infantry set off, "a little hunched, with gritted teeth, and with an indescribable look on their faces, features taut after a frightening, sleepless night." Isolated in his command post, Mangin waited anxiously for news.[16]

When the whistles sounded the young French conscripts left their trenches with cries of "*Vive la France*" and set upon the enemy with gusto.

The assault groups, rushing forward from jumping-off saps dug towards the enemy's line behind "an impressive and precisely regulated barrage,"[17] emerged as terrifying spectres from the mist that still clung to the river valley. They were into and through the enemy's front position in just twenty minutes: the few surprised defenders of Bois Y put up little resistance; the German counter-barrage was feeble. While the grenadiers and bayonet-men following behind mopped up enemy survivors and consolidated the captured lines, the leading waves, hardly touched, pushed on. Mangin spotted them through his binoculars as they climbed up the far side of the ravine, and telephoned the good news to brigade headquarters. By mid-morning XX CA had taken all their objective, capturing nearly 2500 prisoners.[18]

On most of the front the defenders, men from 6 Bavarian IR, had been quickly overwhelmed and taken prisoner. As they passed dejectedly through the French rear areas the gunners abandoned their work to gawp at their prize. Pockets were searched in the hunt for souvenirs; the Germans' *Krieg-brot* was black and hard, but their biscuits were tasty and they had good cigars, noted Charles Barberon of the 121 *Regiment d'artillerie.* The prisoners looked exhausted after their ordeal, but not unhappy. They had survived, and the *poilus* treated them well. "It's surprising," added Barberon, "but the soldier who has suffered the enemy's fire does not show the same hatred for the enemy as civilians." Now they were defeated comrades, victims of the same cruel war, not enemies. "[W]hat good would it do to damn these poor fellows when the real guilty parties are well protected and can't be got at?"[19]

That the French army could walk over the German lines on 1 July was due to the French gunners, men such as Barberon, whose methodical, carefully targeted bombardment had destroyed the enemy's defences. It was a vindication of Foch's attacking methods, and Fayolle's cautious and precise preparations. The artillery had done their job so well that the assaulting infantry encountered little resistance in the enemy's front system: deep dugouts had been closed by well-directed fire, trenches had been flattened and machine-gun nests destroyed, and enemy artillery batteries all but silenced. The few pockets of resistance were quickly overcome with the fire and movement tactics that the French high command had prescribed after detailed study of their 1915 defeats. Only on the flanks that the bombardment had not reached, on the left among the tree stumps of Bois Favière in the rear of the reserve line and on the right in the fortified village of Curlu, was there organised resistance.[20]

ON THE FRENCHMEN'S left, the men of 30th Division, mainly volunteers from Liverpool and Manchester, were making equally rapid progress. Not for them a slow, steady march across no man's land, splendid yet suicidal: their commander had instructed them to leave the front trench before zero-hour and advance across no man's land at the double, with the following waves close behind to escape the counter-bombardment. From the rear an artillery officer observed the spectacle:

> At 7:20 a.m., rows of steel helmets and the glitter of bayonets were to be seen all along the front line. At 7:25 a.m. the scaling ladders having been placed in position a steady stream of men flowed over the parapet, and waited in the tall grass till all were there, and then formed up; at 7:30 a.m. the flag fell and they were off, the mist lifting just enough to show the long line of divisions attacking.[21]

With this head start the assault battalions from 21st and 89th Brigades were into the enemy's front line before the defence had time to orga-nise. "It all seemed so easy—much easier than when we had practised it behind the line," Lance-Corporal Quinn of the 20th King's wrote home afterwards.[22] Leaving it to the following waves to clear the many prisoners from their dugouts, they pushed swiftly on into the support and reserve lines. So rapid was their advance that they had to pause for the supporting barrage to lift ahead of them. By 08:30, 89th Brigade had seized Dublin Trench, its first objective, which was unoccupied. "Fritz had made a 'strategic retreat,' " Quinn gloated, "that is to say he had run like the devil a few miles back."[23] The assault troops consolidated their gains and waited for the supporting brigade to come up and push on to Montauban, silent and exposed.

On 30th Division's front too an effective bombardment had been the key to success, and indicates what might have been achieved if such well-coordinated artillery and infantry cooperation had taken place all along the front. The commander of the defence, Oberstleutnant Bedall, noted glumly in his diary that his 16th Bavarian IR "had sustained severe losses from intense enemy bombardment, which had been main-tained for many days without a pause, and for the most part were already shot to pieces" when the attack began.[24] Wire-cutting by the trench mortars and 75s lent by XX CA had been particularly effective, so the men were not forced to bunch in order to get through gaps in the wire. Formation and unit cohesion were maintained in the all-important first minutes of the attack. Carefully targeted counter-battery fire, closely observed from the air, had silenced most of the enemy's guns in the sector, and so the retaliatory barrage, which was designed to stop supporting troops crossing no man's land, was barely noticeable. Heavy howitzers had pulverised the enemy's fixed defences. Dublin Trench "was so battered by artillery that it was unrecognizable"; its conquerors had to dig a new trench between the shell-holes fifty yards in front.[25] The lifting barrage fired the assault waves onto their objectives before the defenders had time to organise.[26] "We have done better than expected," wrote Private Laidlaw, "and the Germans are putting up a pretty poor fight . . . Of course we are all very excited, as we feel that the end of the war is now near."[27]

However, the assault was not going so well on 30th Division's left, where, despite Maxse's thorough preparations, 18th Division had not been as quick or as successful. As a consequence 21st Brigade's flank was exposed. A single machine gun, located in a German reserve trench, Train Alley, and rifle fire from "The Warren," a network of support trenches on 18th Division's front, caught the brigade's leading battalion, the 18th King's (Liverpool) Regiment, in the open: over half its number quickly became casualties. Its supporting battalion, the 2nd Green Howards, were caught in no man's land and could not get across until Lieutenant H. C. Watkins from the King's led a small party into Train Alley and bombed out the defenders, a deed that was to win him the Military Cross. It was just one of the many individual acts of bravery that were to influence the course of events as the day unfolded. After Watkins's action the brigade resumed its advance, reaching its first objective, Glatz Redoubt, at 08:35.[28] There were "bodies everywhere, in all kinds of attitudes, some on fire and burning from the British bombardment. Debris and deserted equipment littered the area and papers were fluttering around in the breeze." Their first job done, the victors set about looting the hastily abandoned German dugouts for souvenirs; at least until one unlucky man, who had survived the rush across no man's land, was shot dead by a trigger-happy soldier from the relieving battalion who came across him parading around in a captured pickelhaube.[29]

Two tactical problems, which were to bedevil the British advance on the rest of the front, were encountered on 21st Brigade's front. Isolated machine guns and pockets of resistance, missed by the barrage, could take a disproportionate toll of the advancing infantry. Unless a coherent line was maintained, and progress was steady and consistent along the whole front, flanking fire would wreak havoc in neighbouring sectors. Together these would slow the advance appreciably and upset the carefully timetabled coordination between artillery and infantry.

Maxse's 18th Division had just such an experience. At 07:27 mines were blown at Casino Point, the apex of the enemy's front line, and the attackers surged forward, only to find themselves enfiladed by German machine guns sited in the old Carnoy mine craters in no man's land. "Contrary to expectation we were not going to have things all our own way," Private Robert Cude, a runner with the 7th East Kent Battalion, recorded. "He has plenty of machine guns, and is making a frightful carnage." The advance was checked long enough for the enemy garrison, who had abandoned their front line before the bombardment, to take up positions in the support line. But the attackers' blood lust was up: "I long to be with Battalion so I can do my best to bereave a German family. I hate these swines!" Cude continued. When the defences at The Warren were turned by 30th Division's advance up Train Alley, 55th Brigade, on the right flank, whose men had gone to ground between the German front and support lines, could push on to its objective. Cude relished the moment:

> It's a wonderful sight and one that I shall not forget. War such as this, on such a beautiful day, seems to me to be quite correct and proper. A day such as this, one feels a keen joy in living, even though that living is to say the least of it very precarious. Yet men are racing to certain death, and jesting and smiling, yet wonderfully quiet in a sense, for one feels that one must kill, and as often as one can.

An intense and bloody fight among the enemy's maze of trenches ensued:

> No quarter is asked or given in a good many places . . . The reason is not far to seek, for tens of thousands of our men are lying low, never to rise again . . . but even the wholesale slaughter of a beaten but not disgraced enemy is, or grows, obnoxious, and so from 10 o'clock prisoners began pouring in. They had even caught a General with members of a big staff.[30]

By mid-morning, the enemy's resistance, although more stubborn, was also collapsing on 18th Division's front. Its central brigade, the 53rd, assisted by the Casino Point mines, was making steady progress through the successive lines of German trenches, although not without a vicious fight. The defenders were ill-at-ease: rushed forward two nights previously to replace the front-line companies who had been practically wiped out by the bombardment, they were disorganised and

German cavalryman rides down the Somme valley in August 1914 (Corbis)

French dead in the fields near Quesnoy, September 1914
(© Paris—Musée de l'Armée, Dist. RMN/Pascal Segrette)

12 September 1916: at the height of the offensive, Sir Douglas Haig tries to explain to David Lloyd George how it is being conducted. Albert Thomas (behind Haig) and General Joseph Joffre (between Haig and Lloyd George) look on. (Getty Images)

Marshal Foch in his study (Corbis)

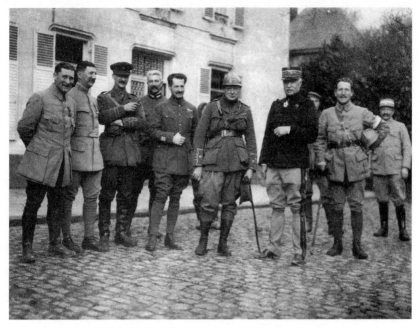

Fayolle and Churchill *(third and fourth from right)* with staff officers at Fayolle's headquarters, January 1916 (Getty Images)

Hindenburg *(left)*, the Kaiser *(centre)*, and Ludendorff *(right)* ponder their next move (Imperial War Museum, IWM)

General Sir Henry Rawlinson
(Australian War Memorial)

General Sir Hubert Gough
(IWM)

General Fritz von Below
(Ullstein Bild/Topham)

General Erich von Falkenhayn
(Ullstein Bild/Topham)

The Kaiser reviews German troops behind the Western Front, October 1915
(Australian War Memorial)

German dugouts on the Somme front (Getty Images)

British troops march into battle past the basilica in Albert, with its famous leaning virgin statue (Australian War Memorial)

A bloodied but triumphant British soldier on the Somme (IWM)

Indian cavalry wait to go forward on 14 July 1916. A good propaganda photograph, but hardly reflective of the situation on the ground. (IWM)

French infantry attack on the Somme (*L'illustration*)

The "fireworks" on the Somme front, viewed from behind the allied lines (IWM)

The bombardment near Montauban, viewed from the air, June 1916 (IWM)

unfamiliar with their badly damaged trenches. Nevertheless, when the British came over, they did their best. "Our men attacked them with hand-grenades and eventually succeeded in holding them at bay," Leutnant Busl of 6 Bavarian RI recorded. But weight of numbers would eventually prevail: Busl and two-thirds of his platoon were already casualties.[31] On the left flank 54th Brigade charged the reserve Pommiers Trench on time, at 07:50. But the defenders fought well, inflicting heavy casualties and taking the wind out of the British attack, thereby gaining time for the second position to be organised. With their flanks now turned it was time for the survivors to pull back and join their comrades to the rear.

Between the reserve line and the second position were three key intermediate positions, XIII Corps's second objective. On 30th Division's front the village of Montauban, with its isolated brickworks, La Briqueterie, a cluster of factory buildings and chimneys that the enemy had turned into an underground fort, was half a mile beyond the first objective. In 18th Division's sector Pommiers Redoubt marked the corps's left flank objective. Thirtieth Division's reserve, 90th Brigade, "pals" of the 16th and 17th Battalions of the Manchester Regiment, were to lead the assault against Montauban. As they advanced from the British front trench a solitary machine gun opened up, cutting down both battalion commanders and many junior officers and men. A Lewis gun team was brought forward to silence the troublesome enemy emplacement, and the survivors pressed on, rushing Montauban the moment the supporting barrage lifted. They found a deserted, shattered ghost-town. The largest village on the Somme had been reduced to rubble by a week of fire from French 240mm heavy mortars. The deep dugouts had been smashed and the village's defenders trapped underground. It was the biggest prize to fall on 1 July, its shattered streets empty except for a solitary fox.[32]

After securing Montauban, La Briqueterie was assaulted. Following a thirty-minute hurricane bombardment a company of the 20th King's rushed forward at 11:30, while bombers worked round to the north to cut off the garrison's retreat. The bombardment had completed its murderous task, and resistance was minimal. The headquarters of two German infantry regiments was overwhelmed. Oberst Leibrock, commanding 6 Bavarian RIR, was astonished:

A little later they reported that the British were indeed behind the dugout. I went out myself to take a look. Hardly had I put my head up when I received a burst of small arms fire from the rear. I could not be absolutely certain about the situation, but it was at least clear that it was not our men standing around firing from the Second Trench, but rather the British digging in.

The occupants of the command bunker, a disparate group of staff officers, clerks and stragglers, made ready to defend themselves. Then there were several explosions, ear-splitting in the confined space; the British were throwing hand-grenades down the steps of the bunker. The screams of wounded men mingled with the officers' frantic shouts as the situation descended into chaos. Leibrock telephoned for a counter-attack to retake the position, but no reserves could be spared from the intense fight outside. "It seemed to me ever more probable . . . that there had been a British breakthrough. In this situation I felt that further sacrifice was pointless and, because the German artillery was completely silent, that there was no chance of a counter-attack." He and his trapped staff decided to surrender.[33]

To the west, Pommiers Redoubt crowned the Montauban spur, behind the German reserve line. Another battalion headquarters, it was protected by thick belts of wire and strongly garrisoned by a company from 109 IR. It would be stoutly defended. Three British battalions were sent against the redoubt, each being checked at the wire by heavy fire. A more subtle approach was needed, and Maxse's careful tactical training now proved its worth. A Lewis gun platoon from the 11th Royal Fusiliers worked around the left flank and caught the defending riflemen in enfilade. Having silenced the defensive fire, the rest of the battalion rushed through gaps in the wire and set to work with bayonet and rifle butt. But although heavily outnumbered and with their position invaded, the defenders were "little disposed to surrender."[34] It was an hour before the determined men of 109 IR were all killed or captured, and they took a heavy toll on their attackers.

Except for a few isolated points that held out until the afternoon in the middle of 18th Division's sector, Congreve's XIII Corps had captured its objectives by midday. The defending battalions' headquarters had been overwhelmed at La Briqueterie and Pommiers Redoubt, and there was now no coordinated defence between Montauban and the river. Although the German garrison had fought bravely and sold their lives dearly, in this sector the defence had finally degenerated into chaos. Behind Montauban, hundreds of Germans could be seen "fleeing backwards in hopeless confusion, and our men thoroughly enjoying themselves trying to stop their flight."[35] In the sheltered Caterpillar

Valley behind Montauban, frantic German gunners, harassed by machine-gun fire from low-flying aircraft, were struggling to limber up their guns. Men of the 16th Manchesters surprised the artillerymen, who fled without a fight, abandoning three of their guns. These were the first guns captured on the Somme.[36]

SINCE 07:30, FRENCH artillery had been keeping up a steady barrage south of the river to prevent the enemy enfilading the attack to the north. At 09:30, two hours after the main attack, the salvos abruptly lifted off the front line. Three mines were blown under a redoubt at Fay, and moments later the Colonials, singing "La Marseillaise," were swarming over the parapet of the German front trench, under the proud gaze of their army commander.[37] On the right, covering the flank, the middle-aged Breton reservists of 61 DI "rushed at the enemy like young men."[38] A few at least were veterans of their last fight on the Somme, at Moislains in August 1914, and revenge would be in their minds. Two colonial divisions made the main assault between 61 DI's left and the river. Within fifteen minutes they had overcome resistance in the German front line. The momentum of the assault was maintained "owing to the effectiveness and flexibility of our artillery barrage, and above all the excellent manoeuvres of our infantry." When the left-flank attack on Frise was briefly held up by uncut wire, General Mazillier, 2 DIC's commander, calmly called up the fire of the reserve artillery and heavy mortars. The infantry pressed on. By 12:30 all the front system had fallen into French hands, and the leading companies of 2 and 3 DICs were probing the second position between Herbécourt and Assevillers. Over two thousand prisoners were moving to the rear. It had been, Thierry summed up, "a very good day."[39]

ELSEWHERE, IT WAS a different story. Lieutenant-General Henry Horne's XV Corps, on XIII Corps's left, was tasked with pinching out the German salient at Fricourt. The village itself, a heavily fortified cluster of machine-gun nests and dugouts, was to be screened by mine explosions, gas and smoke while 21st and 7th divisions flanked it to north and east. The isolated strongpoint would then be captured by a second assault in the afternoon, after which XV Corps would push on to the German intermediate position in front of Mametz Wood. This sector, the key to the German defensive system south of Thiepval, had been one of active mining and raiding in 1915 and early 1916, and its defenders from 28 RID had turned Fricourt and Mametz to the east "into little fortresses; there were numerous strong machine-gun emplacements, and the dugouts were exceptionally fine, some of them with two storeys, lighted with electricity and, as it was said, provided with every convenience except water."[40]

On 7th Division's front the assault started well. Because XV Corps had disobeyed orders and fired off twice its allocation of shells, the preliminary barrage had been highly effective.[41] The wire had been properly cut, and the enemy's artillery all but silenced. "No Man's Land is a tangled desert. Unless one could see it one cannot imagine what a terrible state of disorder it is in. Our gunnery has wrecked that and his front-line trenches all right," Captain Charlie May of the 22nd Manchester Battalion jotted down as he waited to go over the top. Unfortunately for the attackers, however, "we do not yet seem to have stopped his machine-guns. These are pooping off all along our parapet as I write. I trust they will not claim too many of our lads before the day is over."[42] At zero-hour the supporting barrage commenced, lifting at fifty-yard intervals, with the infantry following close behind.[43] On 7th Division's right 91st Brigade, attacking alongside 18th Division's 54th Brigade, swept forward towards Mametz. The assault battalions met little resistance in the front line, and, keeping up with the barrage, had seized Cemetery Trench and Bulgar Alley, intermediate lines which protected the village, within twenty minutes. However, the enemy had not been entirely demoralised, and as the defence recovered from its initial shock and machine guns came into action in Mametz village and Danzig Alley, the communication trench to its right, the attack stalled. Here Captain May was killed: his body lies in Danzig Alley Cemetery west of Mametz.[44] Nearby, Sergeant R. H. Tawney, the Christian-socialist historian and future Professor of Economic History at the London School of Economics, was hit twice and took cover in a shell-hole, where he would lie until dusk as the fight raged around him. A vicious six-hour struggle for the eastern half of the village ensued. Two further bombardments were ordered, but little progress could be made until the capture of Pommiers Redoubt outflanked the defenders of Danzig Alley. In the afternoon the survivors from the leading waves pushed into Mametz, clearing the village with grenades and bayonets. Its garrison, 109 RIR, fought to the last: it lost almost 2100 men, with only 32 of the defenders making it back to the German second line.[45] Although 91st Brigade had taken its first objective, it had exhausted itself in the process.

To its left, 20th Brigade was in dire straits. On either side of the Albert–Péronne road, the boundary between the brigades, three quiet cemeteries now attest to the chaos of the day. North of the road Gordon Cemetery, with its neat, curved lines of gravestones, is the last resting place of men of the 2nd Gordon Highlanders, the right-most battalion of 20th Brigade. To the south, amid the trees of Mansell Copse, lies Devonshire Cemetery. Facing them is Mametz's small village cemetery, the site on 1 July 1916 of a fortified machine-gun post dubbed "The Shrine." From this spot, the defenders had a commanding view of the gentle slope in front of the village across which the Highlanders and the 9th Battalion the Devonshire Regiment, on their left, were to advance. Starting from their own support trench, the Devonshires had to cross some 650 yards of open ground, filing around Mansell Copse, which obstructed their advance across no man's land. By the time the advancing companies had reached the copse more than half their number had fallen; uncut wire in no man's land had slowed their advance and they had lost the protection of the barrage.[46] The survivors took cover in the German forward trenches. Two companies of the reserve battalion, the 8th Devonshire, were sent up in support, to meet a similar fate. The two battalions' dead share Devonshire Cemetery. Lieutenant William Noel Hodgson MC — whose poem "Before Action," penned as he waited behind the lines, concludes with the prophetic line "Help me to die, O Lord" — lies among them.[47] The Gordons, "going over as on parade,"[48] did better, reaching Shrine Alley, the German reserve line, before their advance stalled in the face of withering machine-gun fire from The Shrine and Mametz. The attack in 7th Division's centre was pinned down, and on the verge of collapse. Only the 2nd Battalion the Border Regiment, on 20th Brigade's left, reached their objectives, and it was hard going. Machine guns in Mametz took a heavy toll, and a close-range fight with bayonets and bombs was needed to clear the enemy's trenches.[49] By 09:15 the battalion had captured the enemy's front lines, but with their flanks exposed to murderous machine-gun fire they could make no further progress. Twentieth Brigade's fortunes hung in the balance, awaiting developments elsewhere on the front.

Twenty-first Division had an equally tough fight north of Fricourt. Here the strengthened bombardment had done its job. Lieutenant Siegfried Sassoon, in reserve with his battalion opposite Fricourt, watched the attack:

The artillery barrage is now working to the right of Fricourt and beyond. I have seen the 21st Division advancing on the left of

Fricourt; and some Huns apparently surrendering—about three quarters of a mile away. Our men advancing steadily to the first line. A haze of smoke drifting across the landscape—brilliant sunshine. Some Yorkshires on our left watching the show and cheering as if at a football match.[50]

On the left flank, 64th Brigade made good progress up the gentle slope of the Fricourt spur, weight of numbers sweeping them through the German defensive system. Men from four attacking battalions fought their way *en masse* towards their objective at Crucifix Trench in the German intermediate position. Eventually, concentrated machine-gun fire forced them to go to ground in the Sunken Lane just short of their objective, but the whole front system was in their hands. Because the battalions on either side were in difficulties, they could do no more than dig in and hold on.[51] Lieutenant Basil Liddell Hart, left behind in battalion reserve, came up in mid-morning through "the gauntlet of enemy snipers and other fire." Such had been the chaos and loss among the officers that the twenty-year-old second-in-command of a company found himself temporarily in charge of a neighbouring battalion. "But the expected counter-attack did not come, and it became evident that the enemy were in as much confusion as we were."[52]

To the south 63rd Brigade had run into very heavy fire and their advance had stalled. The 10th West Yorkshire Battalion, on 63rd Brigade's right, was to suffer the highest rate of casualties of any that attacked on 1 July 1916: some 22 officers and 688 men fell in reaching the German lines, over 90 per cent of those who went over the top. On their left, the brigade's other attacking battalions, the 14th Middlesex and 8th Somerset Light Infantry, were sprayed by concentrated fire from six machine guns in Fricourt and the trenches in front. The bombardment by heavy 9.2-inch guns, designed to turn the village into a wasteland, had clearly failed; many of the fuses of the hastily manufactured shells had fallen out in flight. The brigade's advance ground to a halt in the face of a counter-attack by German bombing parties.[53]

By midday, three of the eight attacking corps, two French and one British, had fought their way to their first objectives and broken the German resistance. In the centre, XV Corps's fierce battle for Mametz and Fricourt continued undecided. On the front of the other four corps between La Boisselle and Gommecourt, where it was planned to take the German second position, it was a very different picture.

"Dull and apathetic, we were lying in our dugouts, secluded from life but prepared to defend ourselves whatever the cost," recollected Leutnant Cassel of 99 RIR, which was holding the line at Thiepval. "The shout of the sentry 'They are coming!' tore me out of apathy. Helmet, belt, rifle and up the steps."[54] All along the western flank of the Somme salient German soldiers rushed to their positions. They had survived the bombardment; now these tired, anxious troglodytes would once again breathe fresh air, move and fight. If they were to see out the day, they must fight hard. Heavy machine guns were hauled up the steep steps of the deep dugouts, with ammunition carriers following close behind. Quickly the guns were assembled, a familiar and much-practised routine, ammunition belts were engaged, and careful aim was taken at the ranks of khaki-clad figures moving towards them. In places the British, who had formed up in no man's land or rushed the short distance between the lines, won the race to the German parapet, and vicious hand-to-hand fighting ensued in the front trenches. With a foothold in the enemy's lines, the forward battalions were able to push on, either bombing along communication trenches or rushing over open ground to leap into support and reserve lines. In this way many lodgements, from a handful of men to a whole brigade, were made in the German front system between Serre and La Boisselle, although these penetrations were not continuous. Elsewhere, where the machine-gunners won the race, their Maxims delivered death at six hundred rounds per minute, against which a frontal advance was impossible. Where no man's land was wide—it varied between 150 and 800 yards—the defenders had vital extra seconds to get to their positions before the enemy was upon them. At some points where the wire had not been properly cut and the attackers were held up looking for gaps or cutting their way through, the assault was easily blunted. Sometimes sheer bad luck took a hand: at La Boisselle a German listening post had picked up news of III Corps's coming attack the previous evening so the defenders were alert and waiting.[55] At many infamous places the attack was shot to pieces by concentrated machine-gun fire, and horrifying casualties were inflicted. Gommecourt, Serre, Beaumont Hamel, Thiepval, La Boisselle and Ovillers, where battalions, sometimes whole brigades, were mown down, are names inscribed in Britain's collective memory. This is the Somme of popular myth, of poems, plays and novels; the graveyard of locally raised "pals" battalions, "Two years in the making. Ten minutes in the destroying," in author John Harris's memorable and oft-quoted conclusion to his novel about the destruction of the Sheffield City Battalion at Serre.[56] Here the volunteer army, and the spirit which motivated it, were shattered.

Yet, if anything, the events of this day enshrined rather than shattered patriotic spirit, engendering a vibrant national myth. One German defender at Ovillers viewed the fight as "an amazing spectacle of unexampled gallantry, courage and determination on both sides."[57] Liddell Hart, after the war one of the most trenchant critics of the battle and the British high command who directed it, could never forget that, against

> the Germans of 1916, most stubborn and skilful fighters . . . July 1st was an epic of heroism, and better still, the proof of the moral quality of the new armies of Britain, who, in making the supreme sacrifice of the war, passed through the most fiery and bloody of ordeals with their courage unshaken and their fortitude established.[58]

The bulldog spirit of the British "Tommy," partially trained and poorly led though he might be, would not be bowed in the face of Prussian professionalism.

WHY DID THIS part of the attack fail so disastrously, and what happened as the two armies came to grips? Traditionally, taking a cue from Liddell Hart, the carnage has been explained in terms of the overambition and poor military skill of the British high command, from Haig and Rawlinson down to the corps commanders, Lieutenant-Generals Sir Thomas Snow, Sir Aylmer Hunter-Weston, Sir Thomas Morland and Sir William Pulteney. It was said their military concepts were out of date, their plans were at fault, and their orders were foolish and inappropriate. Churchill's and Lloyd George's political memoirs popularised this interpretation: here was "the story of the million who would rather die than own themselves cowards . . . and also of the two or three individuals who would rather the million perish than that they as leaders should own . . . that they were blunderers," as the latter dismissed the Somme campaign.[59] Down the years, the battle over how good or bad Haig and his generals were has been endlessly re-fought. By and large, this interpretation denies the professionalism of the British army's higher leaders; in its more extreme form it accuses them of callous, sometimes class-inspired, fratricide.[60] Out of the events of

1 July 1916, it would seem, emerged the class divide that would define British society for generations. Although the bravery of the men could not compensate for the presumed failings of their leaders, Tommy Atkins's commonplace heroism in adversity has generated an extensive literature all of its own.[61]

Although sociological explanations of the Somme have fallen out of fashion in today's less class-divided society, the military decisions of the British high command, and their consequences, remain central to understanding what went wrong. The practicalities of the attack—an inadequate bombardment, leaving wire uncut and defenders untouched; simplistic tactics designed for inexperienced troops; poor command, control and communications leading to poor decision-making on the day—have all been implicated to a greater or lesser extent in the tactical disaster.*

Only very recently has the other determining factor, organised and effective German defence, been examined to give a fuller picture.[62] The British did not fail by their own endeavours alone, but in a gruelling fight with a professional, skilled and determined adversary. German defensive doctrine worked well on this section of the front. The mutually supporting machine-gun posts, the cornerstone of German defensive tactics, proved their worth. Even if the front line was entered, machine-gun posts in the reserve and support lines, such as that at The Shrine, would take a heavy toll of attackers as they sought to press on. The heavily wired and well-garrisoned strongpoints, with which the defensive system was stiffened like the towers in a fortress, did their job: Pommiers Redoubt, Schwaben Redoubt, the Quadrilateral, Leipzig Redoubt and others were the epicentres of the fighting on the first day. Once the impetus of the attack had been blunted and the enemy's ranks thinned out, local counter-attacks eliminated the isolated parties that had gained a foothold in the German system but were cut off from reinforcements and supplies by a curtain of artillery fire. In this way a coherent defensive system would be maintained. It was this method of defence that broke up and then repulsed the British attacks. It took the equally professional French army to expose its deficiencies.

The pattern of events was similar on the fronts of all four British corps north of Fricourt. The infantry set off at 07:30 in waves sweeping towards the German front. The crest of the first wave was soon fragmented by belts of uncut wire and other obstacles: mine craters, saps,

---

*Their origins and nature have been discussed in Chapter 4.

ravines and strongpoints. Machine-gun fire and shelling, dealing sudden death or injury at any moment, interweaved with the natural obstacles. The tide of men splintered, as some companies pushed on while others faltered. The description of one reserve battalion's advance, that of the 12th York and Lancaster ("Sheffield Pals") at Serre, is representative of so many from this part of the front:

> they had to pass through a terrible curtain of shell fire, and German machine guns were rattling death from two sides. But the lines growing even thinner went on unwavering. Here and there a shell would burst right among the attackers . . . Whole sections were destroyed . . . The left half of "C" Company was wiped out before getting near the German wire . . . The third and fourth waves suffered so heavily that by the time they reached No-Man's-Land they had lost at least half their strength . . . The few survivors took shelter in shell-holes in front of the German line and remained there until they could get back under cover of darkness.[63]

Faced with this murderous storm of fire, many thousands got no further than no man's land, being caught on the uncut wire or in the open and shot down. Follow-up waves, moving up by predetermined routes and timetables, often fared even worse than those in front: machine-gun sights were raised and flares sent up to call down pre-registered counter-barrages on the British lines. Heavy casualties occurred to formations that had not even reached their own front line: "For many soldiers the 'race to the parapet' was actually a race to their own parapet, and it was a race that they lost."[64] One infamous image recorded by an official photographer, of men shot down in straight rows as if on parade, records the fate of 103rd (Tyneside Irish) Brigade, advancing up the slope in front of La Boisselle and caught by machine-gun fire more than half a mile behind the British front line.[65]

Many "pals" battalions, going over the top for the first time, were destroyed before they reached the German line. On the northernmost wing of the main attack, men from England's northern industrial cities were gathered opposite Serre. The collected memorials to Sheffield, Barnsley, Bradford and Accrington men in Sheffield Memorial Park behind 31st Division's old front line attest to the tragedy in that sector. Here the tactics, prescribed by General Hunter-Weston, were at fault: Haig's pre-battle concerns were clearly justified. Although the assault troops had moved out into no man's land before the barrage had lifted, in order to rush the German front line as a single concerted wave,[66] this did not avail them. VIII Corps's orders had directed the barrage to lift from the enemy's front trench at 07:20, when the mine on Hawthorn Ridge was to be blown. So in the ten-minute gap before zero-hour the crowded troops lying down in no man's land were sitting ducks. The defenders were at their posts and machine-gunning them before the attack had even started.[67] When their officers' whistles finally sounded, those who were not yet hit rose up and moved off towards the line of gun flashes which marked their objective, only to find their path obstructed by the enemy's wire, which had been only partially cut. One German who watched their doomed progress later recollected:

> When the English started advancing we were very worried; they looked as if they must overrun our trenches. We were very surprised to see them walking, we had never seen that before ... When we started firing we just had to load and reload. They went down in their hundreds. You didn't have to aim, we just fired into them. If only they had run, they would have overwhelmed us.[68]

The German machine-gunners, straining and sweating at their weapons, did their job:

> Throughout all this racket, this rumbling, bursting, cracking and wild banging and crashing of small arms, could be heard the heavy, hard and regular Tack! Tack! of the machine guns ... that one firing slower, this other with a faster rhythm—it was the precision work of fine material and skill—and both were playing a gruesome tune to the enemy, whilst providing their own comrades ... a high degree of security and reassurance.

Gun barrels overheated and had to be changed; water to cool the guns ran out so men urinated in the jackets to keep them cool, Unteroffizier Otto Lais's account continued. Belt after belt of ammunition was fired off as the well-rehearsed loading and firing routine was repeated— "Fire; pause; barrel change; fetch ammunition; lay the dead and wounded on the floor of the crater"—until twenty thousand rounds had been fired.[69] In the face of this concentrated fire the British attack on Serre collapsed. Each attacking battalion lost over five hundred men. In 31st Division's sector only a few men from Accrington made it into the German trenches, where they were swallowed up. A signaller, Lance-Corporal Bury, was watching carefully for signs of success from an observation post behind the line:

> We were able to see our comrades move forward in an attempt to cross No Man's Land, only to be mown down like meadow grass. I felt sick at the sight of this carnage and remember weeping. We did actually see a flag signalling near the village of Serre, but this lasted only a few seconds and the signals were unintelligible.[70]

IT IS A military truism that no battle plan survives the first contact with the enemy. But in the carefully paced battle of mid-1916, when coordination and cohesion meant the difference between success and failure, the breakdown of the plan was inevitably disastrous. On the northern sector of the front the plan was in tatters within minutes, and the battle degenerated into hellish confusion: "Bewilderment," as Percy Croney of the 1st Essex Regiment entitled his hazy and fragmented yet still vivid memories of his experiences at Beaumont Hamel.[71] Thousands of men in their own narrow worlds—a section of trench contested with the enemy; a shell-hole in no man's land; an observation post; a machine-gun nest under enemy fire; a command dugout behind the lines—struggled to do their duty in the maelstrom. No amount of meticulous training could have prepared them for what they faced. In such confused circumstances, German military professionalism gave the defence a clear advantage. One German post-battle assessment concluded that British bravery and daring, impressive though they were, were not enough because "the officers and NCOs appointed to command the men lacked experience of war and the ability to react swiftly to new, changing and unexpected situations."[72]

It is impossible to establish how many British soldiers made it to the enemy's front line and beyond. Throughout the day, aircraft and front-line observers, the eyes of the high command, sent back vague and often contradictory reports of groups of khaki figures seen through the smoke of battle: Thiepval had been occupied; the Royal Scots were in Contalmaison; two battalions were in Serre.[73] In fact, the few isolated parties that had survived the first assault and progressed beyond the lively fight in the front lines found the going easier. However, they were cut off from the British lines by the German barrage, or by groups of Germans who had skulked in their dugouts and not been mopped up. They knew that their objective was to push forward, which they did until they met the enemy. Eventually, they found themselves counter-attacked from all sides. Small parties dug in and tried to hold out until reinforcements reached them, but rarely did these detachments—mixed groups from several battalions, some wounded, often leaderless and short of water and ammunition—amount to a viable force. Most were overwhelmed before support arrived, the only clue to their success being the grave markers found in later advances, such as those of a few men of the Accrington Pals buried in Serre.[74] But a few key points in the German front system were seized and held against the inevitable counter-attacks. Leipzig Redoubt, a key salient south of Thiepval, was stormed by 32nd Division in the early minutes of the attack. Its conquerors, the 17th Highland Light Infantry, fitfully reinforced by small remnants of supporting battalions coming up through the murderous fire blanketing no man's land, spent the rest of the day in an intense close-quarter fight for the redoubt, in which individual acts made the difference between success and failure. Here Sergeant James Turnbull won a posthumous Victoria Cross for his defence of one key post: he was killed in a bombing counter-attack late in the day.[75] The redoubt remained in British hands at nightfall.

North of this point there was to be no permanent success, although in two places the German defences were broken into in force. The capture of the Thiepval spur, which dominated the British trenches to north and south, would crown the British attack. This crucial task was allotted to Morland's X Corps, a mini-army of mainly Irish and Scottish battalions. Although 32nd Division's 97th Brigade had broken into the Leipzig Redoubt south of the spur, all attempts by the division's other brigades to seize Thiepval village itself would falter in the face of heavy fire from the ruins of Thiepval château. "Only bullet-proof soldiers could have taken Thiepval on this day," the British official historian laconically concluded.[76] North of Thiepval, 36th (Ulster) Division did break through, seizing the Schwaben Redoubt, a bastion between the first and second German positions atop the western end of the ridge. Ulster Tower stands there today to mark that feat of arms. Here the momentum of the assault was maintained. The wire had been cut and the German front trenches thoroughly destroyed by the bombardment, which made the infantry's progress much easier. Covered by a thick smokescreen, 109th Brigade's leading battalions of Inniskilling Fusiliers rushed the front line, surprising the enemy's machine-gunners while they were still setting up their weapons. The dugouts were cleared systematically with grenades by the mopping-up wave.[77] The Fusiliers surged on into the redoubt and beyond, towards the Mouquet Switch line, their second objective.[78] Although the following battalions fared less well in the face of flanking fire from Thiepval and Bécourt Redoubt to the north, they moved forward briskly in support. By mid-morning, elements of six battalions were established in the Schwaben Redoubt, and a coordinated assault on the second position, the division's third objective, was under way. Here one of the most unfortunate events of the day occurred. As the leading companies of 107th Brigade set off towards the German trenches they were caught by the British barrage—ironically, they were ahead of their schedule—and the attack collapsed when the unoccupied German second position was within reach. A party of fifty men made it into Stuff Redoubt, and found it empty. This was the farthest point reached by British troops on 1 July.[79] Similarly, at Gommecourt, where Third Army had mounted a subsidiary supporting attack to pinch out that fortified village, the assault troops of 56th (1st London) Division broke into the front system in strength and bombed and bayoneted their way forward.

Yet these were isolated successes. North and south of 36th Division, 29th and 32nd Divisions were held up by the fortified villages of Beaumont Hamel and Thiepval. Fifty-sixth Division's partner, 46th (North Midland) Division, attacking Gommecourt from the north, had walked into the enemy's concentrated machine-gun fire. Only a few isolated parties had made it into the German front line, where they were cut off and destroyed. On the western face of the German position the coordinated attack of four army corps on a ten-mile front had broken down into isolated fights for key sectors or short sections of trench. The defence was holding.

WITH CHAOS DESCENDING, the British command had rapidly to rethink. The first tasks facing the commanders on the spot were to reinforce the lodgements that had been made in the German position and to restore cohesion among the attacking divisions. This meant renewing the assault with reserve battalions. Having destroyed the leading waves, the alert defenders turned their attention to these follow-up formations, many of which were moving up over open ground since the communications trenches were clogged with the wounded from the first assault. Concentrated machine-gun and artillery fire wreaked havoc before they had even deployed for the assault, and many reserve battalions never even made it to the British front line. On occasion, brave field commanders, acutely aware of the situation beyond the front line, risked the wrath of their superiors by cancelling supporting attacks. Where they did not, tragedy ensued.

On 29th Division's front opposite Beaumont Hamel village—"a fortress and a masterwork of German brainwork, spadework and ironwork," in Edmund Blunden's recollection[80]—the first attack had been a disaster. As on 31st Division's front, the defenders were alerted by the blowing of the Hawthorn Ridge mine ten minutes before zero-hour. The defenders beat the British to the rearward lip of the mine crater, machine guns were set up, and ten minutes later, when the attack commenced, the whole German line was primed with murderous intent. Very few men from 29th Division made it across no man's land.

In the confusion of battle, 29th Division's commander, Major-General Henry de Lisle, took one of the day's most fateful decisions. Today a proud caribou, staring eternally at the preserved trenches of Newfoundland Park, marks the spot where the colony's tiny army was sacrificed. In support with 88th Brigade, the 1st Battalion the Newfoundland Regiment and the 1st Battalion the Essex Regiment were to follow up the first assault at 08:45. Early reports reaching 29th Division's headquarters suggested, erroneously, that the assault had succeeded (the white flares that the Germans used to call up artillery support were the same as the British signal for success), and de Lisle ordered his reserve battalions forward. The 1st Essex was delayed in the congested trenches, so the Newfoundlanders went forward alone, at 09:05. There was no artillery support for this "isolated and doomed attack" into the enemy's massed machine guns and artillery.[81] Every eyewitness account attests to the calmness of the Newfoundlanders, veterans of Gallipoli, as "in perfect order we waded off through an overwhelming hail of machine gun bullets accompanied by shrapnel."[82] The serried ranks advanced steadily through the firestorm, over the top from the British reserve trenches. Men fell behind the British front line, others before the British wire was reached, and many more as they pushed on doggedly towards the German front trench. A few survivors reached the enemy's parapet, where they were met with bombs, bullets and bayonets. "On came the Newfoundlanders, a great body of men," observed one surviving private from a shell-hole where he was sheltering in no man's land, "but the fire intensified and they were wiped out in front of my eyes, I cursed the generals for their useless slaughter, they seemed to have no idea what was going on."[83] Overall, the battalion suffered 710 casualties, 91 per cent of its attacking strength.[84] All the officers who went forward were either killed or wounded. Alongside the many local tragedies of 1 July 1916, here was a national disaster.

A COHERENT PICTURE of these confused events was subsequently constructed from official reports and survivors' accounts. Those who shared responsibility on the spot had no such clarity of vision. In their headquarters, generals and staff officers waited fretfully for news of the assault. The further back they were, the longer they waited and the less they knew. The slowness and uncertainty of communications meant that reports were generally out of date, incomplete or incorrect, and therefore difficult to interpret accurately. Communications deficiencies were compounded by the over-optimism of the higher command, which tended to latch on to the most favourable reports trickling back from the front line and act upon them. "It is difficult to summarise all that was reported," Haig noted. His own précis, typically laconic as it is, is strictly accurate: "Hard fighting continued all day on front of Fourth Army."[85]

Brigade and divisional commanders such as de Lisle had to take the most immediate decisions: whether to commit reserve battalions, redirect the bombardment, or stop the attack. De Lisle's decision at Beaumont Hamel, committing more battalions to be massacred, was by no means unique, although elsewhere common sense prevailed and reserves were held back, awaiting instructions from the rear. Small-scale local actions continued for the rest of the morning, as supporting waves were sent forward into the fierce German bombardment in an attempt to push through to the parties holding on in the enemy's lines.

The corps commanders further back received a fuller picture from a range of sources, but that picture took time to form, so it would be some hours before they could influence the battle. The news that trickled into higher headquarters was patchy and inaccurate: subordinates wished to give a picture of achievement, not disaster. The best Rawlinson, Fayolle, Joffre, Foch and Haig could do was await a more complete and accurate picture before deciding whether to commit their strategic reserves to the battle. Early reports to G.H.Q. were positive: "Our troops had everywhere crossed the enemy's front trenches," Haig noted at 08:00. Later ones were more mixed, and Haig tried to shrug off his disappointment: "On a 16-mile front of attack varying fortune must be expected!" In the afternoon he visited Rawlinson and Gough; the Reserve Army was to move up in the evening, as "the Fourth Army was getting through its reserves," but no further advance was ordered.[86] Something had clearly gone wrong with the main attack. At Fayolle's headquarters, in contrast, spirits were high. Joffre visited in the afternoon to receive the good news, and "he was beaming."[87] But reports of the British disaster, which became more apparent as the day wore on, deflated his elation. "The British do not yet have the skill," he noted wryly.[88]

By the early afternoon, the commanders were aware that on an eight-mile front between Mametz and Assevillers the German defence had all but ceased to exist. Most of Second Army's immediate reserve battalions, even the divisional depot training battalions, had been committed, and scratch companies of cooks, orderlies, clerks and officers' servants were being hastily formed to occupy the second position. Reserve battalions from behind Second Army's line to the south were ordered northwards to fill the breach.[89] The first reports of the break-in had reached XIII Corps headquarters during the morning. Congreve hurried up to the front to see for himself, then telephoned Rawlinson with the good news. His front had gone quiet, yet he had a fresh infantry division and 2nd Indian Cavalry Division in reserve: should he send them forward against the enemy's second position? Patrols reported that Bernafay and Trônes woods, east of Montauban, were empty. The French on his left had taken all their objectives and were willing to advance. Yet Rawlinson demurred. The plan had not included a breakthrough in the southern sector. The German second line was to be taken north of Thiepval, and Congreve's job was merely to hold the ground he had captured.[90]

On the French front too there was an opportunity to push on. North of the river, XX CA faced the village of Hardecourt and the slowly rising slope to Maurepas. General Nourisson of 39 DI wished to push on with the British XIII Corps, but Congreve's orders from Rawlinson prevented this. The clockwork cooperation of British and French armies, so important to the morning's success, had collapsed. General Vuillemot of 11 DI contented himself with capturing Curlu, which had held up his right flank on the river, in the afternoon. XX CA's two reserve divisions remained inactive, awaiting the renewed assault that Fayolle had ordered for the morning.

South of the river, the Colonials were in triumphant mood after their stunning success. "Our men and officers are keen to progress," noted Thierry, "but the command wisely holds back. Nothing should be left to chance."[91] However, fearing an enemy withdrawal, General Berdoulat pressed Fayolle to authorise an immediate attack on the second German position, which his advanced patrols were already probing. Far from withdrawing, the Germans were struggling to scrape together reserves to defend the second position against the rampant Frenchmen. The Colonials' pause for thought gave them just enough time to rush up their local reserve battalions.[92] Even before Fayolle's assent had been received, Berdoulat had sent his men forward against the villages of Herbécourt, where they met fierce resistance and were repulsed, and Assevillers, which was stormed by the Senegalese *tirailleurs* (light infantry) of 58 *Regiment d'infanterie colonial* (RIC). A counter-attack led by the wounded Leutnant Linder of 60 IR recaptured the village after vicious hand-to-hand fighting.[93] The enemy's defence was far from subdued, and the available artillery support was insufficient to force the position without excessive losses. The attack would be renewed in strength when the artillery had been brought up.[94]

While the corps commanders on the southern part of the front were deciding how to follow up their success, further north the battle for the front system continued. On XV Corps's front the gruelling fight for Mametz and Fricourt persisted. By early afternoon Fricourt was flanked from north and south, although the defenders of the village held firm and continued to take a heavy toll of the attackers on either side. Acting on erroneous and overly positive reports of the situation at the front, Horne decided that the subsidiary operation to take the village itself should be launched in conjunction with a general push all along XV Corps's front. Following a brief and largely ineffective preliminary bombardment, at 14:30 the 7th Battalion Green Howards set off across no man's land. The advance provided further practice for Fricourt's busy machine-gunners, who had already taken such a heavy toll of XV Corps. The assault broke down on the edge of the village. The 7th East Yorkshire Battalion, ordered forward in support a few minutes later, met a similar fate. The two battalions suffered over five hundred casualties each. On the Green Howards' right the 20th Manchesters tried to outflank the village. Sassoon witnessed another brave but doomed assault:

2:30. Manchesters left New Trench...Could see about 400. Many walking casually across with sloped arms. There were about 40 casualties on the left (from machine gun in Fricourt)... Others lay still in the sunlight while the swarm of figures disappeared over the hill. Fricourt was a cloud of pinkish smoke.

Lively machine-gun fire on the far side of the hill. At 2:50 no one to be seen in No Man's Land except the casualties half way across.[95]

Although Mametz village was to fall in the middle of the afternoon after a further sustained bombardment (six hundred prisoners were rounded up out of its maze of dugouts and fortified cellars), the men of 7th Division were too exhausted to push on into the dark, overgrown mass of Mametz Wood behind the German first position; and Fricourt still held out in the centre of XV Corps's front.

Further north, the battle for Thiepval raged on through the afternoon. The Ulstermen struggled to hold on to their gains in the Schwaben Redoubt. Since the divisions on either side had failed, 36th Division found itself holding an isolated tongue of land, "a head and shoulders thrust into the German position,"[96] enfiladed from north and south by machine guns in St. Pierre Divion and Thiepval villages, and under the full weight of the German barrage. Morland strove to reinforce and resupply the detachments in the German position, and to pinch out the Thiepval spur. After a further bombardment, men of the 2nd Royal Inniskilling Fusiliers assaulted the north side of the spur at 13:30. The Thiepval machine guns opened up, and the attack stalled once more. To the south, men of the 2nd Battalion the Manchester Regiment reinforced the defenders of Leipzig Redoubt, but could go no further. The Thiepval front was now deadlocked.[97]

In the woods of the Ancre valley fresh troops, battalions of Yorkshire territorials from Morland's reserve division, 49th (West Riding) Division, waited to go forward. 146th Brigade was ordered to attack Thiepval village at 16:00, after another bombardment by the corps's heavy howitzers. The 1/6th West Yorkshire Battalion led the attack, and became another excellent target for Thiepval's machine-gunners: one survivor, Private J. Wilson, rushed through a gap in a hedge to find "a trench running parallel with the hedge which was full to the top with the men who had gone before me. They were all either dead or dying."[98] Brigadier Yatman advised his superiors that it was impossible to take Thiepval village by frontal assault. Morland ordered a final assault under cover of darkness, but Yatman demurred.[99] Other battalions from 49th Division were ordered to reinforce the Ulstermen at the Schwaben Redoubt, whose position was becoming desperate. They had dug in as best they could under the intensive German barrage, awaiting reinforcement and support on their exposed flanks; there was no senior officer present to order them forward against the second position, which patrols had reported to be empty, so they chose to sit tight. German defensive doctrine dictated that this intrusion be excised. The Schwaben Redoubt was isolated from the British front lines by a barrage of gas shells and machine-gun bullets, cutting off communications, reinforcement and resupply. On learning of this break in his defensive line, Generalmajor von Soden of 26 RID immediately ordered his reserve battalion to counter-attack. Other troops were brought up during the morning and sent towards the breach. It took several hours for them to get across the congested battlefield, but from early afternoon a succession of increasingly strong counter-attacks, spearheaded by men of 8 Bavarian RIR, developed against the apex and flanks of the Ulstermen's position. As the intensive trench battle continued, the defensive perimeter contracted, losses mounted, and ammunition supplies dwindled. The defenders started to wilt under the pressure. At 15:00, Lieutenant-Colonel Bowen remembered:

a lot of men from the 8th and 9th Royal Irish Rifles had broken under Bosch counter-attack from the direction of Grandcourt. We had to stop them at revolver point and turn them back, a desperate show, the air stiff with shrapnel, and terror-stricken men rushing blindly. These men did magnificently earlier in the day, but they had reached the limit of their endurance.[100]

But the German counter-attacks were also weakening, as the field-grey ranks were whittled down in their turn. A final assault, with the disparate remnants of the attacking companies, was organised in the evening by the surviving officers, Oberstleutnant Bram and Hauptmann von Wurmb. Both were subsequently awarded Bavaria's highest decoration, the Knight's Cross of the Max-Joseph Order, for their actions at Schwaben Redoubt. Their motley force pushed forward, making as much noise as possible to hide its small number. They were pushing against an open door. As the attack came forward against his scattered and shattered force, the senior surviving British officer, Major Peacocke, had given the order to pull back to the old German front line. The Bavarians found themselves firing on the retreating British survivors.[101] Thirty-sixth Division's initial success and dogged resistance, which cost it 5482 casualties, had ultimately come to nothing as the supporting divisions could not get forward when faced with an active German defence.[102]

Similar defensive tactics were employed at Gommecourt, where

56th Division's lodgement, exposed to concentrated artillery fire from the south, was isolated, pushed back and slowly worn down by successive counter-attacks. Here again the decision was taken to withdraw the survivors, only five officers and seventy men, under cover of darkness.

THE FRENETIC ACTIVITY of the morning and early afternoon declined as the adversaries fought themselves to a standstill. In the trenches north of Thiepval, under "hellish artillery fire" and on tenterhooks for the next enemy attack, Unteroffizier Hinkel of 99 RIR had "endured an awful time of it. Thirsty, hungry, listless and played out, the long, but largely uneventful waiting time got on our nerves."[103] By mid-afternoon, as attackers and defenders alike lapsed into the torpor that follows adrenaline-fuelled activity, the battle quietened down, except for small-scale attacks and counterattacks, and constant strafing. The Germans held their ground, and Hinkel was able to take his long-awaited revenge: "Wherever a steel helmet showed itself, it was dealt with, just as in a hare shoot. These lads did not seem to know where they were in our trenches and so we allowed some groups to approach us calmly before despatching them with hand grenades."[104]

Characteristically, the English took tea, a risky undertaking in such conditions. Stuart Dolden, with the 1st Battalion London Scottish at Gommecourt, recorded the trials of brewing up under heavy shell- and machine-gun fire near Hébuterne.

> Water had to be obtained from a shell-stricken house, and this proved to be a ticklish job, for as we were carrying the petrol cans to get the water, we suddenly had to throw ourselves flat on the ground, to avoid being hit by a hail of machine gun bullets . . . We were in the act of pouring the tea into the petrol cans when a "Whizz Bang" landed about five yards from the door of the room in which we were working and wounded a fellow who was standing outside. We made all the tea we could, and . . . stood the cans along the wall so that as the troops came in they could help themselves. This was all we could do as, of course, it was impossible to get the tea up to the trenches . . . (We heard later that shortly after we left a shell had . . . scattered all the tins.)[105]

This is one example among many of the labours that follow a great battle. As evening fell the front was a hive of activity. Front-line troops were busy consolidating the positions they had won against possible counter-attacks; new communications trenches had to be dug to link them to the old front line, and wire strung to protect them. Rations and stores were delivered to those who still held their positions. Reserves and guns were moved up on both sides of the line, ready to renew the fight in the morning. Above all, the dead and wounded had to be seen to, a huge task on the British front. Those who witnessed the gory scene behind the British lines knew that something had gone very badly wrong. On "a lovely summer evening" in Meaulte, behind the Fricourt sector, "there was a scene of desolation":

> The main street is strewn with the figures of infantrymen. Some of them have been wounded and, hastily dressed, sent off on foot to the rear. Others are not wounded but, with dreadful pale drawn faces sit or lie exhausted on the cobble stones of the pavements. There is no talking or singing now. Those men have seen death and worse within the last few hours. They are weary, broken and listless. There they lie—the makers of victory. Every few minutes an ambulance whirrs past. Every one is crowded and over-loaded. Men with dreadful wounds in their faces sit on the seat beside the driver, others who cannot sit lie face downwards on stretchers inside—the backs of these last have been shattered by the deadly shrapnel. Then another comes along with men who have lost arms and legs and who are swathed in blood-soaked bandages—a dreadful scene. Those who can walk come limping along on foot or on the limbered artillery wagons which rattle by occasionally.[106]

As night fell the front sprang to life under the enveloping cover of darkness. Men who had spent the day cowering in shell-holes crawled back to their old front line. Medical and burial parties went out in search of bodies and those too weak to move. Sergeant Tawney, still lying exhausted and thirsty in no man's land with a bullet in his abdomen, was tended by a young doctor:

> I knew he was one of the best men I had ever met . . . his face seemed to shine with love and comprehension, not of one's body but of one's soul . . . it was out of the question to get me in that night. But, after I had felt that divine compassion flow over me, I didn't care.[107]

It was a slow, thankless task, less heroic than that of the fighting men, but carried out with equal resolve, and a common humanity. At Serre, Unteroffizier Lais observed the British Red Cross teams active in no man's land: "It is a rare and deeply moving sight in trench warfare . . . Our own first aiders, who are not required elsewhere, go forward to bandage the wounded and deliver the enemy carefully to their own people."[108]

THE SUN SET on a long, brutal day: "deaths in every form—some calm and placid, some blasted and vaporized, some mutilated, one almost burnt to a cinder by me in a dugout," remembered Lieutenant Mulholland.[109] The situation was unclear to both generals and ordinary soldiers. The British command took the decision to pull the few parties who clung on in the enemy's trenches at Thiepval, Gommecourt and La Boisselle back to their own line under cover of darkness; German reserves were now coming up in force and their positions were untenable. The men speculated on what they had done, and what would follow. On the southern part of the front Lieutenant Russell-Jones noted with some satisfaction:

> What a ghastly business this whole affair is, but on the other hand what a success it has all been. The Boches are simply giving themselves up in hundreds. We've captured Montauban, on our left they've got Mametz, and on our right the French have taken Hardecourt. Let us hope we are in sight of the finish. All the Allies are advancing and behind the dark clouds there is just a little ray of sunshine which we trust will mean peace for ourselves, our children, our children's children, aye and even Peace for ever and a day.[110]

As Russell-Jones indicates, over-optimism was not solely the preserve of the high command. A very small gain on the ground was all too easily equated with the victory and peace for which all longed. Further north, however, there was no cause for celebration. Unteroffizier Lais surveyed the devastation at Serre:

> Evening falls. The attack is dead! Our own casualties are severe; the enemy casualties are unimaginable. In front of our divisional sector lie the British in companies, in battalions; mowed down in

rows and swept away. From No Man's Land . . . comes one great groan. The battle dies away; it seems to be paralysed at so much utter misery and despair.[111]

So ended the first round of the long-awaited fight for supremacy; although the misery and despair would continue.

AFTER THE WAR each army recorded its experiences in state-sponsored "official histories." The French history, a monumental 103-volume work, devotes a mere five pages to the events of 1 July 1916, of which a single paragraph covers the British attack.[112] The unfinished German history gives 1 July six short chapters, sixty-two pages.[113] The British history devotes a whole volume to the first half of 1916, culminating with six chapters on the 1 July attack itself, of which a single page covers the French army's success. Rather more coverage is devoted to the German defence. These three official records reveal three very different national perceptions: the French had a victory; the Germans mounted a brave defence; the British had a disaster, for which a military explanation became insufficient.

In his post-war memoirs Joffre delivered a measured, soldier's verdict on the events of 1 July 1916, contrasting the French army's rapid advance—"thanks to the excellent work of the artillery"—with the "less marked" success on the British front. Two factors contributed to their contrasting fortunes:

the Germans did not believe that the French, just emerging from the Verdun battle, would be capable of starting an offensive on the Somme . . . They had therefore taken more precautions along the line facing the British, and this accounts for the more violent reactions which took place on their part of the line. The British also suffered from the fact that their artillerymen were less skilful than ours and their infantry less experienced.[114]

This was a simple military explanation, which Joffre had largely established by 2 July 1916. He felt the British made one key mistake: the German trenches were not mopped up properly by the assaulting waves, so Germans emerged from their dugouts and attacked them from the rear.[115]

Such a practical rationalisation is unlikely ever to satisfy the British:

a straightforward military explanation cannot expunge societal trauma. One anonymous British eyewitness set the tone for their future perceptions:

> Thousands went down that day. I saw from my Post the first wave of troops scrambling out of their Trenches, in the early morning sunlight. I saw them advancing rapidly led by an officer, the officer reached a hillock, holding his sword on high. Flashing it in the sunlight, he waved and sagged to the ground. His men undaunted swept up the mound to be mown down on reaching the skyline, like autumn corn before the cutter. Their story belongs to History, but I have witnessed their deeds with my own eyes, and the sights of that July morning will be ever before me.[116]

Anachronistic sword-waving professional soldiers had apparently led the keen volunteers to unnecessary slaughter. Nevertheless, where there had been success there was much pride in a job well done against huge odds. Major-General Oliver Nugent attempted to explain his division's heroic tragedy to his wife:

> My dearest, the Ulster Division has been too superb for words. The whole army is talking of the incomparable gallantry shown by officers and men. There has been nothing like it since the New Armies came out. They came out of the trenches, formed up as if on the barrack square and went forward with every line dressed as if for the King's inspection, torn from end to end by shell and machine-gun fire.
>
> We are the only division which succeeded in doing what it was given to do and we did it but at fearful cost . . . The Ulster Division no longer exists as a fighting force . . . [It] has proved itself and it has indeed borne itself like men. I cannot describe to you how I feel about them. I did not believe men were made who could do such gallant work under the conditions of modern war . . . I am very proud, but very sad when I think of our terrible losses.[117]

In the immediate circumstances, after proud, virile, brave and fervent battalions had been cut to pieces, it seemed easier to manufacture a myth of heroic sacrifice than to investigate what had gone wrong under the conditions of modern war. Thus the memory of 1 July 1916 would take shape as "an epic of heroism, and better still, the proof of the moral quality of the new armies of Britain, who, in making their supreme sacrifice of the war, passed through the most fiery and bloody of ordeals with their courage unshaken and their fortitude established."[118] But there was also anger at the nature of the job, coupled with a resigned realisation that it was far from over. "Somebody ought to be hung for this show," Lieutenant Billy Lipscome, whose battalion had been cut to pieces at Thiepval, wrote home a few days later.

> The chief fault is that the General Staff sit behind and look at maps of trenches and say "if that is taken so and so will happen," but it doesn't, for trenches are impossible things to judge from maps . . .
>
> This is a rotten letter but we are naturally a bit "down" at present, though of course this is only a local setback and we can't expect to overrun a Boche line when he has had 18 months to fortify it and *has* fortified it too.[119]

Over the ensuing decades, Lipscome's angry disappointment has metamorphosed into collective bitterness. For the British, 1 July 1916 has become a metaphor for futility and slaughter; a national trope and tragedy that defies understanding. The casualty bill alone, 19,240 men killed and another 37,646 wounded and missing, is testimony to the senselessness of the endeavour. (By way of contrast, Foch told Brigadier-General Jack Seely that the French had suffered 1590 casualties on 1 July.)[120] For that reason, the elements of success of that day, in particular the triumph of the French army, more advanced in its tactical methods after its own harsh schooling in 1915, go unsung. While this image took time and much agonising reflection to set in stone, even while the tragically commenced battle raged it was taking root, mediated by the early reflections of those who had survived. Sergeant Tawney, recovering from his wounds, set down his harrowing account of the assault on Mametz, his personal experiences of death and injury, loss and sacrifice, for the eager public. "I suppose it's worth it," he concluded, summing up his battalion's losses, but a note of disillusionment had immediately set in.[121] The "Big Push" had not been expected to change anything, but rather to restore the natural order of things. When it did not, someone had to take the blame, and the collective anguish—a society-wide shell-shock—induced by mass death precipitated pro-

found and unexpected shifts in the nation that had embarked inno-
cently on industrial war. The events of 1 July 1916 started the process.

By way of contrast, the *poilus'* buoyant mood on 1 July 1916 —
which Barberon remembered as "indescribable enthusiasm" and the
expectation of a quick victory[122] — was not to last. As the battle wore
on, it became merely an adjunct to France's prolonged struggle at Verdun, and a consummate success which momentarily held out great
hopes endures only as a brief positive moment in the greater, remorse-
less blood-sacrifice, the price of national liberation.

The other factor, the enemy, and their obvious discomfort at the
smashing of their carefully constructed defences, is usually — and
conveniently — forgotten. Although the allied armies had not achieved
all that they hoped and expected on 1 July 1916, they had momentarily
gained the upper hand. While the German defence had not been broken
completely, it had all but collapsed on a large section of its front astride
the Somme. By the early afternoon, a "broad breach" existed north of
the river, Below later acknowledged.[123] In contemporary military par-
lance, the allies had achieved a "break-in," if not a "breakthrough."
However, this was not the section of the front where the break had been
anticipated, so exploiting it would be a matter of improvisation.

6

Exploitation

FOR THOSE WHO reached the front on 2 July, Picardy's bucolic landscape had been ripped to shreds. Shell-ploughed crop fields, dotted with jagged tree stumps and incised by shattered trenches, stretched for twenty-five miles from Gommecourt to Fay. Hell was their knee-jerk comparator: or Verdun, if they had already visited that affront to nature. "To call these places villages conveys the idea of recognizable streets and houses," noted Australia's official war correspondent, Charles Bean, as he passed through Fricourt and La Boisselle. "They are no more villages now than a dustheap. Each is a tumbled heap of broken bricks . . . Through this heap runs a network of German trenches, here and there breaking through some still recognizable fragment of a wall."[1] These smouldering ruins were littered with the untidy debris of battle: lumps of stone, pieces of metal, smashed weapons; and corpses, or parts of them, scattered pell-mell. The sounds and smells of war and death completed the horror. Between the lines north of Thiepval, wounded men—those who had not succumbed overnight—still moaned, calling out for water and stretcher-bearers. As the midsummer sun rose on another clear, hot day, their cries faded and died. As the day warmed up the unforgettable, unbearable stench of battle—decomposing flesh, explosive gases and lingering chlorine mingled together—invaded the nostrils.[2] Guns and rifles fired intermittently, but there was none of the concentrated thunder of the past week. British patrols and air reconnaissance sent out on the morning of 2 July reported few signs of the enemy. However, after their grisly baptism of fire the previous day, the troops were little inclined to push on. It was a fine summer Sunday, and much of the day's activity in the British sector did it justice. Lieutenant Sassoon spent his afternoon "lying out in front of our trench in the long grass, basking in sunshine where yesterday morning one couldn't show a finger." The enemy's artillery was unusually quiet, so troops could move and work with relative ease. Fricourt was full of British soldiers seeking souvenirs.[3]

South of the river the scene was livelier, as the French prepared to follow up their success. Here territorial soldiers had been detailed to bury the dead overnight. Fatigue parties supervised by artillery and engineer officers had cleared the captured ground of unexploded munitions and other stores while the guns moved forward to the positions already dug for them in anticipation of the second phase of the attack. The infantry had been fed, resupplied and rested as best they could be. Battle had been joined, and it would continue.

For all its horror and subsequent notoriety, 1 July was merely the beginning of the slow, attritional battle that had been planned to break the German army: another 140 days of it, by the conventional reckoning, and nearly five months after that before the epicentre of battle finally shifted elsewhere in April 1917. The General Allied Offensive would take its inexorable course through the summer and autumn. The Battle of the Somme had not been lost "by three minutes," as one anonymous officer later suggested to Edmonds,[4] even if these vital minutes gave the Germans time to set up their machine guns before the British were upon them. If this gave them a temporary advantage on the British front, it did not give them victory. Yet 1 July was not the easy allied triumph that had been anticipated, either. Haig's overambitious plan for a quick breakthrough had been shattered, along with the battalions that had been so enthusiastic to carry it out. Rawlinson's scepticism had proved justified. Nevertheless, as the high command tried to make sense of what had happened and decide what to do next, much still seemed to have gone right. Haig's post-battle summation of 1 July, "the success of which evidently came as a surprise to the enemy and caused considerable confusion and disorganisation in his ranks," while disingenuous, had some claim to veracity.[5] Foch and Fayolle's more careful attack had worked like clockwork, and the French military machine remained ready to spring into action again. The German front had been breached on a continuous thirteen-mile sector, and the second position was in striking distance between Fricourt and Assevillers. However, the breach in the German second position should have come north of Thiepval, where the attack had failed disastrously. The success to the south was welcome, but it was unclear how it could be exploited. Rawlinson was certainly keen to press on: "I want to keep these operations going for at least a fortnight for the Bosch has not many reserves he can bring up and if he does not relieve his frontline they will get exhausted and may crumble."[6] British commanders evidently were not going to let an early reverse get the better of them. The American military attaché reported one English colonel's conversation with Haig, who had expressed "great satisfaction with the situation along the English front, and had stated that they had so far actually accomplished more than they expected." Whether this was over-optimism or calculated dissimulation, Haig was not downhearted. The northern attack had only been a demonstration, the colonel went on to explain, while on the southern sector there were now "only one or possibly two comparatively weak single lines" to get through.[7]

AT BELOW'S HEADQUARTERS in St. Quentin, the night of 1–2 July had been a sleepless and anxious one, although some comfort could be taken from the fact that the Thiepval position had held. The news that the penetration to the Schwaben Redoubt had been driven out finally removed worries on that score. This was just as well, because the Bavarian regiments Below had moved north of the river to meet the allied assault had already been sent forward to reinforce the defence. More reserves were being moved from further back: 12 ID was on its way south from Cambrai, and its leading elements had started to arrive on the afternoon of 1 July to reinforce the defence north of the river. But 5 ID, which Below had ordered forward from St. Quentin to block the French, had been caught by an allied bombing raid while entraining. Sixty ammunition wagons were destroyed, and 180 men killed. One regiment's departure had to be postponed for eighteen hours.[8] It would be three stress-filled days before sufficient reinforcements were gathered to make the second line secure, during which the defenders would have to improvise as best they could in response to allied pressure. There was much to be worried about from La Boisselle eastwards. Fricourt had been outflanked to north and south, and was in danger of being cut off if the British linked up behind it. Orders were issued for the garrison to pull out overnight, which it did without much difficulty, since the British on either side were bloodied and exhausted, and busy evacuating their casualties and reorganising for the renewed effort that had been ordered for the morning. Between Montauban and Hardecourt, where the penetration had been deepest, the newly arriving battalions of 12 ID were detailed for a night-time counter-attack against the British lines across Caterpillar Valley. However, unfamiliar with the ground and harassed by constant shelling, the reserves could not get forward before first light. The counter-attack did not go in until mid-afternoon. After a perfunctory bombardment a whole division went forward in dense skirmishing formation across the Willow stream, under the guns and rifles of XIII and XX corps. Shrapnel, machine guns and rifles cut swaths into the serried ranks. A few Germans reached the edge of Montauban, where they were hunted down with grenades and bayonets. The attack collapsed. The British army did not have a monopoly in tactically unsophisticated assaults.

South of the river, the position was desperate. Scratch companies scraped up from the rear areas had contained French pressure against the second line the previous afternoon, but the remnants knew they could not beat off a properly organised French assault. General von Pannewitz, commanding XVII AK south of the river, received permission to pull back from the Herbécourt–Assevillers position to the river line. Panic set in: "The front is broken! The French are coming!" French civilians took gleeful satisfaction as their once brash occupiers swiftly packed their wagons and limbered up their guns.[9] Large numbers fled back through Flaucourt and over the river, where they would be safe from the rampaging Colonials. The third position, running through Biaches and La Maisonette, was quickly organised for defence. But it was far from certain that there were enough men available to defend it.

Early in the morning on 2 July, the French guns began systematically cutting the wire and destroying the strongpoints in the German second position. Frise, in the river loop, which 2 DIC had been unable to retake in February, finally fell back into French hands.[10] At 16:30 the colonial infantry surged forward, full of fight after their earlier success. An elite division of Prussian Guards would have had difficulty containing them: as it was, only one regiment of 22 RID had been able to reinforce the two battalions from XVII AK manning the second position south of the river.[11] At Herbécourt, the defence was quickly overwhelmed, despite supporting fire from German batteries north of the river. Four RIC proceeded methodically, flanking the village to north and south, linking up behind it before mopping it up. The fight was over in half an hour.[12] At Assevillers, at the other end of the line, the defence was more tenacious: the first French assault was beaten off by artillery fire from unsilenced batteries to the south, and a second with fresh battalions was heavily counter-attacked. More battalions, hastily scraped up from the front to the south, were being thrown in to hold the line.[13] The village was not finally in French hands until 09:00 on 3 July.[14] Following closely the methods prescribed by GAN for the multi-stage set-piece battle, the Colonials demonstrated that early success could be quickly exploited—the capture of one German position could be followed up with the capture of a second while the defence was disorganised. In forty-eight hours the French had broken through on the Somme on an eight-kilometre front![15]

If a similar opportunity existed north of the river, it was on the front between Montauban and Curlu where, covered by 12 ID's costly counter-attack, the defenders were moving guns back behind the second position and rushing up reserve battalions to occupy it. Although Balfourier's men, whose ranks had been little thinned by the previous

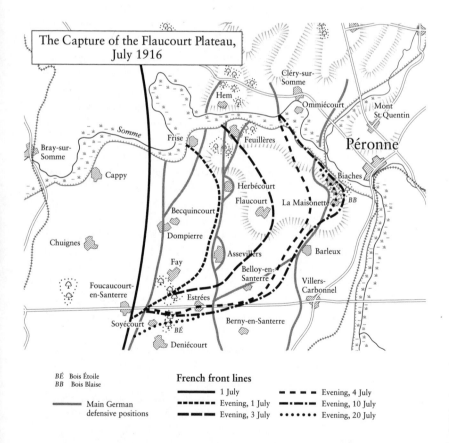

The Capture of the Flaucourt Plateau, July 1916

BÉ   Bois Étoile
BB   Bois Blaise

———— Main German
      defensive positions

French front lines

———— 1 July

▪▪▪▪▪ Evening, 1 July

━ ━ ━ Evening, 3 July

▬ ▬ ▬ Evening, 4 July

━·━·━ Evening, 10 July

•••••• Evening, 20 July

day's exertions, were keen to get at the demoralised enemy, the order was given to dig in and consolidate what they had won to maintain cohesion, rather than try to rush the second position. Moreover, they were already ahead of the British front. German observers had to be cleared from the heights to the north and south or XX CA's regiments would simply be sacrificed in an artillery death-trap. So the Iron Corps's troops waited, momentarily in a surreally quiet sector while the battle carried on to the north and south. Only intermittent German shelling disturbed their repose: 11 DI reported just twenty casualties on 3 July.[16]

Given the chaos and confusion that prevailed in the British front lines, and the need to relieve shattered battalions with fresh ones from the rear, only one set-piece attack could be organised on 2 July, against La Boisselle. III Corps's reserve, 19th Division, came up overnight, and eventually managed to mount a brigade attack in mid-afternoon, some seven hours behind schedule. By nightfall, it had secured the western half of the village.[17] It was not much to show for the second day of a major offensive.

Haig spent the morning of 2 July visiting the wounded, who were in "wonderful spirits." However, his own were turbulent: "A day of downs and ups! . . . The Adjutant-General reported today that the total casualties are estimated at over 40,000 to date. This cannot be considered severe in view of the numbers engaged and the length of front attacked. By nightfall the situation is much more favourable than when we started today."[18] But the reality was that his grand scheme had failed, and he was momentarily at a loss as to what to do next. Should Rawlinson try to redeem the failures north of Thiepval, or exploit the success east of La Boisselle? Their discussions were confused and inconclusive. Rawlinson eventually ordered Gough, who had been put in command of the two corps north of Thiepval, to renew the attack north of the Ancre in the early hours of 3 July, orders which Haig then countermanded.[19] Finally, reluctantly, Haig admitted his bewilderment, and asked des Vallières to arrange a meeting with Joffre and Foch. "Their attempts to attack have produced meagre results," the liaison officer reported. "They will now take counsel, after having received the commander-in-chief's earlier suggestions to visit rather frostily. Des Vallières thinks given the current mood at G.H.Q. a discussion would be fruitful."[20]

· · ·

THE GENERALS MUSTERED at Haig's headquarters talked of exploiting early gains. There was no conception that 1 July had been a disaster: this was only to form in the public mind years later, when casualty figures were revealed, ground won was delineated, and one was factored into the other. Intelligence on the morning of 3 July indicated that, except north of Thiepval, the attack had been effective and the defence was in disarray. Therefore, 1 July had been a clear, if incomplete, success. This dictated continuing the battle. However, there was tension in the air, occasioned by the relative success of the two armies. Just before he left for G.H.Q., Joffre received a congratulatory telegram from Field Marshal Sir John French. "Perhaps the Field Marshal is unwise to address congratulations to the commander-in-chief that he cannot address to his successor," Joffre mused.[21]

How to follow up the success north of the river had to be decided (to the south the Colonials were pressing on to Flaucourt, as planned). An advance through the German second line north of Thiepval to Bapaume was clearly problematic. Although their diary accounts contradict each other, Joffre and Haig clearly had a major row about what to do next. Joffre was bitterly disappointed by his ally's failure (even though Haig had warned him repeatedly of the British army's shortcomings), and he sought to take charge, to tell Haig how to conduct its affairs. He pushed for a renewal of the assault against Thiepval. Haig, who had belatedly settled on following up the success between Fricourt and Montauban, where his troops could work more closely with the French, demurred, and a shouting match ensued. Joffre lost his temper: Foch told Wilson he "simply went for Haig . . . and was quite *brutal.*"[22] Haig claimed he asserted his allied commander's rights: "I am *solely* responsible [to the British government] for the action of the British Army; and I had approved the plan, and must modify it to suit the changing situation as the fight progresses." He concluded his record of the meeting on a triumphant but hollow note: "My views have been accepted by the French staffs and Davidson is to go to lunch with Foch tomorrow at Dury to discuss how they (the French) can co-operate in our operation!"[23] According to Joffre's account, "The commander-in-chief demonstrated the inconvenience of this idea and, not without difficulty and having asserted his authority, General Haig gave up his idea."[24] Both thought that they had won the argument. Supposedly they parted with "mutual assurances of friendship and goodwill,"[25] but these were undoubtedly said through gritted teeth. Next day, des Vallières reported Haig's wounded pride: "This is not the way to treat a gentleman," he had complained. Joffre's outburst had achieved only so much:

> The commander-in-chief had reminded him firmly that the British army was under his orders and must obey his directives. Haig did not demur, although he asserted his right to choose the means. In particular, in the plan to be followed by the British army following its success in the south and check in the north, Haig retained complete liberty. Either he would follow this plan, or do nothing.[26]

It was an unfortunate meeting, with serious consequences for the early phase of operations on the Somme. Neither won the point, whatever Haig's sanctimonious puffing in his piqued account might suggest. After Major-General John Davidson, Haig's chief of operations, visited Dury, he came back with a clearer set of instructions, a workable compromise, which Haig confirmed two days later over lunch with Foch. Haig would now give up the attack north of the Ancre, and focus his efforts between Fricourt and Montauban, where, after securing the intermediate German defences, he would attack the second position between Bazentin-le-Petit and Longueval. Reserve Army would cover this by advancing north-eastwards between Fricourt and the Ancre, to outflank Thiepval from the east via the high ground at Pozières.[27] This did not please Joffre because it shortened the attack front by five kilometres and encouraged divergence between the British and French attacks. Joffre and others at G.Q.G. wanted to exploit the success south of the river, "leaving the British to sort themselves out on their own." But wiser counsels prevailed.[28] The joint offensive would continue, even if it assumed the form of separate, divergent fights. For now, the lines opposite Serre and Beaumont-Hamel would lapse into quietude. Nevertheless, the Battle of Albert, as the British official battlefield-naming committee later designated the fighting from 1 to 13 July, was under way. This would still wear down the German army. And, although Joffre did not know this, Haig would finally abandon the idea of closing down the Somme altogether and shifting British efforts to the Ypres salient, where he could go it alone.[29] He would follow Joffre's broad directives, even if he resented his detailed instructions. Meanwhile, Joffre would stay away from G.H.Q., at least until a success lightened Haig's mood. The British were not yet up to the task; and since they would not accept direct French orders, authority to coordinate the offensive was delegated to Foch, whose more subtle methods might better guide and encourage the touchy officers at G.H.Q.[30] "I am the glue, and that has been so for two years," he remarked ruefully to his wife, "and you would not believe how much patience I have to exercise with the English."[31] Joffre would fulfil his role as coordinator of allied strategy, would see that pressure on Germany was maintained on all fronts, and would focus on balancing the competing demands of France's new battle with her old one on the Meuse.

At Verdun the crisis had passed, because the Somme had quickly achieved its objective of relieving the pressure there. After the capture of Fleury, Falkenhayn had paused. On 24 June Crown Prince Wilhelm's army group on the Meuse front was ordered to reduce its expenditure in manpower and munitions. In the first week of July two divisions and fifteen heavy batteries were placed at OHL's disposal. A week later, on 11 July, Falkenhayn ordered a "strict defensive" at Verdun.[32] But that was not the end of the Battle of Verdun, where operations remained intricately linked to those in Picardy. A German defensive at Verdun was one thing; but an entirely passive front might allow Germany to liberate as many as thirteen divisions for the Somme battle.[33] Therefore, Pétain would have to hold enemy forces there.[34] Joffre committed the French army to a second offensive battle in the second half of 1916, but one whose resources and operations would be referenced to, and restricted by, operations elsewhere. Over the coming months, Pétain and his principal field commander, Second Army commander General Robert Nivelle, would mount a successful if small-scale counter-offensive on the Meuse, wresting Fleury and Thiaumont back from the enemy and keeping a large number of enemy divisions tied down.

WHEN FOCH WAS appointed allied generalissimo after the shattering German offensive in March 1918, he quipped to Clemenceau, "You give me a lost battle and tell me to win it."[35] Perhaps a rueful memory prompted this comment. Foch had warned at the start of May 1916 that if the offensive was scaled down, initial success might produce only a limited tactical victory, and this seemed to be the case. Nonetheless, G.Q.G. remained determined to continue "a long, hard battle on the Somme" as the Anglo-French contribution to the General Allied Offensive.[36] Foch would have to maintain sustained pressure on the German army in the hope that the "rupture" of the front would eventually occur. Such attrition had always been a central component of the allied strategy for 1916, even if Liddell Hart and others' work subsequently established failed breakthrough as the dominating strategic theme in analysis of the "miscalled Battle of the Somme": "miscalled" because what resulted was only "a series of partial actions," not a dynamic, crushing engagement.[37] That is to misconstrue the nature of modern industrial battle. On the afternoon of 3 July, as the allied commanders met at Haig's advanced headquarters at Beauquesne to review their initial successes and failures, reality set in. If it had not stalled entirely, the battle had quickly slowed, as all the models of trench fighting predicted. Foch called it inertia, the process by which the army going forward would slow as its ranks were thinned and became disorganised and its logistics and communications stretched, while the defence, recoiling on its rear defences and deploying fresh reserves, would stiffen and perhaps bounce back. There was little question of the latter on 3 July. At least the principles for consolidating captured ground seemed to be effective. But the real nature of operations was already apparent: "continuing sensible, slow, methodical progress, in which the infantry advance under the continual protection of the artillery, each time taking ground cleared by the guns. The tactics for attacking fortified positions are now set and bearing fruit."[38] To push forward again would require more careful preparation, a new plan, a new build-up and a new set-piece attack. Although at G.Q.G. there was some hope that the cavalry might move forward to exploit the unexpected success south of the river, that was not Foch's way.[39] The battle was just beginning, and it would be a long, hard, costly fight before one side cracked.

Foch had a second problem. He had been given a somewhat ambiguous responsibility, "to act in a secondary role in the battle in which the British army was the main force, but by attacking to support them and pull them along."[40] Even if the British deployed the greater force, des Vallières prescribed that the Somme should remain a French battle "through the energy that our command and troops apply to it."[41] In practice this meant that by September roles would be reversed.

Foch's responsibilities divided naturally into three: the two sectors of Sixth Army's front north and south of the river, closely linked, but in which operations could develop separately, and the British sector, which had to be coordinated as far as possible with the French. Managing this three-front battle would give Foch valuable experience in coordinating operations between armies, which proved invaluable in 1918.

In July 1916, however, he was, if not a novice, still working out how to integrate disparate and divergent attacks into a coherent whole.

Foch had always had a contingency plan, to press on across the river Somme, if things went wrong with the British attack.* Given the initial French success, this seemed feasible, although difficult on account of the narrow French front and the river obstacle they faced. Foch's first idea was to exploit the success on the southern flank. General Micheler's Tenth Army was to be reinforced and extend the break-in to the German defences southwards, after which the river would be crossed south of Péronne. Two more army corps would be deployed north of the river, to push on to the Bapaume–Péronne road to outflank the German defences along the river line. A cavalry corps, acting on foot, would secure bridgeheads over the river, linking the two attacks. It was a bold plan, and growing resistance against the Colonials' advance soon disabused him of the possibility of its rapid success: it was to be a slow, attritional battle after all. But this remained the broad scheme on which Foch based French operations over the coming months.[42]

His task was easiest south of the river, where the Flaucourt plateau was within I CAC's grasp. The morning of 3 July had been somewhat surreal, with no sign of the enemy. Aerial reconnaissance reported the ground ahead unoccupied. Fighting patrols were sent forward to reconnoitre Feuillères and Flaucourt villages. The latter was found to be undefended, and was occupied by midday: 150 German stragglers joined the columns of prisoners, now more than 5000 strong, tramping to the rear. The defence seemed to be melting away, except for enemy batteries firing from the safety of the other side of the river:[43] the mass of guns grouped around Flaucourt were abandoned to the oncoming Frenchmen.[44] During the afternoon the Colonials dug in on their final objective, which it had taken a little over two days to capture. I CAC's cavalry patrolled forward to establish where the German rearguards and next line of resistance lay. If possible, they were to seize bridgeheads over the river. This provided great newspaper pictures[45]—cavalry had not been sent into open country for over a year—but they were not the harbinger of a charge into a demoralised and defeated enemy. Behind them the infantry would follow up, the artillery would be moved forward and re-emplaced, the new front would be connected into the logistics network, and a plan would be drafted for an assault on

*See Chapter 3.

the German third line.[46] In this way, the Colonials advanced seven kilometres and reoccupied the largest tract of French territory since the trench stalemate had set in.

Coming after the trial of Verdun, this advance set France alight from top to bottom. Henry Wilson, visiting GAN headquarters on 5 July, noted:

> Foch ... was well pleased with his attack ... he has captured first and second systems, he has advanced 4000–5000 yards, he has taken 9000 prisoners and sixty guns, and all this at a loss of under 8000 men. It is the finest attack performance of the war, and Foch and Weygand can well be proud of themselves. Result is very high moral tone in his men.[47]

Indeed, his troops were showing all the élan expected of French veterans. A detachment from 24 RIC assaulted Bois 105 on its own initiative on 4 July, silencing an enemy machine gun. Then another group, from 22 RIC, joined in: "The challenge, the difficulty, appealed to them."[48] In truth, after their sudden, unprecedented success, the troops thought that they were winning the war, that the decisive moment had come. Barberon's battery moved forward on to the Flaucourt plateau. Although his comrades were "indescribably enthusiastic" the war-weary and sceptical Barberon himself sounded a percipient note of caution: "No one made a comparison between this attack and that of the Germans at Verdun ... I think we will advance 4, 8, maybe 10 kilometres, like the Germans at Verdun, maybe as far as Péronne."[49] Fayolle was his usual mordant self. With nothing to do at headquarters, he arrived unexpectedly at 2 DIC's command post first thing in the morning on 3 July, and started issuing orders for the morning's bombardment. Thierry remembered:

> [I] remarked humbly that one of the points was 8 RIC's command post, and the other in 24 RIC's rear area. "What," he said, "you are already there? Where are you exactly? I have been very badly informed." And he threw an angry look at his sub-chief, and left in a huff. As a way of congratulations it was a bit thin.[50]

But in his own diary Fayolle noted "a real victory," even if it was incomplete and certainly not decisive. Momentarily it appeared that open warfare had resumed. However, the enemy would surely come back at them, and he was unhappy with the prospects for exploitation. "The plan is simple: cover ourselves with the Somme and fight to the south. But that would mean dropping the British, which we cannot do because they would lapse into inaction."[51] Attention would inevitably shift back to the northern bank, but until then Fayolle was determined to press home his advantage.

I CAC continued to advance over the next few days, although it became clear that the enemy was reinforcing his defences and now appeared determined to fight for them. On the left flank 2 DIC pushed on beyond Feuillères, into the Somme loop. The eastern end of the Flaucourt plateau was reached. From this vantage point a glorious panorama—heart-warming after months staring at the same small piece of enemy line—stretched to north and south. Crops were coming on in the fields of the Santerre, and down below in the marshy Somme valley the trees, barely touched by the shelling, were in full green leaf. Across the river the ground rose to the broad Combles plateau, with the spire of Rancourt church just visible in the distance. Its still verdant fields were pock-marked with white chalk excavations: enemy artillery positions from which intermittent muzzleflashes signalled the incoming barrage. A thicker chalk mark was scrawled along the back of the plateau, alongside a road down which marching men and wagons moved. This was the German third line north of the river, covering the Bapaume–Péronne road, all along which small parties of pioneers were busy raising the chalk parapets and deepening the trenches and strongpoints. It was a fine target for the French artillery. To the immediate front, the roofs of Biaches village could be discerned among the trees; and to its right, atop a steep knoll, the hamlet of La Maisonette. This was the southern extension of the German third position, hugging the western river bank. A mile behind these pretty villages, still relatively untouched by the artillery, the roofs of Péronne, the real prize, seemed tantalisingly close. Here, greater dynamism might have pushed the relatively fresh French units further forward with few casualties. Patrols and air reconnaissance reported the next line of villages was weakly held. But Fayolle, ever cautious, refused to be hurried into a speculative, ill-coordinated advance, despite Foch prompting him to exploit the Colonials' success. Gaining a few more villages would matter little if the attack lost cohesion and the enemy regained the initiative. Better to bring the guns forward, reorganise and resupply the tired and thirsty infantry, and mount a proper attack in the morning.[52]

Over the night of 3–4 July the defenders worked frantically on their third position, linking it through Barleux and Belloy-en-Santerre to the second position at Estrées. The French seized these last two villages on the afternoon of 4 July. Foreign legionnaires from the Moroccan Division assaulted Belloy after a three-hour bombardment. Their morale was excellent. Despite coming under withering close-range machine-gun fire from emplacements hidden until the moment of assault, the legionnaires pressed on into the village. As the bugler sounded the charge, the wounded men in no man's land raised themselves up and cries of "*Vive la Légion! Vive la France!*" could be heard above the gunfire. In the afternoon the legionnaires fought off the first of a series of counter-attacks from the woods to the north-east which 21 RIC had failed to take: the enemy could be seen dismounting from their lorries on the road a few hundred metres behind and immediately joining the attack. The fight went on throughout the night, often hand-to-hand; but the legionnaires kept the village.[53]

Further south, XXXV CA renewed its flanking attack towards Estrées, which fell in the evening. But the enemy soon counter-attacked, driving the French out of half the village in the small hours of 5 July.[54] Most of the village was retaken by that evening, but clearly, as Fayolle had anticipated, the enemy was now reacting with spirit and the advance was losing its cohesion and momentum, breaking down into discrete, small-scale fights. Barleux village, attacked on 5 July, could not be entered.[55] Four new German regiments had arrived to hold the third line, prisoners confirmed. Moreover, the roads were starting to clog up as guns and *matériel* were shifted forward.[56] There would now be an enforced pause as the front was reorganised for a concerted assault against the Biaches–Barleux line.[57]

Over the night of 4–5 July a fresh division, 72 DI, moved into line immediately south of the river and 16 DIC relieved 2 DIC in front of Biaches. The Moroccan Division replaced 3 DIC. The attack went in on 9 July, following a thorough thirty-hour bombardment. Wet weather had forced a twenty-four-hour postponement. A new adversary against which the armies would have to battle regularly in the coming weeks had emerged: "The supply wagons, stuck in the mud, clogged up the rear area completely."[58]

In the centre, south of Biaches, 16 DIC faced La Maisonette. It was the key observatory in front of Péronne, and had been well prepared for defence:

It was protected by four more-or-less complete lines of irregu-larly spaced trenches; the buildings of the château formed a redoubt to the rear of the position: to the north Blaise wood . . . stretched 800 metres to the Somme, to the south there was an orchard and densely wooded park 2–300 metres long. Cellars allowed the defenders to shelter from our artillery fire.[59]

It was Thiepval in miniature. The conscripts of 16 DIC were a credit to 2 DIC's veterans, whom they had relieved. The assault troops, a mixture of European, colonial and Senegalese battalions, attacked at 14:00 on 9 July after an intensive two-hour bombardment. A startled hare shot forward moments before the time fixed for the assault on one company's front: "There's the signal!" one man cried, and they rushed the enemy's front line. The attack met little resistance: the shell-numbed garrison seemed keen to surrender. La Maisonette, outflanked from the south rather than assaulted frontally, was in French hands by 15:15, although machine guns in Bois Blaise stopped the attack to the north. Ahead, trains could be seen getting up steam in the marshalling yard at Péronne. A few French shells encouraged them on their way. Then, under cover of a ruse—a party of Germans pretended to surrender, catching the overconfident Frenchmen off guard—the enemy retook the orchard and château. Two reserve companies drove back the enemy, and the French dug in in the hamlet and orchard. A five-pronged counter-attack was repulsed the next morning. The Frenchmen did not fall for the same trick twice: any Germans who tried to surrender were shot.[60] Thus a legend was cemented. "The Senegalese kill everyone," Fayolle recorded.[61] "Our native troops, having experienced two treacherous acts will not give quarter," their divisional commander noted more accurately.[62] It had been a bloody fight: two French regiments had been decimated, and, whatever the Senegalese's reservations, over a thousand prisoners and six machine guns had been captured. On 10 July, Bois Blaise was occupied. It was only lightly held, but prisoners claimed it was a trap, and its new residents were heavily shelled. Holding the orchard was also problematic: the forward post had been mistakenly abandoned during overnight reliefs, and it had to be reclaimed in hand-to-hand fighting the next day.[63]

Barleux village, in a dip in the plateau and dominated from high ground on three sides, proved an even harder objective. The tall crops in the surrounding fields provided ideal hiding-places for machine guns, which halted 38 RIC's advance on the edge of the village. Without reinforcements, the lodgements that were made in the enemy's defences had to be given up.[64] To the north, though, Biaches had fallen to 72 DI, whose lines now abutted the Somme canal, which ran along the western side of the marshy river valley.

The reality was that the German defence had stiffened, and the Colonials were now caught in a deep, exposed salient, thrust out ahead of the northern attack and the old French front to the south, and therefore vulnerable to artillery fire from three sides. After their enforced quiescence, the enemy's gunners, relatively safe across the river, could retaliate, concentrating their fire and contesting the river line. By the end of the first week of battle, enemy observation aircraft were back in the skies, and tethered balloons were visible in Péronne and to the south. The French deployed their own guns on the Flaucourt plateau to contest with the enemy's, but for the infantry, clinging to shallow trenches, that offered little comfort. The enemy could fire at their trenches from behind, and crossing the open plateau in daylight was impossible. Moreover, in the front-line positions at Biaches, La Maisonette and Belloy enemy infantry were now challenging them for control. After their early elation, despondency started to set in. Their victory seemed to have been wasted, and now they were being forced to endure this attrition while the divisions to the north apparently stood still. Thierry and his staff thought that a new attack could be organised, northwards across the river, to take the enemy holding up XX CA in the rear, something which apparently had not occurred to the high command. But this was not part of the plan, and Sixth Army headquarters was not interested. Fayolle, prudent and methodical by nature, seemed ill-suited to taking quick decisions and implementing such bold initiatives.[65] Thierry's men, now back in the front line, would have to endure while the battle continued to the north.

WHILE THIS PROGRESS was encouraging, it did not promise quick victory. Two vital days had been lost while a new focus was being determined, and shattered British formations relieved and reorganised. By 4 July, however, all were ready to resume operations even if, for the British, ammunition supply remained problematic.[66] With his spat with the French over, and visits to his more successful field commanders providing encouragement, Haig's customary optimism returned. "In another fortnight, with Divine Help, I hope that some decisive results may be obtained," he wrote to his wife. The German defence seemed rather haphazard: "The enemy puts his reserves straight into the battle on arrival, to attack us, thereby suffering big losses."[67] Once again Haig was underestimating the reserves available to throw into the battle. G.H.Q. intelligence staff's assessments indicated there were only fifteen enemy battalions between La Boisselle and Hardecourt, when in fact there were more than thirty, with others arriving every day.[68] But they were correct in pointing out the chaotic state of the defence, with the line "now held by a confused mass . . . whose units seem to have been thrown into [the] frontline as stopgaps."[69]

Foch was less hopeful, as Wilson noted: "On the whole . . . Foch is very pleased with his own advance, and displeased with ours, but does not think that Haig yet understands in what ours failed . . . not nearly sufficient concentration of fire before infantry attack."[70] Getting the British army going again after its visceral shock would prove a challenge. Its main task was to push on across Caterpillar Valley to within striking distance of the German second line, and then organise a new set-piece attack against it; subsidiary to that was the need to cover its left flank by advancing along the Albert–Bapaume road towards the high ground at Pozières. At least on this narrower front the guns could now be targeted on to smaller, shallower objectives, which would redress the main problem inherent in British attacks. But with the front delineated by a succession of separate objectives between the British front and the German second position—Contalmaison village and Mametz, Bernafay and Trônes woods—there would be a tendency for operations to fragment into a number of small-scale, uncoordinated fights. An inherent problem of attritional operations, the grinding struggles for a "jumping-off line" between set-piece attacks, would manifest itself.

How grinding they would be depended on the strength and effectiveness of the defence. While the allies had been debating, Below had been reinforcing. On learning that the defence south of the Somme was collapsing, Falkenhayn hurried to St. Quentin. His meeting with Below and his staff on 2 July was fraught, but at least it persuaded the German commander-in-chief that the real counter-attack was under way on the Somme. Falkenhayn acted with decision, sacking Pannewitz and Below's chief of staff General Grünert who had authorised the withdrawal south of the river, and appointing Colonel Fritz von Lossberg, an acknowledged defensive expert, in Grünert's place. Equally importantly, Falkenhayn agreed to send material reinforcements: four more infantry divisions in the first instance, and thirty-eight heavy artillery batteries, some transferred from the Verdun front. But there was to be no further retreat. "The first principle of position warfare must be to

yield not one foot of ground, and if it is to be lost, to retake it by immediate counter-attack, even to the use of the last man," he reminded Second Army's commander.[71] This notorious order condemned Below and
his men to a long, tough battle of attrition. The German army would
meet like with like. Below's 3 July order of the day summed up the task
that faced them:

> The decisive issue of the war depends on the victory of the Sec-
> ond Army on the Somme. We must win this battle in spite of the
> enemy's temporary superiority in artillery and infantry . . . For
> the present, the important thing is to hold our current positions
> at any cost and to improve them by local counter-attacks. I for-
> bid the evacuation of trenches. The will to stand firm must be
> impressed on every man in the Army. I hold Commanding Offi-
> cers responsible for this. The enemy should have to carve his way
> over heaps of corpses.[72]

Not for the last time, the self-sacrificing task of redeeming the high
command's bombastic martial rhetoric fell on the men. But it would be
several days before new divisions could take their place in the line.
Would it hold meanwhile? Why were the British not coming on, as the
French were south of the river? Was it true, as Second Army's communiqué pronounced on 3 July, that "the Anglo-French offensive had
been halted on the Somme and in places repulsed"?[73]

On 4 July those regiments that had occupied the incomplete field-
works between the first and second positions realised that any halt was
only temporary. Their thankless task was to contest that ground in
order to win time to deploy the artillery and machine guns necessary
for an effective defence. They would have to fight "to the last man" in
improvised defences between Ovillers and the river while new divisions
were moved up to garrison the second position, and, if they were lucky,
relieve them. Many veteran formations, including 3 Guards ID, were
sacrificed. For the next week, outnumbered and outgunned, they
would do their duty, enduring close-range bombardments, fighting
desperately for their shallow trenches, firing until their ammunition
was exhausted, ultimately giving their lives while behind them the
defence consolidated.[74] It was testament to the quality of the German
infantry of 1916 that they fought so bravely and so well; but it marked
the start of a process that would grind the life out of the German army.

Crown Prince Rupprecht noted the chaos that characterised this hurried re-engagement with the menacing enemy: "When at last reserves did arrive it was too late. They arrived in dribs and drabs and had to be deployed immediately to plug gaps. As a result there has been such a mixing of formations that nobody knows what is happening."[75] Falkenhayn had divided the front into three army corps sectors: Gommecourt to the Albert–Bapaume road under XIV *Reservearmeekorps* (RAK) commander General von Stein; from the Albert–Bapaume road to the river under VI AK commander General von Gossler; and south of the river XVII AK under Pannewitz's replacement, General von Quast. But these were corps in name only. Each sector would receive infantry regiments as they arrived, such that recognisable higher formations became hard to identify. For example, reserve battalions had been scraped together from the regiments holding the line down to the river Oise, and so Quast's command included elements from eleven different divisions.[76] Divisional field-gun batteries and heavy artillery would be similarly deployed as they reached the front to strengthen the counterbombardment that was the key to contesting the allied advance. In this way elements of some twenty German divisions were committed to the battle in the first fortnight.

If anything, these were delaying tactics while OHL found its feet after the succession of shocks it had received on Eastern, Southern and Western Fronts, and worked out how to balance its finite strategic reserves between them. It would be two weeks before the defence on the Somme was properly organised. Then Falkenhayn divided the front into two, with Below assuming command of First Army north of the river, and General Max von Gallwitz, now in command of Second Army south of the river, being given responsibility to oversee the joint defence. Naturally, Below resented this reorganisation, blaming OHL's refusal to send him reinforcements in a timely manner for the deficiencies of the defence.[77] But, whoever's fault it was, the allied advance had not been halted, and by mid-July the British as well as the French had penetrated the second position.

WHILE THEIR CHIEFS had been debating their future and their opponents sorting themselves out, Congreve and Horne had been consolidating their gains, without too much difficulty after the initial German counter-attack on Montauban had been repulsed. Twenty-seventh Brigade occupied Bernafay Wood, five hundred yards east of Montauban, without a fight on the afternoon of 3 July. Three abandoned field guns were captured. In the centre, a battalion from 18th Division occupied the long and narrow Caterpillar Wood, capturing five more guns. XV Corps was also making steady, if not entirely unopposed, progress north from Fricourt. Shelter and Bottom woods were occupied on 3 July, counter-attacks from the direction of Contalmaison being driven off with Lewis guns. Over eight hundred prisoners from five different regiments were taken. III Corps finally secured the whole of La Boisselle on 4 July. It was a hamlet of a mere thirty-five houses, but "a rabbit warren, the result of two years' tunnelling by the people holding it who did not mean to be dislodged . . . nearly every house of which was a strong point with dugouts thirty to forty feet deep." The hand-to-hand battle had raged for three days with "big bearded men belonging to a pioneer battalion, who no doubt resented being turned out of their comfortable home." The leader of the British garrison, one-armed Colonel Adrian Carton De Wiart, had to drive off yet another strong enemy counter-attack with grenades, for which he was awarded the Victoria Cross, one of three won by men of 19th Division in the action. Taking La Boisselle cost 3500 casualties: "It was a real 'dog fight' and the Germans never fought like that again."[78] The determined enemy were being mastered, and the traumas of 1 July were starting to fade.

Successful small-scale consolidation helped to straighten the front and hearten the troops, but rather more was needed before the second position could be attacked north of the river. In the French sector two obstacles, the intermediate line that stretched southwards from Maurepas across the Hem plateau to the river, and the village of Hardecourt with a knoll of high ground behind, faced XX CA. The British faced three strong obstacles in their next advance: Trônes Wood on XIII Corps's front; Mametz Wood in front of XV Corps; and Contalmaison village in III Corps's sector. Foch and Haig arranged for a general Anglo-French attack to be launched against the central objectives on 7 July. Meanwhile, to cover the flanks Reserve Army would capture Ovillers and 11 DI would occupy Hem.

The battle for Hem, against an alert and expectant enemy, was expected to be tougher than the Iron Division's first operation. The success of 1 July had been a fluke—the defence had been so overwhelmed that there was no effective response—but now a strong enemy riposte was inevitable. Their objective was another strong, well-wired defensive system fifteen hundred metres in depth, backed by the long but narrow village itself, extending along the river bank to Monacu

Farm. The position was pock-marked with small quarries and dotted with woods. Taking it would bring XX CA within striking distance of the second position, as well as clearing the Hem plateau of guns that were harassing the Colonials to the south. The attack was prepared with customary thoroughness. The assaulting infantry were to "be deployed in great depth, to move forward quickly but without haste, with designated groups, wearing clear insignia, detailed to mop up." Great stress was placed on infantry–artillery cooperation, to ensure that any resistance encountered could be quickly re-bombarded. Hem village was to be outflanked to the north before mopping-up parties went in to clear any surviving defenders. Engineers went forward with the assault troops to wire the captured ground against the inevitable counter-attacks and dig the communications trenches that would link it into the French trench system.

The concentrated artillery of XX CA, augmented by guns firing from I CAC's area south of the river, smashed the fixed defences and obliterated Hem village in the forty-eight hours before the infantry attacked. Thirty-seven RI, supported by a battalion from Mangin's 79 RI tasked with taking the quarries and Hem on the right flank, left the saps they had dug towards the German front under cover of fog at 06:58 on 5 July, two minutes before the bombardment lifted from the German front line. It crept forward with the French infantry following closely behind. By 08:15, the objectives on the northern sector of the attack had been seized, although Hem to the south remained in enemy hands. The village was quickly re-bombarded and assaulted at midday. Most of it was cleared, although parties of the enemy held out in the houses at its eastern end till 17:00, belying a later German declaration that it was voluntarily abandoned on account of an overwhelming bombardment.[79] Finally, Bois Fromage behind the village, bristling with enemy machine guns, was taken by 79 RI at 18:30 after a further bombardment. Only Monacu Farm, another machine-gun nest, remained in German hands. As expected, it had been a much harder fight than that of 1 July, but good coordination between infantry and artillery had kept 79 RI moving steadily forward. Its casualties had been slight, and several hundred prisoners had been taken. One *poilu* was surprised to spot his pre-war boss from the Parisian department store *Samaritaine* among them![80]

Moreover, the fight was far from over. The difficulties of holding on to an unconsolidated line, in practice a series of disconnected positions scraped in the ruins of the enemy's old trenches and insufficiently wired, were demonstrated. Five enemy counter-attacks fought their way slowly back towards Hem over the next twenty-four hours. There was vicious close-quarter fighting in Bois Fromage, de l'Observatoire and Sommet—the latter changed hands four times before artillery fire made it untenable to either side—and in the afternoon the old German front line was reentered. The French front was in danger of collapsing, but 79 RI's reserve company bombed out the infiltrators before they could establish their defence. Its commander, Lieutenant Cordier, won the *Légion d'Honneur* with a fine display of *sang-froid,* calmly indicating the targets to his grenadiers with his cane.[81] Finally, the Germans' fury abated. The French held the disputed ground.

To the north, 39 DI and 30th Division attacked Hardecourt and Trônes Wood. It was a difficult operation, against two sides of a salient. The day before the attack was due to start, a German counter-attack reestablished control of the northern edge of Bois Favière. It had to be retaken by the French before the attack could proceed. So, a day late, at 08:00 on 8 July, the infantry went over the top. After a crushing bombardment, 39 DI captured Hardecourt with customary dash, but they could push no further northwards until the British took Trônes Wood: machine guns were still firing from there because the first British assault had been driven off. At 13:00 they launched a second attack on Trônes Wood. A lodgement was secured at its southern end, but it was to take five more days and six further attacks before the whole wood was in British hands. Thirtieth Division was exhausted, and Maxse's 18th Division had to take its place for the last two attacks.[82] Maintaining cohesion was impossible. Well-placed machine guns took a steady toll of the attackers, disorientated and moving with difficulty in dispersed groups through the thick undergrowth, before local counterattacks forced them back from any ground they had gained. To secure the wood, Lieutenant-Colonel Frank Maxwell VC, commanding the 12th Middlesex Battalion, eventually formed what is best described as a line of beaters out of stragglers from various battalions, and personally led them through "the most dreadful tangle of dense trees and undergrowth imaginable, with deep yawning broken trenches criss-crossing about it . . . with its dreadful addition of corpses and wounded men," shooting ahead and clearing out the remaining nests of Germans as if they were a covert of pheasants.[83]

At the other end of the front the British faced an equally formidable series of obstacles: on the left front Contalmaison village, halfway up a spur jutting out from the forward slope of the main ridge, protected to the west by Bailiff Wood, and almost a mile to its right Mametz Wood. This wood was the largest in the British sector, measuring a mile along its main north–south ride and nearly as wide. It occupied the forward slope of the ridge in front of Bazentin-le-Petit village, with three apexes, pointing towards Contalmaison, Mametz and Caterpillar Wood. Its defenders had intermingled barbed wire with the wood's thick undergrowth and shattered trees, and linked it to the German second position with a communication trench along which reinforcements could pass relatively undisturbed. Contalmaison and Mametz Wood were linked by two curved lines of trenches, Quadrangle and Wood trenches and their support lines several hundred yards to the rear, sited to enfilade any direct assault on Contalmaison and Mametz Wood. The approaches to Mametz Wood itself (from the east side, since the Willow stream in a steep-sided ravine obstructed an advance up the slope from the south) could be enfiladed from the German second position. Although the position was formidable, the forward trenches were in British hands by 7 July. Lieutenant Sassoon cleared one end of Wood Trench with a single-handed bombing attack. It earned him not a medal but a rebuke from his colonel for deviating from the plan and holding up the assault. Many more acts of heroism would be required before the Contalmaison–Mametz Wood stronghold was wrested from its fanatical Prussian Guard defenders, the Lehr Regiment.

Three New Army divisions from Fourth Army reserve were committed to the operation: from right to left, the 23rd, the 17th, and the 38th, Lloyd George's "Welsh Army." The first attack was delivered at 08:00 on 7 July, through cloying mud like that which had clogged up I CAC's advance south of the river.[84] Contalmaison held out until the evening of 10 July, and Mametz Wood was finally cleared on the morning of 12 July. In the interim British battalions made numerous attacks against well-entrenched and skilful opponents. Subsequent analysis has suggested that the British attacks were amateurish, ill-coordinated, made in insufficient strength and with weak and often poorly timed artillery support. They were testimony to the inexperience of the New Army formations that were making them. While not yet of the calibre of their German adversaries, they were at least learning the skills of battle. Although these were difficult positions to take, and tenaciously defended, they were captured relatively quickly, often thanks to the presence of mind of the brigadiers and colonels in the field who took the crucial decisions which made the difference between success and failure.

For those not involved, such as Major Brooke, the fighting made a
fine spectacle:

> On our left we could see Mametz wood, and continual scrapping
> going on for it. One could see the Germans scuttling backwards
> and forwards at the far end of the wood in a continual hail of
> shrapnel ... [We saw] in the distance one of our attacks on Con-
> talmaison; it was a wonderful sight. To begin with the heavy
> Artillery were pounding the village. Great columns of smoke

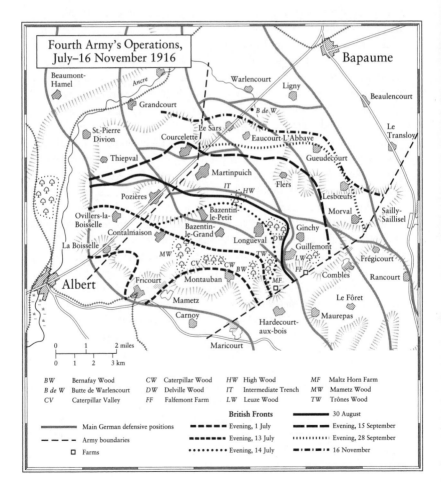

Fourth Army's Operations,
July–16 November 1916

BW	Bernafay Wood	*CW*	Caterpillar Wood	*HW* High Wood	*MF* Maltz Horn Farm
B de W	Butte de Warlencourt	*DW*	Delville Wood	*IT* Intermediate Trench	*MW* Mametz Wood
CV	Caterpillar Valley	*FF*	Falfemont Farm	*LW* Leuze Wood	*TW* Trônes Wood

British Fronts

Main German defensive positions — Evening, 1 July — Evening, 15 September — 30 August

Army boundaries — Evening, 13 July — Evening, 28 September

Farms — Evening, 14 July — 16 November

and brick dust were flying up into the air . . . The whole village was wrapped in a cloud of smoke, lit by flashes of bursting shell. This went on for some time, and then the shrapnel barrage became intense, rows of white puffs of smoke and the ground whipped up into dust by the shrapnel bullets, and away in the distance we could see our infantry advancing in lines towards the village. One felt as if one was in a dream, or that one was watching some extraordinary cinematograph film.[85]

It was a rather more real experience for those in the fight.

The battle began with a night advance by 17th Division against Quadrangle Support Trench. Its commander, Major-General Pilcher, had warned that the objective, open to flanking fire from Contalmaison and Mametz Wood, could not be held. Two brigade attacks demonstrated that fact.[86] For his realism, Pilcher was relieved of his command. (Unfortunately, as Fourth Army's tactical notes had warned, in the hierarchical British command structure "all criticism by subordinates of their superiors, and of orders received from superior authority, will in the end recoil on the heads of the critics."[87]) By the evening of 7 July, men of the Durham Light Infantry from 23rd Division had established themselves in Bailiff Wood, a copse of saplings rather than an impenetrable thicket. After a frontal assault by 24th Brigade had failed — a battalion of the Worcestershire Regiment fought its way into Contalmaison but was forced to retire when its ammunition ran out — Brigadier-General Henry Croft, commanding 68th Brigade, opted to flank the village to west and north, pushing up to and through Bailiff Wood in a series of company attacks.[88] On the other flank, 38th Division's attacks on Mametz Wood from the east came under frontal and flanking machine-gun fire and the attackers were forced to seek shelter in the shell-holes which pock-marked the exposed slope over which they were advancing.[89] Brigadier-General Evans's 115th Brigade made three attempts to cross the exposed slope, but without the promised smokescreen they failed each time. Tragedy turned into farce. Owing to a breakdown in communications, British gunners shelled their own men during the second assault. It was clear to Evans, reconnoitring the ground for a fourth assault, that "no attack could succeed over such ground, swept from front and side by machine guns at short range."[90] The bombardment was too weak, and the infantry attacks were poorly coordinated with it. Evans's splenetic tirade against his superiors as he trudged back to his headquarters acknowledged the huge gap between the practical outlook of officers responsible for operations in the field and the gung-ho attitude of those who directed them from the safety of higher headquarters:

> I spoke my mind about the whole business ... They wanted us to press on at all costs, talked about determination, and suggested that I didn't realise the importance of the operation. As good as told me that I was tired and didn't want to tackle the job. Difficult to judge on the spot they said! As if the whole trouble hadn't arisen because someone found it so easy to judge when he was six miles away and had never seen the country, and couldn't read a map. You mark my words, they'll send me home for this: they want butchers, not brigadiers.[91]

In fact, although Haig seemed to share the opinion that lack of dash, not enemy firepower, was the reason for the check,[92] it was Evans's superior, Major-General Ivor Phillips, who was sacked. A retired Indian Army officer, Liberal MP and under-secretary at the Ministry of Munitions, to whose brother Lloyd George owed a favour,[93] Phillips was a prime example of the sort of home-appointed general whose unsuitability for field command would be exposed on the Somme battlefield. His sacking was probably merited: "As a divisional commander he was ignorant, lacked experience and failed to inspire confidence," one 7th Division staff officer concluded.[94]

Major-General Watts of 7th Division was put in charge of the battle for Mametz Wood, which was attacked again on 10 July. Watts gave his commanders on the ground more discretion over the details of the operation.[95] Now the wood was to be assaulted from the south-east. Although they had to get across the steep-sided Willow stream ravine, the Welshmen would now avoid the flanking fire from the German second position that had cost them so dearly on 7 June. Their artillery fire plan also offered closer support. A false lift (a technique copied from the French, where the barrage lifts off trenches to simulate the moment of an attack but returns to the now-manned trench a few minutes later) cleared the defenders from the edge of the wood, and a creeping barrage protected the advance through it. Four battalions from the division's two uncommitted brigades attacked at dawn on 10 July. Once again the assaulting battalions advanced into a cross-fire of machine-gun bullets and shrapnel. Nevertheless, covered by the barrage, the leading waves of 114th Brigade, two battalions of Royal Welch Fusiliers, forced their way into the left-hand extension of the wood at Strip Trench, although supporting waves were forced to take cover in Willow stream. The enemy fought back strongly, driving the Fusiliers out again. Two more Fusilier battalions then came up in support and, enraged by the sight of Germans bayoneting British wounded, rushed the wood "in a screaming temper."[96] The Lehr Regiment's forward defence was broken. Some men surrendered, while officers ordered the rest back to form a second defensive line across the centre of the wood. Ten more battalions were fed into the engagement, and during a long, hot, horrific day the Welshmen fought their way slowly through the tangled undergrowth and fallen trees. By 18:30 they had broken the rear defensive line and all but cleared the wood. But machine-gun fire from the second German position forced them back from its northern edge, and the enemy re-established themselves in it northern groves.

On the same day, Contalmaison, attacked from the positions established by Croft's men to the west, also fell. Quadrangle Support Trench, which had held out against 17th Division's repeated attacks, was finally abandoned as untenable.[97]

The battle for Mametz Wood did not finally end until 12 July, when the last German outposts were driven from its northern edge. The Welsh New Army battalions had taken a formidable position in one of the most intense close-quarter fights of the war. It had cost them nearly four thousand casualties.[98] Hieronymus Bosch could not have painted a more horrific vision than that in the conquered wood:

> Years of neglect had turned the wood into a formidable barrier, a mile deep. Heavy shelling had ... thrown trees and large branches into a barricade. Equipment, ammunition, rolls of barbed wire, tins of food, gas helmets and rifles were lying about everywhere. There were more corpses than men. Limbs and mutilated trunks, here and there a detached head, forming splashes of red against the green leaves, and, as an advertisement for the horror of our way of life and death, and of our crucifixion of youth, one tree held in its branches a leg, with its torn flesh hanging down over a spray of leaf ... a derelict machine gun propping up the head of an immobile figure in uniform, with a belt of ammunition drooping from the breech into a pile of stained red earth.[99]

"Over all hung the overwhelming smell of corpses, turned up earth and lachrymatory gas."[100]

Three fresh New Army divisions had been used up to capture Contalmaison and Mametz Wood; their casualties totalled over twelve thousand. Overall, the British suffered another twenty-five thousand casualties between 2 and 13 July.[101] It had been a gruesome baptism of fire, a trial for the troops and a test for their inexperienced commanders: Phillips was not the only major-general sent home. How many German divisions were similarly worn out is difficult to gauge. Elements of seventeen regiments, drawn from nine divisions, were engaged in the battle. Three Guards ID suffered heavily and the elite Lehr Regiment (before the war it had been the Kaiser's demonstration battalion at Potsdam) was effectively wiped out. There is much to criticise in the conduct of these operations. Certainly the actions of corps and divisional commanders, unaccustomed to conducting offensive battles, were frequently at fault, and success owed a lot to the self-sacrificing efforts of their men. But the command structure was effective enough to recognise and redress obvious deficiencies. Poor commanders were removed, which opened the way for the promotion of more skilled officers. Regrettably, there were many vacancies: 38th Division lost seven battalion commanders in capturing Mametz Wood.[102] But it was an essential stage in creating a professional army from the civilians in uniform deployed to the Somme. It is moreover important to recognise that these operations were not conceived in the same terms as the set-piece attacks on successive positions. They were tidying-up operations against distinct defended localities, clearing the way for a new set-piece assault that would be conducted with more care and thoroughness, and would pay dividends as a consequence.

AFTER NEARLY A fortnight, the battle seemed to be developing in the allies' favour, but none too quickly. "We have taken 12,000 prisoners and 70 guns, the English 7500 and 24," Fayolle was pleased to record. Joffre nevertheless remained impatient: "He wants a victory; really the victory. He thinks the situation is bad . . . The British have 70,000 men in the ground already and they are still not in contact with the second position."[103] But the British were finally about to give him, and themselves, something to smile about.

The successful British assault on the second position between Bazentin-le-Petit and Longueval on 14 July has often been contrasted with their earlier failures. It certainly surprised the French: at Dury they were talking of "an attack organised for amateurs by amateurs."[104] "This time, grey-clad corpses outnumbered khaki ones on the battlefield," recollected Lieutenant Liddell Hart, who was shortly to be gassed and invalided out of the front line for good: "That sight, and contrast, deeply influenced my future military thinking."[105] Coming only a fortnight after 1 July, it is debatable to what extent lessons had been learned and absorbed. However, the attack demonstrated two features that had been missing from earlier operations: a crushing artillery bombardment and tactical surprise. The most detailed examination suggests that in planning the 14 July attack Haig and Rawlinson hit on these positives by chance rather than through measured judgement.[106] Rawlinson planned to advance his assaulting infantry under cover of darkness. Haig initially cavilled, proposing instead to assault the left flank of the second position from Mametz Wood, and roll it up towards Longueval. After two days of consultation, Rawlinson, backed by Congreve and Horne, eventually convinced Haig. Assembling under cover of darkness within striking distance of the German front line was vital since it was nearly a mile from the British front to the German second position. Massed ranks of infantry advancing up the slope of the main ridge in broad daylight would merely have presented another splendid target to any German machine-gunners who survived the bombardment. That bombardment of the second position, which had been going on intermittently since June, commenced in earnest on 11 July. It was more prolonged and more concentrated than any seen previously. The weight of shell per yard of trench to be attacked was estimated at eighteen times that of 1 July: as one observer noted, "The whole horizon seemed to be bursting shells in front of us, and behind us flashing guns."[107] There would be no pauses in the fire plan to allow the enemy to repair his defences and restring his wire. Four divisions, the 21st and 7th in XV Corps on the left, and the 3rd and 9th in XIII Corps on the right, would assault. Eighteenth Division would support the attack by finally clearing Trônes Wood. The roar of the barrage would cover the sound of the infantry forming up in no man's land. There would also be a cavalry division standing by to exploit success after Longueval was taken.

Late on 13 July, advance parties of Royal Engineers, covered by Lewis-gun detachments, crept into no man's land to mark out the jumping-off line with white tapes five hundred yards from the enemy's front line. The assaulting companies moved out early the next morning, and were deployed on these lines by 02:00. All reached their places undetected by the enemy, and lay down to await the end of the barrage. Some companies decided to crawl forward to close the gap with the enemy's front line. After five minutes of hurricane fire, the barrage lifted and the infantry surged forward. Except on 3rd Division's front in the centre where wire remained uncut, the second position was soon occupied. However, progress beyond was not going to be rapid. Another conglomeration of villages and woods—Bazentin-le-Petit and Bazentin-le-Grand at the western end with their associated woods, and two thousand yards to the east Longueval with Delville Wood to its rear—blocked the infantry's advance. Seventh and 21st divisions took the Bazentin villages and woods relatively easily, although they were soon fighting off a strong counter-attack against the north side of Bazentin-le-Petit Wood. Here the fortunes of war went against the British. The Germans were in the process of establishing a new corps sector under General von Arnim between Bazentin-le-Petit and Longueval, and were relieving the tired troops holding the second position over the night of 13–14 July. A whole fresh division, 7 ID, was coming up to replace the remnants of 3 Guards ID and 183 ID, which had fought for Contalmaison and Mametz Wood. It immediately threw its regiments into the defence of the Bazentin villages. On the right flank, Longueval and Delville Wood, entered by troops from 9th Division, were counterattacked. They were contested throughout the day, and without their capture, progress across the plateau to High Wood became problematic.[108]

On XV Corps's front reconnaissance had suggested that High Wood was empty, so permission was requested to send 91st Brigade, 7th Division's reserve, forward to occupy it. Since this should have been the 2nd Indian Cavalry Division's task, permission was initially refused. With no sign of the cavalry, however, at 15:30 Horne belatedly gave the order for the infantry to occupy the wood. It was to be bombarded until 18:15, but it took so long for the reserve infantry brigade to come forward that the attack did not start until 19:00, without artillery support. One modern soldier judged that "in retrospect it seems ludicrous . . . the most striking feature of the command and staff aspects of the morning and afternoon events is the absence of any sense of urgency."[109] The inadequate state of battlefield communications, coupled with the multi-layered and hierarchical British command system, imposed inevitable delays. Had an opportunity gone begging to push on before the enemy's defence could stiffen? In fact, High Wood was not unoccupied: the Switch Line, a trench along the rear edge of the Bazentin Ridge that had been dug as an outpost line for the German third position, had been garrisoned by a battalion from 7 ID. Faced with intermittent machine-gun and rifle fire, 91st Brigade's infantry entered the wood and dug in in front of the Switch Line, which ran through the rear of the wood and along its eastern edge facing Delville Wood.[110] It would have been suicidal to push through woodland against a defended trench line. The plan to rush High Wood quickly had certainly gone awry, mainly due to confusion and poor communications between front and rear. It was not, however, the cavalry's fault.

Many otherwise sensible historians have an irrational blind-spot when it comes to the employment of cavalry on the industrial battlefield. Robin Prior and Trevor Wilson, for example, have confused Rawlinson's "general idea of future plans" for following up the capture of the second position, which did involve an advance against the German third position with as many as three cavalry divisions, with his operational objectives for 14 July, thereby confusing what the offensive might do with what the attack should do.[111] The cavalry's offensive role was circumscribed by the dominance of artillery and machine-gun fire, which made them vulnerable *en masse* and in open ground. But their mounted infantry role to seize ground quickly and (as the French had demonstrated) their reconnaissance role to locate the retreating enemy were still valuable after a set-piece attack. Second Indian Cavalry Division had been assigned these tasks. It had orders to push across the Pozières–Combles Ridge, dominated by High Wood, along the rear edge of which the enemy had sited an intermediate line. The cavalrymen would then hold this line until relieved by the infantry. Later, if possible, they would push further forward to Flers and Eaucourt l'Abbaye to cover the infantry's deployment for a possible advance on the third position on the next ridge. As conceived, 2nd Indian Cavalry's advance was not a prescription for massed sabre charges, a headlong rush for Bapaume or anywhere else, but a measured scheme to take quick advantage of the enemy's momentary disorganisation once the second position was captured. Haig himself cautioned Rawlinson against the over-hasty deployment of the cavalry, except for a few squadrons to seize High Wood if it was undefended.[112] The error of placing the cavalry under the control of one corps commander and its objectives in the zone of the other was certainly "a sure recipe for confusion."[113] But not for disaster: the fact that rapid exploitation of success did not come off owed as much to bad luck as to bad judgement.

After receiving news of the break-in to the second position, Congreve ordered the cavalry to move forward at 07:40. But the congested roads from Morlancourt, where the cavalry was bivouacked twelve miles behind the front, the maze of unfamiliar enemy trenches to be negotiated, and the churned-up ground all slowed their move forward, albeit not as much as some accounts suggest. The leading brigade were ready to go forward from the old British front line by 09:30, but it was held there because the situation ahead was unclear. When occupation of the wood was finally reassigned to 91st Brigade, cavalry squadrons were tasked with supporting this advance on the right flank. The horsemen fulfilled this role well enough, belying the casual assumption that they were "soon dealt with by German machine gunners."[114] Richard Holmes has clearly demonstrated that one oft-quoted first-hand account of cavalrymen "gallop[ing] up with their lances and with pennants flying, up the slope to High Wood and straight into it . . . horses and men dropping on the ground, with no hope against the machine-guns . . . it was an absolute rout," is false.[115] Nevertheless, such vivid imagery has become an accepted cliché for all that was wrong in British military methods, "precisely what we expect to hear."[116] Crossing the unoccupied western face of Delville Wood was certainly a risk: the troopers came under fire from a machine gun there. But, alerted by a reconnaissance aeroplane, the cavalry squadrons charged and cleared enemy outposts between High and Delville woods. Sixteen unfortunate Germans had been speared by the lancers in the process, and many more casualties were inflicted by the cavalry's rifles and machine guns. After this, they dismounted and dug in, holding a line to the flank until the infantry relieved them. The two cavalry regiments engaged suffered only eight killed and fewer than a hundred wounded on 14 July.[117] As a mobile exploitation force they had accomplished their more limited mission, but they would ever after be categorised as an antediluvian relic that had no place on the industrial battlefield.

HAIG HAD PROMISED President Poincaré that there would be a success on France's national day, and he had kept his word.[118] He finally had his chance to crow to Foch: "He is very pleased with result of our attacks. French openly said their troops could not have carried out such an attack, not even the XX Corps!"[119] But this was scant compensation for the amateurishness of the past two weeks, which the French had regarded with clear disdain. The principal British problem was that they did not use their artillery to conserve their infantry in the French way. Des Vallières reported: "Their infantry is excellent; they take villages with grenade attacks! Yes, but several well-placed heavy shells would prevent heavy infantry casualties. Their morale is fine; they smashed two German divisions which counter-attacked yesterday and today."[120] When Joffre learned that British shell production was only one-quarter that of France, his reaction was predictable: "One is surprised that an industrial nation like England has been unable to do much more. The English preoccupation at the start of the war with commerce and seizing German markets undoubtedly lies behind this state of affairs."[121] Napoleon's "nation of shopkeepers" remained a bane for France, even in alliance! They were so disillusioned with the British performance that G.Q.G. even contemplated assigning French staff officers to British higher formations and artillery brigades to show them how to organise operations correctly.[122]

Such frequent and trenchant French criticisms of British methods inevitably found their way back to G.H.Q., and naturally they did nothing to ease allied tensions.[123] But they were valid concerns.[124] Nevertheless, the Battle of the Bazentin Ridge, as the British dubbed their 14 July success, stands in marked contrast to 1 July and its follow-up operations. Fourth Army was, belatedly, through the second German position, the last of the well-prepared and heavily wired defensive systems that they had faced on 1 July. But it had cost another nine thousand casualties,[125] and that was not the end of the German defences. In the fortnight's grace that the reorganisation of the British army and the slow grind towards the Bazentin Ridge had allowed, the third position had been strengthened and wired, and intermediate lines and redoubts created on the rear slopes of the Bazentin heights. Pozières and Thiepval remained in German hands overlooking the British left flank (although Ovillers would finally fall to Reserve Army on 16 July). On spurs at the other end of the ridge the villages of Guillemont and Ginchy could serve the same function, restricting the British right flank and commanding the ground over which the French would have to advance against the second line adjacent to the river. The time needed to garrison the third position adequately, and deploy artillery and machine guns to protect it, had been won. Meanwhile, the allied attacks had shrunk—the multi-division battles of the first few days were now smaller, brigade or battalion affairs. This combination of stiffened defence and diluted attack would give the next phase of the battle a rather different complexion from its first fortnight.

These two weeks had been a further period of "varying fortune." There were marked successes, particularly south of the river and on 14 July. The German defence had been determined if ill-organised: the hastily thrown in and hopelessly inter-mixed regiments had delayed, but not halted, the allied advance. The French army's early momentum had slowed, and not entirely due to the enemy's resistance, although their ally's temporary hiatus now seemed behind them. But their soldiers' morale remained high, if 79 RI is any example:

> The Somme was a striking contrast with Verdun. In place of persistent defence, of repeated violent German assaults in which we took blows without being able to return them, came a glorious series of attacks. Neither the regiment's skill nor spirit had disappeared, and 1 July was its apotheosis. All were aware of the victory and morale had reached a very high pitch. "We'll have them this time."[126]

A second major shock to the defence, the smashing of the second position north of the river, suggested that the follow-up plan was working. But a universal truth of industrial battle was gradually reimposing itself on the fight, snuffing out any residual hope for a decisive victory. The machines, the guns, were taking control, and the men were once more burrowing into the earth for protection. Fayolle neatly summarised the nature of attritional battle:

> The actual character of the war. Artillery and shortage of munitions . . . The art of war has disappeared. Mechanical means. The importance of airpower. The start of aerial warfare. The German artillery blinded. The troops unaccustomed to manoeuvring. They are numbed by trench war and their outlook is distorted, particularly the artillerymen, who will not cross over the smallest ditches.[127]

He would do his best to guide his troops through it. Now that the machines had taken over the battlefield, the next six weeks would be a much sterner test than the previous fortnight's tooth-and-nail encounters, in which at least human courage, intelligence and fortitude had counted for something. The attritional battle would slow to a crawl.

The Battle of Attrition: August 1916

UNDER A HOT July sun, Private Henry Freeman, a stretcher-bearer in I Anzac Corps, marched south "through the most beautiful country imaginable . . . Wild flowers of every colour and description [were] intermingled with the almost golden corn stretching for miles as far as the eye can see broken only occasionally by a few trees."[1] For a fortnight the great allied offensive on the Somme had been the main topic of conversation along the line, and now the time had come for Australia's sons to do their bit. The glory of a Picardy summer only momentarily cheered the footsloggers, who fell into a more reflective, resigned mood as the rumble of the guns gradually grew louder. Eventually the famous leaning Virgin of Albert came into view. They had arrived. "May it fall quickly!" Freeman would write after a month carrying the wounded from the Pozières battlefield, and his best friend's death.[2]

The Australians joined Reserve Army, and made the acquaintance of Lieutenant-General Sir Hubert Gough, probably the most discussed and vilified British Western Front general (after Haig, at least). "Goughie" had just stepped up to army command, and was to direct the operations on the left flank of the Somme battle in 1916–17. He was commanding there again when the full weight of Germany's March 1918 offensive fell on his Fifth Army. He would also play a predominant role in directing the Third Battle of Ypres, the notorious Passchendaele offensive, the Somme's main rival for bloodiness and battlefield horror. Gough was intelligent, quick-witted and charming, a popular man in the army, both confident and courageous. But he also had all the usual attributes for a public hate-figure. For a start, he was a cavalryman, a "thruster" who, as one of his divisional generals declared, "with the true cavalry spirit, was always for pushing on."[3] His fellow army commander, Rawlinson, certainly baulked at Gough's "Hoorush tactics and no reserves, as they are not sound."[4] Weak-willed subordinates did not last long in his command. This endeared him to Haig, but not to those generals and troops who had to carry out his determined offensives. Indeed, his rapid advancement from brigade to army command owed much to Haig's favour and patronage; and something perhaps to the fact that he was the younger brother of the talented Johnnie, the valued chief of staff whom Haig had lost in 1915. At forty-four years old, Gough was young for an army commander. In July 1916 he was still learning his trade.[5]

The Australians arrived on the Somme in the middle of Gough's push for Pozières, the village that covered the highest crest of the Thiepval Ridge and the northward extension of the German second position. It was an "extraordinarily fine defensive position . . . dominating the whole of the approach up twelve hundred yards of valley which gently rose to the village of Pozières itself."[6] Its capture would cover the left flank of the push forward from the Bazentin Ridge, and outflank Thiepval from the north. New Army divisions had pressed the line forward over the previous days, and a four-day bombardment, "one of the heaviest yet seen on the British front in support of an attack by a single division,"[7] had turned the village into another shell-smashed rubble heap. The 1st Australian Division was brought forward to deliver the *coup de grâce.* As the newcomers awaited their fate,

> we had one occupation of which we never tired. We would sit on the parados of the trench and watch our shells burst over Pozières. Shells of all calibres were used. Shrapnel beautifully timed, huge HE shells, and one shell in especial, a shell that burst with a sheet of flame towards the ground made the life in the German lines a perfect hell. Little did we think that in a few days we would be occupying that position under a similar bombardment from the enemy.[8]

The men sang songs—soldiers' standards and bush ballads from home—and told jokes to keep up their spirits as the interminable hours of waiting rolled on. Whatever happened, they would do their duty:

> though every hour brought more certainly before us the uncertainty of the future, yet this new realization of the instability of existence once over the top did not lessen our desire to make

good or shake our knowledge of the fact that absolute success would be ours. All it conveyed was this: on the morrow when the success had been attained, some of us would not be there. It did not affect our will to do or die. It did not detract one iota from the dash of the charge. It simply gave us knowledge and new thoughts—that was all.[9]

At 00:28 on 23 July, the assault battalions left the saps they had dug forward over the previous days and formed up in no man's land. A shattering barrage hit the German front line. Covered by the rolling shell-fire, the infantrymen advanced into

a terrific thunder storm on a pitch dark night. In the lightning flashes one caught glimpses of phantom figures some with rifles at the slope, some with them at the high port, heads held high in the air striding through the hell that surrounded them . . . the noise overhead apart from the bursting of the innumerable shells recalled the swish of the wings of countless thousands of birds flying above. So closely did the shells seem to move, so great was the weight of metal passing in either direction, that one involuntarily wondered why one barrage did not crash into the other.[10]

There was little opposition; the barrage had obliterated all in its path. The final objective, the line of the Albert–Bapaume road, bisecting the pile of bricks and stones that had once been the village itself, was soon secured. During 23 July the Australians dug in and sent out fighting patrols to occupy the rest of the village. The inevitable counter-attacks were beaten off: 86 RIR was destroyed by concentrated artillery and machine-gun fire on the afternoon of 24 July. The Australians would keep Pozières.[11]

We in the fullness of our conceit and the depth of our ignorance congratulated ourselves on the wonderful success we had made and were not slow in saying that the magnitude of the victory was out of all proportion to the number of casualties. We were young and had much to learn. The next 60 hours taught quite a lot about attacks and the aftermath thereof.[12]

When his infantry could not retake the village, General von Boehn, IX RAK's commander who had just taken over the sector from Arnim's IV AK, cancelled the counter-attacks and determined to make the village untenable through constant heavy shelling.[13] This incessant artillery fire crushed the Australians' spirit. It was supposedly the most intense bombardment endured by them during the war and cost them far more casualties than the first assault;[14] more, indeed, than they had suffered in the whole Gallipoli campaign.

> The class of fighting on the Somme is an eye-opener to all of our men. The intense artillery of both sides tend to unnerve the very best. Curtain or barrage fire was entirely new to oldest of our soldiers. The idea is to establish an impassable wall of steel and shrapnel either in front of our men advancing or behind the country attacked so as to prevent reinforcements coming up. We know from personal experience what German barrage is like. Our own must be terrible as I believe we fire three to the German one.[15]

"This was quite a different kind of fighting to Gallipoli, and one was liable to stop something any minute," remembered Private William Holford: "Had a pretty rough time of it what with whizz-bangs, HE, shrapnel, 5.9s and so we did not know which way to turn and at night time it was always worse for Fritz put them over thick and fast and the marvellous part of it all was that more did not get hit for bits were flying everywhere."

Holford's best mate was killed, and he eventually stopped one himself, a far from unwelcome "Blighty wound" in the ankle that earned him eight weeks of attention from the charming nurses in a Birmingham hospital and a passage back to Australia.[16] Such was the luck of the draw on the fire-swept Bazentin Ridge.

For the next six weeks, the three Australian divisions in I Anzac Corps were to be rotated through the Pozières sector, mounting a succession of battalion- and brigade-strength attritional assaults—nineteen in all, mostly failures—which advanced the line a few hundred yards at a time towards the Mouquet Farm Redoubt ("Mucky Farm" to the Australians), which covered the northern approaches to Thiepval. It was intensive, short-range fighting, for the defenders contested every inch. The strain on men fighting in such a confused engagement was extreme:

> We have all or rather what is left of us been through hell and seen this war and all its attendant horrors. As I look back and try to recall all that has happened I see nothing but blood and dead and

dying men. War—it's nothing but mechanical slaughter. Perhaps a man in all his prime is talking to one, next minute he is in eternity.[17]

When the Australians' tour of duty was over, Mouquet Farm was still in enemy hands. In the meantime, the embryonic Australian army had had its martial spirit sucked out amid the Somme maelstrom. Over 23,000 Anzacs had fallen to secure a square mile of ground (alongside an equally large number of British).[18] The men who tramped the road back to Albert "were very quiet when they came out of the fight— subdued and lifeless—they wanted only to be left quite quiet, alone—to lie and read and write a few letters home."[19] Reflecting on his recent ordeal, Private Athol Dunlop confided to his sister Florence: "Anyway I'm proud of being an Australian and can say without boasting that as fighters they have no superiors and damned few equals."[20] They had endured rather than conquered. But, paradoxically, they were emerging heroes, the Dominion's worthy sons. Thereafter, myth would do duty for experience as the Battle of the Somme started to reshape, and be remade by, those who passed through it. Pozières would become the sacred ground where Australia's divergence from her English heritage took root: the Somme's all-smothering clay would cement the colony's developing sense of national identity.

In a dugout near Pozières, as German shells fell all around, Charles Bean conceived the idea of an Australian War Memorial. Something positive would come out of the Somme. As it took shape in peacetime, the memorial "told a story of test, ordeal and finally triumph."[21] The Australian soldier's spirit of endurance and fortitude, comradeship and individual bravery were the vigorous motifs of national identity taken from the Somme. These central elements of the ordeal were to be enshrined in Bean's official history of the Australian Imperial Force (AIF), "conceived . . . as a solemn and elaborate memorial to the men who fought."[22] The memorial's architects chose to illustrate the AIF's battlefield experiences through a series of dioramas, so that the citizens who stayed at home could get some, albeit plastic, impression of what their fighting men had seen and endured. The Somme is represented by a Lewis-gun team holding an isolated and muddy shell-hole outpost on the "Pozières Heights." Four men fight on, amid the bodies of six of their comrades. Sergeant Twining and his comrades of the 48th AIF Battalion are guarding a flank against counter-attack. "They are deter- mined that the vital ridge, won at the cost of great sacrifice by the men of the 2nd Australian Division and held by their own 4th Division, will not be given up," the explanatory caption reads. "They will remain in their post—alive or dead."[23] There is no sense of a bigger strategic or operational picture in this depiction of the Somme, of the progression or import of the Great War's central battle or of the significance of the fight for Pozières. The diorama represents instead a moral victory, a human sacrifice redeemed by the nature of the ordeal and the soldiers' noble response to it.[24] This should come as no surprise. In the years which followed, the Battle of the Somme was to assume a discrete identity, developing a cultural memory and iconography detached from events on the ground, representative of its centrality as well as its singularity; and not just in Australia. But there is always some generic basis in reality. This was an actual event, even a typical and expected one, recorded by Bean in his official history, which commemorated the individual efforts of Australians as much as the collective endeavour of the AIF.[25]

However, this history was also the basis of "fables, containing morals on which a nation ought to base its ideal behaviour."[26] In jotting down his own "war memoirs" seventy years later, Private Raymond Membrey recounted his Pozières experience. Before being wounded and captured, he had fought with his five comrades in a Lewis-gun team near Mouquet Farm. His sergeant "copped it in three minutes," then another man. With no leadership, the survivors set up their Lewis gun in a shell-hole and fought on, holding their isolated post for two days as the gun team was whittled down. Eventually, only the wounded Membrey was still alive: "I could have retreated at this time but decided to shoot it out, that's the job I was here for, and the funny thing was that I never felt afraid." The next day, with his Lewis gun out of action, he fought six Germans with grenades, accounting for four before being wounded and passing out. Left for dead, he crawled from shell-hole to water-filled shell-hole for five days, hoping for aid. He was nineteen years old, "only a boy," as the German who recovered him exclaimed.[27] Perhaps he was a true Australian soldier; but his memoir was equally suggestive of the soldier he was supposed to be, the mythical Pozières trench-rat, depicted by the memorial. Such represented a new kind of valour—"doing one's bit," fighting with and for one's mates alive or dead, stoic heroism when faced with industrialised slaughter, self-sacrifice in the noble cause—that has come to dominate the memory of the Great War.[28] "I have often wondered how I would stand it," Athol Dunlop wrote, "but after the first sickening sensation of seeing men in all stages of annihilation I got quite steady and did my bit as well as the next."[29]

There had to be scapegoats for all this, of course. These would be the British generals, those who conceived and mismanaged the offensive, who pushed the Australians into the hellfire on that exposed ridge. Bean's Somme history, published in 1929 just as post-war disillusionment was starting to set in, pulled no punches, and "had plenty in it about the ghastliness of war for the ordinary participant and the incompetence of the strategists who sent him into battle."[30] Peter Charlton's 1986 study of the fight for Pozières epitomises a genre of high-command criticism that characterises writing on the First World War and is particularly virulent in Australia, where over the last two decades John Laffin, Denis Winter, Robin Prior and Trevor Wilson have sustained the offensive against the British generals that was originally launched by Bean.[31] The arguments about British generalship on the Western Front pre-date and are not confined to this Antipodean skein. Yet while British, Canadian and New Zealand historians have addressed, and to a certain extent redressed, this viewpoint (admittedly, the "poor" generalship on the Somme is often their reference point for improvements),[32] "down under" they seem particularly enduring, now comprising a key element of Australia's martial myth. Gough was fair game, as were the amorphous hate-figures of "the staff," "stupid overfed fat red-tabs, enjoying their cigars in front of the fire while they drowse and their heads drop over their newspapers," as Bean chastised them after one annoying encounter with the censor's office.[33] Even Lieutenant-General Sir William Birdwood, their respected corps commander who had led the Anzacs to Gallipoli, was whispered to be among the "butchers." "The men have undoubtedly had as much fighting as they want for the present. They are sick of the punishment they have had from the German bombardment. And indeed very few English officers know how heavy it has been."[34] They had fought well, but as Bean recognised, the strain of attritional battle would break the stoutest warriors eventually, and undermine their respect and affection for those whose strident promises of glory and victory were now ringing hollow in the weary survivors' heads.

AT THE OTHER end of the British front line, in the quiet, ordered glades of Delville Wood, one tree stands apart, a Somme survivor that has outlived all the other veterans. In July 1916 Delville Wood was "a thick tangle of trees, chiefly oak and birch, with dense hazel thickets intersected by grassy rides."[35] First South African Brigade, 3153 strong, stormed this wood on 15 July. Just 143 survivors marched out five days later. Over those six days the South Africans endured concentrated shelling from three sides, up to four hundred shells per minute at times, and repeated German counter-attacks from the west, north and east. Although the enemy penetrated the improvised defence on several occasions, the South Africans held on until they were relieved on 20 July. It was another national triumph of heroic sacrifice, of determination, camaraderie and courage, which has left its mark, and its myths. Ian Uys has retold the South African Brigade's shocking and stirring experiences in "Devil's Wood" for modern generations in the voices of its veterans.[36] Among the replanted trees, the serene white-stone colonnades of the South African National Memorial welcome curious and contemplative visitors, inviting them to dwell on that nation's martial endeavours, which began in that now peaceful forest.

Australia and South Africa had particular reasons to create their memorials, either on the Somme or at home. British battalions and divisions, although less celebrated, endured the same gruelling, attritional fight. They fought alongside the Colonials at Pozières and Delville Wood; in the centre of the front they sustained their own, morale-sapping battle for High Wood, worthy of a history of its own;[37] in August and early September two new struggles on the right flank, for the villages of Guillemont and Ginchy, absorbed fresh divisions as the front on the Bazentin Ridge was inched slowly towards a line from which the German third position could be observed. For there was tactical method and purpose, and strategic intention (on both sides), in this repetitive back-and-forth slogging from trench to trench, even if to the casual observer it appears aimless. Jean de Pierrefeu, the French staff officer who drafted G.Q.G.'s daily press communiqué, captured the essence of this phase of the fight:

> The struggle went on with the methodical slowness of which the Germans had shown us an example for four months at Verdun, but with a more skilfully applied use of force and more carefully thought-out manoeuvres . . . The objective on each occasion was some dominant position, some observation post which would give us command of a wide expanse of ground. Once the objective was gained, the advance slackened.[38]

If the offensive were to be renewed in strength, there must be a continuous jumping-off line within striking distance of the enemy's defences from which it could begin. Unfortunately, the methods employed by Fourth Army were unlikely to achieve this quickly and painlessly. But in an attritional battle haste and "careful and methodical preparation," G.H.Q.'s constant promptings for both notwithstanding, were moot, hollow platitudes that gave senior officers some misguided sense of authority and control.[39] There were no simple, effective alternatives for progressing the offensive, whatever armchair generals may subsequently have presumed. As Verdun had done, the Somme assumed its own metallic rhythm, as the two adversaries pounded one another, smashed into each other, like two huge pneumatic hammers striking and throwing off sparks until one or the other shattered.

THE MID-BATTLE SLOGGING match, the wearing-out phase, was as much a consequence of the effectiveness of the defence as of the limitations of offensive tactics, for the dynamic between attack and defence shifted backwards and forwards as the battle developed. By the middle of July, the German defence, initially hurried and improvised, was a much more formidable obstacle. The thin lines and scattered batteries of the early days had been heavily reinforced, and their command reorganised. Fourteen German divisions now opposed the push, backed by a strong concentration of artillery. Heavy guns in particular were now being redeployed from Verdun. If the German army did not have true initiative, at least it could now contest allied assaults, and perhaps contain the Anglo-French advance, if not reverse it. Gallwitz had now assumed responsibility for the defence. He was a German counterpart to Foch and Fayolle, an artilleryman and a thinker. Having served on the Great General Staff and in the Prussian Ministry of War before 1914, he had been ennobled in 1913 after serving as Inspector of Field Artillery. He commanded a corps on the Eastern Front, and then the army that invaded Serbia in October 1915. In March 1916 he took over the "western group" at Verdun and directed the gruelling battle for the heights of Mort Homme and Côte 304. This experience of the artillery battle prepared him for the intensified battle of *matériel* (*Materialschlacht* as the Germans dubbed it) now under way on the Somme.[40]

His first order of the day, issued on 30 July, indicated his purpose:

> The decisive battle of the war will be fought out on the battlefield of the Somme. It is to be made clear to all officers and men in the

front line how much is at stake for the Fatherland. The utmost attention and sacrificial action is to be paid to ensuring that the enemy does not gain any more ground. His assault must be smashed before the wall of German men.[41]

Since Gallwitz had no better tactical method than the same small-scale operations for which the British command have been criticised, he was to engage in a battle of wills. Even if he could have mustered large reserves, getting them across the shell-swept battlefield to mount a general counter-offensive would have been impossible.[42] Nevertheless, judicious commitment of his reserves in penny-packets—a machine-gun section here, a company there, a battalion somewhere else—would sustain a prolonged defence. Hew Strachan has suggested that "attrition came to be about the application or acquisition of material superiority."[43] The principle applied tactically as well as strategically. While overall the allies had great material superiority on the Somme, the dispositions of British forces on the Bazentin Ridge and French forces on the Flaucourt plateau in mid-July—both salients that could be fired into from three sides—gave the Germans local superiority which facilitated a prolonged, attritional defence. The British had unwittingly advanced into a natural amphitheatre. "The shape of the ground is so much in favour of the Germans," one artillery officer recorded, "there is not a single point hereabouts in our possession whence we can see their country."[44] They retained observation over the British positions from High Wood and Guillemont. When he visited the former some weeks after its capture, Rawlinson was struck by the excellent views the Germans had enjoyed over the British lines.[45] The Germans sited their defences to take advantage of natural and man-made features: ridges and spurs, reverse slopes, woods, sunken roads and tracks, quarries and natural hollows. Advantage had been taken of the time gained in early July to thicken their defences: field engineers constructed strongpoints and dug and wired more support and reserve trenches, connecting the villages, woods and ridges behind the second position into an intricate defensive web.[46] One innovation, the construction of "switch trenches" running backward at an angle from the forward lines, and designed to check the flanking moves that the allied infantry were using to pinch out defended positions, would present the allies with yet another tactical conundrum to resolve.

German defensive tactics remained based on the principles of machine-gun and artillery defence and rapid local infantry counter-attack. Such methods worked well on a small scale. Unsilenced enemy machine guns in particular had caused the British significant problems and held the defence together during the critical first fortnight of the battle.[47] Now, with a greater (if hardly sufficient) concentration of artillery, especially heavy batteries, the defence could also contest allied artillery superiority directly, as well as target communications behind the allied front. In this the weather was on the German side. Late July and late August were cold and wet. With terrestrial observation, this was not so much of a problem to the Germans, but allied reconnaissance flights were disrupted by frequent heavy rain, thick, low cloud and fog. Moreover, as new trenches and battery positions multiplied, the German defence became harder to locate. Allied maps, drawn up in the quieter days of spring, became hopelessly out of date.[48] Many of the new lines and strongpoints went undetected by allied reconnaissance until the moment when an infantry attack came under fire from them. As Jack Sheldon has suggested in his survey of the German army's defence, "It was far from a neat and tidy method of conducting a defence, but it was effective and it enabled them to maximise the experience of their men, even when their command structure was shaken by heavy casualties."[49]

The German defenders also had an advantage in logistics. Retreating, they could fall back along established supply lines into prepared positions while the allied armies had to pause in order to extend theirs into the devastated battle zone. Roads had to be repaired and light railways extended before bulk supplies could be moved forward for another general attack. On a smaller scale, the essential infrastructure for conducting an industrial battle—communications trenches, dugouts, battery positions, dumps, telephone networks and rest camps—had to be created anew. Fayolle's Sixth Army, which was fanning out northwards as it advanced and now had two corps in the front line, needed to take over the communications in the Maricourt sector from the British in order to keep its expanding front supplied.[50]

FOR THE INFANTRY, this phase of the battle meant a long, miserable grapple for woods and villages under the metal rain from artilleries competing for dominance. The British struggles are well documented. But those south of the river were as intense and prolonged, if now long forgotten. By mid-July, early French progress had been checked. While Haig had given the French a welcome Bastille Day surprise on the Bazentin Ridge, the following day the Germans gave them a rather less pleasant one at Biaches, where storm-troop companies counter-attacked with flame-throwers and re-established themselves in trenches in the eastern half of the village. On that occasion they were driven out, but over the next few days further strong attacks retook positions in Biaches, Bois Blaise and La Maisonette from the recently triumphant French.[51] The villages were to remain contested ground over the following weeks, as French and Germans fought trench-to-trench, bunker-to-bunker, in their ruins.

This did not augur well for the general attack that Foch had ordered following the check at Barleux, in a renewed attempt to push across the Santerre plateau to the Somme. I CAC and XXXV CA attacked between Barleux and Vermandovillers on 20 July. By now, the Colonials had been fighting for nearly three weeks, and continued effort had thinned their ranks and wearied the men. In contrast, five fresh German regiments opposed them. When 16 DIC assaulted Barleux for the second time, the German front line fell easily enough, after the usual intense bombardment. But when they tried to move on to their next objective the French infantry were caught in withering machine-gun cross-fire and the attack collapsed. By the end of the day, violent counter-attacks had driven the outnumbered Colonials out of most of their gains. They had suffered over two thousand casualties.[52] The Colonials were a spent force.[53]

XXXV CA, reinforced with two extra divisions and heavy artillery batteries transferred from north of the river, had been instructed to get the advance moving again by seizing Soyécourt and Vermandovillers villages and the high ground beyond. After this a new army, General Micheler's Tenth, would join in the battle, extending the offensive southwards to Chilly. However, the failure of XXXV CA's attack would disabuse Foch of the notion that the battle could be developed along a south-eastern axis now the defence had stiffened. While, as usual, the French infantry broke into the enemy's defences well enough, taking Bois Étoile and the northern half of Soyécourt village, progress was soon checked by flanking machine-gun fire and counter-attacks.[54] This marked the high point of the battle south of the Somme in July and August.

The reality was that the French advance south of the river was now in a tightly constricted pocket. Mont St. Quentin, a pronounced hill behind Péronne, was a perfect site for German heavy batteries. It gave excellent views across the river to where the Colonials sheltered in shallow, hastily dug trenches. Shells rained into their lines day and night, making movement impossible in daylight and hazardous after dark. Meanwhile, the French batteries bunched on the Flaucourt plateau were being shattered by enemy counter-battery fire.[55] Supply, particularly of water, was problematic. Two DIC began to show the inevitable signs of strain: "Fatigue, heavy losses, bad sanitation, dysentery and loss of enthusiasm" characterised their second tour of duty. The men needed relief and rest after their exertions, but none was forthcoming. Instead, they were ordered to recapture Biaches. They failed. What happened next was a harbinger of more serious events to come. One battalion of 8 RIC refused to go up into the trenches. "They were very agitated, claimed that they were tired out and did not want to be killed in order to correct someone else's mistakes." A company from 4 RIC made the same protest a few days later. A court of inquiry determined that the prolonged strain of trench fighting lay behind this mutiny. Nineteen ringleaders were sentenced to death (although all but two of those sentences were commuted by the President). The investigation at least found no links to anti-war agitation behind the lines, or any criticism of the division's officers. Nevertheless, it was a black mark on the honour of the Colonial Army, and a worrying development for G.Q.G. Victory was clearly still some way off, and growing war-weariness behind the lines, coupled with the possibility that it would contaminate demoralised troops, was giving cause for concern. Joffre issued instructions that the troops should be forewarned of the likelihood of another winter campaign: "The idea should be promoted that victory is certain, but may take a long time."[56]

By the end of July, a generally static defensive line had been established north of the Somme. It followed a wide S-bend, from Thiepval in the north to Mouquet Farm, curving around Pozières and along the reverse slope of the Bazentin Ridge through High Wood and Ginchy to the southern extension of the German second position at Guillemont, looping forward again around Maurepas in the French sector, and round to the river at Cléry, with a forward outpost at Monacu Farm. In the British sector this line was not to change significantly throughout August, although the front was far from inactive. It formed a "zone of friction" that absorbed many British battalions as they inched the front forward, with many German regiments contesting or counter-attacking this slow but unrelenting pressure. South of the river, Tenth Army was checked between Biaches and Soyécourt. It might be concluded that the Anglo-French offensive had been contained, even if another Verdun-style slogging match was bleeding the German army in its turn. But that was not entirely the case. Paradoxically, it was progress between these two fixed points that would get the battle moving again. One active army, the Sixth, continued to move inexorably forward during August, against and through the German second position between Maurepas and Cléry.

Fayolle's method remained the same, carefully prepared bounds

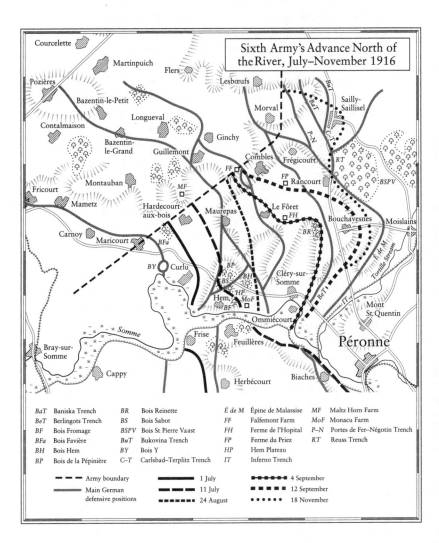

Sixth Army's Advance North of the River, July–November 1916

BaT	Baniska Trench	*BR*	Bois Reinette	*É de M*	Épine de Malassise	*MF*	Maltz Horn Farm
BeT	Berlingots Trench	*BS*	Bois Sabot	*FF*	Falfemont Farm	*MoF*	Monacu Farm
BF	Bois Fromage	*BSPV*	Bois St. Pierre Vaast	*FH*	Ferme de l'Hôpital	*P–N*	Portes de Fer–Négotin Trench
BFa	Bois Favière	*BuT*	Bukovina Trench	*FP*	Ferme du Priez	*RT*	Reuss Trench
BH	Bois Hem	*BY*	Bois Y	*HP*	Hem Plateau		
BP	Bois de la Pépinière	*C–T*	Carlsbad–Terplitz Trench	*IT*	Inferno Trench		

– – – Army boundary	■■■■■ 1 July	■◆■◆■◆ 4 September	
—— Main German defensive positions	■ ■ ■ 11 July	■ ■ ■ 12 September	
	■■■■■ 24 August	●●●●● 18 November	

within the range of supporting artillery fire. Eventually, if pursued "unceasingly, intelligently and methodically" and with vigour, he expected this succession of operations would disrupt the defence sufficiently for a more complete collapse to occur.[57] Such operations would occupy his army for two months. The *poilus* would have to fight for every metre of ground against a strengthened and remodelled defence. Well aware that French howitzers could pulverise a trench line, and their heavy mortars could smash even the strongest concrete emplacement, the defending garrison began to disperse and deepen. The front lines, the obvious target for the enemy's artillery, would in future be held thinly with outposts. Machine-gun positions and rifle-pits would be spread out in the fields behind, often covered by the summer crops that were now several feet high. To overwhelm this defence, the artillery would have to scour much more ground, or the infantry would have to fight much harder. Nevertheless, under Sixth Army's incessant artillery fire the defenders would be whittled down, making their task a desperate and ultimately futile one in the face of elite French infantry. As Oberleutnant Deher remembered after his capture at Maurepas by the *chasseurs alpins* of 153 DI:

> After midday we came under drumfire from all calibres of high explosive and gas shells . . . In a short while we suffered heavy casualties . . . my dugout . . . was quickly filled with dead and dying men . . .
>
> The enemy attacks in dense waves. I counted at least ten of them, at fifty-metre intervals. After a short but heavy pounding with mortars they quickly overran the 9th companies of Bavarian RIRs 22 and 23 . . . To our front, the first wave closed to within fifty metres, but was halted by our heavy fire, flooded to the rear and disappeared into shell-holes . . . at the third attempt they broke through . . . and, thrusting into Maurepas, exploited our left and right bringing down machine-gun fire on us and attacking us from the rear . . .
>
> Our casualties were heavy . . . Finally, the battle was being carried on by only four unwounded men . . . we were soon overwhelmed by the enemy.[58]

In late July and August, Sixth Army operations would take on a regular pattern: an advance of several hundred metres, maybe as much as a kilometre; a vicious close-quarter fight to occupy the objective and then retain it against German counter-attacks; costly small-scale line-straightening operations to retake what had not been held and secure the jumping-off line, with its vital observatories, for the next forward move; and several days of heavy preparatory bombardment before the process was repeated.[59] In this way they surmounted the series of rising and falling bluffs that rose on the north bank of the river.

A new corps, General de Bazelaire's VII CA, came into line south of XX CA for the assault on the German second position, which stretched southwards from Guillemont (where the British were bogged down and so could offer no help by rolling up the line from the north), round the large fortified village of Maurepas, to the Somme in front of Cléry.

The second position was immensely strong, hugging the opposite slope of a steep-sided ravine, and covered by intermediate trenches and strongpoints in the valley itself. And atop the opposite slope sat Maurepas village itself, from where German artillery observers could watch every French move. Sixth Army captured all these objectives during the weeks that Mouquet Farm, High Wood, Falfemont Farm and Guillemont resisted repeated British assaults.

On 20 July the first stage took place—an attack by XX CA on the intermediate line that ran along the bottom of the opposite slope of the ravine. One hundred fifty-three and 47 DIs, which had relieved 11 and 39 DIs, received a warm introduction to the Somme front. On the left, 153 DI seized their objective on the left (although a supporting attack by the British 35th Division on Maltz Horn Farm to the north was driven back). In the centre, however, 47 DI failed to carry its objective, being stopped short of the intermediate line by machine-gun fire. On the Hem plateau to the right, Colonel Maurice Gamelin commanded 47 DI's 2nd *Brigade des chasseurs alpins* (BCA) in its first assault. He had previously been Joffre's chief of operations at G.Q.G. (where he had drafted the tactical instructions that the army was practising on the Somme). Fayolle had congratulated Gamelin on the turnout of his command: "He had not seen 'a better spectacle since the start of the war.'" The brigade's ranks were filled with young recruits of the 1916 class, but they had been imbued with the *chasseur* spirit,[60] which they showed in their attack on the defences to the rear of the Hem plateau. Bois Sommet and Bois de l'Observatoire, and behind them the western half of Bois de la Pépinière, a long, thin wood running along the northern slope of the plateau, were occupied. Although the strongpoint at Monacu Farm remained in German hands, Gamelin's *chasseurs alpins* had advanced the line eight hundred to twelve hundred metres and had taken six hundred prisoners.[61]

The attack was renewed by 11 and 39 DIs on 30 July, after several postponements dictated by indifferent weather, which had compromised the preparatory barrage. This time it was a rare disaster for XX CA. Encountering the new dispersed German defensive methods for the first time, and with their supporting artillery barrage occluded by heavy early morning fog, the French assault units were caught in a devastating artillery and machine-gun cross-fire. French communications broke down—telephone wires were cut by the counter-barrage and visual signals were invisible in the fog—and the continuous assault line broke up into isolated pockets. Almost everywhere the leading waves were forced back to their start line.[62]

After a month at the front, even France's best troops (and commanders) were showing the strain. It seemed to Fayolle that the French infantry were finally losing their élan:

> The actual state of the infantry. The artillery must do everything. Engineers to the front, territorials to the rear, guns all around, in such conditions the infantry consent to mount guard in the trenches. As for going over, they must not run any risks. One can do nothing. If one attacks there are too many trenches. If there were no trenches any more they would not know where to go or where to shoot. Introducing a new unit to reinforce them upsets everything for several days ... one doesn't know how to handle them any more ... They are fed up with getting themselves killed, with no important success, no decision ... How can this war be ended?[63]

Joffre agreed, but he could offer only hollow words of encouragement. On 31 July, after two years of war, he issued an order of the day: "The moment is coming when the common effort will overthrow German military might ... Soldiers of France! You can be proud of the work you have already done; you are determined to see it through to the end: victory is certain!"[64] More practically, if unnecessarily, he advised Fayolle to maximise artillery support to save the infantry.[65]

With eighteen divisions under his command, and two separate battles to control, Fayolle seemed overstretched. Operations south of the river were entrusted to General Micheler, who also received a new corps from GAN reserve, II CA, to insert between the Colonials and XXXV CA. Two new corps on the French front substantially increased the artillery available for the contest against the enemy's batteries, which was becoming the focal point of the battle.[66] This aspect of 30 July's failure would not be repeated.

During August, Sixth Army's advance resumed with more success. Between 1 and 11 August the remaining German strongpoints on the Hem plateau were reduced one by one. After a three-day grenade fight, 79 RI completed its successful tour of duty by seizing and holding Lipa Trench in the centre of the intermediate line, while other key positions, including Monacu Farm, were occupied by other regiments.[67] A set-piece offensive followed on 12 August, "a brilliant advance."[68] Apart from a few isolated sections of trench that held out until mopped up during the next few days, all the German second position from the western half of Maurepas to the river fell into French hands. Gamelin's *chasseurs* did their bit, entering the second position south of Maurepas. Only on the extreme north, alongside the British who were attacking Guillemont once more, had the German defence held. Two further attacks, on 18 and 24 August, mopped up the rest of the second position.[69] Sixth Army was now in a position to assault the next German intermediate line between Le Forêt and Cléry (although XX CA had exhausted itself in the process and would have to be relieved first). Behind that was the third position on the plateau north of Péronne, and then open country.

While the French were making steady progress (except at the point of junction with their allies), the British were still dashing themselves to pieces against the intermediate defences on the Bazentin Ridge.

Anglo-French cooperation was certainly strained as a result of this, although it has been overplayed as a cause of stagnation in the offensive.[70] While the responsible divisional and corps commanders did their best to regulate their attacks at the tactical level, to coincide infantry assaults and integrate artillery barrages, experience had shown that such inter-army cooperation was always rather fraught. Congreve's meetings with Balfourier tended to be long and difficult.[71] And if a joint operation failed there was always an obvious scapegoat to blame. Moreover, with Rawlinson and Fayolle trying to integrate these joint assaults into a bigger plan at army level, local arrangements often collapsed due to revisions of the general plan or postponements owing to poor weather or late reliefs.[72] It did not help that when concerted Anglo-French attacks did take place the British almost invariably failed: "Their assault tactics are infantile," Fayolle groused.[73] Thus it was only at the junction point of the two armies between Guillemont and Falfemont Farm that the German second position held out until September. But this blockage had little impact on the relative progress of British and French forces on either side.

More significantly, repeated British failures revived earlier complaints that the British army was not pulling its weight,[74] which affected the principals' discussions about the broader shape and development of the battle. As might be expected, there was miscommunication and disagreement.[75] Undoubtedly at this point Haig was in an unenviable (if familiar) position. Criticisms of his conduct of the battle by politicians,* pressure for his continued and better conduct of the battle from his allies, the need to teach his troops how to fight this new sort of battle, and the need to provide the wherewithal to fight it effectively all vied for attention and resolution during a turbulent month. For Joffre, 14 July was a short-lived positive element in his ally's otherwise lacklustre performance. Instead of constant partial attacks, which were costly and produced no real result, Joffre wished to resume a general offensive, on the scale of 1 July, against the third position north of the river: "Once again one must lead the British, whose offensive is dying due as much to confusion as to lassitude," he grumbled.[76] This new combined offensive would have two contiguous stages. The first would clear the intermediate defences from Thiepval through Mouquet Farm and High Wood to Ginchy, with the French taking their southern extension between Le Forêt and Cléry. Then a set-piece attack would carry the third position between Grandcourt and the river, clearing the way for an Anglo-French advance in open country against Bapaume and Bertincourt. This would reinvigorate the stalling offensive, which would be integrated into the General Allied Offensive that would take off again when Romania declared war on Germany's allies in mid-August.[77]

Haig was not opposed in principle, but this time he was determined not to launch a large-scale operation before his army was ready. No doubt he had the unfortunate consequences of the last time he gave in to French pressure firmly in mind. Joffre badgered him to bring forward the date of the next British attack, but Haig stuck to his guns. His men had demonstrated that they could beat the best German troops, but they had yet to build up their full strength,[78] so Haig repeatedly stated that he would not be ready to attack before September. In the meantime, the Fourth and Reserve armies would continue small-scale operations in order to establish a jumping-off line for the renewed offensive, and, Haig noted hubristically, would "do all possible to bring the French on my right forward into line."[79] Joffre blamed the British government, who were starting to get anxious about the scale of casual-

---

*See Chapter 8.

ties in the offensive, for Haig's dilatoriness.[80] Haig, conversely, had little idea of how the French were progressing slowly but surely on his right flank. He interpreted one intelligence report, suggesting that the British were wearing out German divisions north of the river at three times the rate of the French, as a sign of French inactivity, taking no account of the fact that Sixth Army was fighting on a front one-quarter the length of the British.[81] But, to his credit, he seemed to be grasping the material basis of industrial battle and was determined not to attack on a large scale until his divisions had adequate artillery support.

To his discredit, though, in the interim he pressed his army commanders to maintain "preliminary" or "subsidiary" operations (as they were dually classified), to gain "small posts held by the enemy within easy reach," pending the "crisis of the fight" that might arise from mid-September. Haig pronounced that this was the "wearing-out battle," but breakthrough might yet occur.[82] Evidently, he adhered to the broader operational conception that had inspired his original plan. In doing so, he condemned his troops to weeks of gruelling small-scale trench fighting. For this, his detractors have castigated him, although his two army commanders, in trying to interpret his confusing and contradictory instructions in a fluid battle situation, certainly share responsibility for the flaccid and unimpressive performance in the battle for the Bazentin Ridge. A skilled, determined defence should not be discounted, but no defence was likely to break when individual battalions were thrown in on short sections of the front against single trenches or villages.

Even on the few occasions when the attack was coordinated along the whole line (it was never possible in August to arrange major British attacks to coincide with Sixth Army's pushes), faulty timing or inadequate artillery support compromised the chances of success. While the Australians were storming Pozières on 23 July, six British divisions were attacking piecemeal against the Switch Line along the rear edge of the ridge between 22:00 on 22 July and 03:40 on 23 July. These discrete attacks were easily defeated.[83] Delville Wood finally fell on 29 July, allowing Fourth Army to turn its attention to Guillemont village, against which XIII Corps and its replacement (from 10 August) XIV Corps smashed its divisions throughout August. High Wood blocked XV Corps in the centre, while on the left III Corps was held up by Intermediate Trench east of Pozières. It was a similar story in Reserve Army's sector, where in more than a month of fighting, II Corps and I Anzac Corps managed to advance the front line no more than half a mile towards Thiepval and Mouquet Farm.[84] III Corps finally secured Intermediate Trench on 30 August after four days of intermittent close-quarter bombing fights.[85] It was the biggest prize Fourth Army had to show for a month of effort.

THE BRITISH OFFICIAL HISTORY states rather blandly of these weeks: "There was undoubtedly room for improvement in method, both as regards the employment of the infantry and the concentration of artillery, trench-mortar and machine-gun fire."[86] In recent years, historians have started to talk of a "learning curve" followed by the British army on the Western Front. Such a regular parabola is probably too simplistic a conception to characterise a complex, up-and-down dynamic, and some historians even dispute the idea's veracity, but it would be churlish to say that the British army learned no lessons from its first big battle on the Somme. It began a developmental process, albeit a jerky and sometimes uncertain one, that turned an army of enthusiastic civilian amateurs led by a cadre of professional but over-promoted senior officers into the skilled fighting force that was to play a major role in the final defeat of the German army. The main lessons acknowledged and absorbed related to the use and utility of individual arms (in both meanings of the word) on the industrial battlefield and their effective combination into what today would be termed a "weapons system." From this, the "combined arms" methods that the British army was practising by 1918 developed.

On the Somme in July and August 1916, the basic principles were being grasped for the first time. The army staff was well aware of the many tactical problems that the fighting troops had encountered, and it was anxious to address them. Who better to teach them than the combatants themselves? In mid-July, Fourth Army operations staff circulated a questionnaire to its divisional commanders, initiating a process that they repeated at intervals during the battle. Simple lessons were being absorbed: the tactical use of the various infantry weapons, how to mop up effectively, and how to deal with isolated enemy machine guns, for example. Appropriate attack fronts and formations—"a series of waves followed by supports in small groups"—were clarified, as were defensive arrangements for captured trenches, woods and villages. More complex techniques, such as how to utilise the supporting barrage effectively, were also being developed, reflecting, if not directly modelling, French practices: "the infantry must learn to be patient, and to go slow and sure," moving forward at a steady walk with a final rush at the instant of assault; they should follow the barrage closely and the timetable should be determined by the infantry's needs. Staff work too was to be improved, as were battlefield communications and supply arrangements.[87] The basic tactical principles of trench fighting were fairly obvious with a little practical experience: as Major-General Shea of 30th Division put it, "Battering of his trenches, shattering of his morale, the most detailed orders well understood, simple assembly, good direction, and a short distance over which to assault."[88] If properly absorbed, Shea concluded, such methods should produce "a first class fighting machine."[89] As new divisions constantly came into the battle they had to learn appropriate methods. As late as mid-August, Rawlinson reported that an attack on Guillemont by the recently arrived 55th Division had failed because:

> the Boches remained in their dugouts until our leading line had passed over them . . . there were not enough clearing-up parties to guard the exits of all dugouts and the Boche therefore came out and gained possession of the front line trench behind our front line. I told the conference that next time we must go slower—follow the barrage closely and ensure that all dugouts are cleared out before we continue the advance.[90]

The other problem identified was that most senior commanders were acting in roles in which they had little practical experience. Haig scribbled irately on one message telephoned from Fourth Army about the confused situation in Delville Wood, "I do not think this indicates sufficient method. In order to win *the first essential* is a Sound System of Command."[91] Certainly many errors and mistakes were made by corps, divisional and brigade commanders on the Somme, and a number were relieved of their commands as a result. Only by such a process of trial and error, however, could the talented rise to the top in the short time available and the expanded army develop an effective system of devolved operational command. Certainly, on the Somme, some of the British army's untried corps commanders proved themselves talented soldiers. Horne and Congreve, whose endeavours were crowned with success on 1 July, were judged "first rate and full of enterprise."[92] In August, Congreve went home after falling ill. (The death of his son in Longueval on 20 July while in action under his command could not have helped his health, although Walter's posthumous Victoria Cross, which matched his father's, must have been some compensation.) It took some time for Lord Cavan, who took his place, to master the mechanics of battle himself, but he was starting to earn Rawlinson's confidence and respect by mid-September.[93] He and Horne both rose to command armies by the end of the war. Others, however, such as William Pulteney of III Corps, the only 1914 corps commander never to be promoted, were consistently poor in carrying out operations.

The fundamental lesson was that this was *matériel*-intensive warfare, and to fight it effectively the tools of the industrial battlefield had to be properly employed. Certainly, the new French light machine guns and 37mm trench guns were proving their worth in the infantry battle, providing immediate firepower to subdue points of resistance.[94] Joffre doubled the order for the former in July,[95] and French infantry battalions were reorganised with integral machine-gun companies, with twenty-four heavy and seventy-two light machine guns per regiment.[96] But the hand-grenade was becoming the principal weapon of trench combat. Waves of grenadiers led the assault, and whoever won the grenade fight would usually keep the disputed trench. Too often in the British sector this was the enemy. The Württembergers of 27 ID held Guillemont against repeated attacks for most of August, using both "grenades and cold steel. Wherever the British were in the rubble of the old village they were finished off in hand-to-hand fighting."[97] Britain's battlefield education came at a high price. Joffre noted at the end of July that the British had suffered 130,000 casualties to date (compared with 24,600 French in the first three weeks); by the end of August, the figure stood at 251,000 (compared with 65,000).[98]

While infantry operations remained the focus, the true defining feature of the industrial battle of *matériel* was the contest between the opposing artilleries. Once the German defence had been organised, it was to be Verdun all over again. "The principal obstacle is no longer the defensive positions of the enemy, but his artillery," Duval noted on 10 July. "Preparing the attack therefore consists of a methodical artillery struggle."[99] Throughout July and August, Fayolle, the former professor of artillery, bombarded his subordinates almost daily with his own salvo of instructions on how to conduct the artillery battle more effectively (some of which were passed down second- and thirdhand from GAN and G.Q.G.): saving ammunition by shortening barrages and targeting them more effectively; reducing wear and tear on the guns and exhaustion of their crews; proper observation of the effects of the bombardment; improving communications between ground and air by using wirelesses; long-range firing; rapid bombardment of targets of opportunity—with a few memos about infantry assaults and improving staff work thrown in for good measure.[100] For Fayolle, effective counter-battery fire was the most important, and most deficient, element in the French offensive system. Over these months the principle that neutralisation—with gas shells to incapacitate gunners and lightning bombardments to force defenders to ground while the infantry attacked—was more effective than trying to destroy everything in the path of the advance was being assimilated. Whether this amounted to a "learning curve" in action or an over-rigorous system of micromanagement from the top down, the French army was certainly not slow or inefficient in noticing the day-to-day nuances of ongoing battle and implementing tactical responses.[101]

If the enemy's artillery could be mastered, the infantry's task was so much easier and less costly. Oberleutnant Deher recollected:

> Until I was captured, I fired an almost unbroken stream of red flares calling for defensive fire, as well as aiming from time to time directly at the swarms of aircraft that came at us from all directions . . . Altogether I fired off the contents of two full sandbags (approximately one hundred rounds). During the entire attack our artillery failed to fire one single round.[102]

Deher was acknowledging the one new dimension of warfare in which the allied armies had a real advantage, and which was essential for making progress while the enemy held the observatories. The Somme was an offensive that depended on air superiority, a new tactical concept. Allied air forces needed to control the skies above their own lines to prevent enemy incursion, and be dominant behind the German front to allow their own observation aeroplanes and balloons to spot unmolested for the artillery.

In the competition for airspace the allies had more aircraft and a more aggressive strategy, which offset the tactical supremacy that the German Fokkers had enjoyed since the second half of 1915. At Verdun the French had developed the offensive tactics suggested by Major-General Hugh Trenchard, commander of the British Royal Flying Corps (RFC): aggressive fighting patrols to engage with German fighters, harass enemy observation aircraft, and bomb and strafe behind the German lines, rather than passive defence of their own vulnerable spotting aeroplanes or their own side of the line. This raised the intensity of air fighting to a new pitch that was costly in machines and pilots for both sides, but vital to ensure that the artillery could do its job. The RFC adopted the same policy over the Somme battlefield. From March, fighting patrols had been actively contesting airspace to enable photographic reconnaissance behind the enemy's lines, while a linear screen of patrols confined the Germans behind their own front. New aircraft—the manoeuvrable de Havilland DH2 and the Sopwith One and a Half Strutter, the first British aircraft armed with a machine gun synchronised to fire through its spinning propeller—gave British pilots a much better chance of defeating a Fokker in a dogfight, and allowed them to dominate the slow German reconnaissance machines.[103] French pilots, and some of the British squadrons, flew Nieuport fighters, which matched the Fokkers in all but armament. In July 1916 the British deployed elements of 17 squadrons, some 185 aircraft, to support Fourth Army. Sixth Army had 201 aeroplanes. Most German aircraft were concentrated over the Verdun battlefield, so there were only 19 fighters and 110 reconnaissance aircraft in Second Army.[104]

Although German airmen tried to fight back, it was a one-sided battle. "More German machines were seen yesterday than at any time previously during the battle," Haig noted at the end of July, "but not more than 20 crossed our line, and none got very far . . . On the other hand we made 451 separate flights over the Enemy's front two days ago, and some 500 flights yesterday."[105] It was an exhausting routine, with pilots making many sorties each day; and dangerous, for as well as dogfights, anti-aircraft fire and mechanical failure were constant threats. Above the Somme, the first British "aces" made their names. Lieutenant Albert Ball began his battle by shooting down a German observation balloon on 25 June, and followed this up with two German observation aircraft on 2 July. He had thirty-one victories, the Military Cross and the Distinguished Service Order and bar, and the Russian Order of St. George, 4th Class, to his name by October.[106] The French redeployed their best fighter squadron, Escadrille III, from Verdun. Among its pilots was Lieutenant Georges Guynemer, "the winged sword of France," who was then the country's leading ace. A four-squadron fighter group with hand-picked pilots was built up around this nucleus: Les Cigognes— "The Storks"—commanded by Captain Félix Brocard of Escadrille III, the squadron that had first borne that enduring emblem.[107]

War in the third dimension was a vital part of the battle, subsequently captivating because it appeared to present a residual chivalry when compared with the mechanical killing of the ground war. Ball reported one dogfight in which both he and his opponent ran out of ammunition without a hit: "There was nothing more to be done after that, so we . . . flew side-by-side laughing at each other for a few seconds, and then we waved adieu to each other and went off. He was a real sport that Hun." But chivalry went only so far. One of his comrades remembered Ball as a "skilled and dedicated killer with no other motive than to use his machine and armament to shoot down enemy aeroplanes."[108]

However, aerial warfare was not all duels between knights of the air. Tethered balloons continued to be targeted, to keep the enemy blinded: new electrically triggered incendiary rockets launched from an aircraft's wings made them easy prey. Raids would be launched against German aerodromes to destroy enemy machines on the ground and the infrastructure that supported them. Railway yards, trains and tracks, headquarters, ammunition dumps and base camps were all targets of opportunity for the airmen's bombs. Guynemer and his comrades in Escadrille III took a particular malicious pleasure in shooting up sleepy German bivouacs first thing in the morning.[109] Only red-cross-marked hospitals were spared.

As well as allowing the allied artillery to compete on fair terms with the German, air superiority posed another danger for the vulnerable infantry. Advancing in a counter-attack on Delville Wood in July, one unidentified German infantrymen recounted that

> enemy aircraft were operating all over the battlefield. Flying fairly low, they gave horn signals from time to time. They were clearly indicating the whereabouts of thickly occupied shell-holes to their artillery, which increased its rate of firing. We decided to act dead if they came our way, in order to conceal the fact that there were fourteen of us.[110]

Added to that, airborne machine-gunners liked to strafe parties of infantry on the ground. On clear days, when allied aircraft would swarm across the skies, German movement was impossible. Men had to bury themselves in the earth or remain absolutely still to escape detection and the attention of hostile artillery or machine-gunners.[111] In the next war French infantry would notoriously grumble that they never saw any of their aircraft in the skies above them; in 1916 it was German soldiers complaining about the absence or lassitude of their pilots.[112]

The positive aspect of this development was that aircraft were now helping infantry to achieve their objectives. Ground–air liaison techniques were still experimental, but they seemed to work: G.Q.G. monitored them closely, and it appeared that the methods of communication, although slow, were appropriate for the steady, methodical operations being mounted.[113] It was the tactical method of the future. At the beginning of August, Joffre placed another big order for observation aircraft and fighters to protect them, and added five thousand men to the French air force.[114] A complex interaction was developing between the various fighting arms: infantry, artillery, aircraft and even cavalry. And one more modern industrial weapon, the tank, which would complete the twentieth-century weapons system, was shortly to make its debut on the Somme battlefield.

Holding on in the face of allied *matériel* superiority and increasingly effective tactical and operational method was becoming increasingly problematic. German guns were outnumbered, overused and breaking down, and the crews were exhausted when they had not been incapacitated by allied counter-battery fire.[115] To save men while maintaining a coherent defensive system, the defence became both deeper and more dispersed. There would not be fixed linear positions, but intricate webs of trenches, thinly held but with many mutually supporting strongpoints, with good observation and taking advantage of every fold of the ground. Isolated positions would be interspersed between these trench lines, sometimes concreted pill-boxes but often only tarpaulined shellholes, distributed in such a way that the allied artillery would have to comb every square metre of ground to be sure of uprooting all the hostile machine guns. The allies' bombardment would therefore be spread more thinly, and there would be fewer obvious targets against which the artillery fire could be concentrated. However, there was an obvious response to this developing defence-in-depth—the creeping barrage, which was designed to sweep the ground in front of the advancing infantry. Fourth Army recognised this in principle at the end of July. But practical problems—particularly wear and tear on the guns that had been constantly in action for six weeks, and the close proximity of the British and German front lines, which made friendly fire casualties more likely—delayed its adoption as standard practice for some weeks.[116] Meanwhile, though, Sixth Army was already utilising the technique. In its orders for an attack on Sivas Trench on 26 August, 46 DI instructed the field artillery to "act in front of the attack in short and violent salvos against the enemy's positions as well as the zones in between (to destroy the enemy's machine guns)." The barrage would creep by fifty-metre intervals every thirty seconds.[117]

For the Germans, the idea that ground could not be given up, that each allied gain should be immediately counter-attacked, still held good, even though their attacks were on a similar company or battalion scale to those of the British and were usually doomed to failure. Among the allies' means of repulsing these counter-attacks were protective box barrages and forward machine-gun posts. In Sixth Army the mopping-up waves were given a second mission: after clearing the enemy's lines, they formed supporting lines bristling with machine guns, so that any enemy counterattack could be broken up in its turn.[118] Hence, in this continuous attrition, the German army contributed materially to its own hecatomb.

FOR THOSE WHO were sent to fight in the attritional battle, the experience would stay with them, if they returned home. A rich literature of trench warfare attests to the profundity of the experience, backed up by reams of unedited contemporary letters and diaries and later reminiscences. Films and photographs, posters and paintings allow us to view the first international war recorded and represented by modern mass media. Notably, the battle was the subject of the first feature-length documentary film commissioned by the War Office to show the British public what their menfolk were achieving in Picardy. *The Battle of the Somme* remains the most-watched film in British cinema history, more than 20 million people viewing it in the six weeks following its opening on 21 August 1916.[119] "After the ordeal you come away mentally and physically exhausted," one viewer recorded, even if having "the realities of war brought home to you" was not quite as traumatic as being there.[120] Testament to its permanent importance, in 2005 *The Battle of the Somme* was added to UNESCO's Memory of the World International Register of culturally significant documents. In the age of colour, sound and elaborate computer-generated special effects, of real-time battlefield reporting and twenty-four-hour news coverage, *The Battle of the Somme*'s silent, jerky, staged black-and-white imagery seems quaint and antiquated. Yet it allows us still to visit those sepia trenches, and we return there every anniversary.

Collectively the emergent mass media have passed a rich, vivid tableau of the trench experience, of its sights and sounds, its smells and texture, its fear and horror, even its humour and pleasure, to subsequent generations. Much of this imagery is generic, evocative of an era and a unique front-line world. The rumble or roar of artillery, the rattle of machine-gun fire, the buzz of aircraft and the clatter of tanks, the cries of dying men, the triumphal yells of their killers, echoed for four years from Switzerland to the English Channel. Rats, lice and flies furnished verminous memories, while the lingering odours of explosions, fire and death hung over that extended battlefield. Men marched in and out, through hot and cold, wet and dry, mud and dust; went into trenches, leapt over the top, floundered among the dead and dying, the maimed and the shell-crazed; trudged out or returned in ambulances crawling slowly along the congested roads that linked the dystopian world of the front with the quiet, peaceful comforts of the rear. Most of the surviving records of trench warfare offer few surprises: a soldier's preoccupations do not change much with time and place. Lucky escapes—the shell that did not explode, the bullet that whistled by—attesting that the hand of fate was on their side, reoccur. These are survivors' tales: Signaller Baurès of 153 RI constructed a whole memoir around the many incidents that defied his aunt Anna's prophecy that he would be killed at the front.[121] Records of basic comforts—a warm place to sleep, dry clothes, good comrades, food and drink (or the lack of them)—fill the pages of journals and reassure anxious kin that all is well in letters home. Changes to the numbing routine might evoke momentary pleasure: on Bastille Day "Special rations. Three extra quarts of pinard, 150 grams of ham, a bottle of Champagne for 4, a cigar. *Vive la France!*"[122] Essentially the soldiers were writing of the repetitious working of the military machine, in which they were merely the functioning parts.

Of course, the Battle of the Somme was like this. But it had its particular character too, one that its survivors brought back with them and then distilled for their relations and their readers. For the Somme stirred the voices of the educated, literate volunteers of Kitchener's New Armies. Siegfried Sassoon, Robert Graves (who was seriously wounded near High Wood on 20 July),[123] Edmund Blunden and their ilk left a rich, reflective, poetical record of the soldiers' experience, a new kind of war literature with a new tone: partly ironic, partly critical, partly celebratory, partly rebellious, partly accepting. Although both had their own celebrated war writers, France and Germany could not really match it. For Britain alone the Somme would be a civilians' battle, an aesthetes' fight, an auditorium as well as an arena. Much of this war literature relates to the world beyond the Somme, and to the shattered relationship between the front and the rear. But some themes distinguished this battle from others. The first thing which struck those who deployed there was its immensity: "From all reports the fighting is on a tremendous scale and that means a lot. How many thousands must fall daily?"[124] That it was war on an industrial scale, with industrial resources and managerial methods, was not lost on its components. Arriving south of the Somme in late August, Captain Humbert observed the panoply of combat:

> observation balloons, munitions dumps, light railways, camps of prefabricated "Adrian" barracks; altogether indicative of long preparation, a business mounted deliberately like an industrial effort; nothing improvised; every possibility anticipated; the front seems to work like a large factory, following a plan no one can derail.[125]

"It certainly was a revelation," Lieutenant Harry Crookshank of the Grenadier Guards scribbled on arrival, "and shows that we really have begun fighting now."[126]

The effect of this war machine on the landscape was equally striking, even to those who had seen Verdun. "From Maricourt we advanced into a shattered landscape. The old German front line had been entirely levelled by our shells, but in odd places reinforced concrete shelters survive. We try to enter them, but a dreadful smell drives us back. The bodies of the occupants . . . are rotting away."[127] Most descriptions of the Somme landscape dwell on the impact of intense battle on quiet countryside, describing the villages reduced to rubble, the shell-tossed earth, the criss-cross of abandoned, broken trenches and wire entanglements, the fly-crowded cadavers, weapons and military stores littering the ground. Comparisons with the undulating surface of the sea, or later the moon, were often made: French bombardments in particular stripped the landscape bare. After a pounding with ten heavy shells per second, the slope rising behind Maurepas to the Bapaume–Péronne road, across which the French infantry would have to advance in September, was "utterly bleak, not a trace of green, just overlapping craters."[128] Such sights soon became unremarkable. A cluster of red-and-blue-clad skeletons, French infantrymen killed by a shell in 1914 and still unburied, was more worthy of note.[129] But throughout these accounts of broken men and earth there persists the perception that reality is not far away. As Alan Brooke noted as he sat in an observation post overlooking the "brown desert" beyond Pozières, "Each time I go up into that bit of ground I feel more than ever that it must all be a dream and cannot be true."[130] To its veterans, the Somme battlefield was an aberration against nature, a nightmare on the edge of reality.

Captain Oscar Viney remembered that after he was wounded on the Pozières heights, "I lay in the sun, there were lots of wild flowers and the birds were singing. It all seemed very incongruous."[131] The satanic war-made scar, gouged by shells and scratched with trenches, contrasted with Picardy's bucolic landscape, which was just out of reach, temporarily left behind or soon to be regained after one last effort. Because of the topography of the battlefield, every time a shattered ridge was conquered there was the invitation of green countryside beyond; a mirage of unspoiled places and happier times. Until that promised land was reached, the metal hell had to be endured.

Unsurprisingly, since the Somme was above all an artillery battle, gunfire defines the soldiers' experience. To observe it was awesome: "It was an arc in which all the individual sounds became one, an immense steel curtain gliding from behind, surging inexorably forward without cease or gap," noted one observer of a French preparatory barrage on 11 September.[132] To be under it was terrifying: the title of the celebrated German writer Ernst Jünger's best-selling war memoir, *Storm of Steel*, pinpointed the centrality of bombardment for the infantryman.[133] The German soldiers called it *Trommelfeuer*, literally "drumfire," the constant percussion of exploding shells, shredding the nerves if they did not shred the body. It was the same for both sides, as the Australians had discovered at Pozières. To witness and survive the competing bombardments unleashing all their fury against a short section of the line was breathtaking, fear and elation mixing in an adrenaline-pumping combination. "Towards evening enemy artillery fire of all calibres increased to drumfire," German artillery liaison officer Richard Sapper remembered. Expecting a British attack, he emerged from his dugout in Guillemont to find that

All hell had broken loose up there! Shells were landing everywhere with deafening crashes, enveloping the collapsing houses with their black clouds of smoke, boring down into the ruins, ploughing through heaps of rubble and sending showers of rock and pieces of metal flying into the air. Shrapnel shells burst with sharp cracks and flashes of light, smashing tiles with their showers of lead. Beams began to burn and above the flames countless white flares lit up the sky, their light casting grotesque silhouettes of the smashed and ruined houses . . . red flares were fired above the white ones calling for defensive fire. Angrily our field guns opened up to our rear and our mortars and howitzers joined in, hammering and banging away. Now at last the whole orchestra was present. I listened with glee to the whistling, roaring and gurgling of our shells which crashed down amongst the British. Great work you gunners. Fire! Fire! Give them everything you've got![134]

Even when there was no attack in progress, the gunners kept up a nerve-racking irregular or steady intermittent firing. Guillemont was subject to the attention of a British "Big Bertha," a 15-inch naval gun that crashed a shell into the ruins metronomically every four to six minutes, shaking everything while Sapper and his comrades waited for the one that would eventfully decimate their dugout.[135]

Such reminiscences relate to the infantryman's experience, which predominates in military anthologies. The artilleryman's participation, while somewhat safer behind the lines except when the counter-battery contest was raging, was equally important for the outcome of the battle. However, gunners' memoirs, being generally turgid, repetitive accounts of fire missions and gun movements, do not make for exciting reading.[136] Amid it all there was a struggle to retain the vestiges of humanity. The lighter side of life, the small absurdities of war, stand out as worthy of note while chaos and death were stalking. Captain Viney, waiting to go over the top near Pozières, had received a reprimand for riding his bicycle without a rear light when in camp on Salisbury Plain in June.[137] It seemed a long time ago, and a long way away.

THE FINAL, HIDDEN dimension of the attritional battle was the struggle for moral ascendancy, less measurable than numbers and rates of casualties, but vital for assessing how the contest was going. Military morale is a delicate balance between mood and spirit (partly *esprit de corps*, partly a deeply rooted patriotic sentiment), the former a fluctuating commodity reflecting day-to-day experiences, the latter a longer-term attitude to fighting and the war. Soldiers are notorious for grumbling: Napoleon's *"grognards"* had their counterparts on the Somme. There was certainly much to moan about in the battlefield routine, the rain, the mud, the fatigues and the uncertain supply situation. Nevertheless, the *poilus'* spirit remained good, as Gaston Lefebvre attests: "Our morale was a lot better than at Verdun, for except when the ration carriers were killed, we could eat when we needed to. Moreover, we felt that here we led the dance, not the German as at Verdun."[138]

In time, however, lack of clear progress threatened to undermine the spirit of the French infantry. When the offensive started, Jacques Play-oust had been willing to bet that Péronne would be liberated by late July, and Lille by January 1917. But by late August, when his time came to leave his quiet sector and move into the inferno, his enthusiasm had waned:

> "Attack attack" for 3 months. Many of us will not see it through. We keep hoping. It would not be so bad if it was the last effort that would be required of us. Gad, people can't imagine the life that we are leading. At time the morale gets pretty low. What we would give for a broken leg, arm now. Many would sacrifice either of them to be finished once and for all. I suppose it is right that the lives of millions should be sacrificed for the general welfare. We are all insignificant units. But from a personal point of view it is atrocious, unfair.[139]

Nevertheless, his belief in the broader success of the allied offensive and a final victorious outcome to the immense battle of wills persisted:

> I still hope to see the end of the war this year. I doubt that the Germans will be able to bear the strain of this general offensive. The Russians are exceeding expectations. The French have not said their last word. The English are putting up a grand fight at last before a very tough opposition. And the Italians are moving too. I think that if I'm still alive and kicking in 6 months I shall have still years to come and annoy you with my hair-raising war experiences.[140]

British confidence was certainly growing as the fight went on, even if Rawlinson's wishful mid-July anticipation of "the great battle of Bapaume" was rapidly deflated.[141] On 10 August, Fourth Army circulated this encouraging message to its troops:

> The general situation has never been better. The Russians continue to gain successes and the enemy's resources are strained to the utmost to withstand the pressure on all fronts. This must be impressed on all ranks and the fact that we are for the moment held up on some parts of the front must not be allowed to cause any slackening in our efforts or weakening in our offensive policy. Determination to win, discipline and a careful study and

application of the tactical lessons of our recent successes will lead to victory in the early future.[142]

Whether such high-command platitudes stirred or washed over the ordinary soldiers stuck on the Bazentin Ridge is impossible to tell. Nonetheless, through taking on the enemy, front-line units were building an *esprit de corps* for the first time, which meshed with the greater skill at arms of blooded troops.

Across no man's land, attrition seemed to be undermining German morale. Constant shelling and fatigue could not but induce war-weariness: "I hope the cruel war is approaching its end, for one no longer has any heart for this sort of thing," one German soldier wrote home.[143] Jünger later wrote that his company approached the battle in a state of great excitement: their morale was high, and they were determined "that positions entrusted to them would only be lost when the last defender had fallen. And that indeed proved to be the case."[144] However, resolve to fight to the last quickly weakened. As early as 17 July, re-emphasising his order to hold positions to the last man, Below had to issue a supplementary injunction against voluntarily abandoning positions without an enemy attack. Any officer who did so would be court-martialled.[145] Nevertheless, through July and August positions were lost incessantly, however stout the resistance, in "a war of *matériel* of the most gigantic proportions."[146] As were the men defending them: Jünger's company had only five of its personnel from the start of the year left by the end of 1916.[147] "We can't really call this war any more, it's murder," one Bavarian war diarist remarked of the fight for High Wood.[148]

Morale had deteriorated rapidly since mid-July, when a German officer, called on to surrender, had shouted back, "I and my men have orders to defend this position with our lives. German soldiers know how to obey orders. We thank you for your offer, but we die where we stand."[149] By the end of the month, Haig was pleased to note, captured officers of the elite 5th Brandenburg ID were declaring, "We are beaten."[150] "German prisoners generally do not seem sorry to have been captured. They all state that they have had enough of this war," Private Freeman noted. It was a sentiment with which he sympathised.[151] On 2 August, after thirty-six hours of continuous bombardment, the commander of a guards unit holding Chancelier Trench between Belloy and Estrées raised a white flag and led a party of survivors over to the French lines.[152] By late August, German deserters were coming over in large numbers, particularly Saxons.[153] All this was not lost on G.H.Q., whose intelligence staff factored evidence of declining German morale, collated from captured enemy documents, into their assessments as the next general attack was prepared.[154]

The material impact of attrition on German manpower reserves was also monitored closely. During the battle's first two months, forty-nine German divisions passed through the Somme sector, almost as many as had fought at Verdun, and it has been suggested that losses in those two months exceeded those of six months' fighting on the Meuse.[155] Allied intelligence assessed German casualties in July as between 130,000 and 175,000.[156] By the end of August, that figure had risen to 200,000, which was actually a significant underestimate: after the war the German official history acknowledged 243,129 casualties in the first two months.[157] The heart seemed to be being torn out of the old army, while its material and human resources were also stretched. G.H.Q. analysts noted a dispiriting report from one soldier's home town: "You should have seen the recruits who were mustered this week; it was like the boys coming out of school, but they have all become soldiers; it hardly seems possible." Older untrained men, "until now unfit for active service, and those who have become unfit during the war," were also to be mustered, "Germany's last hope."[158]

That hope would be marshalled by two new military overlords, Paul von Hindenburg and Erich Ludendorff, because at the end of August, the Battle of the Somme claimed its most exalted victim. Falkenhayn's star had been waning in imperial circles for some time. Gripped in the Entente's tightening noose, the Central Powers seemed to be running out of opportunities. Romania's declaration of war was the final straw that prompted Falkenhayn's resignation (he was sent to the new Romanian front, where he redeemed himself in a field command). His long-standing critics and rivals from the Eastern Front were handed the poisoned chalice of final victory. It was a move of desperation, not of optimism. In his post-war memoirs Falkenhayn reviewed the first two months of the Somme. Because the enemy's progress had been slow, and a skilful attritional defence had been conducted north of the river, he judged that so far the battle "had had comparatively little influence on the further course of the war."[159] Perhaps the German army did win this attritional battle, and Falkenhayn's own strategy was thereby vindicated. However, the Battle of the Somme was not yet half finished.

. . .

BETWEEN MID-JULY AND the end of August, the battle fragmented into a number of separate, gruelling fights. But they remained part of a whole, the wearing-out phase of an industrial battle, problematic but necessary. Both sides suffered heavy casualties, but German morale weakened more than that of the Anglo-French forces. Each monitored the other closely for signs that resolve was weakening: that the German front would break; that the frustrated allies would come to the negotiating table. It was the middle-game, in which the pieces were being manoeuvred for slight advantage.

In the allied camp the false optimism of the early weeks had started to temper, if not dissipate. Joffre's second scenario—of a long, hard-fought campaign—had certainly proved more prophetic. Overall, however, the strategic situation appeared "excellent." The General Allied Offensive had seized the initiative and gripped the Central Powers. Austria-Hungary seemed to be fragmenting internally and falling out with Germany; and Romania was finally going to join the Entente camp, increasing pressure on the Eastern Front. Tentative German peace feelers were also starting to be put out.[160]

Hope of decisive success on the Western Front before the end of the year had not yet been abandoned. In September, a renewed powerful general offensive might yet rout the weakened enemy. Nevertheless, as the offensive slowed, Joffre's political troubles began to mount once more, and he started to look around for scapegoats.* The British were obvious candidates, but he could do nothing about them but moan and cajole (his journal records that he did a lot of both). His subordinates could also be targeted. Fayolle was his usual mordant self, and seemed to be dragging his feet. Joffre considered replacing him with General Humbert,[161] and should he have done so Fayolle would not have objected. Fayolle in turn was at loggerheads with Foch over his step-by-step methods, even though these had been delivering the only appreciable progress during August. "The crafty devil . . . I think he is jealous, he wants to take credit for the victory on the Somme . . . what good did he do, is the artful detail of the offensive battle due to him?" Fayolle griped after a huge row between the three French principals at his headquarters on 18 August.[162] He and Foch simply did not trust each other, Fayolle concluded. Yet only Foch, who seemed (as far as was possible) to have the British in hand and a proper grip on the battle, could be trusted to push things on quickly and with sufficient mass and energy to realise the commander-in-chief's grander intentions.

*See Chapter 8.

Although the battle of attrition continued to work both tactically and strategically in late July and August 1916, there seemed no solution to the operational impasse. With the right method, the allied infantry could make progress, but it was slow and costly. Concentrated artillery fire and a longer attack front speeded it up and reduced casualties, but more was needed for a significant improvement in the moribund offensive's fortunes. Otherwise more powerful heads might start to roll. Such was Foch's challenge for the second half of the battle. Closely watching events unfold and musing on their nature, he was still playing the long game. General Hely d'Oissel, commander of VIII CA, discussed the offensive's progress with him when he called at Dury in early August. Foch expected the "great Battle of the Somme" to continue into September, and to intensify. He explained the "new method" of attack he was using (Fayolle's method!) — taking one German position, moving forward his artillery and then taking another, on a six-to-eight-day cycle. In such a way he had pushed deeper than previous offensives, and he intended to continue his "succession of battles." Fresh army corps would be cycled through the battle—each could mount two advances before it was worn out, Foch had concluded—and he expected VIII CA would take its turn in due course.[163] In this system Foch had found an operational method that could push back the enemy's defences slowly but surely. But there remained two problems. Each successive forward move left behind it a devastated wasteland, so restoring the communications infrastructure and redeploying guns after each step reduced the advance to a crawl. Coordinating the operations of two allied armies was also time-consuming and frustrating. Foch had an alternative: not one big thrust of all four armies under his direction, but a coordinated series of blows, army by army, with each forward push destabilising a portion of the defence and facilitating the advance of the army on its flank, while not allowing the enemy the opportunity to concentrate his artillery fire.[164]

"Repeated actions," Joffre noted, "without let up, against a weakened enemy, in such a way that he has no time to establish new defences."[165] The scale, rather than the split-second timing, of the joint offensive mattered most. In September, such lateral exploitation, which included the extension of the battle to north and south, would get the Somme front moving again; and, as Foch had always anticipated, would push the defence to breaking point.

Despite their rivalry, Fayolle would be Foch's sword arm. During August, Sixth Army had demonstrated its effectiveness, pushing on steadily while Fourth and Tenth armies were held up. Foch and Fayolle's differences abated as the day approached for renewing the general offensive.[166] Both could take a share of the credit if the new plan worked. The epicentre of the battle had shifted to its centre, where the allies' most effective army was driving inexorably towards Foch's original objective: to cross the Bapaume–Péronne road and outflank defences in front of Péronne. But the French military machine was suffering wear and tear, and every effort had to be made to win the battle before it broke down. If the British war machine worked, it would be a bonus, but that could not be guaranteed, so responsibility would fall on Fayolle's world-weary shoulders. As his army had pressed forward, his front had expanded, such that his was now the largest and best-armed formation. He had three corps in line between the British left and the river (I, VII and his old XXXIII CA), while three more (V, VI and XXXII CA) were at Foch's disposal to exploit the next Sixth Army push. With the British held up on the Bazentin Ridge, and Tenth Army making little progress south of the river, Sixth Army would have to break the German centre if there was to be any chance of a significant advance astride the Somme. Fayolle realised his command was on the line.[167] But he was soon to justify himself and his methods in spectacular fashion.

Behind the Lines

O N 4 J U LY, F R A N C E ' S Senate, parliament's upper house, met in
secret session to debate the conduct of the war. Three weeks earlier, the Chamber, the lower house, had done the same. "Perhaps
parliament will have the good sense to understand that at such a
momentous time washing dirty linen, even behind closed doors, is
criminal," Micheler had complained. "It is a sad business."[1] In the
Chamber on 16 June, Deputy André Maginot, one of Joffre's most
trenchant critics, had railed against "uninterrupted partial offensives
which result in no appreciable change of strategic importance, but only
in murderous losses."[2] The filibustering response by General Pierre
Roques, the War Minister, which outlined events at Verdun over two
mind- and posterior-numbing days, showed that industrial war was far
too complicated a business to be reduced to simple formulae. The government won this round, being backed by 440 votes to 97. After the
successful start of the offensive, the Senate vote of confidence in
Briand's government, 251 to 6, accompanied by "acclamation" of Joffre, was even more overwhelming.[3]

The political barometer in France fluctuated wildly, governed by
success or failure. In July 1916 it peaked: Fayolle's advance on the
Somme had momentarily nonplussed those baying for greater parliamentary control of the high command and the war effort. However, as
Esher warned from Paris, "The position would become a very difficult
one in the event of a check; the French being as you well know, subject
to violent revulsions of feeling when their hopes are disappointed." In
that case both Briand and Joffre would be vulnerable.[4] As the offensive
continued even Briand, until now a steadfast supporter of the man he
had raised to be the director of Entente strategy, started to question its
utility and management. "War is a great evil," he confided to Esher in

late August when he heard the details of Joffre's and Haig's latest contretemps, "but it is an even greater one if one leaves it to the soldiers."[5] Such political engagement, interference, fallout and judgement of the Somme offensive were simply further manifestations of the intricate links between events at the front and those in the rear.

ON CLEAR, WARM summer nights the citizens of Amiens could walk out to the hill at Saint Acheul east of the city to watch the "fireworks" in the distance, a concentration of gun-flashes and flares resembling the aurora borealis.[6] Not everyone enjoyed the front-row seats of the Picards (or suffered the disruptions of the khaki- and horizon-blue-clad hordes who camped in their fields and towns). The rest of France, the British Empire, Germany and her allies and fascinated neutrals were reliant on the carefully controlled news from the front emanating from both sides of the line. The First World War was a popular war (in both senses of the word) and its coverage in the emerging mass media, particularly in print but also on film, was something that had to be perfected in order to promote and sustain public support for the war effort. By the time the armies came to grips on the Somme, journalism was changing from a means of conveying official (and carefully censored) information to a news-hungry public to a way of engaging that public with the actions of their men in uniform. If not yet fully formed into the propaganda that blighted later decades, the carefully managed reporting from the front was starting to assume some of its characteristics.

From the start of hostilities G.Q.G. would issue a daily official communiqué, summarising the main military events. The German army did the same. But to sustain public support for the war, more was needed. At G.H.Q., which had issued only cursory official accounts of the day's fighting during Sir John French's command, the importance of conveying a fuller and truer picture of a war in which civilians as well as soldiers were deeply engaged was becoming apparent. Under Haig's command, John Charteris would assume responsibility for the army's media relations as well as its intelligence effort. The first "press officers" were media managers in khaki uniforms.

Neither Joffre nor Haig liked the press. Apart from the obvious threat that reports from the front would leak important military information to the enemy, opinion uncontrolled by headquarters might undermine the commander-in-chief's authority. Nevertheless, as the front stalemated, Joffre started to appreciate that to maintain popular support for the army and the war, France's citizens needed to know something of the activities at the front. Since the desire for information and the curiosity of journalists and newspaper proprietors could not be prevented, it might at least be managed in the interest of the war effort.

G.Q.G. knew the value of good news, especially after four and a half months of morale-sapping reports from the Meuse front. When the scale of the French army's initial victory astride the Somme was realised, Joffre ordered that the "early successes should be very prudently announced." He judged it a good idea to spread the news over several official communiqués, to maximise its positive impact on popular morale.[7] A long-drawn-out summer campaign was just beginning, and good news might not last. France was entering her third year of war, with the enemy still encamped on her soil. This made the patriotic journalist's task easy. The fundamental fact that France was engaged in a war of liberation from the invader, whether it be by defending Verdun or taking back villages on the Somme, provided a simple, overarching theme for French war reporting: the German army had invaded France and *la patrie* would have to be liberated. Yet public support, like military morale, fluctuated in both mood and underlying spirit. While the wish to liberate France would sustain her citizens' patriotic spirit as the offensive went on, they still needed reassuring as shifting fortunes of war unfolded.

Now that the long-awaited counter-offensive was finally under way, it was important to convey the impression that France and her allies were winning, without sounding overconfident. *La France Militaire* took care to analyse the solid statistical basis for the effectiveness of attrition. The war was entering "a new epoch, in which the situation of the adversaries has been reversed, and the advantage passes to our side." Germany's new divisions were her last, and composed of second-class men, while the ranks of her allies' armies were clearly thinning.[8] It was only a matter of time. But that period of time would be longer than anyone expected or wished, and popular morale would have to be sustained in the interim.

Close observers of France in the summer of 1916 confirmed that, despite the gruelling nature of the ordeal, the army and the nation remained committed to the war, to driving the invader from French soil, even if the anti-war activities of the radical left suggested that national resolve was beginning to fracture. Captain Parker, America's official military observer, who joined a tour of neutral military attachés to the Verdun sector in late August, identified the "spirit of absolute confidence and of determination to fight to a finish ... evident every-where." The belief that the allies were now in the ascendancy, at Ver-dun, on the Somme and on allied fronts, prevailed. France and her allies were growing stronger while the enemy was clearly weakening.[9] The counter-offensive was not cracking, but reinforcing Frenchmen's resolve to win.

JULY 1916 WAS a hopeful month for the allies. Their armies were making progress on the Eastern, Southern and Western Fronts. The onslaught at Verdun had been contained. Even the British troops seemed finally to be fighting. Joffre's determined, focused leadership since December 1915 had been vindicated. At this point in the war the allies' political leaders were powerless to intervene to revise the broad parameters of strategy. Joffre must be allowed his chance, Haig his opportunity. But impotence did not suit or satisfy the politicians. If the Chantilly strategy was inviolable for the moment, then they could at least try to find fault with operations and tactics. If the soldiers had emasculated their political function, then they might yet tell them how to manage their military one. The politicians in London had much to ponder during the first weeks of the offensive. Unlike France, Great Britain was not yet used to long casualty lists. The import of Haig's earlier warning to Kitchener—"I have not got an Army in France really, but a collection of divisions untrained for the Field. The actual fighting Army will be evolved from them"[10]—was becoming appar-ent, and Kitchener was no longer there to defend the commander-in-chief. The reality of intensive industrial battle baffled those politicians who had mobilised the empire's young men for the fight. Many minis-ters and Members of Parliament had sons fighting in France. Some suf-fered personal loss. The Prime Minister's eldest son, Raymond, and Paymaster-General Arthur Henderson's eldest son, David, both fell on 15 September. Cabinet Secretary Maurice Hankey's younger brother, Donald, died on 16 October. This added a private justification to Han-key's long-held hostility to the offensive.

The shock of the early losses on the Somme was as difficult for the empire's leaders to absorb as it was for its citizens. There was a natural assumption that something had gone wrong. Momentarily, of course, it had; but battle on such a scale would have produced heavy casualties in any case. The steady attrition that followed the initial shock set in train another slow attritional campaign, waged by politicians in London against Haig's conduct of the offensive. It was the C.I.G.S.'s constitutional role to advise the War Committee on military affairs. In the early weeks of July, Robertson found himself stuck uncomfortably between "the God of War and the Mammon of Politics" as he mediated between G.H.Q. and the Cabinet.[11] He believed, essentially, that "Armies are very human in their nature, and like to feel that they are being ordered to fight by one of their own kind who knows and understands them. Men object to being killed by amateur strategists."[12] This overriding belief that the soldiers should direct the war, and maintain a united front against ill-informed, meddlesome politicians, placed Robertson in a difficult position as news of the heavy losses of early July permeated public consciousness. He had been in France for the start of the battle, and had received full and worrying reports of the disaster from some of Fourth Army's divisional generals. Moreover, the published casualty lists would soon indicate to the wider world that the offensive was going to be a costly affair. In his heart Robertson was sympathetic to such concerns. He believed in a "step-by-step" approach to offensive operations on the battlefield. In a phrase worthy of Fayolle, he suggested to Major-General Sir Launcelot Kiggell, Haig's chief of staff, "It is not a question of men but of artillery until such time as we get through the defences."[13] But it was neither his job nor his wish to tell Haig how to conduct his operations, even if personally he feared that the commander-in-chief had not adapted to the changed principles of attritional warfare. Nevertheless, for Robertson, defending the deployment and use of British recruits in France, there remained a genuine concern that Germany might still win by having the better manpower policy.[14]

This was, however, a longer-term strategic problem. In the short term Robertson had to explain the opening of the offensive and justify its continuation. He was careful not to offer prospects of quick victory to his War Committee colleagues: "I have said that you are getting on very well, but that it would be a slow business," he reassured Haig.[15] But as July passed and the detail of events became clearer, Robertson found that criticism of the high casualties and limited progress was starting to grow. At the Cabinet table Haig had already earned the epithet of "butcher."[16]

Esher, though, judged that Haig's troubles arose from too narrow-minded a perception of strategy. "Can you prevent people at home from fixing their eyes upon the Roll of Honour and gluing their noses to the map?" he pleaded to Hankey. "That is the way, I understand, in which the war is being looked at through club windows. Pray heaven that the outlook of Ministers may be wider!"[17] One ex-minister and ex-soldier was willing to take his criticisms out of the club room and into the highest corridors of power. On 1 August 1916 the Attorney-General, F. E. Smith, presented a memorandum to the Cabinet on behalf of his friend Winston Churchill. It was a forcefully argued critique of the first weeks of the Somme offensive, founded on the premise that the British army was suffering substantially more casualties than the enemy. This basic premise was sound, and borne out by history when the true facts of Britain's early battles on the Somme became public knowledge. Yet Churchill's was exactly the sort of naïve approach to strategy that Esher decried: "In *personnel* the results of the operation have been disastrous; in *terrain* they have been absolutely barren . . . from every point of view, therefore, the British offensive *per se* has been a great failure." The attempt to break the German line and return to "manoeuvre" warfare was not going to work.[18] That, Haig had explained to the Cabinet already, was not the method or purpose of the offensive.

The Cabinet knew Churchill's personality and his amateurish interference in strategy well. At the time they were mooting an inquiry to consider his responsibility for 1915's Dardanelles fiasco: distracting attention at the crisis point of the war "merely to satisfy the vanity of Members of the House," in Esher's opinion.[19] Not surprisingly, the Cabinet dismissed Churchill's memorandum out of hand, and rightly so. Even its sponsor, Smith, in the covering note under which he submitted it, admitted that he was "by no means wholly in agreement with his standpoint, thinking, as I do, that he underrates the importance of our offensive as a contribution to the general strategic situation."[20] Repington, who had visited G.H.Q. in early July and whose tales from the front may in part have inspired it, advised Churchill that "[i]t was a good paper . . . but I did not see the utility of it at this stage. I thought that we were bound to attack and to suffer in the general cause."[21] So did Asquith. He did not pass Churchill's memorandum on to the War Committee, which considered an evaluation by Robertson instead. The Prime Minister was not interested in the maverick opinions of an isolated and discredited ex–Cabinet minister with a grudge, who had been offering determined, if lone and ineffectual, opposition to Asquith's conduct of the war ever since returning from the front.[22] Although at the time his views counted for little, so convinced was Churchill of the validity of his case that he chose to reproduce his memorandum as the central element of his assessment of the Somme in *The World Crisis*.[23] Thus were the parameters for future analysis of the Battle of the Somme established. Ninety years later, Churchill's arguments were still being revived and rehashed.[24] By this means Churchill made himself central to the Somme, and to the conduct of the war of attrition, even though he bore no share of responsibility.

Churchill's untimely and inconsequential intervention in the conduct of the offensive was a minor incident in the increasingly volatile British political climate of summer and autumn 1916. Nevertheless, it betokened the development of renewed controversy over strategy, as well as hostility to the persons and methods of government that were associated with it. The roots of this controversy were well established, and Churchill's maverick intervention was merely an uncontrolled outburst of a more general malaise. In February a gathering of like-minded souls had taken place in Flanders: Churchill, Lloyd George, Smith, Unionist leader Andrew Bonar Law, and Canada's official correspondent at the front, Max Aitken.[25] They were united by their hostility to Asquith's languid conduct of the war: reigning "supine, sodden and supreme," in Churchill's judgement.[26] Over the course of 1916 such opposition would develop, attracting other disgruntled and marginalised figures to the cause: military has-beens such as Sir John French and Admiral Jackie Fisher, and political chancers such as Henry Croft, who resigned his command to resume his seat in the Commons on the prompting of another self-seeking officer, Lieutenant-General Sir Henry Wilson.[27] This motley group shared a desire to see the strategy and policy of the war more firmly directed from London, as well as a contempt for "Squiff," as Wilson cruelly nicknamed the *bon-viveur* Prime Minister. While operations on the Western Front were a part of this complex strategic mosaic, wider issues of mobilisation and manpower, directive institutions, grand strategy and alliance relations, and of course personality all fed into the volatile mix.

The Western Front offensive was actually the least controversial question of strategy. The War Committee had debated, endorsed and re-endorsed the Chantilly strategy before the offensive started, and while all the allies were actively pursuing it, Great Britain could not break ranks without serious, possibly fatal, harm to the coalition. France could not be abandoned. Even Churchill acknowledged the vital need to "come to the succour of our superb Ally with an Army which grows increasingly in strength and power as our latent resources are realised, and becomes a support for all the losses and exertions to which she has been put."[28] By summer 1916, no one would have disputed his pronouncement that Britain was in the war to "the last man and the last shilling."[29] However, that men were not being needlessly wasted and that money was being well spent were fair causes for concern. Towards the end of July, the War Committee threatened to summon Haig home to present his case for continuing the offensive. Haig demurred, feeling that Robertson must take a share of the responsibility and cover his back.[30] So, on 1 August, Robertson presented a full report to the War Committee, in which he put the Somme in its wider strategic context: "I said that we are now engaged in quite a new kind of warfare and that decisive results could not be expected in 24 hours nor 24 days, and that relentless pressure on all fronts was the proper course to pursue and was promising good results by the Winter." German casualties were large and growing. The "impatient" apparently stood corrected. There was no discussion of Robertson's report, and "so far as they are concerned I should say that they are thoroughly satisfied."[31] Haig's own assessment a few days later "pleased [the War Committee] very much indeed."[32] It endorsed Robertson's strategic summary, and indicated his determination to sustain the offensive until bad weather or lack of reserves forced him to stop. And he offered no quick victory: "It would not be justifiable to calculate on the enemy's resistance being completely broken by these means without another campaign next year."[33] The War Committee enjoined Robertson to inform the commander-in-chief "that he might count on full support from home."[34]

The C.I.G.S.'s "damnable ... swines"—Churchill most prominent among them—were now "in bad odour."[35] At least Haig now seemed less profligate in his use of manpower, and more determined to maximise the *matériel* that backed his offensive. In that respect the War Committee's meddling had been beneficial. There would still be casualties, of course, but wastefulness would be reined in as far as possible. Moreover, British casualty rates—which, Robertson posited, "when judged by the standards of this war ... cannot be considered in any way excessive"—were by early August much reduced from the exceptional losses of early July, and proportionately considerably less than those sustained by France and Germany since the start of the war.[36] From then on, ministers were kept informed of developments at the front weekly, and expressed no dissatisfaction with how matters were being conducted, turning their attention instead to the peace terms they would impose once Germany was beaten. In truth, as Robertson reminded everyone, "Our military policy is perfectly clear and simple, and it has the approval of the Government, and we cannot change it every day of the week. The policy is offensive on the Western Front and therefore defensive everywhere else."[37] Since the offensive would have to continue, and victory was clearly not imminent, politicians could but maintain a watchful eye on the development of events.

THE CRITICAL PHASE that Britain had entered in July 1916 was the one for which her first two years of effort had been preparing her, the one in which the mighty imperial forces she had mustered would take on and defeat the beastly Hun. The empire's spirit was high, its mood buoyant. Britain was finally taking her proper share in the allied cause.[38] The empire would need to be united in that cause to win through its greatest trial. The War Office and G.H.Q., which shared responsibility for distributing news from the front, would take great pains to sustain popular backing for the army's effort as the battle unfolded. Early shocks would have to be absorbed, continued strain explained and justified and achievements emphasised. To do this, the media had to be closely managed, accommodated but not unleashed. By such means the information war might also be won.

G.H.Q. followed G.Q.G.'s lead when it came to managing journalistic access to the front. From May 1915, a small group of accredited press correspondents was attached to G.H.Q. (the most prominent were the authors and freelance journalists John Buchan and Philip Gibbs), and their reports would go into a pool for syndication to the British and imperial press. An American press representative, Frederick Palmer, was also included in this group, as were the Dominions' official correspondents, including Max Aitken (later Lord Beaverbrook), representing Canada, and Charles Bean from Australia. With battle getting under way in earnest, the press corps was set to expand. More accredited journalists joined the pool, and many others visited on an ad hoc basis, allied and neutral correspondents among them. Official photographers and cinematographers also went to France to record Britain's great military effort. Their task was to record "the human and personal aspects of the war which must be outside the scope of official despatches."[39] What today would be called human interest stories, tales from the front of courage and endurance, became central to the war narrative, forging emotional links with the front for those whose menfolk were fighting.[40] To this extent, the Somme was the first "media battle" in a mass-media war, in which the home front would experience the soldiers' struggle vicariously, both to improve popular understanding and to sustain public backing for the intensifying national war effort.

At heart, the British press were patriotic, and the newspaper proprietors' desire to support the army and the war effort was an asset that could be exploited. While Haig certainly appreciated the importance of strong media support (in this he was a very modern general) he had initially been reluctant to allow the proprietors to visit G.H.Q.: they would intrude at a busy time and might compromise the secrecy surrounding the coming offensive.[41] However, after the traumas of the first week, the need to rally the media and public behind what was going to be a prolonged, difficult battle won out, and a delegation of newspaper proprietors headed by Lord Northcliffe visited the Somme in the third week of July. The proprietors and editors who were welcomed to G.H.Q. proved particularly useful for promoting and defending its conduct of operations, more so as the battle prolonged. Lord Northcliffe became a regular visitor, "anxious to do all he can to help win."[42] Northcliffe's newspapers, the establishment periodical *The Times* and the popular *Daily Mail*—the latter favoured reading in the trenches[43]— were happy to argue the case for Haig's attritional strategy, even while their proprietor was turning against Asquith's dilatory conduct of the national war effort.

At the same time, the press could prove a catalyst for civil–military tensions. Although Charteris had indicated to Repington when he visted G.H.Q. that the purpose of the operation was "to kill Germans," with the local strategic objectives being a secondary consideration, to an informed observer it was clear that British soldiers were being killed in large numbers too: certainly many more than French, and even, Repington suspected, many more than the enemy, despite Charteris's optimistic figures.[44] While journalists might be valuable to promote G.H.Q.'s cause, they were also trained to seek out candid opinions and ask awkward questions. None of this would appear explicitly in the newspapers—"I am writing two articles on the battle in France, but am delayed by having to send them to Haig to be censored! What will they be worth afterwards?" Repington mused—but it would all feed back into impressions of the offensive at home. On returning home, Repington found "that people were not aware what a terrible battle had been fought on the Somme, or of our losses."[45] But Sir John French, Lloyd George, Churchill and a large number of London society were soon privy to his insights.

Northcliffe and his colleagues were among the extraordinary procession of visitors who passed through G.H.Q. in the summer and autumn of 1916. Some, such as Russia's chief of staff, General Balaieff, who lunched with Haig on 10 July, were important: "a pleasant little man, talks deliberately, but has a small head and seemed rather a mediocrity" was all Haig had to say of this visiting allied dignitary.[46] Others were self-important, such as the King of Montenegro, "a very picturesque old brigand" who arrived with a bag of "impressive-looking" medals to distribute.[47] Many were insignificant, but with enough influence to secure an invitation to the front, such as the Prince of Monaco, an old friend of the Haig family.[48] Some were suspect. The visit of Ben Tillett, "a Socialist Revolutionary" and leader of the dockworkers' union, took Haig into uncharted political and social waters. But Tillett was the right sort of socialist, "determined to beat the Germans," and Haig was happy to charm him with a photo opportunity.[49] A few were unfortunate, such as the Russian colonel who led a military mission around the front line in October and was killed by a stray shell fragment. He was possibly the Russian army's only casualty on the Somme![50] All were anxious to see the great battle now in progress, despite its hazards. Turning the Somme into a sightseeing tour gave G.H.Q. a means with which to maintain political and media support for the ongoing offensive. Some visitors went to find fault, but most returned enlightened, aware of how and why the battle was being fought, if not thoroughly convinced of its utility. Such visits proved the best means of defusing criticism of the British high command: as Robertson reassured Haig, "Everyone who returns from the front speaks in the highest terms of the good work that is being done there."[51]

Of the many distinguished visitors to G.H.Q. and the Somme front, three ranked above the rest: the King, the Prime Minister and the War Minister. George V spent the second week of August touring the Western Front. The royal visit turned out to be a powerful endorsement of the commander-in-chief, as well as an excellent opportunity to sustain public engagement with the battle. Geoffrey Malins, the official cinematographer, filmed the visit, a carefully stage-managed show of majesty, ceremony, human interest and allied solidarity, for the War Office, which released the film in October under the title *The King Visits His Armies in the Great Advance.*[52] King George was received at Beauquesne, Haig's advance headquarters during the campaign, with a guard of honour from G.H.Q.'s household troops. The Prince of Wales, then serving as a staff officer in XIV Corps, was invited to

G.H.Q. to spend time with his father: their affectionate meeting was filmed for public consumption. The royal itinerary included a visit to the shattered defences and dugouts of Fricourt, which Fourth Army had turned into a "battlefield experience" site to which visiting dignitaries, military observers and journalists could be conducted in relative safety, while experiencing a frisson of the battle taking place a few miles further forward (the enemy often obliged by sending a few long-range shells into the ruins).[53] The film showed the King inspecting the ground between Fricourt and Mametz, accompanied by the Prince of Wales, Rawlinson and Congreve. Through a telescope he observed the bombardment of Pozières in the distance. His party paused at a British grave. Like so many later battlefield tourists, the King could not resist picking up a fragment of exploded shell as a souvenir. He was cheered by Anzac and Canadian troops, a nod to imperial efforts. In a stage-managed show of populism he also visited a casualty clearing station and was filmed talking to a wounded soldier.[54]

As the King had stepped from the boat at Boulogne, he had "gravely saluted."[55] It was unprecedented for a British monarch to pay homage to France. The royal visit was a demonstration that Britain was now in the war in earnest and a sop to French public opinion, which before the battle had been carping that the British army was not pulling its weight.[56] At Fourth Army headquarters on 10 August the King decorated Fayolle and Balfourier with the Grand Cross of St. Michael and St. George. French officers who commanded batteries supporting the British offensive received lesser decorations.[57] On 12 August Haig hosted a formal lunch for France's political and military leaders. Malins's film recorded the arrival and departure of the French dignitaries. "Mr Lloyd George arrives late," one caption recorded: just one of his ungentlemanly habits that the generals were wont to criticise.[58] The King sat between Poincaré and Joffre. The only tense moment arose when the *bon-vivant* French commander-in-chief found himself faced with an impossible choice—lemonade or ginger beer to accompany his meal—for the British monarch had pledged to abstain from alcohol for the duration of the war. "Many of us will long remember General Joffre's look of abhorrence, or annoyance," Haig noted gleefully. Nevertheless, the royal visit was a great success: "He expressed himself most satisfied with all he had seen, and with the state of the Army," Haig was pleased to note. On his departure, he handed his commander-in-chief and friend the Grand Cross of the Victorian Order: "He said I had 'his full confidence as well as that of the Cabinet.' "[59]

George V's visit was a high point for both Haig's fortunes during the offensive and Anglo-French relations. By mid-August, the battle, although proving tougher and costlier than anything British arms had attempted before, was acknowledged as the empire's main endeavour; the moment when Britain's army finally engaged that of the main enemy, the point of no return. After some initial dissent, mainly caused by lack of information and understanding, soldiers, politicians and civilians were starting to unite behind the effort: the few dissenting voices remained weak and went unheeded. On his return from the front, George V captured the mood of the moment in a message issued to the troops: the combination of the army's rear services, the men and women working on the home front, Britain's allies and above all the fighting soldiers' "bravery . . . endurance . . . [and] sacrifices" would bring victory.[60]

In the first week of September the Prime Minister came to see things for himself. Asquith was smart enough to realise that he did not know how to do the soldiers' job better than they did. Thus, his visit to the Somme was designed to show solidarity, not to find fault. "Haig is I think doing well," he confided to one of his lady friends, Sylvia Henley, "sticking to his original plan and not allowing himself to be hustled."[61] Haig went out of his way to please the Prime Minister. Raymond Asquith was given leave to meet his father: it was to be their last meeting. The Prime Minister's party received the tour of Fricourt. On cue, the Germans sent over some 5.9s, which burst within fifty yards of the official party, forcing them to take shelter in one of Fricourt's infamous dugouts.[62] His inspection of the full panoply of the army in the field, including an introduction to the new tanks training for the next big attack, reinforced Asquith's confidence in Haig. On his return home, he informed his wife that he was "delighted with his visit to the front, and all he saw . . . The offensive was going amazingly well, munitions satisfactory, the French fighting magnificently."[63] He also felt that Britain's heavy losses had been more than offset by those of the enemy.[64] Asquith was no dupe, and was able to make an informed judgement on the evidence presented to him at this point. He was pragmatic about the undynamic nature of industrial battle, "a question of push, push, push [which] must necessarily be slow work," as he reported to his War Committee colleagues.[65] Haig retained Asquith's confidence to the last,[66] yet such phlegmatic pronouncements were not going to endear the Prime Minister to those Cabinet colleagues who resented his blasé style of leadership, as well as the strategy he endorsed.

Lloyd George's periodic visits were liable to prove more tricky. Robertson judged that the War Minister was vain and could be won over with flattery — "he merely requires to be made a fuss over, and put in the limelight"[67] — but Lloyd George was more inquisitive and stubborn than Robertson realised. While on the surface his first trip to the Somme front, coinciding with the royal visit of mid-August, appeared to pass off well, he clearly had a hidden agenda. Although, much to Charteris's surprise, he did not seem to be worried about the casualty lists,[68] when he visited G.Q.G. he asked Joffre to explain why the British army was suffering such heavy losses, and why the French army was making quicker, less costly progress. Joffre gave a simple, soldier's answer: "One does not make war without losses." Given Joffre's own troubles with political interference, in this at least he was sympathetic to Haig.[69] If Lloyd George did not return to London convinced, he could not set himself against the solid phalanx of generals.

By September, even Lloyd George seemed to be reconciled to the Somme, even if his war memoirs would subsequently bundle that offensive's perceived errors into a broader diatribe against the army's "grim, futile and bloody" attritional strategy and the "narrow and stubborn egotism" of its executors.[70] He visited the front again in mid-September, when the battle was at its height. On this occasion, Haig confided to his wife that "I have got on with [Lloyd George] very well indeed, and he is anxious to help in every way he can"; to such an extent that he wrote after the visit to thank the minister for his "congratulations and good wishes."[71] Even so, Lloyd George did not impress him. The visit — "a huge 'joy ride,' " in Haig's terse summation — apparently vindicated Robertson's judgement. The itinerary played to Lloyd George's vanity. He met reporters and was filmed for the newsreels, "which pleased him more than anything else. No doubt with the ulterior object of catching votes." This jaunt contrasted unfavourably with Asquith's more businesslike tour. "I have no great opinion of L. G. *as a man or leader*," Haig confessed, but Asquith possessed much the superior mind, "even in his cups."[72]

On this occasion, Lloyd George aired his concerns about the British army's relatively poor performance after a lunch with Foch (as usual, the minister had arrived late). Foch patiently explained that the French army had learned from its early mistakes "and were now careful in their advances," perhaps a bit too careful. Loyal to his comrade-in-arms, Foch praised the British artillery, and refused to comment on the skills of British generals. In his own note of the meeting he recorded that he had so impressed Lloyd George with his account of the qualities of the

British commander-in-chief that the minister determined to make Haig a field marshal![73] Inevitably, the details of this discussion quickly reached G.H.Q.: Foch himself reassured Haig that he had met and refuted the War Minister's criticisms. French and British soldiers may have had their differences and disagreements, but they could all agree that politicians were a common enemy. "I would not have believed that a British Minister could have been so ungentlemanly as to go to a foreigner and put such questions regarding his own subordinates," Haig complained.[74] On a more positive note, Lloyd George took away a high opinion of Foch's qualities, and in a rare exception after the war acknowledged his "considerable generalship, which won much ground without undue losses" on the Somme.[75] He might not trust his own commander-in-chief, but now he might rely on his French supervisor to curb Haig's overenthusiasm.

In his war memoirs Lloyd George liked to pretend that once it had become clear that the German line would not be broken on the Somme, he had urged Asquith and Robertson "that the useless slaughter ought to be stopped." In fact, as the responsible minister he could do nothing of the sort, even if he privately anguished over the heavy cost of the battle for which he provided the men and guns.[76] Only after the intensive September battles did Lloyd George start to give any real thought to diversions elsewhere. He urged redeploying military resources to Salonika, to meet another crisis developing in the East. But even then he never suggested that the Somme offensive should be stopped.

Instead, faced with united military opinion and steady if unspectacular progress in France, because Lloyd George could not influence the situation at the front, he chose to agitate at the rear. Asquith's position had been growing more shaky as the year went on. It was never Lloyd George's intention to remove the commander-in-chief and C.I.G.S. However, if control of strategy could be wrested from the soldiers, their excesses might be kept in check. By the autumn, the anti-Asquith cabal was enjoying a growing degree of support on the backbenches. If the Somme did not furnish a reason to topple the Prime Minister, its slow, costly progress was at least grist to the anti-Asquith mill.

Hankey, the Cabinet Secretary, was fond of telling his political masters: "I would sooner have a second best plan in which the soldiers believed than a best plan in which they did not believe."[77] This summation was strategic, and at this time the allies had a strategy that was almost universally supported, and which the generals were vigorously putting into practice; albeit a little too vigorously for some. Churchill's and Lloyd George's criticisms were tactical: the strategy was not being executed as efficiently as it might be in Haig's inexperienced command, and Lloyd George was certainly not above making unfavourable comparisons with the French. Only later did these tactical shortcomings become equated with strategic failure. In fact, as far as the general course of the war was concerned, Esher's suggestion that the early weeks of the Somme marked "the beginning of the end" proved a more perspicacious judgement.[78]

WHILE THE FRENCH people's spirit was deeply rooted in patriotism, the foundations of British spirit were different. For the citizens of Britain and her empire, physically removed from the fighting yet deeply involved, the issues for which the war was being fought were diverse: a combination of the personal, for their husbands, brothers and sons were fighting; the moral, for they were humbling a blustering international bully; and the imperial, for they were united in a common cause and cementing the ties of race that made Britain's empire great. It was pre-ordained and presided over by a belligerent deity, as Bishop Gwynne, the deputy chaplain general, reminded his congregation at the end of the battle: "[He] impressed on the men that this was the trial of the Empire brought about and made necessary by the former shortcomings of the race. It was and is a judgement and we are in the course of proving whether we are worthy to him or not."[79]

Coverage of the early weeks of the offensive in the domestic and imperial press was extensive and positive, and setbacks were disguised as best they could be.[80] To the encouraging stories of Russian forces advancing in Galicia, which had filled the war news columns since June, could be added reports of British forces grappling firmly with the enemy and French troops advancing across the Flaucourt plateau. The context of the Somme was all important, for the British army's contribution to the General Allied Offensive, the rubric under which the press reported the summer's operations, seemed to be modest compared with that of their allies. There was no hiding the lengthy casualty lists that accompanied the British army's effort—they amounted to thousands each day[81]—nor the relatively static lines in Picardy on large-scale maps of the front. In the circumstances, press coverage would have to be careful and balanced; the heavy loss of young British men would have to be justified. If that could not be done in terms of crushing victory, then it would have to be in terms of noble sacrifice.

Florid tales of adventure and derring-do, as penned by such reporters as William Beech Thomas of the *Daily Mail,* rang hollow in the circumstances, drawing deserved scorn and lampoons from the front line.[82] "The ordinary vulgar exaggeration of the battlefield" in the domestic press was yet another factor that would sunder soldier from civilian.[83]

For individuals at home, it was a tense summer. Families scanned the casualty lists or waited fearfully for the postman who delivered the official telegram that brought the worst news of all. There would always be a trickle of bad news; then sudden surges marked the rise and fall of the battle, the engagement of locally raised units in the national cause. Personal losses, which provoked the most intense civilian engagement with events in France, had to be rationalised in terms of the greater cause. Individual suffering and grief had to be mediated through a narrative of purpose and worthy sacrifice, of nobility and bravery amid the tragedy of battle. The belief in King and empire, patriotism and duty, the rightness of the cause and the liberal values with which it was identified provided heady motifs. Local tragedies such as Accrington's and Newfoundland's could be easily, even willingly, subsumed, for the duration at least, into a communal war effort. In 1916 an overarching sense of purpose motivated the British Empire's war effort. "In these great days," Northcliffe wrote, "the breath of war is the breath of life, and the spirit of sacrifice is the spirit of regeneration."[84] The rhetoric and imagery of imperial solidarity and patriotic duty served well to sustain popular morale, in both triumph and tragedy, even if today it is almost impossible to empathise with such antiquated sentiments. But as Churchill identified in an article he wrote for the *Sunday Pictorial* in July, the mood had changed as the attritional battle commenced. The early spirit of adventure had subsided into "a sombre mood . . . The faculty of wonder has been dulled; emotion and enthusiasm have given place to endurance; excitement is bankrupt, death is familiar, and sorrow numb."[85] Now battle was joined in earnest, and the human cost of war was becoming manifest, imperial adventure had to become Bishop Gwynne's imperial endeavour.

As James Douglas pithily put it in his review of *The Battle of the Somme* film for the *Star*: "It is our task to beat the German sword into a ploughshare so that nations may learn war no more." In that Britain seemed to be determined: the warmongers were on the other side of the line, so British public opinion remained broadly in support of the war, and convinced of the justice of the cause for which the empire was fighting.[86] Journeying through Britain on his way to visit the Grand

Fleet at Scapa Flow at the start of September, Bean captured the mood of that portentous summer. Struck by the spirit of the British public, the unity and camaraderie that animated them, he reported it in an official despatch to Australia. Coming on the leave train into London, a city of rich and poor, a metropolis defined by "the separate layers in the big frowning edifice of British society," passing the back gardens of the replicated terraces and villas that lined the railway tracks, he found "one thing that bound them all together just for that moment." As the troop train passed, everyone, from the highest to the lowest, dropped what they were doing and waved to the soldiers.

> I have never seen any demonstration that could compare with the simple spontaneous welcome by the families of London ... For the first time in one's experience one had experienced a genuine, whole-hearted, common feeling running through all the English people ... bound in one common interest which, for the time being, was moving the whole nation. And I shall never forget it.

The purpose of the soldiers at the front was complemented by that of the civilians in the rear, and Bean ventured a hope that these people could not return to the divisive society of pre-war years when the war was won.[87] Certainly, the American ambassador confirmed, despite the all-consuming nature of the war, "people are very cheerful ... it is not only taken for granted—it gives these people activity that brings in some a sort of exaltation, in many more a form of milder excitement ... a depressing monotony of subject and talk and work, relieved by the exaltation born of belief in victory—this is the atmosphere we now live in."[88]

Bean noted privately that something else was happening too: "They were all thinking, and feeling, much more than I realised, about the Battle of the Somme," while Bean himself and the soldiers on his train were, "to tell the truth, very tired—very weary of that fight. Everyone had had enough of it, had come out of a really trying experience and was ready for a little home comfort—and it was a wonderful, wonderful welcome." While civilians could welcome home their triumphant heroes, they could never understand their experience and their world, or really empathise with the combatants. The home front was firmly behind the war, but, Bean lamented, "If only England could wage peace with the unselfish unity of purpose with which she wages war!"[89]

As the battle continued, its reportage evoked debate not just on the offensive but on the rightness of war itself. Front-line soldiers added their voices to the cacophony of moralising and reflection. Sergeant Tawney, recovering from his 1 July wounds, published two articles in the *Westminster Gazette* and the *Nation* setting out one soldier's anguished perspective. The first, published in August, was a matter-of-fact memoir of his experience at the front, a visceral account of the reality of battle and the trauma of being wounded.[90] The second, published in October, was more reflective, an attempt to bridge the "dividing chasm" that had opened between rear and front. "I realise how hopeless it is," he wrote, as he tried to grasp the reasons there was fighting in France, why he had gone to fight there, and the sundering effect it had had on his country and countrymen. The future socialist economist decried the dry accounting ledger of conflict that seemed to define the attitude of the home front:

> You calculate the profits to be derived from "War after the War," as though the unspeakable agonies of the Somme were an item in a commercial proposition. You make us feel the country to which we've returned is not the country for which we went out to fight. And your reticence as to the obvious physical facts of war! and your ignorance as to the sentiments of your relations about it![91]

He was only one of millions grappling with the "dividing chasm," evoked by a failure of the rear to grasp the reality of the front, provoked by the unreal, emotionless media image of war—an image of the nobility, sport and joy of battle rather than its misery and despair, its catatonic emotional catastrophe. Films or words, however stirring or striking, could never bridge that gap. "*The Times*' military expert's hundredth variation on the theme that the abstruse science of war consists in killing more of the enemy than he kills of you, so that, whatever its losses—agreeable doctrine—the numerically preponderant side can always win, as it were by one wicket," did not help the soldiers' frame of mind. Back from the battle, Tawney felt like "a visitor amongst strangers whose intentions are kindly, but whose modes of thought I neither altogether understand nor altogether approve." The battle would never leave him. His England was now "not an island or an empire, but a wet populous dyke stretching from Flanders to the Somme," a hellish, murderous world beyond civilian experience.[92] For

Tawney, the upshot of this alienation from war and warmongers was pacifism and membership of the Union of Democratic Control, committed to a socialist peace.

It would seem that the rear was becoming more engaged with the front just as the front was disengaging from the rear. Until 1916 the war had been censored, through fear of undermining popular morale. Now greater public knowledge was adopted and adapted as the way to sustain national effort. Sympathy and empathy, shock and anger, first-hand evidence of what their menfolk were seeing and doing would rally British civilians behind the expanding war effort to an unprecedented degree.

POLITICS IN GERMANY functioned on rather different lines, and society was mustered in other ways. In the Kaiser's Reich the soldier stood supreme, as is evinced by William II's soubriquet: "Supreme Warlord." If things were not going well, then a military solution, and soldiers to implement it, would be sought. While in Britain and France anxious politicians were trying to wrest some control from soldiers too focused on the main prize, in Germany civilian politicians were being marginalised as the army assumed control of the nation's destiny, presiding over a field-grey populist dictatorship, the age-old solution to acute wartime crisis.[93]

The promotion of Field Marshal Paul von Hindenburg and General Erich Ludendorff at the end of August 1916 was a watershed for Germany. Hindenburg was to embody the state's martial values in what was effectively becoming a military dictatorship. The field marshal was everything the German people expected in a military leader: tall, imposing, with thick white hair cut *en brosse* and a prominent curling moustache, the new Chief of the General Staff epitomised a strong, stern, Prussian character. And he was indeed a scion of Prussia's military aristocracy who had lived a soldier's life, disciplined, ascetic and dutiful to fatherland and Kaiser. By 1916, however, his loyalty seemed to be more to country than to crown.[94] He had first seen active service in the 1866 war with Austria, finally retiring as a corps commander in 1911 at the age of sixty-four. Recalled to command the Eighth Army in East Prussia, his star had waxed after his nation-saving victory over the Russians at Tannenberg in August 1914. Subsequently a personality cult had developed. Wooden statues of him were erected around Germany, and iron and silver nails were sold to be knocked into them: the proceeds went to the German Red Cross.[95] Thus he acquired his less than complimentary epithet, "the Wooden Titan." But the new war leader was more intelligent and shrewder than posterity has generally acknowledged. Certainly, in August 1916 he was immensely popular, and he would remain so.

But Hindenburg was only half of another of those formidable commander-in-chief–chief of staff relationships that lie at the heart of most effective military activity. Erich Ludendorff was a more shadowy figure, a workaholic technocrat and manager supporting Hindenburg's military colossus. To the partnership he brought brains and activity, but also a tendency to excitement and panic, which were well balanced by his chief's calm, reflective nature.[96] In August 1916 he adopted the title of First Quartermaster-General, which covered a wide remit: strategy, offensive planning, logistics, war industries and the management of the press all came within the purview of his multi-dimensional new role, in which he would oversee the intensified mobilisation that might yet deliver Germany from her adversaries' stranglehold. Nor was Ludendorff above interfering in politics and affairs on the home front if he thought this would further the war. Although he was a staff officer of great intelligence and experience, he was not from the usual Prussian noble stock that produced most senior German soldiers: his father was a businessman from Posen. Aged fifty-three in 1916, Ludendorff had been at Hindenburg's side since Tannenberg, and their partnership suited them both. Ludendorff relished the detailed, tedious work that Hindenburg hated. A small, balding man who tended to inspire fear in his subordinates rather than trust, Ludendorff lacked Hindenburg's presence and status among the army and the populace.[97] But he was to become the real driving force of Germany's war.

When Hindenburg and Ludendorff replaced Falkenhayn it marked not only an intensification of the war but the centralisation of the war effort at OHL, where "a new spirit and a new concept of the war" predominated. In place of Falkenhayn's attritional strategy, which had been designed to force the Entente to the negotiating table, the new duumvirate reverted to a "strategy of annihilation," designed to impose a victorious, humiliating peace on the enemy, including territorial annexations and monetary indemnities. In autumn 1916 this policy still had strong popular support, because it was not yet apparent that it would require "more arms and more men to fight more battles."[98] For the Kaiser, however, this meant a further blow to his diminishing authority. The management of the war shifted away from Berlin to

OHL, and the focus from East to West. (In February 1917 OHL itself was moved westwards from Pless in Silesia to Kreuznach in the Rhineland.)

For Bethmann Hollweg, who had been struggling to keep a restless Reichstag behind the Kaiser and his generals, the new regime was a necessary but fateful compromise. Parliamentary and popular agitation for an end to the war started to rise as the German army was properly engaged for the first time. The Chancellor's growing political problem was on the left, as the fragile consensus that had accompanied the outbreak of war started to fracture. The problem with the *Burgfrieden* political truce with which Germany had begun the war was that it was based on a misunderstanding between socialists and conservatives: the former sought to take advantage of it to push their populist agenda, while the latter saw it as an opportunity to assert state authority over domestic troublemakers and to unite the nation in a common expansionist cause.[99]

Before the war, Europe's socialist parties had always professed themselves against any war that would pit worker against worker: the international struggle of the working class against capitalist oppression took precedence over the rivalries of nations and empires. Such principles were generally forgotten or sidelined in the wave of patriotic sentiment stirred up by the arguments for defensive war spun in July and August 1914. But as the war entered its third year, some left-wing leaders started to remember their old values. By the middle of 1916, Germany's Social Democratic Party (SPD) was split over its attitude to the war. Its moderate wing still supported the Kaiser's government, even if they did not believe in the "victorious peace" it advocated. But its militants, led by Karl Liebknecht and Rosa Luxemburg, opposed the continuation of the war, refused to support further war credits in the Reichstag, and agitated for a peace on socialist principles. Their message was propagated as far as the front line. Charteris was pleased to record "one very interesting appeal printed in Germany to the German people to rise in revolt and enforce peace on the German rulers, and thus avoid starvation in their country," found in captured trenches near Pozières.[100] On the home front their message was spreading, as were workers' protests about more mundane concerns. The number of strikes had fallen dramatically with the outbreak of war, but by 1916 they were on an upward trend again, with 240 during the year.[101] Equally worrying, just as Germany's workers were becoming more politically radicalised, many more prosperous middle-class property-owners, civil servants and professionals, also hit by the hardships of war, were starting to sympathise with the anti-war agenda.[102] War versus peace was not simply a question that harked back to pre-1914 class politics: it was growing more relevant and acrimonious as the intensity and fortunes of war started to turn against the Reich.

Faced with growing socialist-inspired anti-war agitation, nationalists who favoured the military's aggressive, annexationist war aims programme were also organising. Bethmann Hollweg, trying to keep Germany's anti-war and more-war factions apart through a series of increasingly desperate political manoeuvres, at this point threw in his lot with the military. The right, increasingly desperate and belligerent, would in future direct German strategy and policy, while in the background the Reich's domestic politics imploded.[103]

GERMANY WAS NOT the only state in which nationalist and socialist views of war and peace were juxtaposed by 1916. But in the more democratic British Empire a more supple, inclusive, political process allowed such ideological and social differences to be negotiated with greater success, if not reconciled.

Responses differed to the bathetic mix of glory and disaster that defined the Dominions' entry into the real war on the Western Front. Newfoundland's response to Beaumont Hamel had been patriotic and self-affirming. On receipt of the tragic news the editorial of the *Evening Telegram,* Newfoundland's leading daily, affirmed:

> The action of July 1st was no defeat. It was the first and greatest step towards the final victory that will crush German power and humble German arrogance. In it our men played no small part, of that we may be sure . . . we know they did not give their lives for nothing. The very size and nature of the casualty list is eloquent. It speaks of an advance, impetuous and fearless, through a hail of bullets. When . . . their achievement in all its glorious detail is told, there will be a thrill of pride throughout our Island such as it has never felt before.[104]

Heartbreaking though that moment had been, the loyal colony's soldiers had won a reputation of which their homeland and the wider empire should be proud. South Africa too could take positives from her brigade's decimation in Delville Wood. The fact that these colonies chose these sites for their national war memorials after the conflict attests to the Somme's central place in national history and the construction of statehood for both. "The Canadians have certainly done very well—got more prisoners than we and advanced much further," the Australian Charles Bean recorded jealously. "[T]heir attacks... have been... part of a general advance... a sort of warfare of which we never had a taste. Ours was of the Verdun type from first to last."[105] Canada had a relatively good Somme, and her army continued to prosper. Her national memorial marks the Canadian Corps's April 1917 triumph on Vimy Ridge.

Australia's response to the Somme, to the long trial on the Pozières Ridge, was more complex and rather different. There was certainly patriotic fervour, coupled with growing belligerence, but it was leavened by a new, assertive sense of statehood and a distinct populist spirit. Australia's proud induction into the pantheon of imperial heroes had taken place at the Dardanelles. But on the Somme she had been let down by the motherland. In future she would take greater responsibility for herself. The loving child was entering a troubled adolescence, in which the values of comradeship and community that Australia's sons absorbed on the battlefield would challenge loyalty and deference as the leitmotifs of Australian citizenship and nationhood.

It was not an anti-war spirit. Australia's populist Labor Prime Minister, William "Billy" Hughes—"that queer combination, a Socialist and Imperialist"[106]—"had no idea of compromise, which I am sure British statesmen have in the back of their heads."[107] He fully subscribed to the noble cause of war. As he pronounced on 25 April 1916, the first "Anzac Day," marking the anniversary of the Australians' landing at Gallipoli:

> The story of the Gallipoli campaign had shown that through self-sacrifice alone could men or a nation be saved. And since it had evoked this pure and noble spirit, who should say that this dreadful war was wholly evil now that in a world saturated with a lust of material things came the sweet, purifying breath of self sacrifice?[108]

Hughes had set himself against Asquith's patrician and apparently easygoing approach to making war, Bean was pleased to note:

> Asquith of course is like most of the others obsessed with the conviction that he knows what is good for the people better than the people themselves ... not my way and the way the British

race ought to be handled, but the way of the snobbery of states-
manship which is the class to which Asquith belongs.[109]

Bean was astute as well as forthright. After Pozières he was also
angry and troubled by the nature of the war he supported. Like so
many others, he became torn between winning the war and avoiding
the sacrifice necessary to do so. This new spirit was immediately evident as Australia engaged with the difficult subject of conscription. The
autumn 1916 debate on the issue was a further manifestation of the
intensification of the war, and of its impact on Australia's soldiers.
Without conscription the AIF could not sustain the losses that Western
Front battle entailed. Yet the moral case against it was that young men
should have the right to choose whether they went to die for their
country. Certainly, by 1916, when the flood of enthusiasm of the early
months of the war had abated to a trickle, manpower was as central a
political issue in Australia as it was elsewhere. Those who had enlisted
voluntarily for "King, country and adventure" had no desire to inflict
real war on the unwilling.[110]

Hughes, who had been opposed to conscription before he had
toured London and the Western Front in the summer,[111] now led the
pro-conscription party, "working hand in glove with Lloyd George,"
Bean noted.[112] The issue, which "provoked the most savage disputes in
Australia's history,"[113] split his own Labor Party, whose trade union
backers forcefully opposed conscription to the point of public distur-
bance. Hughes reinvented himself as the leader of a new National Party
committed to making war "to the last man and to the last shilling," as he
put it, borrowing Churchill's phraseology.[114] Rather than make a parliamentary decision in the face of such hostility, the government authorised a referendum. Now that it was becoming a people's war, Australia
was taking a lead in referring such a central question of war policy, a
matter of life and death, to her citizens. Australia's soldiers were also
citizens, so they would have a say in whether unwilling men should be
forced to fight alongside them: conscription was a lively topic of conversation in the ranks in September and October as Australia's battle-
scarred battalions rested from their recent ordeal.

Canvassing on Hughes's behalf, the journalist Keith Murdoch
toured allied headquarters seeking endorsement for a "yes" vote.
Briand, Joffre and Haig all lent their influence to the cause: the latter
telegraphed Hughes to stress the importance of conscription to keep
the AIF's ranks full.[115] Nevertheless, the conscription referendum, held

on 28 October 1916, resulted in defeat for the Australian government by a margin of 10 per cent, although the troops voted in favour, 72,399 votes to 58,894.[116] A second referendum in 1917 was equally divisive and also rejected conscription.

As elsewhere, the war was reshaping Australian politics and society. Ironically, Hughes advocated the German model of organisation, under a single leader and an all-powerful state, as the way for the empire to win. (Equally ironically, when it came to government, he was very disorganised himself.) Others were not prepared to go that far: in December the rump of the Australian Labor Party started to campaign for a negotiated peace. But Hughes was to lead Australia to victory. In order to do so, like that other "personal adventurer," Lloyd George, he had to split his own party, smash the patterns of pre-war domestic politics, and destroy his own power base in the process.[117]

FRENCH POLITICS HAD always been ideologically motivated and confrontational. Although the immediate parliamentary threat had been checked in July, Joffre's political troubles were not over. Clemenceau's radical press continued to hound the commander-in-chief: Joffre suspected a personal motive, having dismissed General Leblois, the brother of one of the radical editor's leading financial backers. Moreover, there were rumours of a German assassination plot, so the leader of a multi-million-man army had to be given a police bodyguard![118] But ministerial interference in military operations remained his main anxiety. Shortly after the Senate secret session ended, Minister of War Roques and President Poincaré visited the front. Fearing that they would give in to Pétain and Nivelle's pressure to reinforce Verdun at the expense of the Somme just as his new battle was starting to make progress (as had happened the last time Roques had visited Pétain on his own authority), Joffre insisted that he accompany them. At stake was the principle of who directed strategy, for the allocation of guns and troops determined the scale and effectiveness of military operations. In the event, the ministerial visit would be a guided tour rather than an inspection,[119] and it passed off well. Poincaré acknowledged that the troops were in far better spirits on the Somme, where they were on the offensive, than at Verdun.[120]

Fortuitously, when the Somme offensive was reaching a crisis point there was a momentary lull in the day-to-day sniping between high command and government that was the defining aspect of wartime

French politics. Towards the end of August, the Chamber and Senate adjourned until mid-September, and then the long-running sore of the Salonika campaign reopened with Romania's entry into the war on the allied side, distracting politicians' attention from the West, albeit giving Joffre something new to worry about.[121] Major James Logan, an official American military observer with the French army, reported that Romania's entry, coupled with Russian successes and news of hardship on the German home front, "has undoubtedly put the Government itself, and the Higher Command in the best position it has been since the beginning of the war."[122] Such optimism led to loud and heartfelt assertions that the end was drawing near. Realists, however, knew that the war would not be over by Christmas, so attention still had to be paid to continuing hostilities into 1917. Consequently, strategy remained on the agenda. At Saleux on 27 August, Joffre met France's political leaders to review resources and future prospects, with Haig in attendance. The locus of strategic thinking was shifting from operations at the front, which if not decisive were at least proving much more effective, to management of the rear.

The allocation of manpower, a stretched and finite resource now that mobilisation was complete and conscription universal, became central to continuing the war. This was a particular problem for France, which did not have Britain's untapped reserves of manpower. A balance had to be struck between the needs of the army and the intensifying industrial war effort, which was already demanding the release of skilled workers and older men from the ranks. That need was particularly acute, for at Saleux Joffre presented extensive heavy artillery and munitions programmes to provide the wherewithal for fighting the new style of *matériel*-intensive battle in 1917.[123] France seemed more determined now that the war seemed to be turning in her favour. When a papal statement that Germany would be obliged to seek an armistice by October was mentioned, Poincaré declared, "No talk of peace so long as one enemy remained on the soil of the Republic."[124] But at the same time, the politicians seemed anxious for rapid success now that Romania had entered the fray. They badgered Joffre about the purpose and likely outcome of the offensive, and the commander-in-chief was forced to admit that, the relief of Verdun and attrition of the German army notwithstanding, a decisive victory was unlikely. "Briand found this pessimistic," Weygand later recollected. But Foch "did not like to see the government with such false expectations, and dreaded the moment when their illusions were shattered."[125]

While the generals pursued their objective of defeating the German army, all that the politicians could do was to wait and see what happened. The exterior theatres might provide some diversion, but unless something spectacular happened there in the interim an intensification of the attritional war was in prospect for the next campaign.

ROMANIA'S DECLARATION OF war on 27 August initially seemed to be that spectacular event. Her entry into the war, the most significant change in the balance of forces during the second half of 1916, confirms the global impact of the General Allied Offensive. Although localised in Picardy, the Somme campaign was part of a broader war in which other states were engaged. Inevitably, allies and neutrals had views on, and their actions would be influenced by, events in France. Romania had dithered for many months over which side, if any, to support, and was finally persuaded that she stood to gain most from the Entente that was now advancing on all fronts. Her declaration of war was essentially a pragmatic land-grab, seeking to add Transylvania and Bukovina to the territory she had won in the recent Balkan Wars. Unfortunately, she had embroiled herself in a different sort of war altogether.

A new 550,000-man, 23-division army with recent combat experience in the Balkan Wars appeared as an important new weight that might tip the balance decisively in the Entente's favour. The Kaiser, for one, was thrown into a panic: he "completely lost his head [and] pronounced the war finally lost and believed we must now ask for peace."[126] However, "brave and dedicated as they might be," Romania's troops were "poorly trained, poorly equipped and poorly led."[127] The Kaiser's panic soon proved premature: Germany would be able to fight on. Although Russia (Britain and France's principal ally) was fighting hard, General Brusiloff's success against the Austrian section of the Eastern Front had not been followed up with a sustained attack against the northern, German, sector, as Joffre had hoped. Russia seemed to be defaulting to Alexieff's previously rejected strategy of knocking out Austria-Hungary. Of course, this made sense in the context of the shifting military and political situation in the East in the late summer of 1916, but it meant pressure on the German army was not as complete as it might be. Nevertheless, at least another powerful attack was being prepared to coincide with Romania's planned offensive into Hungary.

Similarly, Italy was maintaining pressure on the Southern Front, with a succession of offensives on the Isonzo. The first, in early August, had been more successful than those of 1915, capturing Gorizia, and another was due to be launched in early September. However, although the Italians chose this moment to declare war on Germany too, this was really another struggle with Austria-Hungary, with only an indirect impact on Germany's war-making capacity. The other allies seemed wary of taking on Germany directly. Doubtful from the start that the clash on the Somme would do anything but exhaust the two adversaries and hasten a peace settlement, Belgium's King Albert reflected after two months on the "singular contradiction between the mediocrity of the activity and the immensity of the objective . . . that is the total destruction of the Central Powers."[128] He had been wise to keep his small army out of the hard fighting.

Continued Anglo-French success would boost the Entente, spurring on the Russians and Italians who were fighting hard on their own fronts, as well as tying down German reserves in the West and making allied operations easier and more profitable. Haig subsequently affirmed that one of the strategic objectives of the battle was to "assist our allies in the other theatres of war by stopping any further transfer of German troops from the Western front."[129] Of course, Germany continued to shuffle her reserves between Eastern and Western Fronts, but the number and quality of divisions that could be spared to meet the intensifying crisis on the Eastern Front were undoubtedly reduced by the Somme campaign. German resources were increasingly stretched, even if they had not yet reached breaking point, and Joffre could take considerable comfort from the fact that his strategy was exerting ever more pressure on the Central Powers. In contrast, by the end of August Falkenhayn had dismissed any idea that a German counter-offensive could be launched: stripping any section of the front to muster the necessary reserves for an attack elsewhere, he argued, would risk the long-feared allied breakthrough.[130] This pragmatic approach cost him his job. Certainly, at the end of August 1916 the allies had finally gained the initiative and Charteris judged that the steady pressure on the Somme "prevented the Germans hammering Russia," as had happened in 1915,[131] if not Romania.

Germany's allies were also suffering. Austria-Hungary's army was crumbling: the strength of infantry battalions had already been cut by 20 per cent and troops were now surrendering in large numbers on both the Eastern and Italian Fronts. At home there were bread shortages. A machine gun was deployed in Vienna's Landstrasse to deter rioting. "Extreme pessimism rules here," the U.S. military attaché in Vienna reported. "[A]n Austrian officer said to me, 'We are lost.' "[132] The *Morning Post* willingly reported a Hungarian press correspondent's impressions after visiting the Somme. Their ally was engaged in a battle of exhaustion. For now, their fighting spirit was holding up: machine-gunners were apparently asking to be tied to their guns. But Britain still had two years of manpower reserves to throw into the fray. "Whatever be the case, the German army stands today before the most gigantic task it has had to face during this war, and will have to show the world once again that it is as hard and tough today as it was at the time of its entry to France," the article concluded. "As a matter of fact they are far from being as hard and tough as they once were," the *Morning Post* added.[133] Her ally's prospects of victory were not good, and Austria-Hungary herself had had enough of Germany's war. In December, the aged Emperor Francis Joseph finally died. His successor, Karl, the last of the Habsburg emperors, made peace his first priority.

IN RESPONSE TO their soldiers' harrowing experience on the Somme, Germany's press could find little positive to report. Instead of stories of victory—on the Somme merely holding ground or retaking a lost strongpoint or village were the biggest "victories" the official communiqués could muster—to sustain public engagement with the war and morale on the home front the German press had to take a different line, delivered, in Bean's judgement, in "a notably dramatic picturesque style."[134] Reports from the front emphasised the heroism of the German soldier, his dogged resistance in the face of allied material superiority, the sacrifice being made for the fatherland and the need for the home front to acknowledge that sacrifice, to do it justice and to rally behind the army.

Patriotic journalists such as Georg Queri, the *Berliner Tageblatt*'s official correspondent at OHL, urged and cajoled civilians to back the troops at the front, to accept their privations and sacrifices in the same way that the combat troops were accepting theirs. He reported the human-interest stories from the Somme in grave, admiring tones.[135] These articles, like the most effective propaganda, repeat a simple refrain: Germany could not defeat her enemies materially, so she must break them morally. On the Bazentin Ridge the British were pushing against a wall. The great weight of war *matériel* deployed could not overcome this heroic defence; the survivors' rifle-fire broke the British attacks; the enemy left hecatombs in front of the German positions.

"[A] battle zeal has again suddenly burst forth the like of which could be seen and understood only during the first weeks of this war." This was an attestation that the army remained hard and tough and able to withstand all that the allies could muster. It was a positive and uplifting message. Now both soldiers and society were staking all on the outcome of this battle: "The enemy has gathered all his strength in order to conquer. His ability has reached a high point. He is striving for the final blow and wishes the end and victory. The fighting will last without diminution and the putting in of men and artillery will increase rather than diminish."[136] Germany's soldiers faced a test of endurance, pitting their exhausted bodies against the relentless allied bombardment: British and French guns, American shells. Small comforts could be taken: "The more prodigiously his artillery preparation is developed, the more does the enemy admit that the fighting worth of his troops is less, and it is also probably a fact," Queri suggested. Germany's better soldiers could yet triumph: "Again a handful of defenders took up the fight, and often the far superior enemy was thrown back by a single machine gun, a single little crowd of riflemen and hand-grenade throwers . . . he cannot find his way through."[137] This was heady, jingoistic journalism (under press censorship there could be no other kind), in which words were deployed to reinforce resolve. It suggests a recognition that the crisis point of the war had come, and that Germany's position could only worsen. Nevertheless, if she could never win the *Materialschlacht*, Germany might prevail in the *Moralschlacht*, the battle of wills. All Germany would have to respond: the civilians safe on the home front would have to support the sacrifice of the brave fighters dying for the fatherland with sacrifices of their own.[138]

By late 1916, the home front in Germany was not living up to the effort being made in the field. From spring 1916, monthly reports sent in from Germany's home army districts had charted a growing crisis in civilian morale. Pre-war social and economic divisions were being accentuated by the war. While the rich seemed to be doing relatively well out of it, workers' standards of living were suffering, and there was growing resentment at such perceived injustices when all were supposedly united in the common cause.[139] Debates in the Reichstag about the need for peace were replayed in beer halls and bread queues, while shortages caused by the allied blockade, price increases and calls to subscribe to the latest war loan all contributed to worsening conditions and growing insolvency. Food supply in particular was becoming difficult. By the second half of 1916, meat, sugar, eggs, milk, cooking fats, coal and clothing had joined bread on the list of rationed goods. Moreover, the system of ration cards and registration was complex and inevitably led to long queues outside food shops. Where better to discuss the war, to grumble about the shortages and malign the food controller and other officials who represented the growing power of the state over private lives. Strikes and public disturbances were increasing as a result, and troops were sometimes needed to restore order. All this, nationalist commentators such as Queri could rightly complain, was undermining the wartime sense of community and shared sacrifice that was supposedly holding together the German war effort. In fact, continued censorship of left-wing newspapers, restrictions on socialist meetings and prosecution of agitators under martial law all suggested that the glorified "truce" between Germany's political factions was a mirage.[140]

Germans were growing tired of war and indifferent to warriors: parties of new recruits mustered for the front were no longer cheered on their way. Every six months subscriptions to a new war loan were solicited. That of March 1916 had secured 5.2 million subscribers. But only 3.8 million investors came forward in September to risk their savings on a favourable outcome to the war.[141] Queri might well enjoin them to endure their sacrifices stoically; after all, they were in no way as deadly as those made by the men at the front. Nevertheless, the soldiers were better fed and properly clothed, while their mothers, fathers and younger siblings were wearing rags and having to survive on food scraps and adulterated "war bread" of questionable nutritional value. In Germany the privations at home, rather than a failure to appreciate the grim conditions at the front which were stressed incessantly by the patriotic press, were dividing civil society from the young men in uniform.

In an attempt to engage the civilians once more with their fighting men, Germany took a leaf out of the allied book, producing her own Somme film, *With Our Heroes on the Somme*, released in January 1917. But it lacked the immediacy and authenticity of *The Battle of the Somme*, having been filmed, as German war documentaries usually were, at training establishments behind the lines, rather than at the front. If not panned by the critics—under the military regime such forthrightness was improbable—the film's lukewarm reception was symptomatic of the spirit on the home front as the war entered another year.[142] In Germany it was becoming an impossible task to re-energise the populace with patriotic messages or graphic imagery, not only because Germany's economic conditions were worse than those in

Britain and France, where as yet there was no significant blockade, but also because her political traditions were more confrontational. Left and right could debate war and peace in the Reichstag; but in an empire in which the parliament had no great influence over policy, differences were inclined to spill out onto the streets. The strain of constant, ineffective defence was dividing society between warmongers and peacemakers; the lack of any prospect of victory was starting to tell.

PART OF HAIG'S rationale for continuing the offensive was that it demonstrated to the world that the allies could engage the German army in a sustained battle, shaking the faith of Germany's friends and "doubting neutrals" in German invincibility, while at the same time emphasising British strength and determination to win.[143] To neutrals, the events in France were fascinating and horrifying in equal measure. There was no shortage of foreign press speculation about how the war was being conducted, and the impact it was having on the belligerents. As well as the contradictory syndicated reports emerging from both sides—if not direct propaganda, at least carefully edited and heavily censored—neutral correspondents witnessed the battle at first hand. Such representatives were welcomed, for their reports might promote the cause of one's own side. The Dutch were the closest to the fight, caught uneasily between the main combatants. The effort and losses were staggering, the *Telegraaf* reported. In one three-day period the British army had suffered 1286 officer casualties, 200 more than the whole Dutch professional officer corps. "This gives us a slight idea of the tragedy that is being enacted on the Somme," as well as strongly reinforcing Dutch resolve not to be "dragged into the war."

But the allied cause seemed to be prospering. Despite her early setback, it was felt that Britain had the resolve to fight on, and would bring her full weight to bear on the Somme. Germany, on the other hand, was now tied down on both Eastern and Western Fronts, and was probably making preparations for a strategic retreat in the West if she could not hold on the Somme front.[144] When and how Germany's collapse would come about, Pieter Geyl of *De Nieuwe Rotterdamsche Courant* reported towards the end of August, was a matter of debate at the front. However, "That the defeat is to come is doubted by no one."[145]

Such positive reports in the neutral press undoubtedly fuelled Joffre's determination to battle on. Indeed, Joffre and Haig were anxious to court neutral correspondents who would promote their achievements abroad. James Beck, for example, who toured the Western Front with his English companion H. E. Brittain, was a pro-Entente American lawyer who was seeking material to boost his campaign for U.S. intervention. His journey from the Somme to Verdun served the dual purpose of "vindicating the justice of the Allies' cause in neutral countries, and, on the other hand, interpreting the sentiment of the American people to England and France and to some extent softening a rising tide of resentment against the American people."[146] The visitors were judged important enough for both Haig and Joffre to grant them an audience. "Haig . . . is the beautiful ideal of the British soldier. Strikingly handsome . . . with . . . the most charming personality," Brittain judged, "a brilliant soldier, entirely the master of his task."[147] The admiration was not mutual: "He seemed rather a bounder," Haig noted of Brittain after they had lunched together. However, "I liked Beck!" Joffre's impressions are not recorded.[148] Beck returned with "a quiet spirit of confidence in [the Allies'] ultimate, and I hope and pray not too distant, triumph."[149] Perhaps this was propaganda, but all effective propaganda has some basis in truth.

In early September, with France sensing that victory was attainable as long as the current effort was maintained, Joffre again met with American press correspondents. They left Chantilly imbued with a clear sense of French optimism. While Joffre had been non-committal on the exact moment when the war would end, with another large-scale attack on the Somme imminent he had hazarded a timescale of between a fortnight and six months. Tellingly, the censor ran his blue pencil through this hopeful prediction, reportedly on Briand's personal instruction. Either the meeting at Saleux had convinced him it was unrealistic or he did not want to risk America throwing her weight around at France's moment of triumph. Whichever, the officers at the front reportedly remained confident of victory, even if the general consensus was that it would take another year.[150]

It was important to France to woo the American correspondents, especially if Joffre's hopeful prediction proved unfounded. If America was not in the war, the war was certainly in America: 125,000 flag-waving patriots had marched through New York in May, and such demonstrations were repeated throughout the country.[151] Financiers had invested in the war, factories were producing for it and politics revolved around it. Ideological confrontation was becoming a business venture for the Entente; but for America the war had always been a business opportunity cloaked in moralising rhetoric. For these reasons, the United States was the neutral power whose opinion and actions (not forgetting her money) counted above those of all others.

Unsurprisingly, by 1916, events in France had become inseparable from U.S. domestic politics. America's role in the war pervaded the presidential election that was taking place during the Battle of the Somme. The Democratic incumbent, Woodrow Wilson, was running against the Republican Charles Evans Hughes. While neither advocated American entry into the war (or at least not without good cause), their attitudes to the war, in which America had already made a large financial investment, if not yet a formal commitment, were central elements of their platforms. Wilson, the eventual winner, presented himself as a peacemaker—"the great pacificator," in the British ambassador's rather derisive view. If he could end the war, his Democratic nomination and re-election as President were assured.[152] Ending the gruesome and uncivilised war, rather than fighting for one side or the other, was the predominant standpoint in America, championed by ex-President William Taft's League to Enforce Peace. In 1916 Wilson seemed determined to force the belligerents to the negotiating table, although pragmatic motives lay behind this apparently moral mission, and some of the derision was justified. American liberal capitalism was booming as a result of the conflict, and Wilson intended to sustain the U.S.A.'s financial preponderance and economic prosperity. Faced with the imperialistic closed-market philosophies of the European empires, he chose to champion such liberal ideals as accountable representative government, national self-determination, open diplomacy through a league of nations, freedom of the seas and free trade: principles that might appeal to Europe's peoples if not their leaders. Conversely, if either Germany or the Entente won an outright victory, that would be bad for trade. Wilson might have championed a world safe for democracy, and certainly believed his own rhetoric; but that was also a world of opportunity for Americans after Europe had exhausted itself on the battlefield. Of course, Europe's elder statesmen would not agree to such a new world order without a fight.

Although Wilson's vision might be achieved without American intervention in the war, it certainly required American interference. Restricting German commercial warfare and Entente access to U.S. money were the levers Wilson could use to bring about peace talks. In early 1916 his representative, Colonel Edward House, had toured the capitals of Europe to sound out the prospects for mediating a peace settlement. In Berlin, Bethmann Hollweg had been hospitable enough—

the Chancellor drank a lot of beer, but seemed as befuddled before as afterwards, House recollected of their meeting[153] — but House soon realised that Germany was in no mood to consider reasonable peace terms. He then proceeded to Paris, where he assured Briand: "If the allies obtain a small scale success this spring or summer, the U.S. will intervene to promote peaceful settlement, but if the allies have a set-back, the United States will intervene militarily and will take part in the war against Germany." France apparently agreed not to let the fortunes of war reach a point at which the allies were beyond saving. It seemed then that the Entente was not making any progress in the war, and would need American aid. On the other hand, the French were under-standably sceptical about such an engagement by a lowly emissary.[154] House went even further in London, drawing up a secret memorandum with Sir Edward Grey, the British Foreign Secretary, which pledged that the United States would enter the war on the allied side if Germany refused to attend a peace conference called by the President. Although America's commitment was qualified with the insertion of the word "probably," by early 1916 it was becoming clear where her sympathies lay: on the same side as her investments, not surprisingly. However, Grey insisted that a negotiated peace was not contemplated at that point.[155] As the "Big Push" approached, House accepted that the allies should be allowed to test the endurance of Germany before any peace conference was convened.[156]

The Entente's leaders obviously felt that American entry on their side would seal Germany's fate once and for all. Unless, as the House–Grey memorandum made clear, "the course of the war was so unfavourable to them that intervention of the United States would not be effective," in which circumstances America would "probably disinterest themselves in Europe and look to their own protection in their own way."[157] By September, after the allies had made their effort and Wilson's re-election had been confirmed, it was starting to look as if American intervention was unnecessary and unwanted. Mediated negotiation might yet be avoided. Still, no one suggested that American help would never be needed. Perhaps Briand was simply trying to maintain the premium on the Entente's long-term insurance policy. Grey certainly saw America's diplomatic backing in this way. Even if he always believed in the need to beat Germany for real peace to ensue, and felt that the time had not yet come for negotiation, or indeed even broaching the possibility with the French, he wanted to keep all options open.[158]

The opening of the Somme offensive naturally suspended any consideration of a mediated peace conference. France and her allies would first have the chance to force Germany to the peace table on their terms. Meanwhile, American military and diplomatic observers watched affairs closely, maintaining a steady flow of information on the domestic situations in France and Germany, and the progress of the great battle, which would feed into any decision made in Washington. Certainly, if artillery fire was any guide, the French army now seemed to have the upper hand all along the front, America's military attaché in Paris reported. Nevertheless, despite the many rumours of impending offensives, an end to the Western Front stalemate seemed a long way off in the early days of September. Joffre had been speaking figuratively, not precisely.[159] But Lloyd George too was adamant, as he advised an American reporter on 28 September, "President Wilson . . . must understand that there can be no outside interference at this time . . . The fight must be to the finish—to knockout!"[160]

That America's sympathies lay with the Entente was an open secret in Germany. After the war, Falkenhayn accused the United States of fighting Germany by proxy on the Somme. Allied guns fired American-made munitions, "a slap in the face for real neutrality," while shamefully using international law and threats of becoming a combatant to prevent Germany waging total war against her enemies.[161] Such sentiments were echoed in the German press: on the Somme, "the war industries of the entire world" were being deployed against Germany, which stood alone, "deprived of the terrible, helping hand which stretches itself across the ocean into the arsenals of the war."[162] The world seemed to be conspiring against Germany, even if her leaders were equally scornful of an American-mediated peace. America apparently condoned the allied blockade, and the starvation which resulted, and connived with "England's tyrany [sic] at sea."[163] In May 1916 threats of American intervention had forced Germany to restrict her submarine warfare a second time (the first had followed the infamous sinking of the liner *Lusitania* in May 1915). By the summer, as election fever gripped the United States, Wilson had also started to put pressure on Britain on account of her high-handed breaches of international maritime law and the restrictive trade practices that accompanied a strengthening of the blockade. Nevertheless, despite the President's antipathy to the anti-commercial actions of both sides, by then Germany had lost the "propaganda war" in the United States. "Our news service to America is admirable, and . . . German news has been swept out of the American papers," Charteris was pleased to report.[164] If the House–Grey memorandum "showed whose side America was on," the idea of a mediated negotiated peace in 1916 was premature.[165] Still, America's own interests always predominated in U.S. politics and policy. The United States would not answer the call of Europe without pushing for a new world.

BY AUTUMN 1916, there were two competing political agendas— peace or more war—that pervaded parliamentary politics, and competed for support among a public that was finding its voice as the war continued. Both had their champions, and each would have consequences. Peace was urged by the most powerful neutral state, promoted by an emergent left-wing domestic opposition, and talked of hopefully by men in the line and civilians at home—although these were different peaces: negotiated, socialist or victorious. British, French and German statesmen, those who had taken Europe to war in 1914, all feared a compromise peace. An American or socialist peace would represent the abandonment of much, if not all, of what they were fighting for. It would also be an acknowledgement that their war was unwinnable. Moreover, without victory there might be no peace at all, merely an intermission. It is a fact as well as a truism that many a war is the result of a bad peace.

Europe had gone to war over real diplomatic issues, the Entente and the Central Powers had deep ideological differences that had been highlighted by two years of defamatory propaganda, and domestic political cleavages were only muted, not silenced, by the rallying call of national defence. None of these would be permanently resolved by a compromise peace. Moreover, although given a lease of life by the war, new liberal internationalist or international socialist political forms were still marginal to the capitalist imperialist values of Europe's empires. Those who deplore the fighting of wars, who condemn the Great War for its inhumanity and prolonged stalemate, should not deny its essential rationality for those who were fighting it. More war, despite its now obvious human and financial cost and potentially disastrous consequences, was the only way for ideologically juxtaposed capitalist empires to end the stalemate.

Paradoxically, it was only statesmen who identified with these new agendas and understood the popular will behind them who could win and keep public support for the war effort. During its second half, such radical populist interventionist statesmen, Billy Hughes and David Lloyd George prominent among them, were to take up the baton of war from their old-fashioned predecessors, who were perceived, rightly or wrongly, as having conducted war as a gentlemanly sport, not a national endeavour. Militaristic Germany found her equivalent in the popular field-grey hero Field Marshal Hindenburg. Under this leadership, fighting would continue, and intensify. However much they decried their generals' activities, this meant that politicians would have to support their military commanders.

On the whole, the senior soldiers of the Great War were simple, straightforward, often bluff men who did not work easily with more supple and scheming political masters. However, they were well aware that battle was a political as well as a military event, and they did their best to steer a course through the choppy waters of high politics and public opinion. But even Micheler, one of the more politically minded Somme commanders, was scathing of politicians' deficiencies: "I remain very friendly with the parliamentarians, and the more I talk with them the more I understand how easy it is to get them to believe extraordinary things, on account of their extreme ignorance of all that we are doing. This really explains a lot!"[166] Generals wanted to be left alone to do their job: the job that the politicians had authorised them to do. That simple desire lay at the root of Joffre's and Robertson's constant struggle to stop ministers and parliaments from interfering in the conduct of operations. Robertson, visiting Fourth Army in late September, assured Rawlinson that he was "very pleased with the general situation and with our work on the Somme. He says keep up the pressure and have the best army in Europe in the spring. He made no reference to the casualties but said he was having trouble with LG who was unmannerly."[167] The soldiers knew the offensive was going well, if slowly.[168] They were also aware that it was never going to go as fast as flash-in-the-pan politicians would have liked: destroying the mightiest war machine in Europe was not the work of weeks or months, but years. Cavan certainly decried Lloyd George's expectation of a "knock-out" victory: "I don't think any real soldier considers it possible to absolutely *annihilate* Germany any more than it is possible for Germany to *annihilate* us." At best, with more men and munitions, they could go "a little further and perhaps a little faster than this year" in 1917. Nevertheless, it would take at least two years to finish the war, he feared.[169] Their steadfastness and commitment to the plan that they had agreed upon at the start of the year, that their governments had repeatedly endorsed and that they were seeing through despite the constant external pressures ought to be admired, rather than vilified, as tendentious political memoirs would do subsequently.

Haig complained of Lloyd George: "He never sticks to the same plan for six hours in succession."[170] This was a reference to his travel arrangements to visit G.H.Q., but the same restless spirit animated his dealings with both his political and his military colleagues. Later, when Lloyd George wrote of his mid-September visit to the Somme, he chose to present the enduring image of "squadrons of cavalry clattering proudly to the front." He recollected that he questioned the commanders on the utility of cavalry in the battle. "[B]oth generals fell ecstatically upon me, and Joffre in particular explained that he expected the French cavalry to ride through the broken German lines on his front the following morning. You could hear the distant racket of the massed guns of France which were at that moment tearing a breach for the French horsemen." Nearly twenty years later, this was presented as an image of folly, of the "exaltation produced in brave men by a battle," and the narrow-mindedness of its directors, "quite incapable of looking beyond and around or even through the struggle just in front of them."[171] Lloyd George's account was not fictitious, but he skewed the context to make Haig and Joffre look absurd. On 12 September, the very day when Lloyd George was arguing the merits of cavalry with the generals, the Battle of the Somme reached its climactic point. The long-anticipated breach was to be made that day.

The Tipping Point: September 1916

A T 13:05 ON 12 September, Colonel Adolphe Messimy received a message at his command post, a makeshift shelter hollowed out of the shell-scoured plateau south of Combles. The news was excellent. His *chasseurs alpins* had seized the trenches of the German third position west of the Bapaume–Péronne road. As sector commander, he requested reinforcements. Two fresh battalions from 44 and 133 RIs were ordered forward. At 18:39 they advanced from the third position, over the road where today Foch's statue stands staring defiantly eastwards, across the final line of trenches protecting Bouchavesnes village. Three companies of *chasseurs* followed them across, without orders, but determined to press their advantage. They assaulted a very different sort of objective, a village with houses still standing, with leafy trees and clear roads. It was quickly encircled, then the assault companies closed in and swept through its quiet streets. There was little organised resistance—many of the garrison were found drunk in cellars—and by 19:30, just as the moon rose, five hundred prisoners and ten guns had fallen into the *poilus'* hands. Parties of Germans were seen scurrying to the rear. The Frenchmen allowed themselves a moment to savour their victory: "they hugged each other and danced with joy in the Place de l'Église."[1] For a second time, the French army had broken through into open country.

AT THE START of September the Somme offensive had been resumed with vigour and intensity. By this point, Foch had abandoned attempts to get his four armies to attack simultaneously. Such a tactic had been a managerial nightmare, and did not necessarily produce better results than separate but coordinated attacks. "There has been a good deal of bickering about 'timetables' between the two G.H.Q.s, and a good deal of exaggeration," Esher noted. "But it has now been settled that *exact* timing is not of primary importance, and a very sound decision too. It will spare a deal of bad blood."[2] As Foch outlined in an article published in a Swiss newspaper six months later, in September 1916 the rhythm of the battle became attacks by sector—first left, then centre, then right—to envelop the enemy's fortified positions successively.[3] In this way Foch hoped to increase the scale and tempo of the offensive to such a pitch that the weakening German defence would, as the theory of attrition predicted, finally rupture. His operational method integrated with Fayolle's tactical method: "The mechanism for offensive operations . . . consists of a succession of general attacks. Each of these is completed by numerous localised operations."[4] This method produced steady, sometimes surprising, progress throughout September, and brought the German defence to crisis point. The attritional battle came very close to success: far closer than the customary study of British operations alone would suggest.

Although Haig had stalled and prevaricated throughout August, by the last week of that month Joffre had persuaded him to concert the British effort with that of the French, if not exactly when the French wanted. Their spat of early July was forgotten. At their next meeting on 6 August, Joffre was reportedly "in very good spirits . . . Extremely pleased with everything we had done, and full of compliments." He brought a "peace offering," fifty *Croix de Guerre* to be distributed to deserving British officers.[5] Joffre and Foch intended to renew the attack on the scale of 1 July, with similarly thorough preparation. The main effort would be north of the river, "to pull the British forward."[6] Four armies were to take part. On the British left, Reserve Army would renew the attack on Thiepval, extending operations northwards into the Ancre valley for the first time. To its right, Fourth Army would complete the capture of the intermediate and second positions between Martinpuich and Guillemont and then attack the third position between Le Sars and Morval. Rather than an overambitious single thrust to take both these positions, it was planned as a two-stage advance, coordinated with a similar move by Sixth Army to the south: first against the intermediate line between Le Fôret and Cléry-sur-Somme; then against the third position between Rancourt and the river.

South of the river the battle would be renewed by Micheler's Tenth

Army, which would finally mount the south-easterly attack between Barleux and Chilly that Foch had been contemplating since the end of July. Altogether twelve allied army corps, six British and six French, would be committed to the renewed offensive in the last week of August. Three more French corps were in hand to exploit. The September attack was to be the French army's battle: it would "be pursued methodically, entailing the rapid attrition of the enemy and his extreme anxiety."[7] Once it was in full swing Georg Queri chided Germany's enemies: "The English have submitted themselves to some bad bleedings and . . . the English Army authorities became more modest in their aims as well as in their offensive thought, and the principal burden of the Somme battle has long since rested on the shoulders of the French."[8] However, neither he nor his readers had any reason to be smug about this state of affairs.

The offensive recommenced on 3 September, after indifferent weather had forced a delay. Repeated postponement at least gave both armies time to relieve their tired divisions in the front line. In Reserve Army, V Corps was to attack along the Ancre valley. Forty-eighth and 25th divisions, moved up in mid-August, held II Corps's front at Thiepval. The Canadian Corps was to relieve I Anzac Corps in front of Mouquet Farm after one final attempt to seize it. In Fourth Army seven new or rested divisions (1st, 5th, 7th, 16th, 20th, 24th and 56th) had moved up between 15 August and 1 September. France's best formations were concentrated on Sixth Army's front. Fayolle's old XXXIII CA (70 and 77 DIs) came into line astride the river. I CA (1 and 2 DIs) replaced XX CA. Four new or rested assault divisions (45, 46, 47 and 66 DIs) with colonial or *chasseur à pied* battalions and two independent assault brigades, 4 *Brigade de chasseurs à pied* (BCP) and Messimy's 6 BCA, were distributed to the front-line corps. The new divisions were well prepared and full of fight. For instance, 2 DI entered the battle "in excellent physical and moral condition. Reorganised after withdrawal from the Longueval sector, trained in manoeuvres in camp at Villeen-Tardenois and around Bovelles, re-equipped with every sort of *matériel*, furnished with all the latest weapons, the troops were in excellent condition to give their all." Among these men from the Nord region, fighting so close to their own homes, their occupied towns, "the will to fight was very strong."[9]

The 3 September attack, by twelve divisions in all (eight British and four French), was only partially successful. Mordant as ever, Fayolle had not expected much from this latest attempt at concerted action.[10]

Sixth Army kept going, making "spectacular progress" astride the river, crashing through the next German position.[11] The French supporting barrage was effective as usual, while the retaliatory bombardment was extremely weak, indicating the growing effectiveness of Sixth Army's counter-battery methods. For the first time the commanders on the ground controlled the creeping barrage, regulating its bounds according to local circumstances. A sophisticated system of communications between front and rear was required to do this: as well as line-of-sight, flares, Roman candles, signalling pennants and panels, telephones, optical signals and runners and carrier pigeons would all be used.[12] A machine-gun barrage fired from across the river would strafe Cléry from its open southern flank.[13] Even the Prussian Guard, who held Cléry and the line beyond Maurepas, could not hold up such an assault for long. On I CA's left, I DI pushed forward "with magnificent brio" for more than a kilometre, taking their objectives within the hour. On their right, 46 DI's *chasseurs* overcame more determined resistance, pinching out Le Fôret with the now standard enveloping manoeuvre north and south of the village. Only Sivas Trench in the centre held out.[14] The success on VII CA's front was also marked, if less complete: a few isolated outposts in Cléry resisted into the next day. Nearly 1700 prisoners and a dozen guns were captured in Cléry, with another 2500 men, 14 guns and 60 machine guns being taken at Le Fôret. At the deepest point the French penetrated three kilometres into the German positions. The assault achieved its objective of reaching the enemy's gun line: a tethered observation balloon even fell into French hands![15] " 'Papa' Joffre was beaming," Fayolle was relieved to note. "Thank you, Mother Mary!"[16]

Fourth Army's progress was more mixed. Cavan's XIV Corps did well, if not without a hard fight. Softened up by another heavy bombardment and attacked in strength for the first time by two brigades advancing behind a creeping barrage, the "sea of craters" that had once been Guillemont finally fell.[17] The advance was pressed a mile beyond, with Falfemont Farm, which had held out on 3 September, finally falling two days later, and Leuze Wood atop the spur behind it being occupied. British offensive technique was improving, and the eastern end of the Bazentin Ridge was almost secured. Ginchy, however, held out against repeated attacks by 7th Division between 3 and 7 September. Sixteenth (Irish) Division received its baptism of fire in front of this group of rubble heaps, capturing them on 10 September. The last German outposts were finally cleared from Delville Wood: a two-division counter-attack led by storm-troop companies, the biggest seen for some weeks, had re-established a German line through the eastern side of the wood on 31 August, but this "wonderful victory" was the last to be enjoyed by the Germans for some time.[18]

Elsewhere it was a less happy story. III and XV corps' latest attempt to secure High Wood collapsed, and 4th Australian Division's final assault on Mouquet Farm made little permanent progress: a lodgement made in the Fabeck Graben Redoubt north of the farm was lost by the Canadians a few days later.[19] More small-scale attacks in this sector over the following days started to deplete the ranks of the fresh divisions brought up for the attack on the German third position (old habits died hard). On the extreme right, Reserve Army launched a concerted attack on the Thiepval spur from two sides. II Corps was to continue to press against its southern flank, while V Corps was to try again to push along the Ancre valley towards St. Pierre Divion and the Schwaben Redoubt to turn the strongpoint from the north. Lieutenant Edmund Blunden's battalion, the 11th Royal Sussex, suffered badly in this attack. Their first exposure to the "vast machine of violence" operating on the Somme was unforgettable: "Never had we smelt high explosive so thick and foul, and there was no distinguishing one shell burst from another, save by the black or tawny smoke that suddenly shaped in the general miasma . . . we could make very little sense of ourselves or the battle." Amid the chaos, the enemy replied with "the whole dam [*sic*] lot, min-nies,* snipers, rifle grenades, artillery . . . machine-guns from the Thiep-val Ridge . . . the general effect was the disappearance of the attack into mystery."[20] Although lodgements were made in what was the old 1 July first position, by the end of the day what remained of the assault battal-ions were back in their own trenches. The defence north of Thiepval had not weakened, and "happy valley" was to become another debris-clogged charnel house.[21]

THE REAL SURPRISE for the German defenders was planned south of the river, where Tenth Army's oft-postponed attack began on 4 Sep-tember. General Joseph Alfred Micheler is the forgotten commander of the Somme. His operation, if mentioned at all in histories of the battle, is dismissed as an irrelevant sideshow. Perhaps this is because, as Gary Sheffield (who devoted just one paragraph to the Tenth Army's attack

---

* *Minenwerfers*, heavy trench mortars.

Wrecked British tank on the Pozières Ridge, 1917 (Australian War Memorial)

Baron von Richthofen *(centre right)* standing by his Fokker DRI (IWM)

Lorry convoy behind the Somme front
(Stapleton Collection/Bridgeman Art Library)

Transport negotiates a muddy road behind the British front
(Australian War Memorial)

"Through mud and blood to the green fields beyond": aerial views
of the Bapaume and Chaulnes sectors, with French troops attacking
behind a smoke screen in the latter
(Australian War Memorial/Corbis)

French positions on the Somme (Corbis)

Panorama of the teeming world behind the front lines (Australian War Memorial)

French outpost at Ferme de l'Hôpital, September 1916 (© Paris—Musée de l'Armée, Dist. RMN/Pascal Segrette)

French heavy artillery battery on the Rancourt front, 1916 (© Paris—Musée de l'Armée, Dist. RMN/Pascal Segrette)

The changing landscape of Pozières: before the war, 1914 (Australian War Memorial)

Under German occupation, 1915 (Australian War Memorial)

August 1916, after capture by I Anzac Corps (Australian War Memorial)

Over the winter,
under snow
(Australian War
Memorial)

May 1917, showing
the temporary
cemetery and
1st Australian
Division's battlefield
memorial
(Australian War
Memorial)

September 1917,
reclaimed by
nature
(Australian War
Memorial)

The price of victory: French dead on the Rancourt plateau, September 1916
(© Paris—Musée de l'Armée, Dist. RMN/Pascal Segrette)

The Ancre battlefield, November 1916 (Australian War Memorial)

in his study of the battle) noted, Micheler "had little interaction with the British."[22] Yet the attack is significant, if not for its immediate results, then because it expanded the battle to increase the pressure on a defence already weakened by two months of attrition.

Micheler was a competent, if undynamic, commander, "able but vacillating," according to one French staff officer: "His judgement was marred by extreme nervousness ... On the eve of an operation he swung between the extremes of hope and despair, as his mind dwelt on the strong and weak points of his position. He did not possess the strength of will which enables some men to put aside their worries once they have made their decision." Nevertheless, he was respected at G.Q.G., and had political influence in Paris, so he was able to retain considerable autonomy over his army's affairs.[23] He was also the only French senior commander to emerge from the Battle of the Somme with his reputation unsullied. He went on to receive a key command, of the Reserve Army Group, for Nivelle's spring 1917 offensive. Ultimately, though, he was sacked as an army commander in May 1918 when his defence on the river Aisne collapsed during the German spring offensive.

In September 1916 Micheler commanded six infantry and one cavalry corps, with fourteen infantry and three cavalry divisions. II, X and XXXV CAs were to seize the German defences between Barleux and Chilly and push in a south-easterly direction on to the Santerre plateau, pivoting eastwards towards the river. There was some hope that the crossings over the Somme south of Péronne could be taken quickly during a German panic,[24] suggesting that Haig was not the only optimist. In reserve for exploitation were XXI CA, one division of II CAC and II CC (for now XXX CA held a defensive front south of the main attack).[25] Micheler and his subordinates had studied Sixth Army's methods carefully. His army was to conduct a sustained, four-stage advance behind the usual curtain of artillery fire.[26] Unfortunately, while effective, Tenth Army's first attack was not as spectacular or as shocking to an alert defence as Fayolle's victories in July had been. By September, G.Q.G. was stretched for resources: as well as the expanding Somme offensive, the defence of Verdun was still making demands on a finite amount of *matériel* and manpower. Tenth Army was reliant on its own means for its attack. No more heavy batteries or munitions could be found to support the first assault[27]—these were prioritised for Sixth Army, which, despite Fayolle's pleas for economy, was itself short of ammunition—so Tenth Army's bombardment was not as overwhelming as those fired by its neighbour. Two days before the assault Micheler remained anxious that there were still "huge gaps" in the destructive preparation, especially in X and XXXV CAs' sectors, and that counter-battery fire was not very effective.[28] Moreover, after July the German defences had been strengthened and were on alert for a French attack. The infantry would therefore have more to do, and would be more reliant on their own resources. But these were also weak. The commander of 132 DI noted in his post-battle report that his regiments had come straight from Verdun and had had no opportunity for rest or tactical training since the winter.[29] The same was true of many of Tenth Army's divisions.

They had two lines of German defences to cross. The front position stretched from Barleux to Soyécourt, which had already been the scene of heavy fighting. This joined the pre-July first position, which stretched southwards to Chilly. While this had been knocked about over the previous two months, it had not been subjected to systematic, destructive bombardment. The front position was anchored on a series of fortified villages: from north to south, Barleux, Berny-en-Santerre, Deniécourt, Soyécourt, Vermandovillers and Chilly. Behind, a second, thinner line of defence crossed the Santerre plain, again based on fortified village strongpoints: Villers-Carbonnel, Mazancourt, Ablaincourt, Pressoir and Chaulnes, the last screened to west and north by a complicated, star-shaped group of interconnected woods, part of the château's forested park, from which French preparations south of the Flaucourt plateau could be monitored.[30]

At 14:00 on 4 September Tenth Army attacked on a seventeen-kilometre front, after a six-day bombardment. Infantry from ten divisions went forward with customary élan, against elements from five German divisions.[31] But success against an alert and well-dug-in enemy was mixed. X CA, in the southern sector, took most of its objectives, including Chilly village, and established a foothold in the heavily wooded centre of its front. However, its extreme left flank was checked at Bois Blockhaus, a heavily fortified rectangular copse behind the first position. In the centre XXXV CA pushed the line forward from Bois Étoile. The *chasseurs* of 43 DI seized Soyécourt, and 132 DI entered, but could not hold on to, Vermandovillers. The inexperienced division's tactics were poor, and its casualties were very heavy.[32] II CA did better on their left than their right, where sections of the German front position held out, interrupting their liaison with XXXV CA. In all Tenth Army took 2700 prisoners, but their own losses were heavy.[33] Perhaps such varying fortunes were only to be expected by now.

Over the next few days the line was straightened, with a view to attacking the second objective, but in the face of strong German counter-attacks Micheler halted the offensive on 7 September. It was a muted success, a "half-failure" in the opinion of France's official historians, due (in Micheler's judgement) to an inadequate preliminary bombardment: not through want of trying, if Joffre's note that Tenth Army was firing off its shells with abandon is to be believed,[34] but through lack of accuracy compared with Sixth Army's experienced gunners. Over four thousand prisoners but only four guns had been captured, and the defence remained intact.[35] Foch immediately doubled Tenth Army's daily allocation of heavy artillery shells.[36] It was condemned to a further slogging match before the second German position fell into its hands.[37]

The divisions that attacked on 4 September had already been holding the line for some time, and lacked the experience and skill of the French army's best formations. There was a strong sense of *déjà vu* about 77 DI's attack on Barleux. The plan called for 159 and 97 RIs to pinch out the village, after which the division's *chasseur* battalions would push on to the river. Captain Humbert recounts that at zero-hour the assault troops went forward, in neatly ranged straight lines, marching at a steady pace, calmly and quietly. They advanced into a hail of machine-gun and artillery fire, and came up against uncut wire and unsubdued defences. The first wave nevertheless passed the German front line and pushed on, leaving the garrison to the mopping-up waves. But these were checked at the front line, where the enemy emerged from their undamaged dugouts in strength. As a result, the troops who had passed the front line were cut off and destroyed, while the reserve battalion, sent forward in support, was cut down in no man's land.[38] As was customary on the British front, the disorganisation occasioned by a partially successful attack told against the renewal of a general attack over the following days.[39]

In support of Tenth Army's attack, Sixth Army continued to press its advantage on the plateau south of Combles. On 4 September it took Sivas Trench and consolidated its hold on Cléry. Over the following two days, it pushed to within assaulting distance of the German third position. XXXIII CA moved the line forward in the Somme bend by seizing Ommiécourt in the marshes south of Cléry. Messimy's brigade made its first attack on 4 September, against a cluster of woods in front of the German third position, Bois Reinette, Bois Marrières and Bois Madame. It was a textbook assault. In one rush the leading battalions, 6 and 28 BCAs, took their first objectives, the track from Ferme de l'Hôpital to Cléry, which had been turned into a makeshift trench, and the observation posts on the ridge behind. Despite heavy machine-gun fire from the farm on their left, which 46 DI had not yet secured, Messimy's reserve battalion was ordered forward, in skirmish order, and the brigade pushed on to its final objective in the woods. Within three hours, all its objectives, and 150 prisoners, had been taken at a cost of 670 casualties.[40] The success of Sixth Army's sustained attack provoked a momentary panic in the German lines, which in the southern sector were pulled back hastily to the third position.[41] Foch's hope that a deep drive would force the defence to give ground seemed to be justified. When intensifying machine-gun and artillery fire indicated that the defence had stiffened yet again, the French divisions paused and brought their guns forward for their next set-piece attack.[42]

Foch and Fayolle were keen to exploit their advantage on Sixth Army's front before the defence consolidated.[43] However, three factors were starting to interfere with a sustained advance and forced a six-day delay before the next operation. As usual, one of them was wet weather, which interacted with the increasing logistical problems of getting munitions and supplies into, and exhausted men out of, the area of operations. The recent advance had used up I and VII CAs' immediate reserves, as well as extending Sixth Army's front as it fanned out on the plateau and diverged from the British advance. I CA had to throw out a defensive flank to screen Combles, a large village in the ravine that separated it from Fourth Army, advancing atop the Bazentin Ridge. New divisions had to be deployed, and the reserve corps, V CA, was brought up behind the battle front, with VI CA moving into designated GAN reserve. But, for the first time, Foch gave notice that Fayolle's cavalry corps might soon be brought forward to exploit the infantry advance: Sixth Army had reached the last line of fixed defences.[44] To be of any use, however, the horsemen would have to negotiate their way through the teeming ants' nest of the rear areas, across ten kilometres of shell-pocked landscape, past the mud-bound convoys that ground up and down the few serviceable roads leading forward from Bray and Maricourt, and through the maze of trenches that criss-crossed the featureless, debris-clogged brown ridges.[45] While the generals preached haste, conditions dictated sluggishness. General Guillaumat, commanding I CA, tackled the growing chaos in his lines of communication head-on, ordering his troops to remove abandoned vehicles and clear all the clutter from the roads and tracks, and to begin to resupply the forward zones in daylight, owing to the confusion caused by night-time activity.[46] While Sixth Army sorted out its supply lines there was a momentary calm, at least until the artillery preparation started again. "Petty war" continued in the interim: artillery fire and raiding carried on; isolated points of resistance, such as Ferme de l'Hôpital, were subdued; and jumping-off trenches were prepared for the next, potentially final, step.[47] The calm did not extend to Sixth Army headquarters, where Joffre had to intervene to quell another clash between his fractious subordinates. Foch "does not understand a thing . . . that man never listens to any of his generals," Fayolle complained, "his plan is dangerous and will not work." But he would have to do what his superior ordered.[48] In six days all was ready.

THE 12 SEPTEMBER attack was Sixth Army's biggest to date. Three army corps, with five divisions in the front line, attacked north of the river to seize the third—and final—German position between Frégicourt and the river, and the villages of Bouchavesnes, Rancourt and Sailly-Saillisel beyond. The right flank would then be pushed out to the line of the Tortille stream, a tributary that flowed through Moislains, around the base of Mont St. Quentin, and joined the Somme at Halles. Originally a defensive flank was to be established on the stream, while the reserves were pushed east and north-east in the direction of Bertincourt, but on 8 September revised instructions were issued for XXXIII CA to capture Mont St. Quentin. I CC was moved forward.[49] The German defence seemed to be reaching crisis point after the last attack, so Sixth Army's operations would be intensified to exploit this. Foch's long-nurtured scheme to turn the defences at Péronne seemed about to be realised.

If this operation is mentioned in British accounts of the battle at all, it is generally to bemoan the fact that the French offensive was not coordinated with the large British attack against the third position that was to take place on 15 September.[50] But as part of Foch's plan of concerted but separate attacks, the distinct nature of the French operation makes more sense, especially when it is realised that it was originally scheduled for 9 September, six days before the next large British attack but coinciding with a British attack on Ginchy, with a follow-up attack scheduled for the fifteenth. Moreover, Haig was insisting that the French first capture Frégicourt, from where the enemy had good observation over the right of Fourth Army's front of attack.[51] In the circumstances, in terms of both the wider operational plan and the specifics of cooperation at the Anglo-French boundary, an interval between the French and British attacks makes perfect sense. Unfortunately, a three-day gap (determined as much by the elements as by Anglo-French miscommunication) was exactly the wrong length: too long to allow the British to follow up French success on 12 September, too short to allow the French to give strong support to the British on 15 September. Foch's reassurance to Haig that he would have ample French troops to support any British success on the fifteenth was, in the circumstances, unfortunate.[52]

There was a rainbow over the French front on the morning of 12 September. It was a good portent.[53] Everyone was hopeful. Foch's personal order issued to the troops asserted that the enemy's defences and morale had been shelled to pieces, and the time had now come for bold, decisive action by all.[54] Sixth Army's attack repeated a familiar pattern: early success against the German front lines, with failure in some places where the enemy's machine guns had not been silenced; exploitation of success, consolidation and line-straightening over the next few days. Six kilometres of the German third line were taken, and in the centre at Bouchavesnes the front was pushed forward another three kilometres.

Six BCA's success was both spectacular and uplifting. In a rare positive remark about one of his subordinates, Fayolle noted that "Messimy's spirit was superb." Joffre too was "beaming, and embraced me."[55] Pierrefeu records that when Lieutenant de Brinon, Sixth Army's liaison officer, reported the fall of Bouchavesnes to the staff at G.Q.G., "the break-through was proclaimed; it was believed that on the morrow the pursuit would begin."[56] However, the celebrations were premature.

Two factors gave the defence the opportunity to seal the breach. First, the breakthrough was on too narrow a front, a brigade front of around two kilometres, forming a salient at the apex of Sixth Army's advance. The penetration could be contained by artillery and machine-gun fire from the flanks, from which two natural obstacles — Bois St. Pierre Vaast (a large forest that made Delville Wood look like a copse) and the Épine de Malassise Ridge — controlled the foreground. The other divisions that had attacked on 12 September had not done as well. It had taken all their energy to capture the trenches of the third position. Opposite Combles, 2 DI had taken the Bois d'Anderlu, but they had then been checked at the supporting line of trenches. These were taken over the following days, but vital time had been lost.[57] Forty-five

DI, on their right, were driven out of the trenches they had captured by a counter-attack in the afternoon. Their final objective, Rancourt, was beyond reach.[58] Seventy DI had taken Berlingots and Darfour trenches, which linked the third position to the river, and the Bois Berlingots and Vardar trenches beyond; but the guns on Mont St. Quentin had broken up their attack on Inferno Trench, their second objective on the ridge behind.[59] A local counter-attack forced the French out of both Berlingots Trench and Bois Berlingots overnight. For now, this would form a defensive flank. An attack on the strongpoints and artillery positions of Mont St. Quentin several kilometres away was a major operation that would have to await developments.

The second problem was logistical: "Alas, as always happened, the units required to exploit the breach could not be brought up quickly enough."[60] The breakthrough had come late in the day, and by mid-September daylight was fading by early evening. V CA was ordered to move up overnight and exploit the breach the next morning, with the cavalry ready to move up through Bouchavesnes and push on to Moislains.[61] But in the confusion which inevitably beset the lines of communication after a major push (worsened by torrential showers late in the day) the reserves were far from their assigned positions when daybreak came. I CC's horsemen were stalled among the infantry and transport columns stretching back to Bray, hopelessly lost in shattered and unfamiliar territory. Moreover, I CC's divisions had been stripped of their artillery, infantry and cyclist elements, which were already in the battle. They would not be complete before 15 September.[62] XXXIII CA's reserve division, 47 DI, did not get into place until the morning of the fifteenth, and meanwhile the exhausted men of 70 DI had to fend off further counter-attacks supported by flame-throwers on Vardar and Darfour trenches.[63] Foch blamed XXXIII CA's commander, General Nudant, who was indecisive and lacked drive.[64] Nudant protested, but he was relieved of his command three weeks later. Fayolle was right: his chief had little grasp of conditions on the ground.

As was customary, the offensive continued over the next couple of days—"the battle will be continued without stopping," Fayolle had ordered with uncustomary verve[65]—but with diminishing returns. The strongpoints at Ferme du Bois l'Abbé, east of Bouchavesnes, and the network of trenches around Ferme du Priez, between Combles and Rancourt, were taken, but heavy local counter-attacks had to be repulsed.[66] Inferno Trench continued to resist 70 DI's attacks. Inevitably, the fight degenerated into skirmishes for individual trenches and strongpoints. Nevertheless, intelligence analysis suggested that the defence was crumbling in the same way that it had south of the river in July. On 12 September there had been no concerted response to the French breakthrough. What local reserves there were had been deployed in the open field and had been shot to pieces by the French artillery. There seemed to be no large units immediately available to replace them: once more, individual battalions had been combed up from behind the front and hastily sent to hold the defences north of the Somme.[67] Unfortunately, after three days of heavy fighting, French units were tired and their ranks were thinning. For example, the two battalions of *chasseurs* attached to 48 DI to sustain its offensive power, whose normal complement was over a thousand men, were down to roughly three hundred rifles.[68] Fayolle wanted to pause to replace the front-line divisions and move his artillery forward. Foch demurred. Although he had available plenty of fresh troops—totalling three new army corps—they could not be brought forward in time. Sixth Army must therefore keep up the pressure on the enemy, for at last the British were to attack in force and France must support her ally.[69]

SUCH LOYALTY WAS perhaps misplaced. The British and French offensives were seemingly being conducted on totally different lines, des Vallières had reported wearily on 8 September. Haig was apparently dragging his feet, wilfully ignoring Joffre's directives to exploit success as fully as possible, focusing on short-range objectives and finding excuses to postpone renewing large-scale operations. Stolid British temperament and lack of command initiative were to blame, des Vallières concluded, and not much could be done about it. They would not act until the French had gained the heights beyond Morval.[70] Such was the unfortunate legacy of Haig's political problems in July and early August. For his part, he seems to have had no real perception of the violent struggle going on in the French sector. He noted blithely in his diary on 14 September:

> It is interesting to see how the French have kept on delaying their attacks! Their big attack S. of the Somme which was to have gone in on the 3rd has not yet been launched, and will probably not go in till the 15th. While the attacks of their 6th Army, which were to have gone in with such vigour at the beginning of the month have not yet materialised!

Haig went on to note that French infantry were poor and lacked offensive spirit, their progress was made against a much smaller concentration of artillery than on the British front and the Germans opposing the French advance were "very much inferior in physique."[71] It is a bizarre, erroneous and indeed deceitful summary for a man who always protested that he was doing his best to help the French: especially as he had previously noted the French successes of early September.[72] This may simply have been another short-lived fit of pique on Haig's part. The day before he had had occasion to remind des Vallières, who had just returned to say that Sixth Army would not be able to take Frégicourt before the British attacked, that "I decline to take *instructions* from Joffre or the G.Q.G." This had caused him to question Foch's assertion that the capture of Bouchavesnes heralded the piercing of the enemy's front; which in itself sits uneasily with his own expectation, expressed in his diary the very same day, that he was about to do the same.[73] Such impressions help to explain why the French army's progress in the first half of September—their front line was now some distance beyond that of the British—has all but vanished from history. On the eve of a major battle, perhaps Haig wished to rewrite the Somme campaign to his own credit, to assert his independence in the face of overweening allies and his personal control before the operation that might settle it once and for all. The tone of his most recent letter to Joffre certainly implied that his troops had done the work of wearing down the enemy, and that the French were cooperating with his operations.[74] Now his allies seemed about to snatch victory from under his nose! "The French are indeed very tiresome," Robertson had recently reminded him. But Castelnau's complaints that the British were going too slowly had some justification.[75] While he would not be hurried into it,[76] the time had now come to show France what Haig's army could do.

The Battle of Flers–Courcelette, as the British offensive of 15 September is officially designated, was intended to complete the capture of the intermediate position and push through the German third position. Reserve Army would cover the left flank, while in a four-stage operation Fourth Army would occupy the Switch Line between Martinpuich and Bouleaux Wood (including High Wood), break into the third position at Flers, and then occupy the line of villages that paralleled the Bapaume–Péronne road: Gueudecourt, Lesboeufs and Morval. Sixth Army would cooperate on the right by taking Frégicourt (which had not yet fallen) and Rancourt, thereby surrounding Combles. The newly arrived V CA would push into Bois St. Pierre Vaast.[77] Tenth Army would attack the German first line south of the river. If all this could be achieved, it would be exploited by a rapid advance beyond Gueude-court and Lesbœufs, for which cavalry would be deployed, and the rest of the German third position would be rolled up in the direction of Eaucourt l'Abbaye and Le Sars. Ultimately, Bapaume might be taken.[78]

Rawlinson and Haig repeated their exchanges over just how con-centrated the bombardment, and how deep and dispersed the infantry attack, should be. Haig wanted Fourth Army to push through all three lines of German defences facing it at once, while Rawlinson wished to take them on one at a time (in the French manner), not least because the third line was out of range of most of his guns. Haig's desire for a more ambitious scheme again prevailed.[79] In the final plan Rawlinson's three attacking corps (from left to right, III, XV and XIV) were to commit nine fresh, retrained divisions to the assault, with three tired infantry and five cavalry divisions in reserve. Reserve Army's Canadian Corps was to support the left flank by attacking Courcelette. II and V corps would make a feint attack around Thiepval.

These pre-battle discussions, which seem to echo those of May and June, can again best be explained by the fact that Haig was thinking strategically—his sense of self-importance notwithstanding, he was working within the broader parameters set by Joffre—while Rawlin-son, with wider experience of operations at the front, was thinking tac-tically and focusing on the strengths and obstacles of the defences facing Fourth Army. The ambition of the plan provides grist for the mill of Haig's critics (as well as belying des Vallières's reports of want of drive): once again over-optimistic wishful thinking based on mixed intelligence, which produced an unrealistic impression of declining German morale and shrinking reserves, persuaded Haig to think big. He certainly had evidence that German reserves were almost exhausted, their morale was in steep decline, and their resistance was on the verge of breaking,[80] albeit such assessments were "rather shaky" and empha-sised political and social rather than military factors.[81] The intelligence picture undoubtedly furnished a fair case for increasing the pressure and tempo of the battle. By the rationale of attrition, if the allies pos-sessed fresh divisions and the enemy did not, then the time had come to hit him hard.[82] "Destruction of enemy's field forces" remained a pri-mary objective of offensive operations.[83] Moreover, Haig was fully aware that this was a contingency for which he should be prepared, never a certainty: "The moment might possibly be favourable for cav-alry action... *Everyone* must put forward the greatest effort and be ready to suffer privation for several days in order to reap the fruits of victory."[84]

While Rawlinson paid lip service to Haig's promptings, urging his divisions to go forward as quickly as possible and to take risks, with a view to "forcing the battle to a decision at the earliest possible moment,"[85] it is clear that his own expectations remained more constrained.[86] But as Sixth Army was demonstrating, German defences could be breached. Moreover, broader offensive schemes, designed to develop over days or weeks, should not be falsely conflated with the specific objectives of a particular attack. But, as was becoming a habit, Haig certainly did expand the purpose of the operation at short notice, continue tinkering with the details of the plan right up to the day before the attack, and force a tempo on his subordinates that was too fast for the conditions of the battlefield,[87] none of which made Rawlinson and Gough more likely to succeed on the day.

Like that of 1 July, this attack was conceived as the first stage of a broader offensive, the break-in that might be followed by a break-through. This was not to be a mad cavalry rush, but a "bold and vigorous" all-arms exploitation of disorder in the enemy's defence if the infantry took all four of its objectives on time and there were no enemy reserves behind. The scheme of attack prescribed that the cavalry would "assist the infantry advance," first by helping to establish a "covering force" along the ridge paralleling the Bapaume–Péronne road, and then cooperating in rolling up the German third line.[88] This might take some days. It might in time allow the British armies further north to join in the rout of the enemy, "a success sufficient to repay all the great efforts made during the last two and a half months," although too many uncertain variables remained for detailed instructions for future exploitation to be issued.[89] Since this did not occur, Haig's paper plans to roll up the line from Bapaume, and bring Third, First and Second armies successively into the battle as the enemy's line fragmented, have inevitably provoked scorn.[90] Maps that indicate cavalry objectives as far away as Cambrai do give substance to those who claim that his intentions went far beyond the means and methods at his disposal.[91] However, to argue from this that his ambitions "were now virtually unlimited . . . nothing less than the most grandiose vision for victory developed by any commander since Schlieffen" is overstatement.[92] Rather, they reflected a general belief that the battle was reaching its climax after two and a half months of wearing down the enemy's reserves and morale.[93] In the circumstances contingency planning—in case it was true, as Foch suggested, that there were now only rearguards in front[94]—makes perfect sense. Haig would have been more open to criticism had Fourth Army delivered a crushing blow, on 15 September or afterwards, with no follow-up scheme in place for "after the enemy has been driven from his prepared lines of defence."[95]

What really threatened to curtail the prospects of "pursuit," however, was "the broken state of the ground and roads from shellfire,"[96] and the difficulty of moving troops forward and supplying them over the "high, waterless and roadless plateau" that Fourth Army was expected to occupy.[97] Careful preparations were made to begin restoring communications beyond the old British front line as soon as the attack began, including horse-tracks for cavalry and horse-drawn transport.[98] The cavalry were to go forward in dribs and drabs, to cooperate in troops and squadrons with the other arms. Although there were five cavalry divisions behind the British front, with communications in their present state there was no scope to employ more than a fraction of them in the early stages of the offensive.

The British army did, however, have one new weapon that promised a nasty surprise for the enemy and the possibility of more rapid and complete success than hitherto. The Battle of Flers–Courcelette's place in the annals of war is guaranteed because it saw the first use of tanks on the battlefield. Indeed, such was their novelty and impact that accounts of 15 September, and subsequent controversies, centre on their deployment and use. Winston Churchill claimed a degree of paternity for the idea of armoured caterpillar-tracked vehicles to cross the enemy's trenches,[99] but it had occurred to a number of people around the time that stalemate set in on the Western Front. This is not the place to retrace the story and arguments about the development of the tank. But the controversies about their first use should be mentioned. The principal accusation against Haig is that he committed a new wonder-weapon prematurely. It is a charge that Churchill lays cogently in his memoirs.[100] Lloyd George, Asquith and Hankey had issued such a warning when they visited.[101] Haig's critics argue that it would have been better to keep the secret, to have massed tanks for a large surprise attack (as was to happen at Cambrai in 1917 and Amiens in 1918), than to give it away in a moribund offensive. But this does not hold up under close scrutiny. As Haig rightly pointed out to Lloyd George at the time, the offensive was not moribund but reaching its climax, and in the circumstances everything that could contribute to victory on the battlefield ought to be employed.[102] Only with hindsight was the tank's potential clear. Even by 1918 it remained a one-shot weapon, effective only in combination with the other arms, an important addition to the tactical weapons system of industrial battle, but not yet a battle-winner in its own right. Moreover, a new weapon has to be tested in battle conditions, and appropriate tactics developed. Devoting precious raw materials and factory space to building hundreds when its battlefield utility was uncertain would have been irresponsible.

Haig had fifty of the new war machines at his disposal on 15 September: he had hoped for more but production bottlenecks delayed deliveries. G.H.Q.'s tactical note for employing the tanks—officially "machine-gun carriers" of the Heavy Section, Machine Gun Corps— recognised that they were an untried infantry-support weapon. Their job would be to crush wire—special lanes were to be left in the supporting barrage to allow them to advance ahead of the infantry—to take on unsubdued trenches and machine-gun posts and to protect the infantry once it had reached its objectives. They would be distributed in small groups to the attacking divisions, and assigned particular German strongpoints as their objectives. The Mark I tanks were slow, cumbersome leviathans, which thoroughly justified their designation of "landships."* They advanced more slowly than marching infantrymen across the shell-torn Somme battlefield (and were likely to get stuck in the mud), were mechanically unreliable and exhausting for their crews to operate, and were virtually blind and potentially vulnerable to enemy artillery fire. They would certainly surprise the enemy; they might give valuable aid to the infantry advance; but they would have to work closely with the other arms to fulfil their potential.

That Flers–Courcelette was a far more successful operation than those of the previous two months was due more to effective artillery method than to the commitment of tanks for the first time, although in places the new machines did contribute materially to subduing local defences. It is clear from the detailed orders and careful training of the assault divisions that tactical lessons from earlier operations were being incorporated into planning and preparation.[103] Artillery technique was developing, even if Fourth Army's bombardment was not yet being prepared with the meticulous calculations of shells-per-metre of Sixth Army's gunners. By mid-September, Fourth Army had received many more heavy guns, and stocks of shells were plentiful. The offensive

---

*There were two variants: "male" tanks armed with two 57mm cannons and four machine guns; and "female" tanks armed with four machine guns.

would be supported by 1563 guns: 1157 18-pounder field guns and light 4.5-inch howitzers, and 406 medium and heavy field guns and howitzers.[104] The barrage would be twice as concentrated as that of 1 July, although only half as heavy as that of 14 July.[105] Moreover, artillery technique was becoming much more sophisticated, with high-explosive shells used to cut wire, gas shells (fired by British guns for the first time) for neutralisation of enemy artillery, long-range interdiction fire and a creeping shrapnel barrage—at a slower speed and with a greater concentration of shell[106]—all employed to disrupt the enemy's response and fire the infantry on to their objectives. Survivors' accounts attest to its awesome power:

> A sea of iron crashed down on all the front and support lines . . . The noise was terrible. Impact after impact. The whole of no man's land was a seething cauldron. The work of destruction grew and grew. Chaos! It was impossible to imagine that anyone could live through it . . . It was like a crushing machine, mechanical, without feelings; snuffing out the last resistance with a thousand hammers. It is totally inappropriate to play such a game with fellow men.[107]

At 06:20 on 15 September the infantry went over the top. This time was chosen to allow the tanks to come up under cover of darkness and move across no man's land ahead of the infantry. The intention was to engage the defence two minutes before the infantry assaulted. However, moving slowly in the dark across unfamiliar and shattered ground, many of the tanks broke down or got lost and arrived late. XV Corps's success in the centre, against Flers village, where the tanks fulfilled the mission assigned to them, was not matched on the flanks. Forty-first Division's left brigade cleared its first objective in just twenty minutes, with the help of tanks that had come up on time, and pushed on to its second, Flers Trench, covering the village. The garrison from 9 Bavarian IR panicked in the face of this rapid advance, abandoning the trench. Four tanks then pushed on to Flers, one leading the infantry in while the others strafed the village from its eastern edge and crushed German strongpoints under their tracks.[108] One of the enduring images of this day, of a tank spotted from the air moving up the main street of Flers with British infantry cheering behind, symbolises both the high point of the attack and the morale impact of successful tank action.[109] Flers was in British hands by 10:00. However, such effective infantry–tank cooperation was not repeated elsewhere. To right and left of 41st Division, 14th and the New Zealand divisions' tank support had not arrived on time. Both attacks encountered heavy machine-gun fire, and were able to press on only when the panic on 41st Division's front spread to left and right. These divisions came up on either side of Flers (with their surviving tanks belatedly joining the battle) but, in the face of stubborn resistance from German field-gunners firing shrapnel over open sights from the slopes in front of Gueudecourt,[110] they were unable to reach their final objective. The Flers–Courcelette attack was collapsing due to overambitious initial objectives, and poor timing and cooperation between the infantry, artillery and new tank arm.

The tactics for integrating the tanks with the artillery bombardment and infantry advance went badly awry in XIV Corps's sector, opposite Morval and Lesboeufs. Hundred-yard-wide lanes had been left in the creeping barrage along which groups of tanks would advance towards the strongpoints that had been assigned as their objectives. When the tanks failed to arrive, however, as most of them did on XIV Corps's front, this left gaps in the infantry's supporting barrage, and machine-gun-filled strongpoints which were not under artillery fire. All three attacking divisions, the Guards, 6th and 56th, inevitably suffered heavy casualties, particularly from machine guns in the Quadrilateral Redoubt in 6th Division's sector, which, owing to inaccurate aerial observation, had been missed by the preparatory barrage.[111]

Nevertheless, as one might expect from the British army's elite, the Guards pressed on through the hail of fire, across the German first line and on towards Lesboeufs. They gave no quarter to the Bavarians who opposed them.[112] Raymond Asquith was killed during this confused mêlée. This relentless advance carried the guardsmen forward more than a mile on a mile-wide section of the front. But it was a chaotic advance, with the battalions becoming hopelessly intermingled. Lieutenant Harold Macmillan was wounded in the knee and rear; Harry Crookshank in the groin. The two badly injured future Cabinet ministers were eventually brought in by stretcher-bearers and evacuated to a field hospital. Another future minister, Oliver Lyttelton, advanced with a small party that included a future field marshal, Harold Alexander, on Lesboeufs, the division's final objective. They were forced to ground in an isolated trench ahead of the rest of the division, then eventually driven back to their own lines by an overwhelming counter-attack. They had waited within striking distance of the empty village of Lesboeufs for several hours. If organised reserves had been available to occupy it, then this fleeting chance might not have gone begging, and the attack might have delivered on its high promise.[113] Lyttelton was awarded the Distinguished Service Order for his leadership and enterprise.[114] It was the furthest penetration on the right flank. With the failure of the divisions to their right, and with heavy casualties themselves, the Guards were obliged to dig in well short of their final objective.

On the left, III Corps finally captured High Wood, and Martinpuich village to its left; but only after a hard fight, and a certain amount of farce. In order not to strafe the British front line, which was very close to the German, the creeping barrage in High Wood was to start 150 yards *beyond* the German front trench. Hence the infantry went forward against unsilenced enemy machine guns that took a heavy toll of 47th and 50th divisions' men. Against the advice of his tank advisers, Pulteney had ordered four tanks to enter High Wood to clean it up. But the shattered tangle of undergrowth and debris proved almost impossible for tanks to cross. Belatedly, one tank managed to enter the central ride of the wood; another drove along its eastern edge, machine-gunning the British infantry waiting to assault it. High Wood did fall on 15 September—after two months' fighting—but only after a further intensive trench-mortar bombardment and being outflanked by the divisions advancing through the Switch Line to either side.[115] Major-General Barter, commanding 47th Division, who had been highly critical of the details of III Corps's planning, was sacked for "wanton waste of men" but really to save his corps commander's face.[116] Fifteenth Division on the left flank did better, taking Martinpuich. It did so without useful assistance from the tanks, which came up late.

Reserve Army's attack on Courcelette was a great success. Here the tanks were assigned to follow the infantry and mop up the village, so there were no gaps in the creeping barrage. When the attack went in, there was little organised resistance in the front position. Overnight the enemy had mounted a counter-attack against the Canadians' right flank, which had been beaten off with heavy losses; troops were mustered in the trenches opposite preparing for another attack when the Canadians' supporting barrage hit them. Following the moving wall of shrapnel closely, 2nd Canadian Division's assault companies overwhelmed a trench full of dead and wounded men before pushing on to their final objective, the ruins of Courcelette's sugar refinery and the appropriately named Sugar and Candy trenches to left and right. These fell after a short, sharp fight. By 07:00 the division could wire back that it had secured its objectives. The tanks arrived too late to do anything but mop up a few isolated pockets of resistance. Although they had fulfilled their mission of establishing a defensive flank west of Martin-puich, the Canadian Corps had orders to push on into Courcelette village if opportunity presented itself. It did, and in the afternoon 2nd Canadian Division's reserve brigade was deployed. The assault battalions, the 22nd (Canadien Français) and the 25th (Nova Scotia Rifles), set off from the old British front line. In an impressive display of field-craft and small-unit leadership, they marched forward three-quarters of a mile in well-dressed lines through the German counter-barrage, wheeled half-right at the new front line and followed their own barrage into Courcelette. The assault battalions pushed through to the northern edge of the village while the 26th (New Brunswick) Battalion, in support, mopped up. Some parties put up fierce resistance, but the Canadians were equally ferocious and quickly prevailed. "There is not much fight in the Germans," Private Gordon Silliker said derisively:

> They are all right if they can get in some safe place where you cannot see them and they can snipe at you or put a machine gun on you . . . It's a different tune they have when you get them on the point of the bayonet and especially when they get up against a Canadian. Whenever they see the Canadians coming they as a rule meet them with their hands up.

Second Canadian Division, which had not done very well in its previous engagements in the Ypres salient, had initiated the process that would turn the Canadian Corps into first-class shock troops.[117]

Courcelette's defenders, a newly arrived battalion from 212 RIR, wilted under their first experience of a concentrated Somme bombardment. The situation behind the German front was critical: G.H.Q. intelligence staff's estimation that the German army's reserves were exhausted was accurate. Future German Chancellor Franz von Papen, operations officer of 4th Guards ID, wrote that in the late afternoon he

> hurriedly collected batmen, cooks, orderlies and clerks from our own divisional headquarters and from a few neighbouring formations, and with them produced signs of activity as if fresh reserves had arrived. In fact, there was not a single reserve company for scores of miles behind us. A complete tactical break-through, the achievement of which the enemy dreamed, had taken place, although they did not seem to realise it. A few dozen administrative personnel was all that stood between the enemy and a major victory.[118]

Germany's official history acknowledged that the defence was almost completely broken by the end of the day.[119] The 15 September attack must be judged a qualified success, even though as on 1 July the quickest advance had been on a flank, rather than in the centre. Four thousand yards of the third position were seized, and all the Switch Line that had been holding up the British for two months was occupied. It endorsed the principle that powerful general action was more effective than small-scale, localised attacks. But if it had come near to achieving Haig's ambition for a breakthrough into open country, Papen was right that G.H.Q. were oblivious to that fact: the cavalry stayed in their assembly areas and the infantry consolidated their objectives rather than pushing beyond them. The caution inculcated by previous failures may have precluded a real triumph. Moreover, the tanks had not made a particularly promising debut. Although in many places they lent valuable assistance to the infantry, especially in subduing hostile machinegun emplacements, they proved very vulnerable.[120] During the course of the battle their number was whittled down: of the 50 delivered, 49 were deployed for the attack and 36 made it to the front; 27 reached the German front line, either before or after their own infantry, but only 6 got as far as the third objective.[121] Enemy shells (and on occasion friendly fire), but more often mechanical failure or ditching on the hummocky, trench-scored battlefield, accounted for the failures. Only three tanks remained serviceable on 16 September. Such an attrition rate indicated that they still left much to be desired in terms of reliability and use. Careful thought would have to be given to their future tactical handling, and more attention paid to their armour and mechanics. They were accepted as a useful addition to the armoury, but it was recognised that they remained "an accessory to the ordinary method of attack, i.e., to the advance of the infantry in close cooperation with the artillery."[122] Nevertheless, Haig immediately ordered another thousand.

IF THE TANKS' debut on the battlefield had been inauspicious—lending credence to the view that such novel weapons needed to be tested in real conditions to ascertain their tactical capabilities—their reception by the public was rapturous. The newspapers competed with hyperbolic accounts of these new land leviathans, of the murderous assistance they brought to the infantry, and the terror they inspired in the enemy. H. G. Wells's "Land Ironclads" or William Heath Robinson's "Terrible Machines" of war, imagined in 1914 and 1915, had come to life. In the absence of photographs, sketch artists, Heath Robinson among them, let their imaginations run wild when depicting the landships: huge, trench-crushing metal monsters, bristling with flame-spouting guns, with hysterical Huns fleeing from their grinding tracks.[123] It was media hype that belonged to a later age. The reality was that few British troops saw the tanks in action, and those who did regarded them with amusement rather than awe; and soon with concern, after they realised that they were best avoided since they drew the enemy's fire.

The German press was more muted in its response: it would not do to stir up an excitable and potentially fractious population with tales of war-winning weapons. The "devil's coaches," the *Düsseldorfer Generalanzeiger* reported, caused only momentary awe in the front-line soldiers. As soon as the new war machine fired its guns it was just another weapon: "The men came back to their senses; their vigour and tenacity returned."[124] By now, German soldiers were familiar enough with the contraptions and surprises of industrial warfare not to be too thrown by the tanks' appearance. Furthermore, anti-tank defence was easy to improvise. The thin-metal boxes were vulnerable to field guns and the armour-piercing bullets used against British snipers' shields.[125] Even machine-gun and rifle fire, which caused splashes of red-hot metal when they hit the tanks' armour-plate, could incapacitate the crews, if the fumes from the vehicles' own engines had not done so already.

So on the battlefield the tanks were soon just another weapon in the expanding arsenal of trench warfare, although "tank fever" persisted on the home front. One man who had been impressed, Geoffrey Malins, gave the new war machines star billing in his third film, *The Battle of the Ancre and the Advance of the Tanks,* which opened in January 1917. Tanks became the subject of music-hall songs and the centrepiece of variety sketches, making a mockery of the bloodbath going on in France. As Siegfried Sassoon opined in his February 1917 poem "Blighters,"

> I'd like to see a tank come down the stalls,
> Lurching to rag-time tunes, or "Home, sweet Home"
> And there'd be no more jokes in Music-halls
> To mock the riddled corpses round Bapaume.[126]

ON 16 SEPTEMBER the usual attempt was made to follow up. Rawlinson urged his subordinates to push on "with the utmost vigour" to "complete the enemy's defeat" while the defence was disorganised.[127] But the hard fight of 15 September had been costly for Fourth Army: the losses were on the same scale as those of 1 July, over 29,000 casualties (although there was much more to show for them), and Rawlinson's divisions were blown.[128] So only a few small-scale, poorly organised, line-straightening operations took place on the sixteenth, with, as might be expected, heavy losses and meagre gains. With the onset of further heavy rain there was a pause. The guns were moved up and the front-line troops relieved. However, despite its incomplete results, Flers–Courcelette was finally a feather in Haig's rather scuffed cap. Friendly congratulations were exchanged between G.Q.G. and G.H.Q., and King George sent a special message congratulating his troops.[129] Privately, Haig gloated. His army seemed to be taking on more Germans, and he was beating his ally as well as his enemy now. A bizarre dig at French "folly" in thinking that they could push through the German defence in one continuous sweep suggests that Haig had grasped the true nature of the offensive, if not his ally's methods.[130]

One single blow, no matter how well executed, was never going to decide the battle. Foch knew that. Nevertheless, a faster tempo promised more than intermittent blows or small-scale operations because it would allow the enemy no time to recover. For this reason, by mid-September the renewed offensive was clearly delivering positive results and, if sustained, it had the potential to break the defence once and for all. Even Fayolle allowed himself a small, qualified hope. "We may well be on the eve of the decisive period of the battle of the Somme," he judged. "The Boche is really disorganised in front of us. He is suffering terribly."[131] As Below acknowledged, 15 September was "a very heavy day, with serious losses, even by Somme standards." Many defending battalions were annihilated, suffering more than 50 per cent casualties. Morale inevitably suffered.[132]

There was more to come. After the twin punches of 12 and 15 September had knocked the defence off balance, Foch intended this to continue at regular intervals for the rest of the month. With Fourth and Sixth armies rearranging their fronts—as Fayolle had warned, Sixth Army's hastily mounted 15 September attacks with tired troops and inadequate bombardment had made only meagre gains[133]—the focus shifted once more to the flanks. Between 15 and 18 September Tenth Army renewed its offensive, taking Berny, Vermandovillers and Déniecourt, along with several thousand prisoners.[134] On the eighteenth, in a hastily organised but relatively smooth and bloodless line-straightening operation, 6th Division captured the Quadrilateral.[135] Another general attack was arranged for the twenty-first, with Reserve Army joining in with another attack on Thiepval. But the weather intervened once more, and the resumption of operations was delayed. Thereafter, though, with four days of sunshine from the twenty-second to the twenty-fifth, the Anglo-French armies were to prepare their most powerful combined attack since 1 July.

Fourth Army was to complete its 15 September operation, renewing its attack on the Gueudecourt–Lesboeufs–Morval line and pushing on towards Le Transloy. Cavalry squadrons were to be attached to XIV and XV corps headquarters to allow for more rapid reaction to favourable developments, but again their role was really that of mounted infantry, patrolling forward and seizing empty ground in cooperation with the other arms. III Corps was to cover the flank by pushing into the network of trenches that covered Eaucourt l'Abbaye and Le Sars.[136]

Sixth Army had been heavily reinforced: VII CA had been relieved after ten weeks in the front line, and V, VI and XXXII CAs had come into the line between I and XXXIII CAs to fill the expanding front, some twelve kilometres longer than before the 12 September advance. Heavy batteries were redeployed from Verdun and Tenth Army's front to support the renewed attack, and Sixth Army was allocated an additional daily allowance of heavy munitions from GAN's carefully managed stocks.[137]

Reinforcements entering the Somme battle were both hopeful and apprehensive. In early August, Commandant Courtès's 46 RI had left the Vauquois sector, at the western end of the Verdun battlefield, where they had been engaged in a close-range mining and counter-mining struggle with the enemy for the previous sixteen months. Two weeks at Mailly training camp followed, where they updated their fieldcraft: "evolutions, combat formations, rapid advancing, assault waves, mopping up of trenches, use of the new weapons—light machine guns, rifle grenades, 37mm cannons—it's intensive instruction, preparation for the immediate future. Will it be Verdun? Will it be the Somme?" Neither was a welcome prospect, but the Somme was seen as the lesser of two evils. Still, "Don't go there," soldiers shouted from troop trains passing to the rear as 46 RI moved up, "don't go there!" In camp on the banks of the Somme, along which barges filled with British wounded moved, the rolling thunder of the guns could be heard in the distance. The barrack huts' thin walls reverberated with the concussions of the barrage, and the firework display of shells and flares could be seen on the horizon, a "very impressive spectacle." Convoys were passing unceasingly towards the front: "Oh, the roads of the Somme. Such an impression of intensity and febrile activity, relentless work, but with a will. But also, what mud." Lines of men and trucks moved along the crests, undisturbed by enemy shells. Rumours from the front line passed through the throng of expectant reinforcements: we have started the breakthrough; the British are doing well; Messimy's *chasseurs* have taken Bouchavesnes; we are to go up to reinforce them; the enemy is in disorder, they cannot hold on; our artillery are working miracles; the front is ready to break. "A waft of victory seems to pass over everyone. It would be well if that were the case . . . Is it true that the enemy is routed? Or is he playing dead, regrouping his disarrayed forces?"[138]

Sixth Army's intention was to renew the attack along its whole front from Combles to Mont St. Quentin; seven divisions would participate in the first stage, pushing north-westwards. I CA would take Frégicourt and Sailly-Saillisel, while XXXII CA would take Rancourt, the western half of Bois St. Pierre Vaast and Saillisel. V and VI CAs would take the rest of the wood and the Épine de Malassise. If all went well, reserve infantry and cavalry would move on Rocquigny, Mesnil-en-Arrouaise and Etricourt.[139] Haig's oft-planned cavalry exploitation might attract knee-jerk scorn; but by this point Foch felt the time might be near when he could launch cavalry against deep objectives in a defeated enemy's rear areas.[140] To the south, VI and XXXIII CAs would push east and south-east, establishing a defensive flank along the Tortille with a view to manoeuvring against Mont St. Quentin and Péronne.

However, during this brief lull the enemy had been busy too. A new Switch Line had been excavated on the reverse slope beyond Frégicourt and Rancourt—Prilip, Porte de Fer and Négotin trenches—linking Morval to the western apex of Bois St. Pierre Vaast and covering Sailly-Saillisel. Trenches were dug along the north-western and south-western edges of the wood, with the line extending along the Épine de Malassise and down to the Tortille at Allaines. The enemy, Fayolle complained, were better field engineers than his own men. In the battle with picks and shovels they could construct a defensive line faster than his own troops could prepare the gun emplacements, jumping-off lines and communication trenches needed for the next assault.[141] German engineers had also been active opposite the British. Air reconnaissance reported a fourth position being constructed along the Bapaume–

Péronne road between Bapaume, Le Transloy and Sailly-Saillisel.[142] The advance would be contested every metre of the way, if only to gain time. Reports from agents in occupied France indicated that the Germans were laying out a new defensive line some thirty-five miles behind the Somme front: the possibility that they would conduct a strategic retreat if allied pressure became too great had to be considered.[143] Germany's reserves were stretched. Although Ludendorff had recently created fifteen new divisions, this did not represent a real accretion of strength since it had been done by stripping regiments from existing divisions and combing Germany's depots.[144] Three of these green divisions, 212, 213 and 214 IDs composed of *Ersatz* reservists, were brought up to replace the shattered formations opposite Sixth and Tenth armies.

Fourth and Sixth armies' combined offensive, which for almost the first time since 1 July began at exactly the same time, 12:35 on 25 September, had mixed results. In the centre good progress was made either side of Combles. Fourth Army, attacking more limited objectives under a more concentrated bombardment than on 15 September, occupied Morval and Lesboeufs, and came within striking distance of Gueudecourt, although its garrison repelled attempts to enter the village. To the south, 42 DI's left flank regiment, advancing behind the creeping barrage "as if on an exercise,"[145] took Rancourt within the hour, but its right could not push on to the machine-gun-filled edge of Bois St. Pierre Vaast. Here 94 RI's 2nd Battalion, attacking Jostow Trench, lost two-thirds of its strength in just a few minutes, "one of the bloodiest checks of the war."[146] I CA completed the capture of Frégicourt in the early hours of 26 September, and pushed on with 42 DI's *chasseur* battalions to the new line of trenches behind, gaining footholds at either end of Porte de Fer Trench, and in the western apex of Bois St. Pierre Vaast.[147]

The *poilus' esprit de corps* remained excellent. Ninety-four RI still wished to go forward, even with no officers: "It is always the *chasseurs* who get all the glory."[148] British and French armies finally seemed to be working in harmony; even, at their point of junction, in tandem.[149] Combles, now virtually surrounded, was pinched out in a joint operation by Fourth and Sixth armies working from north and south in the early hours of 26 September before the Germans could complete its evacuation.[150] Gueudecourt fell later that day. British tactics were becoming more sophisticated. After a preliminary bombardment, a tank went forward and straddled Gird Trench, which blocked the

British advance, raking it with machine-gun fire. Bombing parties followed behind. Artillery-spotting aircraft flew low and strafed the beleaguered garrison: 370 German survivors threw up their hands in surrender. Two enemy battalions had been overwhelmed. The attackers suffered just five casualties. The infantry pushed into Gueudecourt, and cavalry patrols were ordered forward to scout the apparently empty ground towards Ligny and Le Transloy. Artillery fire forced the horsemen to dismount, so on this occasion there would be no exploitation of disorder in the defence, but the signs were growing that it was imminent.[151] Fourth Army's large gains with relatively light casualties in the Battle of Morval were testimony to how far the British army's tactical methods, especially in the vital area of infantry–artillery cooperation, had come since July. Robin Prior and Trevor Wilson, who have delivered the most detailed critique of British fighting methods in the offensive, were willing to concede that "in Somme terms it had been an outstanding success," even though they insisted that this was achieved more by accident than design.[152]

However, the day was not a total triumph for the attackers. On Sixth Army's front south of Bouchavesnes, where the guns around Mont St. Quentin dominated the ground, progress was disappointing. Gamelin's *chasseurs alpins,* back in the front line, attacked Inferno Trench, concealed on a reverse slope. Without good observation the French bombardment had done limited damage, and Gamelin's men could make no progress against concentrated machine-gun and artillery fire.[153] Further north, Courtès's 46 RI, dug in along the Bapaume–Péronne road, had a bloody baptism of fire. Moving up through the litter of German corpses, the men were smiling. But once they reached Bouchavesnes, "the village on which the daily communiqué focuses its attention," the monstrous reality of the Somme, "an atrocious battle of artillery," quickly dawned on the newcomers. 210mm shells erupted from Mont St. Quentin and crashed down along the desolate road, its surface smashed by shelling, the trees on either side reduced to a double row of skeletal brushes, and its side ditches filled with bodies. "The road seems to be the edge of the last cycle of hell; on the other side, one enters the unimaginable." On 25 September Courtès led his men across that road, and into the inferno, back into the fields where General Ebener's divisions had been cut to pieces in 1914. At midday two battalions went over the top, astride the road from Bouchavesnes to Moislains. If the enemy had been in disarray, he had now recovered. The German barrage crashed down, while aircraft flew overhead, strafing the assault waves. Previously undetected machine guns hidden in shell-holes between the lines opened fire. The attack gained only three hundred metres. Courtès's men added their own to the jumble of cadavers and limbs that stretched from Bouchavesnes back to Curlu: field-grey German corpses, horizon-blue *chasseurs,* colourful *Zouaves,* even disinterred blue-and-red *poilus* and bleached bones from the desecrated village cemeteries. "Vauquois was a crest, Bouchavesnes is a hole," remarked General Valdant sardonically.[154] Between the guns of Mont St. Quentin and the machine guns of Bois St. Pierre Vaast, Sixth Army's southern flank was contained. Renewed assaults on 26 and 27 September pushed 10 DI's line to within striking distance of the south-eastern edge of the wood, but only by adding hundreds more bodies to the heaps already spread across the plateau.[155]

ON 26 SEPTEMBER, RESERVE Army renewed its attack on Mouquet Farm and Thiepval. British tactics had developed significantly since 1 July, and the Thiepval garrison's tenacity could not defeat overwhelming firepower. The bombardment was the heaviest yet fired by Reserve Army gunners. It was to be supplemented by an indirect machine-gun barrage, which would sweep the German rear areas as the assault troops advanced. The British battalions, assisted by tanks, made steady progress behind their creeping barrage. Mouquet Farm fell in the late afternoon, after British pioneers winkled out its last few defenders from their dugouts with smoke grenades.[156] The assault along the Ancre valley secured lodgements in Zollern, Stuff and Schwaben redoubts, and patrols pushed on to the edge of St. Pierre Divion. Thiepval village was finally surrounded. Inevitably, the German garrison resisted this final assault on their strongpoint: it was a matter of pride to the Württembergers of 180 IR who had held Thiepval against all comers since 1914. A vicious close-quarter fight continued through the night as the garrison was mopped up. But by 08:30 on 27 September, Thiepval was finally in British hands.

It was appropriate that this *coup de grâce* was delivered by Maxse's 18th Division, a New Army formation that had done well on 1 July, and had now returned for its second tour of duty on the Somme. Maxse was justly proud of the fact that his men carried out "a deliberate assault and the capture of a considerable depth of intricate trenches defended by stubborn German regiments who had held their ground against many previous attacks."[157] Although Maxse's pre-battle preparations had once again been thorough,[158] on-the-spot leadership and individual acts of heroism made the difference between success and failure. Three Victoria Crosses were won in 54th Brigade, while Lieutenant-Colonel Frank Maxwell VC, who organised the scattered remains of the assault battalions into a coherent defence once the village had been occupied, probably deserved another. Instead, he was soon after rewarded with command of a brigade. Haig came to congratulate Maxse personally on his division's success; Clemenceau too. But the warm welcome extended to the "conqueror of Thiepval" by the local country doctor in

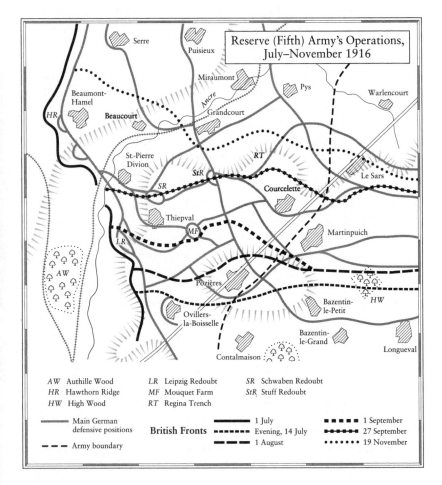

Reserve (Fifth) Army's Operations, July–November 1916

AW	Authille Wood	LR	Leipzig Redoubt	SR	Schwaben Redoubt
HR	Hawthorn Ridge	MF	Mouquet Farm	StR	Stuff Redoubt
HW	High Wood	RT	Regina Trench		

Main German defensive positions
Army boundary

British Fronts　1 July　Evening, 14 July　1 August　1 September　27 September　19 November whose house Maxse was billeted touched him even more.[159] On 28 September, 18th Division followed up its success with an advance into the Schwaben Redoubt, although it could not repeat its earlier triumph. An eight-day bomb fight ensued before the division was relieved. At that point small parties of the enemy still held out in the northern edge of the redoubt.[160]

The fall of Thiepval, which had tormented the British since July, was a decisive moment: it was "absolutely crushing . . . every German soldier from the highest general to the most lowly private had the feeling that now Germany had lost the first great battle."[161] General Palat, one of France's official historians, acknowledged that this victory marked the end of the British army's apprenticeship; it could now take on and beat the professional German army.[162] Certainly, British and imperial divisions, of which Maxse's 18th is a prime example, had come a long way in their tactics and leadership during three months of intensive fighting. Before the end of October all but a handful of the British divisions serving on the Western Front were to endure this harsh but effective battle school. If not yet the veterans that they would become, by the end of September 1916 British junior officers and other ranks were no longer naïve or casual in the military arts.

GALLWITZ AND BELOW seemed to have no effective answer to this relentless hammering. German defensive tactics had become predictable, and the regular counter-attacks were easily beaten off with artillery and machine-gun fire, sacrificing the army's best combat units in the process.[163] A succession of counter-attacks between Bois St. Pierre Vaast and the river between 20 and 23 September, supposedly the heaviest yet delivered during the battle, failed to regain any of the ground lost since 12 September.[164] The thick attacking waves were destroyed by French artillery fire: losses that could not easily be replaced. It has been suggested that in September the German army suffered its heaviest rate of loss since the battle began. More worryingly, only 10 per cent of the drafts required to fill the depleted ranks could be found.[165] Men of the 1917 class and reclassified men from home garrisons, support and *Landwehr* units were being sent to the front as replacements. With an increasing rate of attrition, divisions could not be rested sufficiently between tours in the line: their standard fourteen-day rotation meant that a new formation had to be fed in once a day. The defence once more became reactive, hurried and disorganised, like that of early July. Although it had not yet ruptured (or at least only briefly and not terminally), the loss of key defensive positions, the increased rate of attrition and sustained pressure throughout September was bringing it to the point of collapse, both morally and materially.[166] By the end of the month, the German army was wilting, and there was justification for some gloating. Brigadier-General John Gelibrand, whose 6th Australian Brigade had earlier been on the receiving end of the worst the German artillery could manage at Pozières, informed his brother on 26 September:

> At this date things are moving a bit and I guess it will be a pretty melancholy ... Xmas in the Fatherland ... Their cry that the Somme is not war but murder is of course the comic relief of the piece — and whilst they had the superiority of guns and rounds they thought it was quite a good war. Things do look a bit different when the boot is on the other foot.[167]

Certainly, in September, as Romania's entry and attacks on other fronts were coordinated with the reinvigorated Somme battle, the strategy of the General Allied Offensive seemed finally to be producing results. Germany's new war leaders were hard pressed. "September was an especially critical month," Ludendorff admitted. "The battles [were] among the most fiercely contested of the whole war, and far exceeded all previous offensives as regards the number of men and the amount of material employed." The attacks from 25 to 28 September were "the heaviest of the many heavy engagements that made up the battle of the Somme."[168] But the huge armies of 1914–18 were resilient: hard to control, but also hard to break. The crisis could be weathered.

Hindenburg and Ludendorff, who had spent the previous two years fighting poorly armed Russians on the Eastern Front, were certainly surprised by the situation they encountered in the West: "an evil inheritance," Hindenburg termed it.[169] The policy on the Somme front had to be reviewed within the context of a multi-front war. The first decision taken was to adopt a strict defensive at Verdun. Combat divisions, heavy guns and aircraft from the Meuse sector would have to go north to the Somme, whose exhausted formations would replace them. Other divisions would have to go east to counter the new Romanian threat: Falkenhayn would command them.

On 7 September a conference of Western Front commanders and chiefs of staff was convened at Cambrai. Out of these discussions, not for the last time, came a tactical response to what was essentially a strategic problem; perhaps there was no strategic solution. Ludendorff was highly critical of the defence on the Somme. Both its tactical principles and its combat methods were at fault. Rather than trying to hold ground at all costs, and counter-attacking every Entente success, which led only to excessive casualties, Ludendorff prescribed a more "elastic" defence, "broader and looser and better adapted to the ground," which would absorb the shock of enemy attacks without wasting precious lives. The defence would become deeper and there would be less reliance on fixed positions and dugouts, which, given allied artillery superiority and thorough mopping-up techniques, were both death traps. An outpost line would mark the German front, but most hard fighting would be done behind this, after allied assaults had been drawn into and broken up within a lattice of mutually supporting strongpoints and concealed machine-gun nests. Moreover, Ludendorff prescribed a greater reliance on firepower rather than manpower in future, an oblique acknowledgement of the superiority of allied *matériel*-intensive tactics over the "flabby" German defensive response. German infantry formations were to follow the allied lead and re-equip with light machine guns, trench mortars and grenade-launchers. The rifle was to be revived as the main weapon of defence, in place of the hand-grenade. Attacks might then be broken up at a distance, avoiding the bloody close-quarter trench fights that were taking such a heavy toll of the infantry. The Somme front was to be supported by greater artillery and air concentrations: to meet like with like, the guns and aircraft were to go on the counter-offensive. Responsibility for coordinating artillery support was to be devolved down to divisional commanders to improve the flexibility and speed of response to allied attacks.[170] Finally, the higher command in the West was restructured. Three German army groups were created. Gallwitz reverted to command of Second Army, and Crown Prince Rupprecht, newly promoted to field marshal, took over the Northern Army Group, responsible for the Somme sector.

This short-term response was a clear acknowledgement that Germany had lost the initiative. If she could hold on until the winter, then she might regroup her forces effectively, review her strategic priorities, and fight on into 1917. But for now she had to contain the allies' superiority in numbers and *matériel*. One final decision—to lay out a new, deep defensive system in the rear of the central section of the Western Front, to which the army might subsequently make a strategic retreat—

was taken in September 1916. This new line, the Siegfried Stellung ("Hindenburg Line," to the allies), was to be constructed over the winter. It was an admission that Joffre's attritional strategy on the Somme was working, and that the German army did not have the reserves for another attritional campaign on its current length of front. In the meantime, it was "imperative" that the Somme front hold, and divisions were to be stripped from the rest of the Western Front to ensure this.[171]

It would take a few weeks for this new policy to have an impact in the field, but by the end of September a clear change in the methods and effectiveness of the defence was apparent. Although counter-attacks did not stop—as mentioned above, those around Bouchavesnes in mid-September were particularly costly for the attacking divisions[172]—other aspects of the defence improved. Increasing artillery strength allowed a firepower-based response to allied infantry attacks, which could be shot to pieces before they reached the German front.[173] The deployment of machine guns hidden in shell-hole positions between the German trench lines also gave the attackers a nasty surprise, and a new tactical conundrum to resolve.[174] As we have seen, Courtès's men were on the receiving end of these new tactics on 25 September.

Nor were the allies still dominant in the air. Allied pilots, in constant action, were beginning to suffer from fatigue and nervous exhaustion. Their replacements did not have the veterans' experience, while their opponents were more deadly. Captain Ball had to be sent home to rest at the beginning of October. While the famous hero entered a "media circus" in Britain, his comrades had to face the first "flying circus." At the end of August, German ace Captain Oswald Boelcke was sent to First Army, where he developed an offensive doctrine to counter that of the British and French air forces. He claimed his first "kill" on 2 September. His own formation, *Jagdstaffeln* (or "Jasta") 2, the most famous of the newly formed German fighter squadrons (of which four deployed to the Somme in September and October), gathered together a group of hand-picked combat pilots, including Baron Manfred von Richthofen, the "Red Baron." Jasta 2 flew the latest Fokker, the DIII, and the new Albatross DI, which was better armed and faster than any allied aircraft: it deserved its nickname, "The Shark." In the second half of September, British pilots were outmatched in the air for the first time. On 17 September Boelcke's squadron shot down five machines, with Richthofen claiming his first victim.[175] The Red Baron had added another ten British aircraft to his score by the end of November, including a British ace, Major Lanoe Hawker VC. By the end of the year, Jasta 2's pilots had scored eighty-six victories.[176]

Allied squadrons still outnumbered German (although with the assumption of a strictly defensive attitude at Verdun, more German squadrons were moved northwards, and by mid-October 540 machines, nearly two-thirds of the German aircraft on the Western Front, were supporting First and Second armies[177]), and they maintained ascendancy over their side of the line. But once they ventured into German airspace they were guaranteed a warm reception. It was some compensation that Boelcke was killed in a crash-landing after a collision with one of his own squadron. The RFC sent a wreath.[178] But between September and December 1916, 311 allied aeroplanes were brought down on the Somme front.[179] Haig immediately requested a reinforcement of fighter squadrons.[180] Joffre's response was to re-organise French fighter squadrons into four "wings" that could be deployed as needed.[181] The Storks, who had commanded the skies over the Somme, were officially recognised as the first permanent fighter group. In October they were joined over the Somme front by Squadron 124, the famous "Lafayette Squadron," whose aircraft were flown by American volunteers. Equipped with new Spad single-seater fighters, the French aces were able to fight back with some success. Georges Guynemer and Alfred Heurtaux, the first two pilots to fly Spads, each shot down two Albatrosses in one sortie in the bright, sunny days of late September.[182]

However, increased German competitiveness in the skies offered only marginal assistance to the hard-pressed German infantry on the ground. Although another change of tactics might for now check the relentless allied advance, the longer-term impact of the slow, grinding struggle was starting to manifest itself. "The German of 25 September 1916 is no longer the German of 25 September 1915," noted one French report. Prisoners were in a pitiable state, and both officers and men were quick and happy to surrender. Even the officers loudly expressed how fed up they were with the war. Only the elite machine-gun companies seemed committed to sustaining the defence.[183] Rates of shell-shock and desertion, coupled with large-scale surrenders, the mental and moral symptoms of prolonged strain, were growing among the combat troops: "the feeling of absolute superiority of the German soldier was lost," one staff officer observed.[184] The observation of Captain von Henting, a staff officer in the Guard RID, that the Battle of the Somme was "the muddy grave of the German field army," is frequently cited.[185] Less well known is his continuation, "as well as of our confidence in German supremacy, crushed by British industry and its shells. The German supreme command which entered the war with great advantage was beaten by the technical superiority of its adversaries, and obliged to throw division after division, unprotected, into the cauldron of annihilation."[186]

HERE HENTING ACKNOWLEDGED the importance of the home fronts, without whose continuing production efforts the Battle of the Somme could not have been sustained. Haig did the same in his end-of-battle despatch, which recognised "the efforts and self-sacrifice" of the men and women in Britain to sustain his army at the front.[187]

Defensive reorganisation might have counted for little if other factors had not contributed to slowing the assault. However, in testimony to the back-and-forth dynamic of industrial battle, just as the defence was reviving, the offensive was beginning to run out of steam. The allied armies' own supply of fresh divisions was starting to dry up, and finding the guns and shells with which to support them was turning into an administrative headache. By late September, Tenth Army had lent heavy artillery to Sixth Army and was forced to stop its own advance south of the river due to a shortage of fresh reserves and shells. Moving munitions and supplies forward to the troops along the few congested roads and distributing them across the shattered battlefield remained a complex logistical challenge.[188] But these problems were common to both sides. What helped the defence to hold out was the unconquerable force of nature. President Poincaré had intimated at the end of August that the weather in northern France broke around the time of the autumn equinox.[189] As the battle ground on towards a cold, grim winter, the fight against the elements would gradually supersede— if never entirely replace—the fight against the enemy.

THE SEPTEMBER BATTLE on the Somme was not, as it has usually been depicted, a renewed British attack to pull along a weakening and dispirited French army. Nor was it an overambitious flight of hubristic fancy on the part of the British commander-in-chief (although reading his own account in isolation does give grounds for such a conclusion). Examined individually (and normally only the British attacks on 15 and 26 September are examined in any detail), the operations in September 1916 appear to be merely more of the same, more extensive but still relatively localised attritional fights with modest results—some villages won here, a few trenches taken there. Seen collectively, however, Foch's renewed offensive represented something new; and Haig's plans make sense in this context. For these reasons, a true picture of the French army's repeated efforts during September, more and larger than those of the British, is a vital missing piece of the Somme jigsaw puzzle. As far as was possible with such a diverse and crude instrument as the allied armies of 1916, Foch had them working as a fairly well-oiled war machine: if not a truly combined offensive, by September it was at least a concerted one. His battle reached its climax with a succession of blows—piston thrusts—as four Anglo-French armies successively struck against the German defences. On 3 September, Sixth Army took the German position between Le Fôret and Cléry-sur-Somme. Fourth Army finally captured Guillemont, and Ginchy soon after. On 4 September, Tenth Army extended the battle on its southern flank: Soyécourt and Chilly fell. On 12 September, Sixth Army struck a very heavy blow north of the river at Bouchavesnes: by doing so it had doubled its penetration of the German front in ten days. Three days later, Fourth Army, using tanks for the first time, broke into the third position at Flers, also taking Martinpuich, Courcelette and the previously impregnable High Wood. On 17 September, Tenth Army struck again, south of the river, securing Berny-en-Santerre, Vermandovillers and Deniécourt. On 25 and 26 September, Sixth and Fourth armies took Rancourt, Frégicourt, Morval, Lesboeufs and Combles. On 27 September, Thiepval finally fell to Reserve Army. This was how an attritional offensive was supposed to proceed, and it brought the defence to crisis point.

These attacks constituted a push for victory, which France really needed. Her war-weary troops did not want another winter in the trenches, so they gave their all to drive the invader from their homeland.[190] Whether better weather would have allowed Foch to force a decision will never be known for sure, although probably not, because attrition was taking a toll on the attackers as well. Strategic victory in a battle of attrition was undoubtedly a chimera—hoped for, planned for, even on occasion apparently imminent, but unrealisable. The truth, as Castelnau recognised, was that "offensives with limited objectives were offensives which would certainly succeed, but yield no return." There would always be another ridge to take; at least while there were men enough to defend it.[191] As Joffre admitted on 21 September, he did not now expect the German line to be pushed back significantly before the spring. Nevertheless, on all the measures of attritional warfare—manpower, *matériel,* morale—he remained confident of victory. He even found words of praise for Haig and the British army: the longterm weakness in France's strategic hand had finally been discarded.[192]

Hitherto, the essential purpose of offensive operations had been to make the war mobile again, to resume big battles and thereby defeat the whole enemy army in the field in quick time. This had not happened; perhaps it could not happen in the circumstances of 1916. Haig's battle plans are customarily criticised for this overambition, but they are frequently misunderstood, being operational schemes not tactical directives. It was the commander-in-chief's job to think big, even if all operations would begin with the storming of the enemy's next defended position. By September, it seemed that Haig's army, corps and divisional commanders could in favourable circumstances be trusted to deliver real results in the field, as Foch's had been doing all along. Despite their steady forward progress, however, on the Somme the allies found themselves in the same situation as the Germans had encountered at Verdun. The scale of the battle, and improving tactical and operational methods, meant that in September they had progressed further and faster, and caused more damage to the enemy (and to themselves in the process). Nevertheless, until operational art improved, and battle developed a faster tempo, more supple command methods and more flexible logistics, tactical triumphs would not be translatable into strategic victory.

"This damned war will not finish quickly," one *poilu* wrote home. "I believe we will be here for ever."[193] It was a justified conclusion in 1916, and the strategic stalemate was far from over. The Battle of the Somme had relieved the pressure on Verdun, restored the initiative to the allies, worn down the enemy's manpower and morale and, as part of the General Allied Offensive, stretched German resources dangerously thin. In September it had reached a pitch that caused the German defence acute problems. With the onset of autumn, however, the Somme offensive would ultimately prove abortive. Nevertheless, a new way to fight a big battle that would defeat the whole enemy army — not in the open but along the defensive lines where he was encamped — was emerging from the ruins of the old one. Eventually, Foch would use it to end the damned war.

Muddy Stalemate: October–December 1916

Captain Tristani's battalion awaited its turn in the corridor that "serves as the entrance and exit to the furnace," his men apprehensively listening to the rumble of the guns and watching the light show in the night sky as so many thousands had done before. What had they to fear? They had already survived Verdun. A chapter title in Tristani's unofficial history of 32 RI's 3rd Battalion acknowledged what they had to face next: "The Mud of the Somme."[1]

The battle still had another seven weeks to run when Tristani's battalion embused for the front line on 1 October. The final month and a half of the offensive appeared to follow the pattern of the previous three: French and British attacks repeated at regular intervals; slow German retirement; gruelling trench fights under the steel umbrella of artillery fire. But the scale and cadence of October and November's fighting were different. Allied attacks, although still frequent, were less powerful and their immediate objectives more restricted. The defence yielded less ground. Although Haig intended "to go on until we cannot possibly continue further either from the weather or want of troops,"[2] the momentum of the offensive, the pressure on the defence and September's intensive attrition could not be sustained. Although pressure on a weakened enemy could and would be maintained, after a final, unsatisfactory, general attack on 7 October the idea of a breakthrough, if not entirely abandoned, effectively disappeared.[3]

The push-me-pull-you nature of the advance precluded a decision so late in the year. The battle was shrinking inexorably as winter came on and there was nothing that the high command could do about it. While Haig's smug note that Foch's "grand attack had gradually resolved itself into an attack by one Division on a village (Sailly-Saillisel)!" was hardly fair,[4] by mid-October it was certainly much less grand than it had been a month before. Joffre tried to reverse both the lethargy that seemed to be endemic in French divisions and the return of British short-sightedness. Local fights for tactical advantage had to be replaced by battles for operational superiority, G.Q.G. enjoined: attacks should be made more often, and aim for the enemy's gun line, not the trench immediately in front. The only consequence, inevitably, was that the British commander-in-chief restated his customary objection to being ordered about by a Frenchman![5] This forceful repetition of old mantras could have little real effect, because it was not the troops or the methods, nor even the intransigent Haig, but conditions on the battlefield that had slowed progress to a snail's pace. Joffre was merely window-dressing as his own vaunted ambition became bogged down, and political agitation against his untrammelled power once again welled up. French troops had done well enough in September to suggest that they knew their job, and British soldiers were also finally manifesting comparable skill. Thus, under the high command's stern but naïve instructions, the troops struggled on, game but increasingly disillusioned, through the autumn. But in the face of wind, rain and shortening daylight the military machine virtually seized up. The logistical network behind the allied front, always fragile, clogged up with mud and lost, damp, morose men. The September offensive had held out great promise, had pushed the defence towards breaking point. But urge, order and bully as they might, Joffre, Foch and Haig could not press their advantage to a decision. The battle passed beyond human control once nature started fighting back against the desecration of her works.

LARGE-SCALE OFFENSIVE OPERATIONS were renewed on 6 October. The pause had been longer than usual, and longer than Foch wanted, because rain had turned the roads into rivers of liquid mud and kept the artillery's spotting aircraft on the ground for several days. Mud had been an intermittent difficulty, whenever it rained, since the battle had commenced. Near-continuous downpours would now make it an overwhelming problem. The mud of the Somme was something special. When the chalky subsoil mixed with the thin brown topsoil, a "liquid yellow grey mud . . . extraordinarily buoyant, like quicksilver," resulted.[6] It stuck to everything, caked men, clogged wheels and

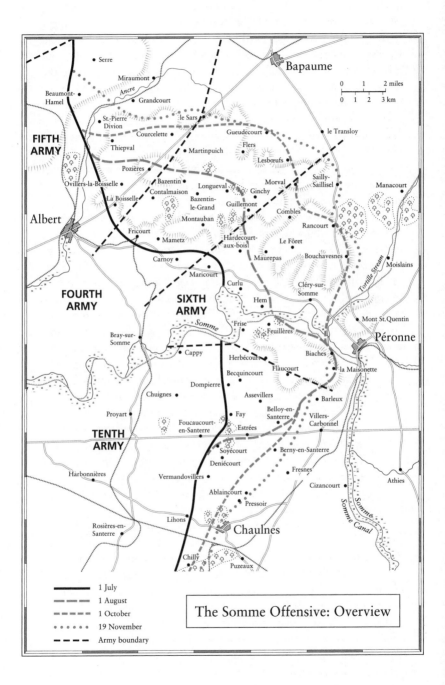

The Somme Offensive: Overview

Serre
Bapaume
Miraumont
Ancre
Grandcourt
Beaumont-
Hamel
St-Pierre
Division
le Sars
le Transloy
FIFTH
ARMY
Courcelette
Gueudecourt
Thiepval
Martinpuich
Flers
Lesbœufs
Pozières
Sailly-
Saillisel
Ovillers-la-Boisselle
Bazentin
Longueval
Morval
Manacourt
Albert
Contalmaison
Bazentin-
le-Grand
Ginchy
Guillemont
Combles
La Boisselle
Montauban
Rancourt
Fricourt
Hardecourt-
aux-bois
Le Fôret
Mametz
Carnoy
Maurepas
Bouchavesnes
Moislains
FOURTH
ARMY
Maricourt
Curlu
SIXTH
ARMY
Cléry-sur-
Somme
Hem
Somme
Frise
Mont St.Quentin
Bray-sur-
Somme
Feuillères
Péronne
Cappy
Herbécourt
Biaches
Flaucourt
la Maisonette
Becquincourt
Chuignes
Dompierre
Assevillers
Barleux
Proyart
Fay
Belloy-en-
Santerre
Villers-
Carbonnel
TENTH
ARMY
Foucaucourt-
en-Santerre
Estrées
Soyécourt
Berny-en-Santerre
Harbonnières
Deniécourt
Fresnes
Vermandovillers
Cizancourt
Athies
Ablaincourt
Pressoir
Somme Canal
Lihons
Chaulnes
Rosières-en-
Santerre
Chilly
Puzeaux

0   1   2 miles
0   1   2   3 km

————— 1 July
– – – 1 August
- - - 1 October
• • • • • 19 November
– – – Army boundary

jammed the mechanisms of guns and rifles. Moreover, it was every-where. On the roads it sucked in the wheels of transport lorries and wagons, congealed round the hoofs of horses and pack-mules and gripped the boots of marching men, often to the knee. Away from the roads the going was almost impossible. Shell-holes were filled to the brim with it, forming a quicksand that sucked in men and animals. It was, one history remarked, the worst mud which the *poilus* encoun-tered anywhere along the front.[7] Recurring images of the later fighting on the Somme are of men waist-deep in the mud, immovable and rav-ing; of animals sunk in to their bellies, which had to be shot. It seems astonishing that anything could move at all, let alone that a huge battle could be conducted in such conditions.

The unrecorded heroes of the Somme are the technicians and work-men who kept the whole battle flowing, if not smoothly, then at least effectively. Military engineers oversaw the construction and upkeep of the logistical network, maintaining and improving roads, building light railways as the front advanced, making new "corduroy" roads out of logs and railway sleepers and laying duckboard tracks across the empty landscape to guide troops forward. Transport officers worked out the complex timetables and routes for getting men, supplies and ammuni-tion up to the line, and wounded and tired men out of it. Every time the front lurched forward the whole system had to be rearranged: the roads and railways extended; the guns repositioned; new supply dumps established; new bivouacs and command posts built. It fell upon the many battalions and companies of line-of-communications troops, pio-neers and territorials, military policemen, doctors and nurses of the Army Medical Corps, drivers, blacksmiths and mechanics of the Army Service Corps, even labour companies organised from German prison-ers, to operate the infrastructure: to clear and repair the roads; drive the convoys; direct the traffic and unblock the traffic jams; evacuate the sick and care for the wounded.

All this effort was expended to keep the guns firing and the infantry creeping slowly forward. Yet, paradoxically, it was constant forward movement that had brought the road network to breaking point: the extra few miles to the front and back from the railhead, over the worst ground, along the most rudimentary roads, rivers of mud after heavy rain, crippled the battle. More lorries were needed to cover the extra distance, and they moved more slowly in the inclement conditions. In Heilly, just south of the main Amiens–Albert road, Bean noted, a con-voy of motor lorries passed constantly along "the muddiest street in all France . . . like vast black beetles crawling through the mud."[8] Many broke down and had to be shunted aside, to wallow in the sumps of liquid mud that had been dug alongside the carriageway in a vain attempt to drain it. Unsurprisingly to Tristani, as his battalion's transport crawled forward in the traffic jam, all the effort seemed futile. The panorama resembled the engravings he had seen of seventeenth-century sieges, not a modern battlescape.[9]

It was a valid comparison. Despite this ceaseless effort, the truth was that the communications infrastructure was inadequate for such a sustained struggle in a relatively small area. The expectation had been that the battle would move forward quickly. By the end of August, however, it was apparent that the existing infrastructure was insufficient to sustain a relatively static battle like the Somme. Nevertheless, during September, still heavier demands were made on it. Even the French logistics service, which had experience of managing a similar battle at Verdun, was floundering. Sixth Army's lines of communication were always hampered by the fact that the main roads north of the river were assigned to the British army. The forgotten *voie sacrée* along the Roman road south of the river, passing from Amiens through Villers-Bretonneux to Proyart, kept both Sixth and Tenth armies supplied. At the Proyart crossroads Sixth Army's transport columns branched off northwards, across the river to Bray, while those of Tenth Army continued on to Estrées. A two-way lorry traffic, and one-way horse traffic, moved constantly along the road. On 30 September this road carried 38,000 men and 3700 tons of *matériel*, twice as much as passed along the road to Verdun on its busiest day: at greatest intensity vehicles passed every four seconds on the Somme, every five at Verdun. Between 5 July and 10 December the *Commission Régulatrice Automobile,* which managed this vital supply line, dealt with 371,000 tons of *matériel* and 2,064,000 men.[10] However, as more French corps were fed through this bottleneck as Sixth Army's front fanned out, so its supply lines were stretched to breaking point. There were not enough labourers to keep the expanding transport network beyond Maricourt functioning.[11] In February, Foch had envisaged that he would need three steamrollers to keep his roads in a good state of repair. By November, GAN had 49 steam- and petrol-driven road rollers and 34 mechanical diggers, 8340 French labourers and 1450 prisoners of war at its disposal.

Together they shifted 671,000 tonnes of aggregates, and built 23 new roads into the battle zone. But only the Amiens–Proyart road could be properly maintained. The smaller roads that branched off it were broken up by the constant traffic which passed along them.[12]

While the French were preparing to attack Bouchavesnes, Haig was sitting in a conference room at G.H.Q., discussing light railway and roadstone requirements with Lloyd George and Sir Eric Geddes, managing director of the North Eastern Railway and now one of Lloyd George's "men of push and go" who had been sent to France to report on the army's lines of communication. Roadstone was now pouring into Fourth Army's area at the rate of 1500 tonnes a day, but it was difficult to find the transport and labour to move it. There were no mechanical means available, so loading and unloading had to be done by hand, while trucking the heavy roadstone around was simply adding to the mass of vehicles clogging the roads and helping to break up the surfaces that the roadstone was needed to fix! Geddes had toured the rear areas of Fourth, Reserve and Sixth armies, and had been particularly impressed by the network of light railways that were used to deliver shells to the heavy batteries behind Fayolle's front, saving road transport and manpower. Light railways could also be used to move roadstone. Geddes proposed a similar system for the British front, because industrial battle needed modern transportation to function. By early October, light railway tracks were extending on to the Pozières Ridge (the rails and locomotives were begged from the French).[13] However, it was too late for this to have much impact in the current campaigning season.[14] Haig noted sanctimoniously after the conference that he had anticipated the need for a light railway network from the moment he had assumed command of an army in January 1915, but back then the War Office had cavilled at the cost.[15] Whoever was ultimately to blame, it was certainly true that the roads and railways behind the Somme front could barely cope in good weather, and they were the main drag on maintaining the tempo of operations in wet periods. Ironically, now the British troops at the front were doing rather better, the services at the rear were failing. As well as roadstone, the British army was short of timber, rolling stock and labour by the last quarter of 1916. Although the supply of shells was now adequate, transporting munitions to the right place at the right time was increasingly difficult, especially as heavy shells (which were now the main munitions loads) required much greater transport capacity than the smaller, lighter field-gun ammunition.

The army's quartermasters and transportation officers had hitherto managed its increasing transport and supply needs with a series of ad hoc expedients that dealt with particular problems as they arose but did not create a functional infrastructure for an army of several millions. Port capacity, railway lines and motor and horse-drawn transport, as well as procedures for timely supply and maintenance, were all lacking.

Ian Brown, in his evaluation of the British army's logistics, concluded that:

> The transportation system, in spite of the best efforts of the administration to patch it up, could not properly sustain the Somme. Until a system had been created that could, the [British army's] offensives would invariably have the same character as the Somme—a prolonged drive into the German lines, using ever-increasing quantities of ammunition and increasingly damaging the transportation infrastructure, until the offensive could no longer be maintained.[16]

Geddes's immediate appointment as G.H.Q.'s director general of transportation services—a civilian with the honorary rank of major-general—was proof that success in modern battle depended on good management as much as hard fighting. It was a business enterprise, which needed civilian management skills, as much as a military one. Over the winter, Geddes and his subordinates would overhaul the army's transportation systems, establishing an "administrative military institution" much better suited to fighting long, attritional battles.[17] Not enough could be done in the remaining months of 1916, however, to alter the fact that constant wear and tear was bringing the military machine to a halt.

Despite these ongoing difficulties, throughout October and November the allies did maintain their *matériel* superiority—First and Second armies' rear areas were, of course, also clogged and chaotic—even if it could now be brought to bear only on narrow sections of the front. One airman flying above British infantry advancing on Eaucourt l'Abbaye on 1 October noted:

> They do not seem to be the target of much enemy shell fire. The enemy barrage . . . bore no relation to the wall of fire which we were putting up. I should have described it as heavy shelling of an area some 300–400 yards in depth from our original jumping off places. Thirty minutes after Zero the first English patrols were seen entering Le Sars. They appeared to be meeting with little or no opposition, and . . . no German shells were falling in the village . . . To sum up, the most startling feature of the operations as viewed from the air was:
>
> 1) the extraordinary volume of fire of our barrage and the straight line kept by it. 2) The apparent ease with which the

attack succeeded where troops were enabled to go forward close under it. 3) The promiscuous character and comparative lack of volume of enemy's counter-barrage.[18]

This advantage meant that the battle could and would be sustained, with the resolve if not the results of the previous months.

For German regiments on the receiving end of allied firepower the purgatory of battle continued. On 2 October, 16 Bavarian RIR—the so-called List Regiment in which Corporal Adolf Hitler served as a despatch runner—deployed to the Somme. For three months now they had listened to the distant rumble of the guns, read the press despatches, and heard the hair-raising accounts from *Sommekämpfer* recuperating in their quiet sector in Flanders: "Their stories strengthened the impression that something completely out of the ordinary awaited us." The presence of its soon-to-be-infamous other rank has led to this regiment's wartime experience being better studied than that of other German front-line units. It was on the Somme for only ten days, but that was quite long enough for it to be muddied, bloodied and eviscerated, having suffered 1177 casualties, including Hitler himself, wounded in the groin. The Bavarian official history suggests that the regiment had only 350 effective riflemen by the end of its tour. With their morale still buoyed after smashing an Anglo-Australian diversionary attack at Fromelles in July, the Bavarians were blasé about what awaited them. But after "its first actual experience of true modernist warfare," the regiment's historian John Williams judged, 16 Bavarian RIR, and 6 Bavarian RID, to which it belonged, suffered a blow from which it never recovered.[19] Its defeated opponents at Fromelles, 61st and 5th Australian divisions, went on to glory. In contrast, 16 Bavarian RIR was "dragged along in the darkness and night [and] spewed out in ruins" on the Somme, like so many other German regiments.[20]

AT THE START of October there was still great hope among the allies. After the rapid advances of late September, Major-General Currie, commander of 1st Canadian Division, felt that "the Germans are beaten."[21] Haig himself was planning to commit the Cavalry Corps north of the Ancre, and expand the offensive northwards to Third Army's front, to exploit the apparent demoralisation of the enemy vigorously.[22] The troops, Fayolle noted, were in high spirits: "Our men know that victory is really coming." Fayolle himself was in a surprisingly good mood too. On 2 October Poincaré, Joffre and Roques had visited his headquarters to appoint him Grand Officer of the *Légion d'Honneur*. His portrait appeared in *L'Illustration*. He was France's most successful field commander. But, as he noted the same day, "it was raining." An attack on Bois St. Pierre Vaast had been postponed. And his troubles with Foch, "that devil of a man who talks of time and space, while never considering the ground," had returned. At least in church Notre-Dame-des-Victoires smiled down benevolently.[23]

However, Fayolle's victories would dry up during October, as Sixth Army's operations shrank and slowed. Not just the weather, but the troop, munitions and logistics problems, as well as the local geography of the front north of the Somme, meant that Sixth Army's advance would be constrained into a narrow arrow-head between the British right wing and Bois St. Pierre Vaast. Moreover, Duval, "the dynamo that drove the French Army on so successfully during the battle of the Somme," had gone on sick leave in the last week of September, slowing the tempo of French operations even more.[24] After a general attack on 6 and 7 October had produced meagre results—in the wet and dull conditions the bombardment had been ineffective—Sixth Army's right flank south of Bouchavesnes was halted. There would be no crossing of the Tortille or capture of Mont St. Quentin. Péronne would stay German for the time being. The battle to the north, alongside the British, would continue, but without the common purpose of September. British attacks were orientated northwards, towards the "holy grail" of Bapaume. Sixth Army's advance diverged east–north-east, as it sought to extend its grip on the Bapaume–Péronne road and outflank Bois St. Pierre Vaast to the north by occupying the twin villages of Sailly-Saillisel and Saillisel.

Fourth Army's October operations, officially designated the Battle of the Transloy Ridges, represented a reversion to August's attritional fighting. The putative British objective was now the fourth German position, which stretched along a low ridge running north-westwards from Le Transloy towards Ligny, in a dip at its apex, covering Bapaume. During the month a series of attacks was mounted, interspersed with the usual smaller-scale, line-straightening operations. There was little to show for all this effort by November. Conditions on the ground certainly contributed to the slow progress, but the nature of the position cannot be discounted. Although the British official historian strove to make this fighting comprehensible by outlining attacks on particular trenches, in practice the positions that Fourth

Army's corps (and the Canadian Corps supporting its left wing) attacked were a confused lattice of trenches, old gun positions and strongpoints that made little sense to the troops on the ground, whose maps were hopelessly out of date.[25] Now that the battle was reorientated northwards, the main advance was sweeping up between the remains of the old German positions north of Flers, along the Flers and Gird lines. The communication trenches between them, hastily organised for defence, furnished a succession of switch lines from which the advance could be contested. What appeared on a map as a neat red line of trench was, after constant bombardment, no such thing in reality. Fresh troops arriving to take up positions in the German front line, used to taking over the neat, revetted trenches and tidy dugouts of quieter sectors with due military ceremony, were aghast at their new positions. When the Marine Infantry Brigade, fresh from Flanders, took over Regina Trench, beyond Courcelette, on 30 September, "there was no trace of either position or trench! Smashed and buried dugouts, wrecked camouflage, a tangle of assorted equipment lying around intermingled with all types of weaponry, told us in very stark terms a tale of destruction and intense fighting ... we had entered into a completely new phase of large-scale warfare."[26]

Capturing this maze of trenches, interspersed with machine-gun nests, was every commander's nightmare—slow, difficult and exhausting. It was made worse by the fact that Fourth Army was moving down from the Thiepval–Pozières ridges into a natural amphitheatre overlooked from the Le Transloy and Warlencourt heights to right and left. In its centre sat the prominent Butte de Warlencourt, an ancient burial mound from which the ground to front and sides could be observed and commanded. From here the German garrison could watch the British ranks breaking on the strongpoints around them.

XIV and XV corps' attacks on the right, against the intermediate trench line beyond Lesboeufs and Gueudecourt, made very little progress during October. The line advanced only a few hundred yards, and the fourth position along the Le Transloy Ridge remained invulnerable. Whether set-piece attacks (on 7, 12, 18 and 23 October) or battalion- and company-strength trench operations, concentrated German artillery and machine-gun fire usually forced the British pushes towards Le Transloy to ground very quickly.[27] One notable exception was the successful capture of Hilt Trench by the Newfoundland Regiment, attached to 12th Division, on 12 October. The Newfoundlanders' return to the Somme three months after their heroic yet disastrous assault on 1 July was likely to be traumatic. Their ranks were filled with the recovered casualties of the earlier attack and keen new recruits anxious to emulate their comrades' heroism. Newfoundland's tiny army had something to prove to the watching empire. Although insignificant in the context of the battle or the war as a whole, for Newfoundland the capture of Hilt Trench went some way towards redeeming the earlier failure. While attacks to either side failed, the Newfoundlanders took and held their own objective and part of that of the neighbouring battalion.[28]

III Corps on the left did much better initially. The farm complex at Eaucourt l'Abbaye, between the Flers and Gird lines, was taken by 47th Division in a preliminary operation in the first days of October; and on the 7th, 23rd Division captured Le Sars village on the Albert–Bapaume road.[29] This was to be Fourth Army's last major victory. Over the next four weeks the Butte de Warlencourt withstood successive attacks by five different divisions.

WHILE FOURTH ARMY was trying to take the last ridge in front of Bapaume, Sixth Army's left wing was engaged in the worst fighting that it was to see in the whole Somme battle. Today, the fight for Sailly-Saillisel, if not completely forgotten, is viewed as a meaningless engagement at the end of a long, strength-sapping battle. But it rivalled the fights for Guillemont, Ginchy and Thiepval in its duration and intensity. If Sailly-Saillisel, north-west of Bois St. Pierre Vaast, could be taken, the huge wood might be outflanked to north and south and pinched out, rather than taken by a costly frontal assault.[30] Then the offensive might gain momentum again. For seven weeks, this T-shaped complex of villages,* dominating the shallow dipping valley between them and Le Transloy, held up two French corps. In atrocious battlefield conditions, and against reinvigorated German defenders, progress was to be frustratingly slow and exhausting. It was not the way to pursue decisive victory.

To attack Sailly-Saillisel required more space. Bois St. Pierre Vaast hemmed in the right flank of XXXII CA, which held the line between Rancourt and Frégicourt, so in late September Sixth Army extended its left wing by taking over Fourth Army's positions in front of Morval.

---

*Sailly-Saillisel lies along the Bapaume–Péronne road, with Saillisel almost at a right-angle along the road that runs behind Bois St. Pierre Vaast to Moislains.

Two corps could now attack along a four-kilometre front: I CA would push north-eastwards from Morval to cross the Bapaume–Péronne road and take Rocquigny beyond, while XXXII CA on its right would move on Sailly-Saillisel. On the left flank the objective consisted of a line of trenches, delineated as Bukovina and Jata-Jezov, in the German fourth position south of Le Transloy, which ran along the western side of the Bapaume–Péronne road and into Sailly-Saillisel's northern apex. They sat in a concealed dip at the top of the shallow valley between Le Transloy and Sailly-Saillisel. In front of them the sunken road linking Le Transloy and the Morval–Sailly-Saillisel road had been fortified, christened Baniska and Tours trenches by the French. Sailly-Saillisel village was the right-flank objective, covered by two lines of trenches: the Prilip–Portes de Fer–Négotin line immediately beyond Rancourt, which abutted the western apex of Bois St. Pierre Vaast at Reuss Trench, and Carlsbad, Terplitz and Berlin trenches two thousand metres behind. A redoubt around Bois Tripot and Sailly-Saillisel château protected the southern approaches to the village. No stronger defensive position had yet been encountered.

The battle began with a bombardment of Sailly-Saillisel by heavy artillery, 270mm, 280mm and 370mm mortars.[31] The infantry assault was timed to coincide with Fourth Army's in what would be the last general offensive of the battle. On 7 October I CA's left wing, alongside the British, made little progress. The corps was exhausted after leading the advance since late August. It would be relieved over the next few days by the fresh IX CA. On 12 October IX CA's 18 DI made its first, unsuccessful, attack on Baniska Trench. It was repeated, to equally limited effect, over subsequent days, and Fayolle took the new corps to task for these failures. Its commander, General Pentel, was sacked.[32] Fayolle seems to have caught Foch's disease, urging progress—after all, such obstacles would not have been insurmountable in August or September—but with little perception of the true conditions at the front. Pentel had taken care to train his formations in current tactics and manoeuvres before they went forward,[33] and IX CA was struggling as best it could in the circumstances, but a combination of battlefield conditions and enemy tactics rendered its efforts almost hopeless. Eighteen DI, which did two tours of duty in the front line from 10 to 21 October and 4 to 21 November, subsequently outlined the reasons for its failure to take Bukovina Trench. Terrestrial observation posts had limited views of the trench, and the atmospheric conditions—a combination of fog, wind and rain—made both aerial and ground observation difficult.

But even if the final objective could be seen, crossing its two-thousand-metre glacis—a featureless, shell-hole-pocked and waterlogged slope, over which the enemy had distributed concealed machine-gun posts, and which was dominated by his artillery and aircraft—was near impossible. The French infantry's own light machine guns, clogged with mud, could not give fire support when they advanced. During 18 DI's first tour, when the infantry had gone over the top they had struggled forward up to their knees in liquid mud; by their second tour, the quagmire reached their thighs. Not surprisingly, progress, if made at all, was measured in metres. The capture of an individual machine-gun post was considered a good day's work.[34]

On 17 October it was Tristani's battalion's turn to go over the top, against Baniska Trench, four hundred metres away: "The usual ceremonies took place slowly—artillery preparation; enemy counter-barrage on our trenches; putting down of rolling barrage in front of the assault wave." The omens were not good. The French supporting barrage fell on Tristani's men, assembled in a jumping-off trench that they had dug a hundred metres beyond their own line. The assault company went forward into a cross-fire from machine guns hidden in the shell-holes in front of Baniska Trench. The German counter-bombardment caught the supporting waves in their own front line. The survivors went to ground short of their objective: the battalions attacking alongside them had made no progress, leaving the intermingled companies of 32 RI in an exposed salient ahead of the French line. The isolated men—nothing but "muddy slabs" by now—then came under fire from their own howitzers. The regiment suffered 130 casualties for little effective gain.[35] It was not Sixth Army's finest hour. But it was typical of the operations taking place in this phase of the battle. It took IX CA almost a month to take Baniska Trench, which finally fell on 1 November, due more to good fortune than careful preparation. General Andrieu, commanding 152 DI, had applied directly to Fayolle to postpone that attack on account of the condition of the ground, but Fayolle had refused.[36] It seems that the trench's garrison, who did not anticipate an attack in such atrocious conditions, were caught in their bunkers.[37] Bukovina Trench was never reached.

Progress was better on the right wing, where the advance pushed forward a further two kilometres. In the first week of October, in a series of hard-fought trench fights, I CA established control of the Prilip–Portes de Fer–Négotin trench line.[38] On 7 October its 56 DI pushed forward twelve hundred metres on to the slopes west of Sailly-

Saillisel village, taking Carlsbad Trench and entering the Bois Tripot strongpoint behind it. From their new vantage point they could look over the plain rising to the Bapaume–Péronne road and Sailly-Saillisel, dotted with German stragglers fleeing to the rear. The mirage of a breakthrough, of victory, momentarily revived.[39] On the right, battalions from 40 DI took Terplitz and Berlin trenches, and established the right flank on the south-western edge of Bois St. Pierre Vaast, although a foothold established in Reuss Trench could not be held.[40] It was the most successful part of the general attack north of the river, but fell short of the ambitious objective of pushing through Sailly-Saillisel to Rocquigny.

But this was an increasingly rare success in a stalling advance. In the second week of October General Marie-Eugène Debeney's XXXII CA took over responsibility for the sector. Debeney was a former *chasseur* officer, who had been professor of infantry tactics at the *École de Guerre* before the war. Like Fayolle, he believed in the value of fire-power, and it showed in the relative success of his operations when compared with those elsewhere in October. Or perhaps it was the influence of his new chief of staff, Colonel Mangin, from 79 RI. After vicious bayonet fighting in the ruins of the château, XXXII CA's first penetration into Sailly-Saillisel, on 12 October, was driven back like IX CA's attack on its left on the same day.[41] 66 DI, the conquerors of Cléry in August and the closest the French army had to a storm division, returned to do the job. On 15 October two battalions of the elite 152 RI—the so-called Red Devils—and 68 BCA took the rest of the Bois Tripot position and the ruins of the château that covered Sailly-Saillisel to the south-west, and fought their way into the village. They managed to infiltrate the hitherto solid defence through gaps that had been made in the German defensive line by the crushing bombardment. The Prussians blamed the Bavarians, and vice versa. But the defence was far from spent. A six-day street-fight, ruin-by-ruin, cellar-by-cellar, shell-hole-by-shell-hole, ensued.[42] 66 DI held its positions around the Bapaume–Péronne road crossroads against repeated German counterattacks for a fortnight, while on the right flank XXXII CA's *chasseurs* fought their way into Reuss Trench.[43] "This XXXII CA is really very good," Fayolle recorded in a rare positive comment.[44]

So good was it that one of its successful defensive actions—the defence of the church and crossroads in Sailly-Saillisel by 94 RI (dubbed the "Bar le Duc Guards") on 29 October—was used as an example in a post-war textbook on infantry tactics. It should not be forgotten that Sixth Army's success was also a product of effective defensive tactics, which defeated innumerable German counter-attacks, as well as offensive skill. Good junior leadership, careful management of local reserves to sustain the firing line, maintenance of weapons, tactical use of the ground and high morale—the survivors of the 94 RI's 5th Company, whose weapons were all clogged with mud, even picked up stones to hurl at the advancing enemy—were identified as the fundamentals of this successful defensive action. It is even more noteworthy as this was the battalion which had been cut to pieces in front of Jostow Trench only one month before.[45]

A planned attack on the eastern side of the village, as part of a further general attack in late October, failed to materialise because of more atrocious weather. On 5 November the attack finally took place, even though a storm was still raging, but progress was limited. It was renewed the next day with equally meagre results. The last "houses," reduced by now to rubble heaps, finally fell to XXXII CA on 12 November.[46] This was the anticlimactic finale of Sixth Army's offensive, which had been checked by the same sort of active defence, based on village and wood positions linked by well-sited trench lines, that had served the Germans so well during the attritional phase of the battle. Although the XXXII and V CAs had established themselves on the edges of Bois St. Pierre Vaast, the outflanking manoeuvres to north and south were blocked. The only alternative, taking it by frontal assault, was mooted—Fayolle thought Foch was crazy to suggest it in face of the revived German defence[47]—and the Iron Corps and Messimy's *chasseur* brigade were brought forward to attempt it. Messimy's men attacked it from the south-west on 5 November. The leading companies waded through knee-deep mud into the wood, but they could not hold it, and withdrew to their start-line.[48] The attack was not renewed. All that effort and the tactical result was stalemate.

IN OCTOBER AND November the allied armies enjoyed more success on the wings than in the centre. Here lines of communication were less extended or congested, and there was a shorter expanse of shattered ground to cross, making operations more feasible, if still difficult. Tenth Army's offensive continued in a series of fairly regular attacks during October, which secured the next line of villages in the centre. On the flanks, however, at Barleux and Chaulnes, the defence remained solid.

Reinforced by the fresh XXI CA and 2 CAC, on 10 October

Micheler's centre began the battle for Ablaincourt. It was another large-scale attack, on a ten-kilometre front, and penetrated the German second position around Ablaincourt village, taking nearly fourteen hundred prisoners. On its southern flank, however, the attack was contained by the woods that covered Chaulnes from the north. On 14 October three divisions, including 10 DIC (commanded by national hero General Jean-Baptiste Marchand, who in 1898 had led the French expedition which encountered Kitchener's Anglo-Egyptian army at Fashoda), attacked again on the northern end of Tenth Army's sector, seizing the next complex of trenches and taking nearly a thousand prisoners. Over the following days the French hold on Ablaincourt, which the enemy had been fiercely contesting, was consolidated. Marchand himself was wounded by a shell splinter but insisted on staying at his post. Tenth Army had maintained its momentum a little longer than the Sixth, although repeated German counter-attacks had slowed its progress. Nevertheless, despite preparations to push on to the Butte de Fresnes and cut the Péronne–Chaulnes railway line, in the second half of October its advance slipped into torpor too. Once again the exhaustion of the troops, the state of the ground, the continuing bad weather and the solidity of German defences were the causes.

The situation on Tenth Army's flanks was crucial. To advance on the northern end of its front against the high ground around Villers-Carbonnel and Fresnes (the next line of defended villages) required Barleux, which had held out since early July, to be captured.[49] This was now in Sixth Army's sector, where the Germans were actively fighting back. Fighting flared up again in Bois Blaise, between Biaches and La Maisonette, in the last weeks of October.[50] On the 29th the French were to receive a rude shock, when after an intensive, eight-hour bombardment 359 IR recaptured La Maisonette. One battalion of 97 RI was completely overwhelmed, and a large gap was made in the French defensive system. The situation remained precarious for some days and the planned attack on Barleux had to be abandoned.[51]

On the southern flank the fighting for the Chaulnes woods, especially Bois 4, was vicious, akin to that in Delville Wood back in July and August. Tenth Army's 10 October attack had secured a foothold in the woods,[52] but the exhausted 51 DI failed to complete their capture the next day. That night they were counter-attacked by German storm-troops with flame-throwers. A battalion of 25 RI was destroyed.[53] On 21 October the French attacked again, surprising the enemy, who in turn counter-attacked the next day.[54] It was the start of a vicious short-range fight that would last for several weeks. The bare details disguise the character of the fighting, which was as grim as any that had taken place in the woods north of the river.

ALTHOUGH ALL FOUR allied armies remained active throughout October, there was none of the startling progress, or concerted effort, of September. Trenches and villages fell, were recaptured, fell again, but there was no repeat of the shocking attacks or deep advances that had caused momentary panic in the defence in the past. The only gains of the October and November fighting on Fourth and Sixth armies' fronts worth mentioning in the official communiqués were the villages of Le Sars and Sailly-Saillisel. Elsewhere, merely "the odd yard of coagulated mud . . . [was] wrested from the enemy . . . at prohibitive cost."[55] The German troops endured the same atrocious conditions as the British and French, but these favoured the defenders, who had less mud to traipse through to the front line—"merely a narrow strip along the front. The slopes leading down to it were green"[56]—and generally retained the advantage of high ground over the allied infantry trying to fight their way up from the boggy valleys. Mud assisted the defence in other ways too. The bombardment lost some of its potency. Shells falling into this swamp would not explode. The near-miss is a staple of soldiers' memoirs, and those of autumn 1916 describe shells landing with a splash in the many shell-hole pools that dotted the landscape, or plopping into the liquid ooze at their feet, rather than exploding on contact with hard ground. Weapons, equipment and uniforms covered in mud were useless. French *Chauchat* light machine guns were particularly prone to jamming, but rifles and heavier weapons also had to be cleaned regularly and carefully. Mud would coat the long skirts of the French service overcoat to such an extent that the men found themselves restricted in their movements by the extra weight. Scottish soldiers encountered similar problems with their kilts. Many Scottish regiments marched out of the battle with bare legs. Such conditions did nothing for the men on either side. Even fifty years later, "the mud of the Somme remained one of the cruellest memories of the French army."[57]

By this point, the battle was being fought by divisions that had been through it once (sometimes twice) already, or less experienced formations that had been kept out of the fight. Here the allies had an advantage. Returning British divisions had learned lessons during their first tour, and would do better because of them. German divisions had less time for rest, and newly formed units such as the 211–214 IDs had to be deployed even though they were ill-prepared for the full horror and intensity of the fight. By contrast, the Marine Infantry Brigade, having gained experience in Flanders, gave a good account of itself when defending Regina Trench. It held up Reserve Army's right wing for much of October, indicating that the arrival of a fresh formation with high morale and good discipline could be locally significant. But it paid a heavy cost. When paraded in front of the Kaiser after coming out of the line, the brigade could muster only two composite battalions from the nine that had gone into battle.[58] Fresh French divisions could still be found throughout October, even though they did not possess the élan or skill of the colonial and *chasseur* formations that were returning for their second or third tour. In the atrocious conditions of the battle for Sailly-Saillisel and Bois St. Pierre Vaast success or failure was determined as much as anything by the amount of time a unit had already spent in the line, and the rapid *matériel* and moral deterioration that front-line duty occasioned. Tired, depleted formations would generally be overwhelmed by new battalions with filled ranks, whether French or German, attacking or defending. Attrition was decisive at the tactical level as well.

By autumn both sides were having trouble keeping their front-line ranks filled. The German army had created fifteen new divisions in the final months of 1916, but these, composed of garrison troops, older classes and *Ersatz* reservists, used up the last depot reserves. By the end of the year many of the 1917-class recruits were already in the trenches. In the French army the quality of replacements was similarly in decline: 151 RI's ranks were being filled up with "shirkers" who had so far avoided front-line service, workers sent back to the front for disciplinary reasons and youngsters from the 1917 class.[59] Meanwhile, the first conscripts had started to arrive in the ranks of British infantry battalions. They lacked the enthusiasm of the earlier volunteers, and it would take time to train them in the newly mastered skills of battle.

British soldiers were starting to show signs of strain. On the eve of the 25 September attack, 150 men in the 10th Battalion of the King's Own Yorkshire Light Infantry in 21st Division had tried to go sick.[60] A few weeks later Britain's most notorious military execution, of Private Harry Farr, took place, *pour encourager les autres*. Farr, a pre-war regular soldier who had exhibited symptoms of shell-shock earlier in the battle, had gone absent without leave on the eve of an attack by his battalion against the notorious Quadrilateral. He was quickly tried by a field court-martial and found guilty of cowardice in the face of the enemy. In the circumstances—deserting one's comrades on the eve of an attack—the military authorities judged that commutation of the death sentence was not possible if military discipline was to be maintained. He was shot at dawn on 16 October.[61] Although the New Army men seemed to have established a moral ascendancy over the enemy, clearly towards the end of September battle fatigue was starting to take its toll, and Rawlinson recommended that Fourth Army's tired divisions really needed rest.[62] However, the battle still had seven weeks to run.

THE BATTLES OF late autumn generally degenerated into dozens of tiny combats around strongpoints and sections of trench, where individuals or small groups of men tried to hold on to their lives and their lines. If not understanding exactly why, they appreciated that their personal efforts were somehow vital to the outcome of the Great War. One German NCO's vivid account of yet another French attack near Sailly-Saillisel is typical of what they faced:

> The horizon glowed blood red with the muzzle flashes of the guns and hundreds of flares rose all along the enemy front as far down as Gueudecourt... "That is the French signal for the assault." For a few moments all was still, then our flares soared upwards as well: the artillery was to bring down defensive fire. Within a few seconds, shells were crashing down in front of the enemy's trenches. The French men joined in and a truly hellish concert got under way ... Soon bullets were whistling past us. Left and right, behind us and in front shells came down, causing us casualties. Machine guns began their Tack! Tack!
>
> After only ten minutes, the Battle of the Somme was working away like a giant machine. Everything operated with a terrible rhythm. I could not make out a single patch of ground where things were not exploding. There was no time for fear. There were the enemy. Here we were and it was a matter of sticking it out. We shot ceaselessly, until our barrels were hot ... "Fire low!" I bawled at my neighbour. Now we were firing without counting ... Splinters clattered against our steel helmets, but we took no notice. An attack absorbs all the senses ...

After an hour the attack had been smothered all along the line. Gradually things quietened down. The battle resumed its dreary look of boring desolation . . . "That was great!" said my neighbour, "Let's hope it always goes as well" . . . He had come through his baptism of fire well.[63]

Unteroffizier Feuge lived to fight another day. Others were not so lucky. A German corpse spotted among the many bloated, black cadavers around Combles, with one hand in a bag of grenades and another clutching a prayer book, suggests that once in combat these were all desperate men, fighting for survival in an uncontrollable hellfire with their last reserves of strength and sanity.[64] Perhaps this was the only way left for the ordinary soldier, lost in the huge mechanical battle, to assert his individuality. Captain Fritz Wiedemann of the List Regiment certainly gave that impression in his post-war regimental history. The regiment's commander was a drunken incompetent, and responsibility devolved down to battalion and company commanders, as the "shattering muddle" of battle disintegrated into individual actions by the "few" men still with the will or strength to fight. Most, however, gave up the struggle, and cowered, sleep-deprived, hungry and ravaged by dysentery, in shell-holes and dugouts, hoping and praying merely to survive: "a destructive sense of abandonment began to take hold. Otherwise peaceful and rational men became irrational . . . despair . . . crept into the hearts of brave men."[65] The German army was cracking under the strain. Corporal Hitler, wounded on his third day in the line, escaped such mental and moral collapse.

The ordinary solider also started to realise that he had more in common with the man whom he fought and killed on orders from above than he had with those who issued those orders. Friendly "Fritz," the man in the trench across the way, was a very different individual from the hegemonic "Huns" and the beastly "Boche" who collectively threatened the peace of the world. Lieutenant Philip Chattaway mused as he sheltered in a recently captured German dugout:

As I sit here I like to think of the Boche not as a bullying professional soldier but as some quiet homely fellow called up like ourselves to take his part in the war which is so distasteful to him. His letters and postcards from home lie in the rack over his bed, his walking stick and clothes brush lie on the shelf and the batmen are even now cooking some of his tinned food for our dinner. So, as I picture it, must he have sat at this table when off duty and perhaps have speculated just as I am doing now on the feelings of his enemy in the trenches opposite. I have taken quite a fancy to him this enemy of mine.

Common humanity might still intrude into industrial war, but its grim reality dominated: "I don't know how he died; very few got out of this death trap whole but I can only hope he is not one of the festering twisted corpses lying in the shell-holes outside . . . May your soul rest in peace, Musketeer Arnant of the 66th regiment of the line."[66] He was probably dead or wounded: 66 IR suffered over fourteen hundred casualties in the early days of October.[67] Chattaway's own soul was to join Arnant's within the week.

Inevitably, in these conditions and circumstances, effort declined and morale suffered. The veterans of 86 RI had not forgotten "the terrible mud of the Somme" when they commemorated their own small part in the battle fifty years later. On 17 September the regiment had participated in 120 DI's capture of Vermandovillers, attacking with "a superb élan." By mid-October, however, in the operations against Ablaincourt, "a vast lake of mud" had become the main enemy, exhausting the men "despite their extraordinary qualities of resistance and tenacity."[68] Just getting about in the forward zone was becoming increasingly difficult. Survival instincts prevailed over martial ones. When Tristani led his men forward in early November to take over the recently captured Baniska Trench, the journey from the reserve position behind Bois Bouleaux, although only a couple of kilometres, was akin to the harshest polar trek, with the ever-present prospect of sudden death thrown in. Tristani had trouble rousing his catatonic men from their holes in the ground; they seemed literally frozen into them, resembling living rags. A solitary guide from the forward battalion led them over the featureless Morval plateau, through the shell-holes, bodies and debris, "past craters into which a whole house might disappear." Every twenty seconds a 150mm or 210mm shell whistled in: each time they threw themselves face down into the mud. Once in the line, endurance took over. Tristani's new billet turned out to be a filthy formerly German bunker, with rivers of mud running through its shattered entrances, filling the bottom with knee-deep, glutinous clay. Fifty men crammed into this rat-hole designed for a dozen, but at least it kept them out of the corpse- and debris-filled stream of mud that had been Baniska Trench. The unlucky ones had to shelter in the water-filled shell-holes behind the trench. Frostbite cases soon appeared. Rain persisted for twenty-four hours, the ration parties never arrived, and several men sank into the mud, screaming piteously for rescue. A few German stragglers wandered into the French positions. It was impossible to know whether they were deserters or simply lost. After five days Tristani's men hobbled out. Most had rotten feet, and had to hitch rides on empty trucks to the rest area. They got lost in the fog while recrossing the plateau and were caught in a German gas barrage. Their gas masks were too clogged with mud to be of any use. Two men were killed and many others suffered badly.[69]

As the front-line journalist "Paul Loti" (the alias of Commandant Julian Viaud) reported towards the end of November, convoys of blue-painted lorries still rolled along the Amiens–Proyart road, their drivers swathed against the wet and cold, "taking up the thousands of proudly smiling victims for the great, sublime holocaust, or bringing out the human debris which survived." But, he acknowledged, the rain was putting out the fire of the Somme, "night was falling on the infinite desolation . . . and Death seemed at last to be falling asleep."[70] Clearly, by now, the offensive was all but spent, as were the troops who were caught up in it, "stuck in the mud" literally and metaphorically. The Foreign Legionnaires who returned to their old trenches in front of Belloy in November looked back fondly to the sunny days of July when they had previously held that sector. Now it took two hours to advance one kilometre through the ooze. When time came to pull out, rescue parties had to be organised to recover the many men sunk up to their waists in the mire. If they came out, the mud would suck off their boots, so they had to make the chilly trudge out barefoot.[71] Those who had done their time were glad to see the back of the bleak, featureless landscape, "the big, naked ridges, the far horizons, the terrible mud into which one sinks." Courtès was happy to bid *"Adieu la Somme"* as he led out the remains of his command after their second tour of duty in the line opposite Bois St. Pierre Vaast. They were shattered. Yet, even in this debilitated state, he could not help but wonder at the larger meaning of their ordeal: "Only later would we understand what the gigantic battle of the Somme was, even if our losses then and there indicated its importance to us. For the Somme was the artillery battle *par excellence,* the artillery battle in all its horror, a deluge of fire in the middle of a quagmire."[72] Surviving it was a victory in itself, no matter which side you were on. "Our spirits were not exactly high," Unteroffizier Feuge remembered, when the remnants of his company were finally withdrawn from their shell-holes near Sailly-Saillisel, "we were too hungry, thirsty and tense for that. But we were certainly relieved." As they drew away from the fighting an uncertain sense of elation overcame them. "We're saved!...We had escaped the destruction. The roaring hell behind us concerned us not one bit."[73]

The mood of the troops was starting to turn. Weeks wallowing in mud and shivering in damp trenches, reduced to cold iron rations because of the logistical breakdown and the inability of ration parties to locate the confused front line in the dark,[74] were bound to lead to complaints, even if, after relief, a hot meal, a strong drink and a clean, dry bed—"blankets on an attic floor" in Feuge's case—soon wiped away the worst memories of the ordeal.[75] Fighting in such conditions, however, was another matter: now the rare dry days were welcomed not for the chance to fight, but because both sides could get out of the trenches to dry their sodden greatcoats in the weak autumn sunshine. The soldiers, it seemed, were just waiting for the end of the campaigning season.[76] There would, after all, be no victory in 1916. Tristani reflected on his battalion's disappointment in front of Baniska Trench: "We suffered a rude check, despite our hopes for success. We get the impression that battle is no longer possible in this mud-field...No one is happy— losses have been suffered for no appreciable results." Staff work had been poor: their own gunners fired on them rather than the enemy's defences. There had been rumblings of discontent when Tristani's men were trucked back to their rest billets and had to wait five hours in the rain while the cantonments were prepared. Unfairly, their divisional and corps commanders were dismissed.[77] The *poilus'* harsh judgement on the workings and competence of higher authority were to find mutinous expression when the French army resumed the offensive unsuccessfully in 1917.

Fayolle, the most perceptive of the senior commanders, judged that by the last week of October the battle was no longer going anywhere. If it still had a point, it was merely to "pull along the British and oblige them to fight." For Fayolle, "exhausted men, shattered ground, a strong enemy, and winter" signalled the end of his long, hard task. The best he could contemplate was to "settle down quietly over the winter and organise a battlefield which can be utilised in spring or summer."[78] Rawlinson agreed, although he was determined to maintain an active front over the winter, while training his army for a resumption of the offensive in the spring.[79] Joffre and Foch still looked to take Bapaume and Haig, of course, shared their determination. But even the rationale that good tactical positions had to be secured before the winter, which had been creeping into operations schemes since mid-September, cannot fully explain the later fighting. The whole Thiepval Ridge was in allied hands by mid-October. By attacking the next line of German positions beyond the Ancre and across the valley from Le Sars and Lesboeufs, Haig condemned his men to a wet winter in low ground beneath the German observatories. But this was the final ridge, beyond which the tantalising spires and roofs of Bapaume could be glimpsed. This was the first prize of the battle, and one last gasp might still secure it before the year was out. If one is inclined to criticise First World War generals, in the events of late October and early November 1916 one has reasonable grounds. Foch continued to press, Joffre to bluster, and Haig to scheme, which condemned their troops to battle on in appalling conditions to what one anonymous veteran identified as "the limit of endurance."[80] This battle ought to have been stopped before the end of October. In mitigation, in the wider battle of wills that encompassed the battle for ground, Haig was anxious not to compromise the moral and *matériel* ascendancy that his troops had finally gained over the German soldiery. President Poincaré, who had come to congratulate him on the success of the British armies, was informed that the Anglo-French armies had

already broken through all the Enemy's prepared lines and . . . now only extemporised defences stood between us and the Bapaume Ridge. Moreover the enemy has suffered much in men, in material, and in *morale.* If we rested even for a month, the Enemy would be able to strengthen his defences, to recover his equilibrium, to make good deficiencies and worse still, would regain the initiative![81]

Therefore, Haig was not going to stop his battle; and Joffre did not dare to stop his.

However, it would be stopped for them. In November, for the first time, front-line commanders began to protest that their men could not fulfil their designated tasks in the prevailing conditions. They also criticised the futility of any attacks that were carried out. Lieutenant-Colonel Roland Bradford of 1/9 Durham Light Infantry, soon to become the youngest brigade commander in the army, was privately very critical of yet another vain assault on the Butte de Warlencourt, which his battalion was directed to mount by the staff warmly quartered well away from danger.[82] Thus an enduring image of ignorant, detached staff and suffering soldiery started to take root, forming in the minds of those who were being pushed to the limit.

It is ironic, then, that a staff officer helped to bring a long-overdue conclusion to the battle. In the first week of November Cavan's XIV Corps had been ordered to assault Le Transloy yet again. Cavan formally protested to his army commander: his men could not be expected to attack in such atrocious conditions.[83] The attack took place regardless, with predictable results. But Rawlinson had referred Cavan's protest to G.H.Q.,[84] which promptly sent a junior staff officer to investigate conditions at the front. Major the Viscount Gort, a future field marshal who would command the next expeditionary force Britain was to send to France in 1939, inspected 17th Division's sector, adjacent to IX CA. His report, which went right to the top, made worrying reading. The troops could struggle from their trenches only if the men pulled each other out of the morass. Even then they seemed reluctant to move forward, going to ground a short distance beyond their own trenches. Perhaps it was not the solidity of the defence, but "physical exhaustion owing to mud and water which the men had been standing in and the resultant reduction in morale" that ultimately checked Fourth Army's advance. This was not so much a fighting as a logistical problem: it had taken almost ten hours for one "fresh" battalion brought up for the most recent attack to struggle forward the three miles from Ginchy, after which they had no energy for the assault itself.[85]

Rawlinson reinforced this mood of "enough is enough" when he delivered a sternly worded memorandum to G.H.Q. protesting against Haig's intention to maintain large-scale offensive operations throughout the winter. His divisions were tired, their ranks were filled with new recruits who needed training, and to push them over the winter would undermine the great improvement that they had shown during the battle.[86] Rawlinson was closing down Fourth Army's offensive on the Somme. His corps commanders were instructed to improve their line for the winter, strengthen their defences in case of a heavy German counter-attack, and get their divisions trained and fit for renewing the offensive in the spring.[87] G.H.Q. agreed.[88] Rawlinson was belatedly directing his battle.

Although the later stages of the battle had badly affected the mood of the front-line combatants, on the allied side of the line spirits remained high. Pierre Petit, in the trenches at Sailly-Saillisel, opined on 31 October, "the Somme has become a nightmare for our men. In place of the dreams of glory which fired them up, they found only rain, mud and an unassailable enemy. Their own artillery kills them; the water and mud freeze them, grip them so that they can do no more." Nevertheless, their unexpected success the next day restored their smiles.[89] Although the *poilus* had not finally defeated the Hun, they had clearly mastered him on the battlefield. As one anonymous soldier in II CA bragged, "I can assure you that here we have the distinct impression of our constant superiority, more marked and continual than that which the Germans had over us in the battle of Verdun."[90] The British soldier's cheeriness in the face of adversity was also noteworthy: "I think it is this delight of his in being what he should call a 'care not' that makes him march to his death singing about 'the Bells of Hell' and his ladyfriends," recollected one admiring officer, on the eve of his own demise.[91] This good humour was grounded in Tommy's success and prowess. Partly this was based on obvious *matériel* superiority. While he was being somewhat hyperbolic when he suggested that the allied artillery fired a hundred shells for every German one, Gunner Keith Buchanan knew that "the steel band round the Hun grows tighter and tighter."[92] "Where the British shells flew over in sheaves," Charles Bean remembered, "the Germans fired single shells or salvos."[93] Moreover, British battalions were beginning to develop a sense of their own martial skill and ability to match their opponents in ways that had not been possible back in July.

But even if allied soldiers believed they were winning, their wish for peace remained strong, and their reasons for fighting on were shifting. Charles Barberon, returning from a short period of leave in which he had had to endure the alien surroundings and empty speculations of the home front, inevitably discussed the prospects of victory and the nature of a peace with the comrades he met on the leave train. Some still thought victory was not far off, although most now did not. One corporal felt that France must win a complete victory, or else she would have to fight again in five years' time. Barberon himself wanted France to seek immediate peace, "but if we cannot get an honourable peace we should continue the war without remorse . . . to the bitter end." Hate it though they all did, they were realising that the war was changing its nature and purpose. It must be fought to the death to vanquish the militaristic Hun, but also to conquer war itself, in all its industrial barbarity. And to rein in the warmongers: "those who sell shells only ask one thing, that this continues," Barberon railed as he returned to his billet under the incessant shell-fire. If the socialist pamphlets found in captured dugouts were any guide, Fritz seemed to share this ideal.[94]

IN ENGLAND TAWNEY was equally enraged by those who remained safe far from the front lines. However vivid the imagery served up daily by the media, he cautioned, war was not the civilians' spectacle but the soldiers' "state of experience." "When men work in the presence of death," he cautioned, "they cannot be satisfied with conventional justifications of a sacrifice which seems to the poor weakness of our flesh intolerable." Unless the civilians renewed their faith in the cause, the soldiers' burden was not to be borne. For now, they stood on a cusp, some "denouncing the apostle of war, yet not altogether disinclined to believe that war is an exalting thing, half implying that our cause is the cause of humanity in general and democracy in particular, yet not daring boldly to say so lest later you should be compelled to fulfil your vows."[95]

"I do hope that England realizes what is being done out here by her troops," Charteris had worried as the battle reached its climax. "[T]hat we have effected already more than any previous attack either by the French or ourselves, that we have beaten a great portion of the German Army in battle, and that we may still force them to peace this year."[96] Clearly, however, the mood changed between July and October. In July, the U.S. military attaché in Paris had reported: "Even the much heralded secret session of the Chamber of Deputies . . . and the present secret session of the Senate have not caused any excitement among the people. Their determination to win and their confidence in final victory are as unshaken as ever."[97] By October, when Tawney and Barberon were writing, soldiers, people and even politicians were starting to question why they were fighting, and whether victory was worth such a price.

IF THE BATTLE of the Somme was going nowhere by October 1916, its continuation is understandable in the context of the war as a whole. Foch identified it as "the relief of the Verdun of the Eastern Front."[98] Joffre's broader strategy remained in operation, even though there were signs that the General Allied Offensive was also starting to dislocate.

The Italians were still attacking on the Isonzo, although their operations were suffering from the onset of autumn rains as well. Heavy fighting was continuing in Romania, and one of Joffre's main justifications for maintaining pressure on the Somme front was to prevent more German divisions going to the Eastern Front, where Falkenhayn was mauling the inexperienced Romanian army. However, the practical coordination between Romanian and other allied operations was indifferent. Russia preferred to continue her offensive into Galicia rather than go to the aid of her southern neighbour, and by the time Brusiloff sent some divisions it was too late to prevent Germany occupying Romania. A belated offensive on the Macedonian front was making some progress against the Bulgarian army, but it was too little and too late to save Romania. There was some justification in complaints about "the slow offensive at the Somme and elsewhere on the Western Front" reported by the British military attaché in Russia, Colonel Alfred Knox, rather than a concerted action against Bulgaria in the Balkans.[99]

The Somme's drain on German resources at least allowed Joffre and the French army one final—in Joffre's case self-defeating—coup before the campaigning season came to an end. While Fayolle and Micheler had been making progress astride the Somme, the French army had been pursuing a second attritional battle on the Meuse. From mid-July to September, General Nivelle's Second Army had the initiative, and in order to tie down German divisions it conducted an active defence and local counter-attack around Fleury on the eastern bank. It was cruel, intensive *grignotage,* just like that on the Somme: the Thiaumont Work, a commanding outpost northwest of Fleury, changed hands sixteen times between June and September.[100] Thus Verdun had assisted the Somme at the same time as the Somme—as had been its short-term purpose—had relieved Verdun. As Messimy's *chasseurs* advanced on Bouchavesnes, Joffre was on the road to the fortress. On 13 September, in a solemn ceremony in the crypt of the shattered citadel, the heroic defence was acknowledged. President Poincaré presented the city with the *Légion d'Honneur.* Great Britain awarded a Military Cross.[101] Verdun had not fallen.

Throughout these months Nivelle and his army group commander, Pétain, had been pressing Joffre for strong reinforcements in order to mount a proper counter-offensive. Their objective was to regain the lost forts at Douaumont and Vaux, more symbolic for France than any number of Picardy villages. The Somme offensive remained Joffre's priority for as long as it made steady progress, absorbing the bulk of men and guns, while Verdun received only enough reserves to maintain the active local offensive. But it was clear that the German defence on the Meuse was slackening. Divisions and heavy guns had gone north to shore up the Somme front, while tired units that had already been through the northern battle had replaced them. As a result, Second Army finally had superiority in men and modern heavy artillery, so when the Somme offensive bogged down Joffre finally gave in to Pétain and Nivelle's importuning.[102] On 24 October General Mangin's III CA advanced behind the first intensive creeping barrage fired at Verdun and recaptured Fort Douaumont.[103] France rejoiced. By 3 November, Fort Vaux was also retaken. For the first time at Verdun the artillery was conquering and the infantry occupying; the French had established moral ascendancy over the enemy on the Meuse as well as on the Somme. As Crown Prince William, who had overseen the Verdun battle, admitted in his memoirs, "I now know for the first time what it is like to lose a battle."[104] France's long agony on the Meuse had been redeemed in glorious style by a dynamic army commander apparently practising a new style of unstoppable offensive. Although Nivelle, in Ian Ousby's estimation, was a man "with a gift for disguising old ideas in new formulae,"[105] it was to be to his great good fortune that politicians had short memories.

WHILE NIVELLE WAS winning glory at Verdun, the battle on the Somme was playing out its final, inconsequential moments. GAN and Sixth Army's resolve had not faltered, even if common sense started to prevail. The fresh troops brought up to assault Bois St. Pierre Vaast spent their time consolidating the defences in front of Bouchavesnes. Tenth Army planned for a resumption of its advance across the Santerre plateau if atmospheric conditions and resources allowed. To that end, line-straightening company- and battalion-strength attacks persisted against salients and strongpoints opposite the French front during November. The long-standing tactical sores—Sailly-Saillisel, Biaches, Barleux and Chaulnes—continued to weep. With a certain degree of irony, the slowing of the French offensive meant that the last victories on Sixth Army's front would belong to the enemy. On 29 October La Maisonette had been lost, along with over 450 French prisoners. Between 14 and 16 November German storm-troops re-established a foothold in Sailly-Saillisel and recaptured Reuss Trench. Foch was livid; Fayolle unsure whether to blame his chief for wearing out the troops in constant small-scale attacks or the 9th *Zouaves* for not being bold enough to hold their positions.[106] Either way, the French army's battle was inexorably going the same way as that of the British.

Although the recapture of La Maisonette was of only minor tactical significance, it was a godsend to the German army's press office, which had spent four months trying to disguise the steady series of defeats astride the Somme, and a small compensation for the loss of Fort Douaumont.[107] But such modest German triumphs were inconsequential when compared with the final shock that Gough's army delivered north of the Ancre in mid-November, an attack that confirmed who were the winners of the Battle of the Somme.

Fifth Army had spent October and early November consolidating its hold on the northern end of the Thiepval Ridge.* The slow slog that became known as the Battle of the Ancre Heights testified to the revived German defence. It took Gough's divisions more than a month to conquer the Schwaben and Stuff redoubts and the lines that extended eastwards over the rear slopes of the ridge, Stuff and Regina trenches. They advanced little more than half a mile, and British and Canadian losses were heavy. It might have been considered a more than adequate effort at this late stage. Haig and Gough, however, looked beyond the Ancre. The going would be easier on the flanks, and the cavalry might yet be allowed to have a gallop. From early October, therefore, a new thrust north and eastwards to destabilise the German lines beyond the Ancre was prepared.[108] Once conditions improved, Gough would throw his forces against the villages and ridges that had defied VIII Corps on 1 July 1916.

It has been suggested that the Battle of the Ancre, as Fifth Army's final offensive was dubbed, need not have been fought; that it was a political battle to allow Haig to go to the pending allied strategic conference with a recent triumph in the bag. Political criticism of his conduct of the offensive had revived, and a striking victory would give him some ammunition against his critics and add weight to the case for continuing the Somme offensive in 1917. It would satisfy his demanding allies, too: Joffre had not yet formally halted the battle or the General Allied Offensive. Finally, it would be a moral victory. If Serre and Beaumont Hamel, unconquered villages that were imprinted on the army's psyche, could be taken, then 1 July's humiliation might finally be absolved, and the fact that Bapaume was still out of reach mitigated.

---

*Reserve Army was formally redesignated Fifth Army on 29 October 1916.

The British army's final offensive on the Somme can be usefully contrasted with its first. On 13 November Fifth Army attacked with five divisions, with two brigades covering the flanks. A few tanks were available, but ground conditions rendered them ineffective. It was the biggest operation seen in the British sector since September. The preliminary bombardment lasted seven days and was twice as heavy as that of 1 July. The wire was effectively cut along most of the front, and German fixed defences destroyed, but the deep dugouts in the villages generally survived. The assaulting infantry had the assistance of a creeping barrage, which they had been trained to follow very closely, and a machine-gun barrage designed to sweep German machine-gun posts off the ridge behind Beaumont Hamel. The elements offered both help and hindrance. Heavy fog covered the initial advance; but many attacking units could not follow the barrage because of it. Deep mud also slowed the assault. The outcome, as on 1 July, was mixed: some real successes, but not complete victory; a steady infantry advance behind the barrage but no opportunity for cavalry to exploit. On the northern flank, 92nd Brigade of 31st Division, tasked with establishing a defensive flank, by accident or design attacked its 1 July objective of Serre again. On this occasion the assault battalions made it to the German support line but were obliged to retire late in the day because 3rd Division, attacking to the south, did not get as far. At this point the topsoil was loam rather than chalk, and the mud was waist deep. Third Division's assault waves bunched up on uncut wire in no man's land and its attack collapsed.[109] To their right, 2nd Division's left also failed where German wire was still intact opposite the strongpoint known as the Quadrilateral; but its right wing, following the barrage closely in lighter mist, reached its first objective in the third German line within half an hour. In the confusion, once the attack on the northern flank had broken down, an effective assault on its second objective could not be organised.[110] In the centre, 51st Division fought its way into Beaumont Hamel: machine guns brought up from its undamaged dugouts had to be overcome, but the village was pinched out from two sides and the defence subdued. The division's second objective beyond the village remained out of reach. On 51st Division's right, 63rd (Royal Naval) Division, in its first engagement on the Western Front, pushed towards, but could not take, Beaucourt village. South of the Ancre, 39th Division took St. Pierre Divion.

On 18 November fresh divisions delivered the second stage of Fifth Army's attack, pushing the line up the slopes north of the Ancre and consolidating the hold on Redan Ridge. Beaucourt, finally taken by 63rd Division, was the last village to fall into British hands. Serre, where 31st Division's "pals" battalions had been cut to pieces on 1 July, remained German. On the same day the first snow of winter 1916 fell. It marked the formal end of the Battle of the Somme.

British infantry and artillery cooperation had improved immensely since 1 July, and barrages were now heavier, more concentrated and more carefully targeted, with destruction, interdiction, neutralisation and infantry support all incorporated. On 13 November the entrances to the dugouts in St. Pierre Divion were each assigned their own 4.5-inch howitzer to keep the garrison underground until the attackers reached the village (a practice the French army had employed on 1 July). Mopping up had also been greatly improved. Two platoons were specifically designated to occupy two German battalion head-quarters in Beaumont Hamel, which had been identified from a captured document.[111] But timetables were still inflexible, and when they broke down an attack would stall. Casualties remained heavy—around 22,000 men in the six-day battle—but now there were tangible results to show for them. Although German casualties are not recorded, Ludendorff noted that the Battle of the Ancre was "a particularly heavy blow." Over seven thousand German prisoners were taken. The two divisions holding the Ancre sector had to be replaced after 14 November, and again after the 18 November attack, having been very badly handled.[112] The British army's military skills, if not yet perfected, were much improved. For Edmund Blunden, who fought there, the Battle of the Ancre was "a feat of arms vieing [sic] with any recorded. The enemy was surprised and beaten."[113]

Although the capture of Beaumont Hamel, Beaucourt and St. Pierre Divion meant nothing in strategic terms, it signified something more profound. "I was with the infantry . . . when they took those places," Lieutenant Arthur Waterhouse remembered, "and could not help feeling at the time that we had got level with the people who had repulsed us when we last had a go at Beaumont Hamel. Under these circumstances the Germans could not hold on."[114]

WHILE THE BATTLE had ground inexorably on, seemingly adding little to the military balance other than further casualties, the politicians had declined to interfere. Robin Prior and Trevor Wilson have argued that at the start of October the British War Committee took no action on a memorandum Haig submitted which presented his arguments for maintaining the attritional offensive, simply disregarding his "flights of fancy." Certainly, as they say, "No authorisation was given"; but no authorisation was needed.[115] At that point the battle still had some potential. Towards the end of October an allied strategic conference at Boulogne proved equally reticent about the offensive, focusing instead on Balkan affairs. Nobody challenged the strategic consensus. Lloyd George was maybe too pusillanimous to speak up, although he was restive and rancorous: "I am the butcher's boy who leads the animals to be slaughtered," he complained to Max Aitken, "when I have delivered the men my task in the war is over."[116] Certainly, Lloyd George always had qualms about pursuing the Somme battle with the resources Britain had at her disposal, and he had confided to Repington early on that he thought it was unlikely to prove decisive. He kept on delivering the men throughout the offensive; yet he had admitted at its start that he could not deliver guns on the French scale until November, and that these were the key to "the grinding process" that must go on if Germany were to be beaten. For Lloyd George, attrition in 1916 was inadequately resourced and premature—"by next year we should have a lot more men and guns, and the Russians the same; we should begin breaking the Germans down"—but not wrong.[117] If he strongly opposed the offensive for which, as the responsible minister, he provided the men and the guns, he should have tendered his resignation. Apparently, at the start of November, he did threaten to resign if the attritional strategy continued into 1917.[118] But this was mere political sparring: the life of a disgruntled backbencher, Churchill's life at the time, was not for him. In autumn 1916 the Chantilly strategy and attrition were central to all politicians' conceptions of how to defeat Germany, the main objective. Even after the death of his son, Asquith continued to back his commander-in-chief, thereby keeping any lingering criticism of Haig's tactics in check.[119] Subsequently, the politicians' reluctance or inability to stop the offensive have been presented as craven, but at the time the constraints of coalition, civil–military relations, public opinion and increasingly fraught domestic politics gave them little room for manoeuvre. But they also had much less desire to do so than their war memoirs, compiled after years of reflection and with the benefit of hindsight, would later pretend.

Both Asquith and Briand had failed to deliver victory after more than two years of war and their positions were shaky, but not because of the Chantilly strategy. Rather than the Somme, it was long, heated arguments about how to save Romania, in which the politicians and their military advisers were also engaged, that precipitated Lloyd George's first open attacks on Asquith and another secret parliamentary session in France.[120] Such arguments, while heartfelt, were mere distractions while attrition did its job. Over Romania the gloves finally came off in the battle between the War Minister and his C.I.G.S.: both men knew how to fight dirty.[121] Robertson mobilised his allies in the press in his ongoing struggle to control strategy.[122] He was quite candid when he confided to Northcliffe that "The Boche give me no trouble compared with what I meet in London."[123] But by now even the Northcliffe press, which had always supported G.H.Q.'s attritional strategy, was turning against Asquith's dilatory conduct of the national war effort. Attrition should not be abandoned; rather, it should be pursued more determinedly.

Haig's critics from the beginning of the year seemed to have been won over by the later success of the offensive. Both Arthur Balfour and Reginald McKenna visited G.H.Q. and gave their firm backing to the commander-in-chief.[124] Haig even claimed to "get on very well with Lloyd George" when they met at the Boulogne conference. The War Minister tried to make excuses for his earlier discussions with Foch; although he did not hide his differences with Robertson and the War Office military staff.[125] However, despite his attempt at reassurance— "it is important that soldiers and politicians work together in this war. It is only by the most complete understanding and cooperation between the military and civilian elements that we can hope to win"[126]— the situation remained uneasy. There was still an ongoing backbench conspiracy focused on Churchill, F. E. Smith and Sir John French, who had recently been sent out by Lloyd George—apparently on Joffre's suggestion—to review the French army's artillery practices.[127] Lloyd George wanted to rein in the War Office staff, and the others wished to oust Haig, the *Morning Post*'s editor H. A. Gwynne told Rawlinson.[128] By then, Haig had become inured to the intrigues of politicians, and he was not unduly perturbed. "I expect that neither will succeed in doing much harm," he informed King George, especially given Clemenceau's paeans to Field Marshal French on "the skill of everyone and the accuracy of the artillery etc. etc."[129]

Despite all that had happened since July, the early judgement that attrition was costing Britain more men than Germany, while less valid by October, still held the field among the dissenters in London.[130] But Lloyd George did not openly criticise the Somme or attrition until early November, when the battle was almost over. After a month of arguing over strategy and authority with the C.I.G.S., Lloyd George arranged an informal meeting of the War Committee to discuss Robertson's latest endorsement of the attritional strategy, from which the latter was deliberately excluded. In Robertson's absence the War Minister put his confused, disingenuous, amateur case against the offensive. His argument harked back to July's concerns:

> To be successful in the plan pursued up to the present time it is essential that the anvil should be more damaged than the hammer . . . Further it was essential to keep up the spirit of the people at home, and this might not be easy where there were no definite victories. The people of this country firmly believed that the Somme offensive means breaking through the German lines, and there would be great disappointment when they discovered that this was not likely to happen.[131]

Breaking the German army was not the same as breaking through their lines, something which Lloyd George, who had been told often enough, ought to have realized by this point. "I remember at one time when I was in France the public were frequently disappointed by hopes being held out to them in the communiqués which were not realized," Robertson reminded G.H.Q. at the end of September: "Whatever you do you must not disappoint the public." Robertson acknowledged that G.H.Q.'s official communiqués were now fuller and more effective in describing the work of different units, "claiming nothing to which you were not entitled and holding out no hopes which were not likely to be realized," and that as a result the reporting from the Somme was supposedly having a good effect on public opinion at home. Nevertheless, the most important civilian remained determined to be disappointed: it would seem that Lloyd George was equating his own disappointment with public opinion.[132] Moreover, Charteris suspected he was also deliberately manipulating the War Office censorship in order to undermine his rivals, concealing from the public the German army's declining morale. It would not do if the army or Asquith were seen as successful, "so I suppose the British public will not have it impressed upon them that their army has won a great battle."[133] Lloyd George, who feared the public would blame the politicians rather than the generals for the apparent lack of progress, was finally making his move against Robertson and potentially against the Prime Minister who backed him.[134]

At the allied politico-military conference that met in Paris in mid-November to discuss the strategy for 1917, Lloyd George made a bid to revise strategy (ironically, Asquith was his mouthpiece). This stratagem was more symptomatic of the intensifying rivalries between politicians and soldiers than evidence that Lloyd George (or anyone else, for that matter) had a realistic alternative strategic policy. He called emotively yet impotently for "victories," not more morale-sapping casualties: the draft conclusion of the earlier War Committee meeting could only suggest a vague desire to "seek for some more promising plan to end the war."[135] But, as clear-thinking strategists recognised, the defeat of Bulgaria, Lloyd George's panacea at this time, was no more likely to bring Germany to her knees than the loss of Romania would destroy the Entente.

So divided were politicians and soldiers at this point that separate military and political conferences met on 15 November. At the plenary session that followed the next day Briand endorsed the Western Front as the main theatre, and Joffre and Haig's proposal for a renewed Somme offensive, potentially starting in February 1917. Asquith had reassured Haig at dinner the previous evening that he was "very pleased with our success, and said that it had made a very great difference to their relations with the French and other Allies":[136] although with another son recently wounded in the offensive, the Prime Minister was finally starting to have qualms about Haig's tactics, if not Joffre's strategy.[137] Lloyd George, then a lone dissenting voice among the senior war leaders, had been stymied by the proponents of attrition once again: "crushed," Haig was pleased to note.[138] Nevertheless, Rawlinson suspected that Lloyd George, unfazed and now emerging as Asquith's main rival for power, sympathised with Churchill's cabal and certainly wanted Robertson's head, if not Haig's. He was, after all, a politician: "he can't help intriguing."[139] Lloyd George had clearly not been satisfied with Foch's explanation for the heavy British losses on the Somme, or the offensive strategy that had led to them. Still, even if Haig was only, in Lloyd George's oft-quoted jibe, "brilliant to the top of his army boots," there was no obvious replacement in the higher echelons of the British army, and the commander-in-chief had patrons in high places, not least the King. Nevertheless, the human cost and indecisive ending to the Somme were certainly factors behind the confrontational attitude Lloyd George would take to the army command in London and France after he wrestled the premiership from Asquith at the start of December.

In November both Asquith and Briand had the option of blaming their commanders-in-chief, and obliquely their armies, for the costly strategy of attrition, and the lack of victory, even if in autumn 1916 the Western Front strategy was not the central issue of contention that it was later made out to be. That would mute, if only for a while, the growing parliamentary clamour against them. But such a self-serving course would not endear them to the ordinary soldiers who had been championing a cause that their governments had persuaded them was worth fighting for. Asquith chose to fall on his sword rather than scape-goat his commander-in-chief. After eight years in Downing Street his energies were waning, while the material, financial and manpower demands of war were intensifying. Victory seemed no nearer than it had been a year earlier.[140] His critics in the Cabinet and Parliament would not be appeased, so he resigned on 7 December 1916. Briand, on the other hand, chose to cull the high command that had brought France close to victory on the Somme.

THE WINDING DOWN of the battle came as no surprise. Even while the muddy, bloody actions of October and November had been ongoing, thoughts of closing the 1916 campaign and preparing to renew battle in 1917 had crept into the commanders' thinking. In late September Haig had advised Rawlinson and Gough that their battles should at least secure good positions for the winter. (Nevertheless, as we have seen, pushing on in October and November lost many of the positional advantages gained in the September fighting.) In early October Foch and Fayolle had enjoined the corps holding the flank astride the Somme bend to prepare deep, multi-line defensive positions to contain an enemy counter-attack. Joffre anticipated the British army extending its front southwards, at least as far as the river, relieving French divisions for rest and training, and to provide a fresh reserve for sustaining the offensive in the spring. By the end of December, the British held the line down to the river. The shattered Somme battlefield was to be their responsibility and they were soon to face "the cruellest winter for fifty years."[141]

Military operations, and misery, did not end with the formal suspension of large-scale fighting. The armies could not simply pack up everything and head home. If battle was to be recommenced in the spring, as was Joffre and Haig's intention at the end of November, then jockeying for advantage must continue. Petty war—trench raiding, sniping, strafing, shelling, air reconnaissance on days when the sky was clear—persisted, if on a much smaller scale than previously. "The game of hide and seek with Death" went on for those who held the line during the freezing winter months.[142] I Anzac Corps was tasked with holding the central sector of the British front, between Le Sars and Gueudecourt, "the worst sector of the sodden front . . . the hardest trial that ever came to them."[143] The AIF now had five divisions rather than the original three (four in I Anzac Corps on the Somme front), but in this expanding national army the majority initially had little real battle knowledge, and little *esprit de corps*. The Australians' second tour of duty on the Somme was to be as harrowing as the first, albeit for different reasons. Now the main battle was against the elements, with the enemy a few hundred yards away to add occasional distraction. Nevertheless, this time the experience was altogether more positive, more rewarding, than the first. Out of adversity came comradeship, *esprit de corps,* and renewed confidence in their martial skill, which had been so badly undermined by Pozières. Come the spring, the Australian infantry were shaping up to be audacious trench-fighters.

The Australian battalions had departed from their quiet sector in the Ypres salient with heavy hearts: when the men of the 12th AIF Battalion were asked by Lieutenant-General Birdwood if they wanted to return to the Somme he "got a straight 'no' from one end of the trenches to the other."[144] At the end of October they came into the line opposite Bapaume, whose clock-tower was visible just beyond the next ridge. It would taunt them for the whole winter.[145] Their ordeal began with a failed assault. First and 7th Australian brigades were ordered to attack Gird and Bayonet trenches. After several postponements the assault was finally delivered on 5 November with little enthusiasm. Bean witnessed "a long line of men . . . going very deliberately forward, they seemed to be looking at their feet. I dare say they were, to avoid the shell-holes. They were certainly going slower than any line of Australians I had ever seen."[146] The ground was in an appalling state: it had rained overnight, and in places the attackers had to wade forward through thigh-deep mud.[147] Although they entered Gird Trench in places, the Australians were bombed out by 4 Guard ID's counterattacks once their ammunition ran out. It was a commendable achievement in the circumstances, but futile, and cost over fifteen hundred casualties.[148] The men who came out of the line bore the mark of their experience:

> not demoralised in any degree, but grey—drawn faces—and very very grim . . . It is the first time I ever passed an Australian bat-

talion without seeing a single smile on any man's face ... those faces wouldn't look any different if they were dead ... as worn and haggard as ever I have seen a British Tommy ... There could be no doubt that these men had been through very severe hardship and were feeling it very heavily.[149]

It smacked of men remembering a previous hell,[150] and entering a new one, "in which the awful conditions under which we struggled during those November days dwarfed the slaughter of the artillery to almost nothing in comparison."[151] Indeed, the shelling had none of the intensity of Pozières in what was a torpid and mud-bound battle. But life in the trenches over the winter of 1916–17 was certainly harsh, and existence in the scoured Somme landscape difficult. The earth itself was depressing:

It's a soft undulating piece of country of gentle slope and shallow valleys; but what an abomination of desolation it is! One can almost imagine that it is as things were in the beginning, a torn shattered landscape, a leaden sky that reaches down and hugs the earth, as tho' the breath of some foul, clammy monster had breathed on it and the roar and flash of the guns like some unseen power rending the silences of this corner of a primeval world. A puny, insignificant atom moves across the face of it; it is a man, and it is he and his kind who, with their lust and passions, have turned a fair land into a wilderness and bathed a world in blood and tears. Round about is strewn the litter of a battlefield ... The land is rent and torn, shell-holes are everywhere, and old German trenches, shattered and blown to pieces by our artillery, zigzag up one hillside and down another. In between shell craters are many unkempt mounds of earth, with a rude wooden cross or rifle thrust muzzle first into the ground, under which sleep so many of our brave dead.[152]

On a largely featureless battlefield, certain landmarks stood out. Near Delville Wood a shattered tank marked the way forward to the Australian front lines. A party of enterprising signallers were sheltering in a hollowed-out cavern beneath its protective bulk (excavated by Tristani's men some weeks before) for their winter quarters.[153] Such rudimentary cover was vital to survival on an exposed upland across which blew "a bitter wind ... that nearly cuts us in two."[154] Most of the natural or man-made shelters had been swept away by the tornado of shells. "A trench is a gash in the ground that will hold plenty of water and mud, give you a bon attack of trench fever if you stay in it long enough, and give you just enough room to duck when the 9.2s lob," Sergeant Arthur Matthews recorded.[155] In time, however, the new front was engineered to provide more comfort. Dugouts, barrack huts, cook- and bath-houses colonised the landscape as the army settled into winter quarters. Bean was charged with editing a new trench newspaper for the Australian soldiers, *The Rising Sun,* and the Australian Comforts Fund organised hot coffee and sausage rolls in canteens behind the front lines: one was set up in the porch of the ruined church at Fri- court.[156] Tristani's battalion, returning to the line south of Boucha- vesnes in mid-December, found their positions at least habitable, if not yet complete or commodious. Intermittent fire was exchanged across the lines, but with no real purpose except routine. "The sector, if not entirely quiet, seems relatively calm," Tristani noted. "The enemy seem to have accepted their defeat and they work like we do."[157]

It took time to adjust to the winter cold. Earlier problems of static trench warfare—trench foot, frostbite and exposure, bronchitis and rheumatic fever—ravaged the men as the muddy swamp turned into an eerie, frozen wasteland. Issues of warm winter clothing (sheepskins and mittens so big they could be used as foot-warmers), rubber waders, braziers and winter fuel (previously the remnants of the wooden houses of Flers had been burned for warmth) provided some comfort to those who occupied the front-line holes.[158] Men found their small corner of shelter, their dugout or gun-pit, their very own "Shell Hole Villa, Mud Valley,"[159] and hibernated as well as the ongoing military routine would allow. As one Australian gunner recorded:

After about a week here seven of us got hold of a good dugout. We have a bunk each and a fireplace at the end; we are very cosy having a table and all; we live pretty well having a mess of our own. I do not think there was a night we missed our rum and milk, condensed milk of course. The rum undoubtedly is a great health preserver.[160]

Nevertheless, tea and bread froze solid. On the colder nights men did too: sentries died at their posts from hypothermia, and individual trav- ellers or even whole ration parties might be found days after they were due to arrive, frozen in the empty landscape after they lost their way.[161] Yet the cold had its advantages. The swampy ground solidified, so men and animals could get across it more easily, even if slipping on ice replaced getting stuck in the mud as the new hazard, and shells started to explode again when they hit the hard ground. At times "nature . . . disgusted with the everlasting mud," sent a thick blanket of snow which turned the fields white, hiding the worst ravages of the fight and giving all a fresh, new feel.[162] When this happened there was almost a holiday pause:

> For a day or two the fire slackened on both sides, and for little spaces, measured in seconds, the silence seemed more pronounced and impressive than ever. Like a lot of children, we indulged in energetic snowballing one another, the officers and conspicuous persons such as the sentry and cooks and batmen getting a little more than their share.[163]

Temporary escapes and more homely distractions became sublime. Matthews remembered the camp at Ribemont fondly: "the beer and champagne is the best there, or there are more mademoiselles knocking about, or one can always get a decent 'feed' there, and so on."[164]

For those accustomed to the warm Antipodean sun, the front could have proved a grim, morale-sapping environment. Surprisingly, however, as the Australians adapted to their icy world, renewed martial vigour manifested itself. "There is a certain fascination in doing one's job up there," Lieutenant Norman Heathcote wrote home, "somehow the personal factor does not seem to come into it. We do more grumbling out of the line than in it."[165] *Esprit de corps* could be built more effectively in groups of men who were sharing adversity, yet were not being decimated by constant shell-fire or costly and demoralising military operations. For example, the 53rd AIF Battalion rose to the occasion by developing a sense of pride in their appearance, despite the inclement conditions. They had seen a Guards unit cleaning themselves up before marching out of the line in parade-ground order, and determined to imitate them. Whale oil, issued to the troops to rub on their feet to ward off trench foot, proved more efficacious for polishing steel helmets and military equipment. Thus buffed up, the 53rd would march out of the line in dressed ranks in full battle order. For the rest of the war, they proudly bore the unofficial title of "The Whale Oil Guards."[166] November's "regiment of old crocks" was metamorphosing into a proud national army.[167] "The men were very fit in the cold weather," Bean recorded in February 1917, "and on the whole liked it . . . they aren't the same depressed men as in October and Novem-

ber . . . the winter has made them a much more serious fatalistic lot."[168]
When the 3rd AIF Battalion went back into the line in the last week of
January they put an end to "the mutual sitting-down-and-looking-at-
each-other stage" that had been the norm while the Northumberland
Fusiliers had been in the sector.[169] The Australians would in future take
the fight to the enemy, with the insouciance born of experience. Soon
they were exhibiting a casual pride in their well-executed trench operations: "On Monday night we assisted the people adjoining us in a bit of
a 'stunt' straightening out one of the 'crinkly' bits in the line. They did
it all right and got about a hundred Fritz to send back as souvenirs."[170]
When the battle resumed in earnest in the spring, the revivified I Anzac
Corps—trained, re-equipped and shiny—would commence its ascent
to the pinnacle of battlefield skill and reputation.

THE THIRD CHRISTMAS of the war passed with the trenches still
deadlocked from the Channel to the Swiss border. The "festivities"
reflected the changing nature of the war. Against a background of grey
skies, sleet and snow, on 25 December, when there should have been
"peace on earth and goodwill amongst men . . . the heavy guns are
booming in the same old way," Lieutenant Ulric Walsh wrote home to
his brother in Australia.[171] The high command ordered a strafe of the
German lines for 11:00 on Christmas Day; presumably, thought Bean,
to interrupt their Christmas dinner.[172] For those in the front line it was
no different from any other day: "What a Christmas! Standing in rain
and mud eating Xmas dinner consisting of muddy bread and meat."[173]
In the rear the troops celebrated as best they could, with food, drink
and thoughts of home.

Across no man's land Christmas was sombre, one of shortages and
despondency. Leutnant Mandl of the Bavarian reserve artillery
recorded the mood: "Among the men I find the same feeling of home-
sickness which makes one feel choked. One must pull himself
together." The Christmas trees that had decorated the front-line
trenches in previous years did not arrive, although Red Cross parcels
did, plus a Christmas bonus of two marks and two litres of beer from
the canteen: "I force myself to feel cheerful, and joke a little and observe
with pleasure that most of the men are in a better mood than I am. In
the evening we sit together in the [officers' mess], speaking mostly in
monosyllables. Finally most of us write letters home. From the front
thunders unbroken cannon fire."[174]

Mandl had good reason to feel depressed. On 12 December he had received a telephone call in his dugout. His superior relayed the news that the Kaiser had offered the allies terms of peace. The Kaiser's triumphal rhetoric rang hollow:

> Soldiers! With the feeling of victory that you have gained by your bravery, I have by consent of the Allies made a peace proposal to the enemy. Whether it will reach the destined goal is doubtful, you have with the help of God held back the hand of the enemy and conquered him.

"In my opinion it is hopeless," Mandl's informant concluded.[175] The news that the Kaiser had made a peace offer, however welcome, was received with cynicism and disbelief on both sides of the line. It was a hubristic, hollow offer, typical of the leadership and mind-set that had plunged Europe into war, not a genuine attempt to bring an end to hostilities. "If only he'd admit that militarism was wrong or had shot its bolt—that would be half of what the world's fighting for and we could forgive a great deal," was one Australian's comment.

> But the beggar must announce that he offers the world peace because he has smashed it—that because might has won, and because he has crushed little countries and defied the world successfully—therefore he is ready to make peace—and he puts the question of peace out of the question by his bloody preamble!
>
> The Kaiser always boasted that when Germany was in a position to give her peace to the world—as a strong man making an ordinance for a weak one—she would grant it. It is a very clear move—just now he is at the height of his success. Of course it is meant to split the allies.[176]

Peace on such terms was no peace at all.

For the German soldiery, huddled in their own equally cold and muddy shelters across no man's land, the peace offer was dismissed as too good to be true. Mandl's men fell contemplatively silent after he put down the telephone receiver:

> Their eyes shine for a moment, then they shake their heads and say "it is too early to hope to rejoice. Nothing will come of it."

They have been too often disappointed, it might seem still another disappointment, and what is the use of a hopeful future, however near, when one does not know if he will live through the next hour. Only the present is certain. The future is too uncertain to allow oneself to rejoice; we shall not rejoice until peace is quite certain.

In fact, the Kaiser's peace offer, its note of German victory notwithstanding, was a recognition that the allies were strong, united and slowly but surely mastering German militarism. At the ragged front Mandl's men had unequivocal evidence for this: their reflection was cut short when a huge shell blew in the door of their dugout.[177] France met the Kaiser's initiative with another successful counter-attack at Verdun; Britain with a dynamic new government headed by Lloyd George, resolved to fight on to victory. The 1916 campaign had ended, the war was far from over, but after the Somme there could be only one ultimate victor.

"Victory Inclining to Us"

ON 5 JANUARY 1917, FAYOLLE's old chief, Pétain, showed him around the Verdun battlefield. "Basically he did not show me anything interesting," Fayolle commented tartly.[1] The most successful Somme commander had good reason to be bitter. In the general reshuffle of French senior commanders that had occurred at the end of 1916, Fayolle's Sixth Army passed to General Charles Mangin, Nivelle's fiery front commander in the Verdun counter-attack: "Nivelle judges him a better [attack] commander than me. In that he is mistaken." Just as bad, Micheler, "who did much less than me... only carried out limited offensives at wide intervals, which is easy," was given command of the Reserve Army Group for Nivelle's anticipated breakthrough. Fayolle was undoubtedly right when he scribbled that Nivelle's successes at Verdun were "a consequence of the Somme offensive, [and] the result of the methods used on the Somme reconfigured by Mangin." He recognised that behind all this there was "a desire to lower the Somme to make Verdun shine with a brighter light."[2] It remains so in France to this day.

However, Fayolle was relatively lucky. Joffre was forced out by Briand in order to secure his own shaky seat for a few more months. He was given a marshal's baton, an office in Les Invalides, a post as chief strategic adviser to the government, but nothing to do. Foch was also moved, the day after he was decorated with the *Médaille Militaire* for his direction of the Somme offensive. On the trumped-up grounds of his exhaustion—few men had as much energy—Foch was rested from army group command and placed at the disposal of the new commander-in-chief, Nivelle.

Once the Somme slowed and the reality that the allies were facing another year of war sunk in, the upheavals that had been bubbling under while the battle was allowed to run on to its boggy denouement were no longer containable. Joffre's sacking was perhaps overdue. He was old and increasingly tired, and his authority among the allies, once the bedrock of his front against his own government, had seriously weakened. The General Allied Offensive had not delivered victory; indeed, it had even failed to stop the total defeat and occupation of France's newest ally, Romania. That damaging distraction when events in France had reached crisis point was probably more Romania's fault than Joffre's, but the politicians could not ignore it. As a consequence, another secret parliamentary session discussed the failure of France's Eastern strategy and prompted Briand to move against his commander-in-chief. This was merely another event in France's volatile affairs of state and did not mean that Joffre's strategy of attrition and strangulation of the Central Powers had been wrong. Rather, it would simply take much longer, and require much greater effort than had been possible in 1916. Unfortunately for Joffre, his familiar strategic prescription for 1917, "attack together and on all our fronts,"[3] was not welcomed by the politicians who had to shoulder the responsibility for lengthening casualty lists. Cachet from his recent successes at Verdun, modest though they were by Somme standards, was enough to bring Nivelle to the attention of the short-term-focused politicians. The poisoned chalice of breakthrough and liberation was handed to him. Euphoric, he made a rash, immodest promise to deliver rapid victory where Joffre had failed. He would break the Aisne front in forty-eight hours, he boasted. It was just what the politicians in London and Paris wanted to hear after 1916's bloody attrition. Experienced commanders were more guarded. "He thinks big, but I do not know if his method of execution is right," Fayolle commented after he met the new commander-in-chief.[4]

Briand's "devious stratagem," he was not ashamed to tell Joffre, had been designed to save his government: thereby the "supple and opportunistic" President of the Council of Ministers saved his own skin, albeit temporarily, although his majorities in parliamentary votes were in rapid decline.[5] His ministry and the direction of the war were re-organised at the same time. At the *Ministère de la Guerre*, Roques was replaced by firm-handed General Hubert Lyautey, nicknamed "The Dictator,"[6] who had made his reputation as a colonial governor in North Africa. Briand wanted to wrest control of military strategy from the front-line commanders. Nivelle would merely direct the North-east Front, while a new five-member ministerial war council would oversee the general conduct of the war.[7] Fayolle attributed these unwelcome changes to a conspiracy by "the Colonials" who had served in France's overseas forces working hand-in-glove with the politicians. Specifically, they accused Fayolle of letting the decisive victory slip through his fingers by halting I CAC's advance in July 1916. His response was robust: "[W]ith the river in the way, in front of Péronne? With the British to the north, at right angles? In those circumstances, no chance. And what did the Verdunois do in October and on 15 December? Did they break through? Cretins!"[8]

In Britain Asquith had culled his naval staff, owing to the worsening U-boat situation; but that was not enough to disarm his critics on the backbenches and in the popular press, or to thwart Lloyd George's intrigues.[9] When he became Prime Minister in December, Lloyd George was determined to translate the principles with which he had hitherto conducted the country's war effort into its politics and strategy. There would be no compromise peace with Germany. Instead, a reinvigorated nation, led by a small War Cabinet of dynamic, talented individuals, would fight the enemy to the finish. After Britain's inexperienced New Army had "beaten the greatest army in the world, day after day, battle after battle," the rest of the nation must follow its lead. Such was his inspirational rhetoric. He was proud of his army, even if he skirted round the question of strategy.[10] Indeed, on that subject civil–military relations would remain fraught.

The British generals fared rather better than their French counterparts after the Somme. As Lloyd George had intimated to Foch, in the New Year's honours Haig was promoted to field marshal. The irony was not lost on the more successful French general.[11] Rawlinson was promoted to full general but "rested" from army command during the course of 1917, while Gough prospered in command of Fifth Army. (Their fortunes were to be reversed in 1918 when both led their armies back to the Somme.) Robertson retained his place as the government's chief military adviser. However, after Lloyd George assumed the premiership his long-running arguments over the strategic conduct of the war with the forthright, uncompromising C.I.G.S. worsened. Naturally, they were apprehensive at G.H.Q. The Prime Minister was liable to seek "short cuts to victory . . . with all the advantages of the Somme thrown away."[12]

Lloyd George had much sympathy for Briand's sweeping changes in the high command and his centralisation of military policy in ministerial hands. But a new prime minister could not be so bold. Nevertheless, if he could not sack Haig, the commander-in-chief might be brought under closer control. When they met at 10 Downing Street in January 1917, Lloyd George finally confronted Haig over the Somme. Haig was defensive, but had no real answer to Lloyd George's assertion that "the French army was better all round, and was able to gain success at less cost of life. That much of our losses on the Somme was wasted [*sic*], and that the country would not stand any more of that sort of thing." Haig answered disparagement and misrepresentation with dissimulation: the French infantry was no longer what it was—it "lacks discipline and thoroughness"—and he had been obliged to sustain his attack to keep pressure off Verdun.[13] Such was the uneasy relationship between the Prime Minister and his commander-in-chief. The former was reportedly "sketchy and goes into nothing thoroughly. He only presses forward the measures which he thinks will meet with popular favour."[14] He certainly seemed "very bitter against Haig,"[15] was reportedly hunting for an excuse to sack him and remained determined to curb all his generals' power. Lloyd George's Machiavellian follow-up— formally to subordinate Haig to Nivelle for the duration of his offensive, a move that was finalised in conference at Calais in February 1917—proved a disaster. As well as permanently souring relations between London and G.H.Q., it badly damaged what had been developing into a reasonably cordial and functional relationship between Haig and the new French commander-in-chief. The latter had an English mother, so the English language and the idiosyncrasies of the English character were not alien to him, as they were to Joffre. Haig even liked him personally, finding him "a most straightforward and soldierly man," both more energetic and more gentlemanly than his predecessor.[16] But Lloyd George's move soon undermined one positive development which came out of the winter's reshuffle.[17]

Such were the immediate consequences of the Somme: restructuring of the French high command; a new British government; reorganisation of, if no improvement in, civil–military and allied relations; and a strategic rethink. But this was a short-term response, with quick-fix outcomes. The new, bold strategy to win the war with a single attack (in effect a return to the old one that had been discredited in 1915) denied the reality of industrial war. Tactical finesse, which the French army had perfected in 1916, could never convert so easily into strategic victory. Foch and Fayolle were to have the last laugh: "Those Colonial asses . . . can only carry on for 48 hours. They have shown that they cannot pursue an operation . . . justify themselves by saying Ah! If only they let us do everything on the first day. Who will stop them?"[18] By the spring, the French government wished to stop them but proved too craven. By then the rationale for, and practicality of, Nivelle's offensive was moot. Fearing another 1916-style mauling, before Nivelle could launch his attack the German army commenced a strategic withdrawal on the Somme front.

IN NOVEMBER AND December 1916 the battle of the Somme had subsided into a cold, muddy winter campaign of observation, waiting for the sun and the daylight to spring it back into life. If not over, then the Somme was in suspension, while the participants took stock and the next campaign was planned. The prospect of final victory in 1916 was always a chimera. In that year the war was still expanding, both in terms of the number of nations engaged and the resources they were committing to the conflict. By 1917 it had become clear that before there could be peace there would have to be more war. What sort of peace it would be and what sort of war was needed to enforce it were determined by the Somme offensive, the point at which ideals changed and fortunes turned, and during which both sides realised that much greater effort was required to avoid defeat. At the time, the campaign's worth and impact seemed clear, although as the war ground on over two more years both became occluded. Henry Wilson, who had given up his command and was back in London, acting in an unofficial capacity as Lloyd George's military adviser, summed up the previous year's fighting as 1917 arrived: "The last day of a year of indecisive fighting. Verdun, Somme, Greece and Rumania all indecisive, both sides claiming victory; on the whole victory inclining to us, and the final decision brought nearer."[19]

This verdict is echoed repeatedly by the battle's observers, as well as its directors. Clearly something important had happened on the Somme. There was a consensus that, if it had not ended the war, the Somme campaign marked *The Turning Point,* as *The Times'* correspondent Perry Robinson titled his 1917 anthology of despatches from the battle.[20] The troops could appreciate this: that summer, Lieutenant Waterhouse reassured his mother that the Somme had marked "the turning point of the war. The battle of the Somme showed [Germany] for the first time that their opponents could break through their hitherto impregnable fortress."[21] There was still much to be done before victory was confirmed. Nonetheless, John Masefield, the first author to sum up the battle for the British people, could legitimately claim that on the Somme "the driving back of the enemy began ... It was the starting-place. The thing began there."[22] Haig's despatch ended with the confident (and ultimately justified) appraisal: "The enemy's power has not been broken, nor is it yet possible to form an estimate of the time the war may last before the objects for which the Allies are fighting have been attained. But the Somme battle has placed beyond doubt the ability of the Allies to gain those objects."[23] He was not alone. The outgoing Minister of Munitions, Edwin Montagu, recorded in December that "there is not a soldier who I meet who does not believe that this German force is a beaten one."[24] There were no such self-assured predictions on the German side at the end of 1916.

While Haig could confidently predict ultimate victory through attrition, as the year turned others were still seeking victory under false pretences. Back in July, Esher had decried those "ministers ... who count up the losses, which they exaggerate, and measure advances on the map, which they minimise."[25] But that did not stop them. On the balance of territory lost and won it appeared that Germany was the clear victor of the 1916 struggle; and in Germany comfort could be taken from the fact that a couple of hundred square kilometres won back on the Somme were negligible in relation to the 44,000 square kilometres of France and Belgium occupied by Germany in 1914.[26] However, that was applying an outmoded pre-twentieth-century concept of strategy, in which occupying ground was the measure of victory, to modern industrial war, in which applying superior resources in order to destroy the enemy's ability to fight (after which you could occupy as much ground as you desired) was the way to win. Those couple of hundred square kilometres had, German commentators were pleased to report, cost the enemy some half a million casualties; what defending them had done to Germany went unacknowledged.[27] This dichotomy between ground and strength was apparent throughout the planning and conduct of the battle, as Rawlinson had explained to Esher in July: "Even if we fail to break the Germans this autumn, we shall, he believes, find ourselves next spring superior to them in every respect: men, material and morale. The end is assured absolutely, in his opinion, although the date is still doubtful."[28] Therefore, it is not surprising that it echoes through contemporary evaluations of the Somme and subsequent discussion of its impact. The Somme's territorial gain appeared negligible, at least for its cost in human life, even though the offensive had liberated the largest tract of France to date, and much more was to come.

Certainly, it seems reasonable to question the value of one village more or less, reduced to brick-dust and at the cost of thousands of men killed and maimed: Sailly-Saillisel and Beaumont Hamel made little difference. However, less tangible factors must also be weighed up in the strategic balance, as they were all along. These were quite familiar to those following the battle on the home front. On 3 July the Newfoundland *Evening Telegram*'s strategic analyst had laid out a series of objectives for its readership. Would the offensive break the front, or would it be another Verdun? Would the Central Powers' interior lines allow them to contain concerted allied pressure? The first purpose of the offensive was to prevent the Germans reinforcing the Russian or Italian fronts: the enemy might be broken there if not in France. The British effort would at least inspire a French counter-attack at Verdun, and provide "a lesson to Germany of the decline in her power greater, perhaps, than actual advances by the British or French elsewhere. Germany is at last being made to feel the real meaning of the 'brutal arithmetic' of man-power which is our greatest ally."[29] These unquantifiable strategic advantages were those that Haig reviewed in preparing his 1 August memorandum for the War Committee as a rebuttal to the casualty counters. He re-emphasised them in the despatch he submitted to London at the end of the offensive: Verdun had been relieved, German troops had been held in the West and worn down considerably. The first two had been the objectives of June 1916, the third the outcome of attrition, although the heavy autumn rains "prevented full advantage being taken" of the situation. "Any one of these three results is in itself sufficient to justify the Somme battle . . . They have brought us a long step forward towards the final victory of the allied cause," Haig believed.[30] These intangibles represent the underlying achievements of the Somme, and made one village more or less immaterial. Fayolle echoed Haig's conclusion. While the Somme battle had begun "with no clear purpose," he had found one. He had disengaged Verdun and drawn in the enemy's reserves, achieving to date "the most success with the minimum of casualties." Decisive victory had never been possible with just two corps, as his colonial detractors were suggesting.[31]

Haig presented the Somme as an important step towards victory; one that brought the strategic initiative to the allies.[32] How and when final victory would be delivered, however, no one, not even Haig, could predict at the end of 1916. But the hope was soon, and certainly before the next winter. Rawlinson, for example, accepted Charteris's optimistic assessment that "the Boches are on their last legs," and expected them to be forced to the peace table by July 1917.[33] At the front, though, two further years of fighting were anticipated.[34] All were agreed that the success of 1916 had to be rapidly exploited; before the German army could recover, before Hindenburg could seize back the initiative. Joffre and Haig's slow, steady policy no longer suited. Nivelle's bold boast that he could win quickly, could beat the German army in the field with one big blow, caught the mood of the moment.

AS THEY STARTED to plan for 1917, both sides were aware that 1916's attrition had had an impact, even if then it was difficult to quantify exactly—full casualty statistics would not be collated until hostilities were over—and was probably far from finished. This was acknowledged in the title of Haig's end-of-battle despatch: "The Opening of the Wearing-out Battle." Attrition was designed to wear down reserves of manpower: both immediate battlefield reserves to facilitate a return to more mobile operations; and strategic reserves, which kept the ranks full. Although Haig's intelligence staff had extrapolated from published German casualty lists that the offensive had cost the enemy some 600,000 casualties by November,[35] Germany's reserves were far from exhausted. But there were indications that they were now in decline. One neutral estimate in December 1916 suggested that the Central Powers had 7.45 million men in uniform, compared to 14.66 million allied troops—incidentally, a significant shift from a 1 July 1915 U.S. War College estimate of 9.65 million and 11.87 million, respectively. The German figure had been adjusted down from 5.4 million to 4 million. The Turkish and Habsburg armies had also shrunk, as had the French and Italian, while the British and Russian armies had grown.[36] Part of such decreases is explained by the redeployment of armaments workers from the front into key war industries, but casualties were the predominant factor. At the end of a year of intense fighting some armies were still growing, and there was still more civilian effort to be harnessed to the escalating war effort. The process of attrition, as Haig recognised, had only just begun; but on those figures the Entente was always going to outlast the Central Powers. But the development of more efficient, *matériel*-intensive methods was essential if allied losses were not to be prohibitive.[37]

In fact, Germany still had 2.85 million men, some 134 divisions, on the Western Front at the end of 1916, out of 5.1 million men in uniform, following a further expansion of her army. She was able to raise thirty-three new divisions in the last quarter of 1916 and the first quarter of 1917, although this was as much a response to strategic overstretch as an indication of large reserves of fresh manpower still to be tapped. Germany was having to juggle her declining manpower resources. Men of the 1918 class were called up three months early, in September 1916—most of the 1917 class were already at the front—and older recruits, home-front workers of military age and men previously designated for garrison duty were incorporated into the ranks. It must be remembered that Germany began the war with proportionately a much greater pool of unconscripted manpower than France. By the end of 1916, this was starting to run dry, although this final levy, along with recovered wounded men, gave Germany 1.3 million reserves with which to sustain her army through the next year's fighting.[38] At the same time, in September 1916, 1.2 million workers had to be released from the ranks for the intensifying home-front war effort, so there was no real net gain in fighting strength.[39] In the same month only 10 per cent of the replacements needed to fill losses in Prince Rupprecht's army group could be found.[40] Although there were more German divisions, they were shrinking: while the number of divisions on the Western Front increased by 18.3 per cent between March 1916 and March 1917, Charteris calculated that there were only 3.6 per cent more infantry battalions,[41] and these were generally well under established strength. Documents captured by the British army indicated that German battalion establishments were reduced twice in the early months of 1917: first to 800 (including 100 men fit only for garrison or labouring duty), then to 700 (with 50 garrison troops).[42] The dilution of veteran units caused by the deployment of so many inexperienced recruits also undermined the cohesion and fighting effectiveness of existing formations.[43] Battalion complements of light and heavy machine guns were doubled, however, indicating that greater firepower would be used to compensate for reduced manpower.[44]

While Germany was clearly starting to have manpower difficulties by 1917, it should also be recognised that British and French divisions were short of men by the end of the 1916 campaign. For Britain, this was potentially a temporary shortage. A lean period had been anticipated in August and September before her recently raised conscripts could be deployed and trained to fill the ranks thinned by battle.[45] Resupply of manpower to Haig's army became one of the long-running disputes between G.H.Q. and the government, partly because Lloyd George tried to constrain Haig's offensive tendencies by restricting his supply of "cannon-fodder," but also because of the competing demands from industry and transportation as the war effort intensified during 1917.[46]

France had made her most intensive mobilisation effort in 1914 and 1915, and it was recognised at the start of 1916 that she could keep her army's ranks filled only if the British army did more of the hard fighting. Although this happened on the Somme, British inexperience had obliged the French army to assume the major role in the combined offensive, while maintaining the Verdun battle. French tactics, operational methods and *matériel* superiority meant that by the second half of 1916 her army was finally able to give better than it got on the battlefield, with a corresponding reduction in the rate of casualties. Nevertheless, the damage done in 1914 and 1915 could not be reversed.

Although accurate figures of German losses on the British and French sectors of the Somme front cannot be established, it is reasonable to assume that the French army inflicted as heavy (and probably significantly heavier) casualties on the German army on its section of the Somme front than it sustained. G.H.Q. intelligence staff calculated at the start of October 1916 that the British had taken on and defeated fifty-eight German divisions, and the French sixty-one (including divisions that had passed through the battle for a second or third time). The balance had tipped in France's favour during the intense September fighting.[47]

The French had committed 44 divisions to the Somme offensive (doing 70 front-line tours in total),[48] to the British 52 (96 tours),[49] and had taken 41,605 prisoners and 178 guns, compared with Britain's 31,396 men and 125 artillery pieces.[50] Together, they fought 95.5 German divisions (completing 133 tours).[51]* In the battle both British and German armies sustained losses at a hitherto unprecedented rate, if for different reasons. In contrast, the French infantry, well protected by their guns, were now capable of sustained combat. Seven of the eight divisions in XX CA and I CAC fought throughout July and well into August. Judicious rotation of assault divisions through corps sectors, and filling gaps in the ranks with new and re-formed recruits while in rest areas kept French regiments effective. As a result, although the intensive fighting of 1916 thinned the ranks of France's regiments, she

---

*French and British divisions normally had a larger infantry element than German divisions (usually twelve battalions to nine), but German divisions generally spent longer in the front line than allied, and suffered a higher percentage of infantry casualties before relief.

was not obliged to deploy her 1917-class recruits until the final months of the year. However, French units were 140,000 men under establishment by then (compared with 92,000 at the start of the offensive),[52] and the 1918 class was in the process of being called up, with a view to filling gaps in the ranks that would increase when active operations recommenced in the spring.[53]

If, after two years of war, bodies were starting to run short, *matériel* was starting to take their place. At the end of 1916, Joffre restructured his infantry divisions on the nine-battalion model adopted by the Germans in December 1914. Instead of one artillery and two infantry brigades, French divisions would in future have an infantry element of three regiments, and a field artillery element of three groups. This allowed 101 active divisions (10 of them newly created) to be kept in the field for the 1917 campaign. Although smaller, they packed more punch. Infantry battalions were reduced from four to three fighting companies (the fourth became a divisional depot training company), while their complement of heavy weapons—light machine guns, grenade-launchers, 37mm guns and Stokes mortars—was increased. Each battalion gained an integral heavy machine-gun company. There would also be more heavy artillery assigned to corps to support them.[54] The French army of 1917 would be leaner but fitter, having assimilated the principles of modern infantry combat that had been practised successfully on the Somme.

On the Somme the New Armies "learned many valuable lessons which will help them in the future."[55] Although British divisions were not restructured, infantry battalions were reorganised and new tactical instructions, which distilled the lessons of the Somme, were issued and practised over the winter months, when training once again became the priority.[56] Two G.H.Q. pamphlets published over the winter— "Instructions for the Training of Divisions for Offensive Action" and "Instructions for the Training of Platoons for Offensive Action"— codified the wide variety of Somme experience into a systematic battle doctrine, establishing the basics of the small-unit and combined-arms tactics that British forces would employ for the rest of the war. The infantry platoon—now divided, like the French equivalent, into specialist sub-sections of riflemen, bombers, light machine-gunners and rifle-grenadiers—would in future be the tactical formation on which battle doctrine was focused.[57] While rarely acknowledged, the new tactics owed much to France's methods,[58] as well as to Germany's defensive tactics. Divisional artillery was also reorganised, allowing corps and army artillery to be strengthened.[59] The first stage of the British army's learning process, the development of effective battlefield method, was complete.

While small-unit tactics for the industrial battlefield were being perfected on the Somme, modern operational doctrine, which in time would break the Western Front stalemate, was also being formulated. Killing Germans on the Somme would have allowed Haig's army to reach Bapaume and beyond in the spring: Foch judged that steady sapping forward over the winter had made the German front untenable.[60] But such front-on slogging lacked sophistication. Moreover, the offensive had demonstrated that more grandiose schemes of manoeuvre were unfeasible on a congested, industrialised battle front. Although French infantry pierced the German defences twice, between Herbécourt and Assevillers on 2 July and at Bouchavesnes on 12 September, neither of these breaches could be exploited, being either in the wrong place, or too narrow. Moreover, French method was no longer directed to achieving the "breakthroughs" that operational doctrine had conceptualised in 1915. Moreover, if cavalry, or even infantry and artillery, had gone through, they would have been stopped soon enough on the next natural line of defence: probably the ridges and woods beyond the Bapaume–Péronne road, where the offensive ground to a halt with the onset of winter. Trenches and barbed wire could always be overcome by well-motivated troops with sufficient artillery support. But enemy reinforcements deploying along shorter, less congested lines of communication, supported by fire from the flanks, could always seal a breach.

A better solution was to exploit not forward but laterally, in order to extend the fight and kill more Germans, thereby destabilising the whole front. The concept was difficult to grasp. Castelnau had explained to Repington in July that the object was "first to pierce the line and then to enlarge the gap by continually bringing up fresh troops on the flanks."[61] This remained Haig's conception of battle. Foch's conception was more appropriate to the circumstances. Debeney, XXXII CA's successful commander at Sailly-Saillisel and subsequently one of Foch's army commanders in 1918, succinctly summed up the Somme's importance for what was to follow in a speech he gave in 1926, when unveiling the statue of Foch on the Bapaume–Péronne road. The Somme, he pronounced, was "the first of the great mass battles, in which we asserted our tactical superiority over the enemy . . . After the Somme [Foch] began to abandon the simplistic idea of obtaining success by breaking a short section of the enemy's front, replacing it with the more fruitful idea, which was to give us victory, of progressively dislocating the various sectors of the front."[62] This method would push back the enemy's line in successive, coordinated stages. Tried in microcosm on the Somme in September, it would be the operational basis of Foch's successful counter-offensive in summer and autumn 1918.

Foch's personal study of the offensive recognised that, while over four months the armies he directed had taken many kilometres of enemy positions, thousands of prisoners and hundreds of guns and other weapons, and had used up 120 enemy divisions, the slow cadence of the scientific battle had allowed the enemy time to repair the damage and to replenish his strategic reserves; just enough time to contain the allied advance given the tactical advantages the defensive enjoyed. However, if 120 German divisions could be reduced to half their strength in weeks rather than months, put out of action simultaneously for several weeks, the whole German army in the West could be beaten. For this to be achieved, the attack front had to be much longer. Speed, scale and *matériel* would be the decisive factors. In 1916 it had been necessary to plod side-by-side to train the British but now that they had mastered the art of attack such supervision would be unnecessary. In future, the allied armies should operate independently but with a common strategic purpose, battlefield method and reciprocal trust.[63] This new operational method, which Foch dubbed "general battle" (*bataille générale*), could be applied along the whole front. It was attrition and manoeuvre combined and adapted to the constricted space of the Western Front, yet more supple and speedy than Joffre's "*bataille de rupture.*" The entire line would be pushed back but never broken through. Eventually, it should fracture completely. Here was a way to capitalise on the hard fighting of 1916, which had gained allied supremacy over the German army. Foch, France and the alliance took something away from the Somme, a vital lesson for the future conduct of the war, the key to decisive battlefield victory. This was the most important, if now long-forgotten, intangible outcome of the Somme campaign: forgotten because Foch was not to get the chance to put it into practice for eighteen months.

THE SOMME OFFENSIVE'S impact on military morale is unquantifiable, yet needs to be considered alongside its material and doctrinal consequences. Undoubtedly such close-range, intensive infantry fighting was stressful for combatants on both sides. A huge increase in the new psychiatric phenomenon of shell-shock was just one manifestation of the strain of industrial battle. The horror of the trenches is rightly recognised for what it was, and that horror was particularly acute on the Somme, made worse towards the end by the onset of winter (even though this paradoxically mitigated the worst of the battle). The immediate, changing impact of ongoing battle on soldiers' morale has been evaluated in discussions of the fighting. Certainly, by its end, soldiers on both sides were pleased that it was all over, although shared hopes for an end to the war were disappointed. Assessments of the outcome of the fight, from the highest to the lowest, indicate which side had better endured the trial of 1916.

The tenacity and effectiveness of the German army's defence on the Somme should be acknowledged. After the war, Falkenhayn paid well-earned tribute to the *Sommekämpfer*:

> Once again the unsurpassed fighting qualities of the German soldier were shown up in the strongest light. Always inferior numerically, he gave ground step by step before the fury of the enemy artillery only where a stand had, in fact, become impossible. He was always ready to win back lost ground from the enemy, and to take advantage of every weakness.[64]

The extracts of German war diaries and personal accounts anthologised by Jack Sheldon in his survey of the German defence on the Somme add substance to this claim.[65] The brave, self-sacrificing actions of individuals and small units, doggedness in defence and boldness in counter-attack all come through in these survivors' tales. (Although this is hardly surprising, as fear, cowardice, funk, shirking and desertion do not make good soldiers' stories.) The army facing the allies in 1916 was certainly a formidable fighting machine, even if probably not as homogeneous or perfect as the paeans of praise heaped upon it by its leaders and admirers might suggest. But after the "monotonous mutual mass murder" of the Somme there can be little doubt that it was no longer the force it had been at the start of the year.[66]

Haig was criticised for sending his troops forward against unsilenced German artillery and machine guns. But for months Germany's troops had to be sent forward under guns that could never be silenced. The casualties and experiences of the Somme and Verdun decimated its human resources and undermined its moral fibre. The loss of an irreplaceable cadre of experienced junior officers and NCOs is frequently cited as especially significant.[67] By the end of the year, German ranks were increasingly being filled by the young and the old, who were inexperienced and physically inferior to the front-line troops of earlier years. Charteris's conclusion in autumn 1916 was astute: "the German, though he is far from being a demoralised enemy, is undoubtedly not of the same calibre as he was this time last year. The offensive has shaken him up."[68] This reduction in the fighting capacity of the German army, while impossible to measure precisely, is the other side of the equation, and the one on which arguments for a Somme success are usually grounded.

One individual will have to do service for the thousands of dispirited, psychologically damaged Somme veterans who lived to fight on in 1917, to sacrifice themselves in the attritional battle which the allies still imposed upon outnumbered and outgunned German forces. Leutnant Mandl, who returned to the front from an officers' training school in February 1917, was surprised at the peace of Flanders: "the pace seems to soldiers from the Somme absurd. The men's dugouts are like toy houses." But even in a quiet sector he no longer had spirit for the front: "I can't stand as much as I used to. My nerves have been bad since 20 December . . . I often think if the bullet had got me instead of Vogel, all my worries would have been at an end."[69] The cruel winter conditions, the loss of comrades and the nervous strain brought on by intensive combat induced a fatalistic torpor in Mandl: he did not live to see 1918.

Moreover, this once mighty army was being overmastered by the determined, experienced, well-armed veterans in the French ranks and the young, enthusiastic and intelligent fighters in the now well-tempered British divisions. With a certain degree of personal and patriotic pride, Haig reported that his army of semi-trained civilian soldiers had matched themselves effectively "against an Army and a nation whose chief concern for so many years has been preparation for war."[70] Fourth Army's tactical notes, prepared in May 1916, posited:

> It has been rightly said that this war will be won by superior discipline and morale. We undoubtedly started with the disadvantage of pitting an undisciplined nation against a disciplined one; but this advantage is rapidly disappearing, thanks to the self-sacrifice of the best element of the nation. This self-sacrifice is not enough to ensure success unless we attain a high standard of military discipline.[71]

The self-sacrifice of 1 July 1916 still hangs heavily on the nation. The disciplined, confident, conquering army of November goes less remarked.

Their training was far from over, however, and their challenging task had still to be completed. After the Somme, Charles Carrington suggested, "The German Army was never to fight so well again, but the British Army went on to fight better."[72] In future British troops would approach combat with less brio, but greater pragmatism born of experience. One Somme medical officer, Captain J. C. Dunn, who was later to compile an influential analysis of the day-to-day war experience of his battalion, the 2nd Royal Welch Fusiliers, commented on G.H.Q.'s opinion: "The German is not what he was, but his falling off seems, on contact, to be no greater than ours. Without our superiority in guns where would we be? The French seem to be far ahead of us in recent attack technique, formation, and the co-ordination of rifle, grenade and automatic fire."[73] *Matériel* superiority was certainly important. Fritz von Below implicitly acknowledged that battle was now an industrial process when he identified "the perfected application of technical means" as the reason why German defensive methods had been bested by September.[74] But so was growing confidence. For young volunteers such as artillery lieutenant Arthur Waterhouse, although "it was a hard time . . . it was the making of us. We came out of that furnace truer steel. We were B/58 of the 11th Divisional Artillery, and we were proud of it."[75] They were now serviceable components in the bigger military machine. Colonel Treadwell of the United States Marine Corps, who was guided round the Somme front in late December, reported on the shifting martial balance:

> The general feeling with regard to the Somme offensive seems to be one of disappointment that more was not accomplished, but also of satisfaction that marked superiority had been shown to the Germans in much of the fighting, and that the German moral and prestige had been lowered . . . The British were generally thought to now have the superiority over the Germans in aviation, artillery, trench mortars, and bombing, bayonet work, and morale, while the Germans were conceded superiority in organization, standardization, trained officers, entrenchments, and use of railroads. In machine guns in which Germany has so long had the predominance, conditions are now about equal.[76]

Foch was happy to acknowledge that on the Somme "the fresh British troops had familiarised themselves with the process of attack. They acquired spirit."[77] This was all to the good since, like the German soldiery, the *poilus* were sustaining attrition with their minds as much as their bodies. As the mutinies in 2 DIC suggest, their resolve was starting to falter. It is a recurring motif among observers of the French infantry of late 1916 that while they had developed tactical skill, they now lacked the élan of earlier years. Foch admitted as much to Lloyd George in September; Haig used it as an excuse for intensifying British efforts.[78] In Berlin Georg Queri could snipe that the French infantry no longer showed the spirit of 1914 or 1915 in their attacks.[79] But his fellow Germans at the front knew that their enemies now had many more bullets and shells.

In his final order to his brave *chasseurs* Messimy expressed his pride and expectation: "on the wide ridges of the Somme, where the tragic grandeur of the gigantic battle unfolded, you excelled." The cost was heavy, 71 officers and 3000 men, but their sacrifice would not be in vain. "In a few months you will resume the fight, will finish what you have started so well. Your battalions' pennants will flutter joyfully in the wind of victory; you will win it at the charge."[80] As 1917 began the French army still had one big battle left in it, one final chance to deliver France by its own efforts before it reverted to the role of "brilliant second."[81] In fact, it faced two more years of gruelling combat before there would be any victory parades. There would be no charge, but the sturdy *poilus* would see it through to the end. Although war weary, they were veterans; their cause was just. Their mood might waver, but their spirit, their resolve to free *la patrie* from the invader, would not falter.

The victors of the Somme had seen horrors, lost friends, empathised with enemies, turned against staff and commanders. "We haven't forgotten the familiar faces of those who were not allowed to pull out," Lieutenant Waterhouse wrote, "nor shall we ever forget the tall white crosses: 'Killed in Action, Courcelette, November 1916.' "[82] The battle provided a personal cause to supplement or supplant the national one: they would fight on to justify and avenge the sacrifice of their comrades in arms, their friends. Although the process of defeating their determined enemy would still be long and costly, the Somme had given British and French soldiers the methods and the confidence, and good reason, to vanquish the German army.

For Britain, this was not only a military victory. Haig was happy to acknowledge "the obligation of the Army in the Field to the various authorities at home, and to the workers under them—women as well as men—by whose efforts and self-sacrifice all our requirements were met."[83] War was now a national undertaking, in which the maintenance of civilian morale was as vital as developing military *esprit de corps*. In fragile France, home-front solidarity was now becoming an issue. When General Thierry went home on leave everyone was talking about the Somme: "All deplored the stagnation: they hoped for more from it, they foresaw the decisive 'breakthrough,' they were deceived. I saw that at the end of the summer of 1916 morale in the interior had notice-ably fallen."[84] Jacques Playoust noted a similar decline in civilian morale while he did duty in a quiet sector: "He is beginning to feel the pinch of war. In Paris, coal, sugar and potatoes are hard to get. Every-thing is getting frightfully dear."[85] Shops and restaurants had recently been instructed to close early, to save fuel: Parisian life, which had hith-erto continued in all its gaiety and splendour, was now having to adjust to the reality of war.[86]

If the Anglo-French armies had secured an ascendancy in the battle of wills played out at the front, the rear also had to hold steady for the advantages won in battle to be pushed through to victory. The twin dis-appointments of the defeat of Romania and the atrophy of the Somme offensive "with comparatively slight gains in territory at a great cost of men" meant that after the optimistic peak of September, France was now languishing under a wave of pessimism.[87] The war would be drawn out and difficult, and the spirit of the home front would in future need to be nurtured along with that of the soldiers. The mood in France, grumbling at the inefficient conduct of the war despite the most concerted and effective campaign to date, was certainly worrying, even if calls for its conclusion remained muted.

The other vital yet intangible element of strategy—maintaining the cohesion and commitment of the alliance—also had to be ensured. The General Allied Offensive had demonstrated a remarkable degree of allied solidarity, even if the extent to which Anglo-French action had helped the Russians by holding German forces in the West, or vice versa, can never be accurately assessed. Eleven German divisions did go from West to East during the second half of 1916 (they had to because of the crisis on the Eastern Front), but nine of them had been eviscer-ated on the Somme or at Verdun beforehand. Six relatively fresh divi-sions came the other way.[88] Concerted action on all fronts at least constrained Germany's freedom to send troops where they were needed. For the first time, Britain had made a sustained effort in support of France, even if there were Frenchmen who continued to believe that this (and Russia's and Italy's endeavours) was insufficient when compared to the human sacrifice France was making.[89] Atavistic fears of allies, always lurking below the surface, occasionally bubbled over. Clemenceau, for one, worried that Britain would expect due reward for her sacrifice on the Somme: colonial territory; or even the 350-year-old bone of contention, Calais.[90] But such rhetoric merely reflected the volatile, confrontational nature of French politics. Briand's changes managed to check this pessimistic mood, for a time. A new wave of optimism would sweep through in the spring, with the new commander-in-chief offering the next sniff of victory. But after 1916 had not produced victory, the harsh truth was that more effort was needed from France herself: "Above all, should we not reply to the German civil mobilisation by the civil mobilisation of all our forces?" the radical Socialist Deputy Victor Hervé asked in *La Victoire*.[91] Root-and-branch change, with all its potential for political conflict and social turmoil, was the inevitable consequence of attritional stalemate, the necessary precursor for victory—or defeat.

AT THE END of the battle, the first voices calling for peace were being raised in the allied nations, not least because President Wilson was stepping up pressure to negotiate an end to the conflict: following his re-election, in December 1916 he had asked the belligerents to state their war aims with a view to a peace conference. A flurry of non-committal peace notes issued from Europe's capitals, but there was not enough common ground for the initiative to prosper, especially now the Entente seemed to hold the advantage: France's answer to the Kaiser's note was "capturing guns by the hundred and Germans by the thousand" at Verdun, Captain Dunn noted.[92] Nevertheless, in future war aims and peace terms were factors that had to be considered in strategic policy, especially since there was a growing belief on both sides of the front that victory was impossible and that the war would end only through exhaustion and compromise. The pacifist socialists were increasingly vocalising their belief in a new sort of peace, one without annexations and indemnities, that reflected the values of an international workers' community rather than the interests of hostile, capitalist Great Powers.

In 1916 such voices were still on the fringe of politics, with most socialists, while horrified by its conduct, continuing to believe in the justice of the war. The radical peacemongers in both camps would become more vocal as the war dragged on inconclusively.[93] Neverthe-less, the views of those whom Robertson forthrightly dismissed as "cranks, cowards and philosophers" were unlikely to prevail now the war had entered a new stage.[94] Although, by the start of 1917, Cap-tain Percy Chapman's hope that "rifts seem to be appearing in the heavens, so perhaps the dove of Peace will find her way to earth once again" was shared by most,[95] the ideological juxtaposition of liberal-ism and authoritarianism, and the clashing territorial and financial agendas of Great Power diplomacy, rendered German and Entente peace terms irreconcilable. The grinding down of Germany would have to continue.

If the size and permanence of France's expanding armaments works provided a fair measure of her war effort, then "there was no intention on the part of the French government other than to see the war out to a finish."[96] Britain too was finally reaching the point where she was becoming "a nation in arms." After Lloyd George's ascent, further measures in manpower control involving "the most complete limitation of individual liberty" would strengthen that trend. Here was "evidence of the Englishman's growing appreciation of the magnitude of the task in front of him and his determination to see it through . . . so far as war on land is concerned, England's war power is steadily increasing and . . . her most formidable effort is still to come."[97]

In the early months of 1917 everyone's overriding concern was how and when the monstrous war that had assumed a new and all-encompassing form in 1916 would finally end. Would it be through one great battle or via a negotiated peace? Could German militarism, or Entente obstinacy, be broken soon? "G.H.Q. is absolutely confident that we are going through this year," Charles Bean recorded.

> And [Philip] Gibbs told me two days ago that the Germans very sincerely wanted peace. It is the blockade killing their youthful population, he says, which is frightening them. I wonder—it's a bestial war, as everyone thinks who hears of this. Still, if they get fats in Germany they turn them into ammunition, and they are trying for all they are worth to starve us. Good God—what can you do in a struggle like this? What *is* the right thing?[98]

Bean's reproof was aimed at the war itself, as much as those who waged it. The new year was commencing under a cloud of fear; and new ways of annihilating civilians and soldiers, unrestricted submarine warfare and new types of poison gas, promised only to intensify the barbarism. To justify carrying on, the allies published their own war aims in response to the Kaiser's abortive peace offer. "A1," Bean thought.[99] "Restitution, reparation, guarantee against repetition," Lloyd George had told Parliament when he came to power.[100] In January this statement of principle was elaborated in detail in a formal reply to President Wilson's note: restoration of Belgium, Serbia and Montenegro; evacuation of French, Russian and Romanian territory and financial reparations for damages; self-determination for Italian, Slav and Romanian minorities; the break-up of the Ottoman Empire and recognition of the rights of Habsburg national minorities; restoration of Alsace-Lorraine to France; and a league of nations to oversee post-war international relations.[101] This was not a programme of compromises, nor a cynical ploy to win over neutral opinion and keep the men fighting. Such realpolitik peace terms indicated that the Entente powers now believed they were winning, and so could propose, if not yet dictate, terms. Furthermore, they showed that there was still much to be achieved, and that their soldiers were willing to fight on in order to achieve it.

BY CONTRAST, IN Germany the nation's resolve to fight the war to the bitter end was starting to crack. The small crumb of comfort that Germany could take from the 1916 campaign was that she had held out. Although the late victory over Romania's second-rate army gave reason to crow, this was a sideshow to events in France, where success at Verdun had been reversed following her army's long mauling on the Somme. The future Chancellor Prince Max of Baden noted that the year "ended in bitter disillusionment all round." The best he could say was that the war remained "deadlocked" and that, despite all the bloodshed, "neither we nor [our enemies] had come one step nearer to victory."[102] If "holding on" was all that the high command could offer the German people as an objective for their continuing effort and increasing hardship,[103] then growing hostility to the conflict was only to be expected.

The alternative was a negotiated peace. The Kaiser's positive response to Wilson's peace initiative was an admission that the imperial regime would have to secure peace before the empire's growing internal fissures tore it apart. In practice, dogged defence, however heroic, was not going to force the allies to the peace table, collectively or individually. German expansionism certainly would not. Even defeated and occupied Romania had not sued for peace. The field army might hold Russia and France, but behind them lay Germany's real enemies: Britain, backed by American finance. The "Anglo-American race" would sustain the Russians with guns and money, and the French with men and munitions. Different means were needed to bring Britain, the supposed linchpin of the Entente, to her knees. So in 1917 German strategists would take new, desperate steps to break the backbone of the alliance. If the army could not end the war, the air force and the navy just might, by taking the war to England. The aerial bombing campaign—with Zeppelins now backed by long-range Gotha bombers—would be stepped up to undermine morale on the British home front. More importantly, in February 1917 Germany would resume unrestricted submarine warfare. "There is no antidote against U-boats and Zeppelins," some optimists believed.[104]

But this was a sign of desperation. Unrestricted submarine warfare had been suspended in 1915 in the face of neutral, especially American, pressure. Its resumption only strengthened the hostile coalition. It was anticipated that the United States would join the allies. Not everyone welcomed that, however. Lieutenant Ronald McInnis recorded:

> We wonder whether all the talk of neutrals joining in will come to anything. We reckon we can quite well manage to settle this "dog-fight" ourselves now, but if they come in it will certainly end it quicker. We grudge them very much their share in our victory. They are like snarling timid curs who hang round the fight until the big dogs have really decided it, and then snap at the already beaten enemy.[105]

Two months later, McInnis had to swallow his reservations: America entered the war in April 1917. Germany's latest stratagem to break the Entente's asphyxiating grip had succeeded only in ranging the world against her. Ludendorff gambled that Britain could be forced to the negotiating table before America's latent military strength reinforced the industrial and financial power with which she was already sustaining the Entente. But it was a forlorn hope, for when America flexed her muscles Germany stood no chance. Meanwhile, the German army, and the home front, would have to hold out against the relentless twin pressures of attrition and blockade that the Anglo-French armies and navies would maintain throughout 1917.

Allied intelligence carefully monitored worsening conditions in Germany as the blockade started to bite. It was a powerful extra limb for the strategic stranglehold squeezing the Central Powers. Conditions on the home front were deteriorating as a new year of war began. Leutnant Mandl's depression had been brought on in part by his visits home in early 1917. He missed the camaraderie of the front. Although he saw his sweetheart, that pleasure was offset by the dispiriting experiences of the rear area and home life. The staging post of Metz was filled with overdecorated administrative officers, "bored and cool in their manner," who had no grasp of the reality of the front. Germany was starving as a consequence of the allied blockade: "Had a mighty meal myself to prepare for the fourteen days' starvation at home," he recorded before going on leave.[106] The winter of 1916–17 became notorious in German folk memory as the "turnip winter" (in fact rutabagas, which were usually fed to cattle). With the failure of the 1916 potato harvest, and access to the international grain market cut off by the blockade, the staple of the workers' diet became the ubiquitous watery and tasteless root vegetable. Christmas 1916 was a joyless occasion. The usual cakes and sweets were not available because there was no sugar or cooking fat. Coal shortages meant that the electricity supply was intermittent. The war had even subverted that holy festival's generous spirit. The most common presents for young boys were toy soldiers and a model U-boat that fired a torpedo which "sank" a model British battleship.[107]

With the new military regime determined to intensify Germany's war effort, the home front was undergoing radical transformation. If the army could not force a victory, maybe this was due to insufficient civilian effort. Such was a natural assumption in a state that was rapidly becoming a military dictatorship, one in which the army did not trust the people. At the very least more was needed from industry to help the hard-pressed army meet the enemy's *matériel* dominance on more equal terms. The Somme therefore prompted major changes in the management of Germany's war effort. In a regime in which authority and obedience were cardinal values, this meant applying military discipline to the home front. On assuming command in August 1916, the commander-in-chief had lent his name to an armaments programme that was designed to triple weapons and munitions output. Germany would meet the *Materialschlacht* by developing her own industrial effort, economising the lives of her men. Implementing the ambitious Hindenburg Programme required a more thorough mobilisation of the industrial front, which would challenge the political and social status quo of imperial Germany. The Patriotic Auxiliary Service Law, which followed in December, extended conscription to an unprecedented degree. In future all males aged seventeen to sixty would be liable to some form of national service, either at the front or in the war economy. This required a significant compromise with the representatives of organised labour, for such far-reaching reforms could not be implemented without the cooperation of socialist political parties and factory trade unions.[108] For the first time, the ultra-conservative high command had to acknowledge that it could not conduct the war without the active involvement of the left, but at the same time it was not prepared to go so far as to offer genuine political concessions in return for the workers' effort. As a result, while soldier and worker, army, government and society were mutually dependent, they became increasingly polarised as war was prolonged. The consequences were to be felt immediately after defeat in 1918; and the repercussions would resonate through German domestic politics and European affairs until the final defeat of German militarism, in its bastardised Nazi form, in 1945.

As 1917 began, Germany was hurting from the effects of war, yet mobilising more intensely to try to force the allies to the peace table. However, prospects of victory, in the pre-1914 sense of the word, were slipping away, unless a new strategy could be found. The army seemed unable to win on the battlefield, and it remained to be seen if it could hold its positions until the navy could win the economic and material war. Full-scale home-front mobilisation, which the French press likened to the *levée en masse* that the revolutionary republic had summoned in the desperate year 1793 following the first Prussian invasion, was Germany's last resort. The whole nation was now at war.[109] This change of strategic tack attests to the effectiveness of the General Allied Offensive in checking the German army and winning military initiative.

Obliquely, Haig's and Foch's verdict on the Somme was confirmed in Falkenhayn's memoirs. In face of the allies' "immense superiority in men and material," which had become manifest during the 1916 campaign, Germany had to win the battle of wills, the psychological and moral fight: she had to break her enemies' resolve. For Falkenhayn, attrition had been the means to do this, but Verdun had failed, and attrition on the Somme, rather than weakening the allies' resolve, had hardened it and given them confidence in their military methods and ultimate victory. At the end of 1916, he acknowledged, even if the position was not yet desperate, Germany was "now engaged in a struggle in which the very existence of our nation, and not only military glory, or the conquest of territory, was at stake." Germany and her allies would have to "hold out," to sustain their resolve to win longer than their opponents.[110] In this death-or-glory scenario Germany's uneasy social peace and the political structures of the imperial regime were expendable. The army was now the master, and it would use any expedient to triumph.

AT THE END of February 1917 Leutnant Ernst Jünger rejoined his regiment, now holding the line near Villers-Carbonnel. The Somme front was still active. He had to run the gauntlet of British shell-fire, across the badly damaged bridge that was the only link from the east bank of the river to his company's exposed position: three hundred metres away, the enemy were "full of curiosity and enterprise." But the day-to-day trials of trench warfare were overshadowed by a heavier anxiety: "There were already rumours of some vast impending '*matériel* battle' in the spring, which would make last year's battle of the Somme appear like a picnic."[111]

As part of their review of German strategy, in order to blunt the enemy's unrelenting attrition which would sooner or later crack their army's powers of resistance, Hindenburg and Ludendorff had resolved to forestall the next allied push with a strategic retreat. It might just win enough time for the submarines to do their work. Since the previous September military engineers, labour battalions of Russian prisoners of war and pressed French civilian labourers had been constructing an elaborate new defensive position across the base of the Noyon salient. Designed by Colonel Fritz von Lossberg, Second Army's defensive expert on the Somme, the Siegfried Stellung ("Hindenburg Line" to the allies) was a system of great strength and depth. The line's forward position, arranged in a jagged, saw-tooth pattern, was thickly wired and dotted with concrete artillery and machine-gun emplacements, and sited so as to be difficult for allied artillery to target. A similar position was being constructed 500 to 1500 metres behind it, from which the first could be commanded by machine-gun fire. The two positions together comprised a "battle zone" more than a mile deep, which was to be backed by further positions several miles to the rear.[112] If the allies intended to attack in the centre of their front, after the retreat they would have to begin the slow process of photographing, mapping, emplacing and registering their guns and installing their infantry all over again.

Between mid-February and early April the German army withdrew to this new line, straightening out the Noyon salient between Arras and Soissons. The defensive line was shortened by around forty kilometres, and fourteen divisions were released to form a strategic reserve for sustaining attritional warfare on the Western Front.[113] There was now no question of strategic counter-attack. Retreat—acceded to reluctantly by Hindenburg and Ludendorff but essential as battle on the Somme resumed in the spring[114]—might gain vital weeks for the submarines to do their work before the allies started hammering again.

Jünger's company was his battalion's rearguard. As they moved up he witnessed frenetic activity as the withdrawal was prepared:

> The villages we passed through on our way had the look of vast lunatic asylums. Whole companies were set to knocking or pulling down walls, or sitting on rooftops, uprooting the tiles. Trees were cut down; windows smashed; wherever you looked clouds of smoke and dust rose from vast piles of debris . . . With destructive cunning, they found the roof-trees of the houses, fixed ropes to them, and, with concerted shouts, pulled till they all came tumbling down. Others were swinging pile-driving hammers, and went around smashing everything that got in their way, from the flower pots on the window-sills to whole ornate conservatories.[115]

If they had to give up ground Germans had sacrificed their lives to hold, it would be restored to France as wasteland.

"You people at home have no idea how eagerly this Spring has been looked forward to," Gunner Keith Buchanan wrote at the start of May.

> The winter was one long, wet, cold dreary night and far away on the edge of the night was a tiny glimmer, Spring, that seemed ever so far away. I'm sure none of us ever longed for the warm weather as we did this past winter in that sea of mud, the Somme; often I wished I was a grizzly that I could feel oblivious to it all by hibernation! It's over a week since any rain fell, such an unusual occurrence. There's no mud anywhere, and the roads are in excellent order . . . It's a treat to get enveloped in a cloud of dust and to feel it grit on one's teeth![116]

Buchanan was recording two victories: one over the elements; the other over the foe.

In the middle of February Fifth Army's patrols reported that the German trenches opposite were very lightly held. Gough ordered his divisional generals to prepare for a general advance. On 17 February attacks on the enemy's front line met with heavy opposition. But a week later the forward companies walked into the enemy's front positions astride the Ancre, and by the next day they had occupied the villages of Warlencourt, Pys, Miraumont and Serre. The enemy had melted away; barely a dozen German stragglers were captured. After its hammering of the previous year, the German army was beginning its retreat with a tactical withdrawal. The Battle of the Somme had sprung back into life: "the cavalry was moving up by the thousand at the time and everybody seemed to think that we should have them on the run within a few weeks."[117]

The first stage of the retreat, straightening the line in front of Bapaume in late February, was modest. But it showed that British trench operations on the Ancre Heights over the winter, constant small-scale raiding and harassment, had made the salient between Gommecourt and Le Barque untenable. Rumours spread quickly that it was the start of something bigger, fuelled by the palls of smoke that could be seen rising from the next line of villages behind the German line. For some months, allied intelligence had been monitoring the construction of the Hindenburg Line;[118] but whether that was merely a precaution in case the allies broke through in the spring or presaged a voluntary retreat could not be determined.[119] Poor winter weather and German airmen's determination to harry prying allied pilots made it difficult to confirm the reports of prisoners, agents and refugees from the occupied zone by aerial reconnaissance.[120] When the large-scale retirement began in mid-March it took G.Q.G., focused on its own plans, somewhat by surprise, despite indications from January onwards that the German line was thinning and that aerodromes, base camps and hospitals were being dismantled and relocated behind Cambrai and St. Quentin. Perhaps it was simply not part of Nivelle's plan. It certainly threw preparations for his offensive into disarray, for it removed the enemy from a large part of his intended front of attack. The officers of the Operations Bureau, charged with preparing the offensive, refused to believe that the enemy would voluntarily pull back, for "it would be an avowal that they had lost all hopes of beating us."[121] It was certainly an avowal that, as the balance of forces stood, they could not expect to beat the allied armies in the field on the Western Front.

The columns of smoke behind the German line as stores and standing crops were burned finally convinced G.Q.G. that the retreat was happening. Nivelle determined to take moral, if not military, advantage of it. GAN's divisions were ordered to follow up, while Pierrefeu was ordered "to convey the impression in the communiqué that the enemy had retired under constant pressure from us, and that we were energetically pursuing him, engaging his rearguard and harassing his troops."[122] In fact, after capturing enemy orders which indicated that a retreat was being prepared in early March, General Louis Franchet d'Esperey, Foch's successor in command of GAN, had wanted to launch a powerful offensive using France's new tanks for the first time in order to catch the withdrawing Germans in the open. Nivelle demurred: it was not part of his wider offensive scheme. The French army would follow the German army, not push it.[123]

One Somme victor at least could relish the late proof of his success. Fayolle paid due homage to Notre-Dame-des-Victoires, because his reward for the German retreat was command of a reserve army in the coming offensive, with the prospect of an army group in the future.[124]

On 18 March British and French cavalry patrols, scouting open country, met in Nesle,[125] fifteen kilometres beyond the old front line. The next day they reached the river Somme. But its bridges had been destroyed, so it would be some days before the allies managed to cross that emotive waterway. The bigger prizes of Bapaume and Péronne would fall to the British army, which had extended its front southwards across the Somme in preparation for Nivelle's offensive.[126] The follow-up was vigorous but careful, and marked a brief return to more mobile operations.

The Australians opposite Bapaume welcomed the new spring, and the new challenge. In the spirit of Haig and Nivelle's agreement to maintain pressure on the enemy pending the main offensive, active trench operations had been continuing north of the Somme since late January: "daily extensive assaults, faultlessly carried out and successful in securing for them numerous prisoners," actions deserving of acknowledgement even in the French official communiqué.[127] Such small-scale harassing attacks on the enemy's forward lines, for which Rawlinson coined the euphemism "peaceful penetration," and in which the Australians became the acknowledged masters, affirmed allied soldiers' tactical dominance over their forlorn enemy. But progress was slow, and the Australians were keen to get moving again, to reach Bapaume, which had lain tantalisingly just out of reach beyond the next ridge all winter. The prospect of moving forward into green countryside lifted the spirits: "after months of solid drudgery in the mud and snow to know that we are soon to leave all this behind for fresh fields and open country without bogs, mires and shell-holes was great news indeed."[128]

Something big was clearly in the air: training for mobile warfare had recommenced. Despite the rumours, however, the concrete intelligence that the German army was about to pack up and move back was limited and inconclusive until a few days beforehand. Astride the Somme petty-war—raiding, sniping and strafing—had continued through January and February: the defending regiments themselves had no inkling that they were soon to give up those hard-dug and harder-held positions voluntarily.[129] Orders to prepare for a retirement were issued to the German troops on 5 March. Guns and *matériel* were to be sent to the rear, and what could not be salvaged was to be destroyed. Artillery fire was to continue until the last minute by guns left behind to fire off shell stockpiles that could not be removed (although the diminution in heavy shells fired from January onwards had been one indication that something was up).[130]

The opportunity to push forward more rapidly came in mid-March. With the evidence of a withdrawal increasing, Australian fighting patrols started probing the trenches north of Le Transloy, across a no man's land still littered with the bodies of those who had fallen in the failed attacks of November. Eventually, on the night of 12–13 March, they found them empty. The enemy had withdrawn in front of Bapaume to forestall a large-scale attack by Fifth Army, which had resumed active operations with the coming of spring.[131] Captured German orders confirmed that a withdrawal was commencing.[132] Infantry patrols and Australian Light Horse squadrons went forward to probe and harass the German rearguards; the going became firm as the boggy shell-fields were left behind. Australian infantry marched into Bapaume on 17 March 1917, unopposed except for a few outposts and snipers who were flushed out like rabbits. Lieutenant Arthur White of the 30th AIF Battalion raced one of his NCOs to be the first man to enter the town.[133] The offensive's original prize was finally in British hands, eight and a half months after the first assault. Even though the retreating enemy had left the town a smoking ruin, it was a moment of genuine triumph: "to reach Bapaume" had become a mantra among the Australian forces during the previous months.[134] As curious men flooded in, the newly captured town became "like Blackpool on a bank holiday," in one ex-Lancastrian Aussie's judgement;[135] or "Manly on a

Sunday" to a Sydneysider.[136] But the war was not far away, and tragedy quickly returned. Eight days later, time-delayed German mines levelled Bapaume's town hall, one of the few buildings left standing, and former headquarters dugouts under the town's ramparts. Two French parliamentary Deputies who were visiting the newly liberated town were among the dead.[137]

Such atrocities were common. Churches often seemed to be the only buildings left standing, too large to burn or knock down, but these too periodically became death traps when German booby-traps exploded. To slow the allied pursuit, crossroads were mined and bridges blown up. The land had been cleared of crops, wells were poisoned, flood-plains inundated and anything useful for the war effort — timber, farm vehicles, machinery and metal — carted away. In Vraignes a mutilated piano in the middle of the main street added a surreal note to the usual chaos.[138] Only the roadside calvaries seemed to be beyond violation,[139] although at Bapaume Cemetery the "retreating Germans have erected another cross made from two saplings lashed together, on it they have hung an old soiled shirt, as though in mockery."[140] Engineers would have to comb the evacuated ground for booby-traps. So many "souvenirs" left for the pursuing soldiery caused death and injury when picked up that strict orders had to be issued against that age-old military practice.

It was difficult to ascertain how far and how fast the Germans were going, or where they would stop. Moreover, by the time it was clear that a retreat was under way, the Germans had gained several days' valuable head start. Fourth and Fifth armies hastily organised all-arms forces to follow up. Cavalry squadrons, infantry and cyclist battalions and artillery batteries, occasionally supported by armoured cars, moved forward together, the precursors of later combat groups. Brigadier-General Seely, commanding the Canadian Cavalry Brigade, which was allocated to Fourth Army, improvised one such of his own, commandeering infantry from XV Corps to capture Equancourt village. General Du Cane's wrath was assuaged by the timely arrival of the commander-in-chief's congratulations on the exploit.[141] The surge forward was welcomed after the long winter in the trenches. "We looked forward to warfare of a new kind," Lieutenant Waterhouse recalled, "and were congratulating ourselves on having a good time, with much loot, in the near future. But it was not to be."[142]

It was a retreat conducted by a skilful enemy, not a rout; a slow, measured retirement to carefully prepared defences, through a number of intermediate defence lines. At its deepest, the withdrawal covered twenty-five miles. Well-placed rearguards, local counter-attacks and the total destruction of the communications and domestic infrastructure in the Germans' wake would slow the pursuit. For the allies, getting guns and supplies across the devastated roadless zone that delineated the two static front positions (some two miles deep) was difficult enough. The mud, one gunner recalled, remained "a far more formidable foe than the Germans."[143] It took the first British batteries three days to cross.[144] Beyond it, the infrastructure of industrial battle—roads and railways, barracks and gun positions, water pipelines and telephone systems— had to be constructed afresh. The 13th AIF Battalion found itself "roadmaking, cable-burying and carrying" for a week, rather than fighting.[145] Moving forward was unfamiliar. Bean recorded a three-hour, 1200-yard-long traffic jam on the Pozières–Bapaume road, with no traffic control officers to sort it out.[146] It was the engineers and military policemen more than the infantry, cavalrymen or gunners who were busy during the advance. But there was a very welcome change once forward progress started. "The mud hell had its limits, and now we came across fresh country, where we found green grass and trees and other wonders of a like nature," Waterhouse wrote home. "We went mad in that beautiful green country, and the process of harrying the Hun in his retreat was certainly well done."[147]

If a morale boost was needed, the German retreat provided one. "The beginning of the end is at hand," Geoffrey Malins judged, "the triumph of Christianity over barbarity, of God over the devil."[148] For the French infantry, moving forward again, crossing German trenches and liberating French countryside was an emotional experience. War seemed a human activity once more: "the infantry advanced in line, as on manoeuvres," one 53 DI artillery officer recorded. But military reality soon returned. Entering Lassigny, all was chaos and confusion: traffic was jammed in all the streets, and any truck that tried to skirt round the obstructions got bogged down in the mud. The artillery officer's batteries were stuck. The rations did not arrive until two o'clock in the morning. Accustomed to the routine of the trenches, much went to pot when the army found itself on the move again: some feared that the Germans would take advantage of the confusion to counter-attack. It was lunchtime the next day before the column finally started moving. But the sight of the French tricolour, hung out by welcoming civilians, made it a poignant moment. Meanwhile, the military bands playing French tunes brought tears to the eyes of the liberated.[149]

Yet behind the momentary happiness lay another human tragedy. Civilians of labouring age were evacuated to work in the German occupied zone before the retreat; old men and women and young children, often parentless, were left behind.[150] The unproductive population of St. Quentin was abandoned in the devastated villages in front of the new German position: fewer hungry mouths for them to feed and another problem for their adversaries.[151] The advancing troops had to find food, water and shelter for these helpless refugees, as well as their own men, in a countryside that had been stripped bare. Malins's fresh bread and sausages were very welcome to the abandoned villagers of Bouvincourt, who had been subsisting on American Red Cross food relief for some months. The cinematographer was able to capture some excellent scenes for home audiences in return.[152] By this point in the war, the allies had gained considerable experience in supporting refugees and displaced persons, and restoring communications after German armies had passed through.[153] Such a situation had been anticipated and the prefecture had stockpiled grain to feed the residents of the occupied territories after their liberation.[154] But it placed another burden on an already overstretched transport system, as military trucks had to be diverted to move civilian foodstuffs.

"Do not be annoyed—only wonder" was the haughty notice that greeted the liberators of Péronne.[155] But the devastation that the retreating enemy left in its wake was also a boon: moral capital could be made out of this new manifestation of Hunnish vandalism and atrocity. At a time when military and public morale was at a low ebb after a long, cold winter, another desecration of France's peaceful fields and display of immorality furnished a timely example of German "*Kultur.*" The advancing troops were suitably shocked by what they found. The 55th Battalion AIF was billeted in the woods of Velu château, built in the early eighteenth century but now reduced to rubble: "Fritz has utterly and purposelessly ruined it, with his usual thoroughness . . . the wanton, needless destruction makes one's blood boil," Lieutenant McInnis noted.[156] With the troops went the official photographers, recording the "ruins heaped up with barbarous refinement, and particularly the felled fruit trees," for distribution to the press, "and in the general anger the mistake of the Command was forgotten."[157] Malins's final feature film, *The German Retreat and the Battle of Arras,* released in June 1917, brought the British public vivid imagery of the burning ruins of Péronne and the debris of departed battle—the deliberate destruction of homes and felling of trees, and the Somme's shattered bridges—

which contrasted with the glistening river and the verdant country beyond it. The film was crowned by recording the first British patrols crossing the symbolic river on a makeshift bridge, and cavalry and cyclists hastening forward in pursuit of the beaten, retreating enemy.[158] Such reportage reminded soldiers and civilians why they had gone to war, and what they were fighting against, an antidote to the tide of opinion that was starting to flow against the conflict. Yet this film did not attract the audiences of the previous summer:[159] the public had seen enough of war. In the short term, however, the retreat was both a great victory and a moral uplift for the allies, and the significance of the manoeuvre was not lost at G.Q.G.: "All regarded this retirement as the beginning of the end . . . it was clear that the retreat was the logical outcome of the Somme, and [Nivelle] reaped the fruits of it while Joffre and Foch, to whom all the credit should have been given, were in disgrace."[160]

WHILE OPERATION ALBERICH, as the German army code-named their withdrawal, was undoubtedly "a most impressive feat of military engineering, deception and discipline," it was also an admission of disadvantage, an acknowledgement of the unsustainable rate of loss of the Somme and Verdun and the fragile state of front-line morale.[161] The impact of the retreat on morale was far from positive. Voluntarily to give up lines that they had shed so much blood to defend the previous year was bound to lower German soldiers' spirits while strengthening the growing perception that Germany was in dire straits and that a compromise peace was the best that they could now hope to get for all their efforts. Even if the command tried to represent the withdrawal as a strategic manoeuvre on Hindenburg's part,[162] the front-line soldiers knew that it was a desperate move to gain time. Their gallows humour was evident in the messages left in the abandoned trenches at Le Transloy: "Welcome to the Conqueror. Much pleasure . . . Capture our next line, and you beat Germany . . . It's a long way to Tipperary, but further to Berlin."[163] Some men were left behind to cover the retreat, but after capture many admitted to "dereliction of duty and not carrying out the orders given by their officers."[164] Prisoners taken by Fayolle's First Army reported that "there was a general lassitude among the men of their company." One, who had been on leave in December, also confirmed that there was great misery in his industrial village and popular dissatisfaction was growing each day.[165] There was real fear among the rearguards left to cover the retreat that, if captured, they would be subjected to reprisals for the devastation that their army had wreaked. However, despite the current situation, they still apparently believed that Germany would ultimately win.[166] Seely surmised that the scorched-earth policy was designed to show the French that further advances would lead only to ruin, and so might act as a deterrent. It had quite the opposite effect: "Parties were sent from every regiment to view the devastation. Indeed, I took some of them round with me. The result was that they returned to spread renewed ardour among their comrades. They did not 'ponder,' but they were exceedingly angry."[167] Even to many Germans—not least Prince Rupprecht, who had protested in vain against OHL's order to destroy everything in his armies' path, to the point of threatening to resign[168]—the vandalism took German efficiency one step too far. For now, they would fight on, more out of habit than from confidence in their final triumph. But trouble was being stored up for the future.

ALTHOUGH GERMANY'S STRATEGIC expedients were a clear indication that the General Allied Offensive had done her great harm, she had still to be finished off. That meant that 1917's plan had to be appropriate and rapidly executed. In the more liberal Entente states politicians believed that they had reasserted control over strategy after the indecisive 1916 offensive. Yet there remained little doubt among the military about the utility of attrition. Although Henry Wilson lamented that 1916 had been "a very disappointing year,"[169] he remained fixed in his belief that the only way to win the war was to defeat Germany: "if we beat the Boches, all else follows." His method was "two Sommes at once" in 1917. Such advice, of course, was anathema to Lloyd George and his ministers: "we shall never beat the Boches," he whined when Wilson suggested it.[170] Brock Millman has suggested that after Lloyd George came to power, British strategy was dominated by pessimists who believed that the Somme had demonstrated that the war was unwinnable, or at least not worth winning at such a cost. The best that might be gained was a peace with advantage, a strengthening of Britain's global position in anticipation of the renewed conflict that would inevitably follow a compromise peace. Haig and Robertson nevertheless remained optimistic that their strategy of hammering the German army in the main theatre would eventually bring victory.[171] Foch shared their confidence, if by now he had a more sophisticated conception of how to do so.

What the Battle of the Somme had confirmed—as some had come to believe before it began—was that the German army would have to be pummelled into submission, not felled with a single knockout punch. A multi-million-man army was too resilient to be broken quickly. Prince Rupprecht's 17 December 1916 order of the day acknowledged his soldiers' "heroic courage" in an unequal battle of *matériel*: "The greatest battle of the war, perhaps the greatest of all time, has been won." If the allies had planned a breakthrough, or to take more than "a narrow strip of utterly ruined terrain," then his troops had certainly contained this.[172] However, Germany's official historians subsequently acknowledged, "It would be a mistake to measure the results of the Battle of the Somme by mere local gains of ground."[173] Since the allies were pursuing attrition, the German soldiers' sacrifice was in the allied cause. In the second half of 1916 on the Western Front the German army had continually lost ground, and much more besides, as its pounding commenced. "We must save the men from a second Somme battle," Hindenburg affirmed as Germany reviewed strategy for a new year.[174] At the same time Germany was raising "more arms and more men to fight more battles,"[175] for the army could never escape battle entirely. Consequently, in 1917 it was to lose much more, first by voluntarily surrendering a large tract of occupied French territory, and then in further heavy losses and involuntary territorial surrenders as the Anglo-French armies sustained the attritional battle.

At the end of 1916 the German army was far from beaten, even if those who directed it and those who opposed it all knew that Germany was now losing the war. As Colonel F. B. Mildmay was proud to assert in response to Lloyd George's opening speech to the Commons as Prime Minister, at the front "there is in the air, and prevalent in the atmosphere, a sort of conviction that we have got the Germans; that we have got to the turning point."[176] Nonetheless, the German eel would still squirm and pull on the allied hook for nearly two years before Foch's and Haig's armies finally reeled it in.

Although unable to report dynamic operational success, Haig's December despatch accurately summarised the lengthy campaign's strategic significance in the context of the war that Britain and her allies now found themselves fighting: a national war of attrition. Haig's measure of victory was not a few villages more or less in Picardy, but allied armies across the Rhine: he had pronounced as much to Poincaré in May 1916. But the Somme had undoubtedly furthered that objective. Such was attrition: relentless, bloody, undynamic, yet decisive. That strategy was contributing to victory, and had to be continued and intensified, although everyone knew it would be a slow, drawn-out business. Yet Haig's December 1916 assurance of final victory was hedged privately with one caveat: "provided the Russians hold."[177] Before the allies' *matériel* and operational superiority was definitively asserted, and British troops marched, flags flying, into Cologne, Russia had collapsed into revolution in 1917, and the allies had repulsed a ferocious German counter-attack on the Western Front in spring 1918.

Foch, who had always doubted that the Somme would bring about a decisive victory, privately echoed the British commander-in-chief's verdict. By the end of 1916, Germany was contained on both fronts, if not yet beaten. Her victories in 1914 and 1915 had been reversed. The Somme had been "a battle which worked, always victorious, beating the Germans, pushing them back. We should continue in this vein as far as we can, denying them any freedom of action and opportunity, continue to beat them." To defeat Germany, however, powerful, concerted allied action was still required: much more powerful offensive action was also Foch's prescription for victory. He rightly identified that the real obstacle to pursuing this successful attritional strategy to its denouement was not now the enemy, but allied politicians. In Nivelle, who reportedly "invariably says yes to anything you may say to him,"[178] they seemed to have found a commander-in-chief who would respect their authority and bow to their strategic imperative.[179]

Remobilisation

NIVELLE'S DELAYED AND controversial offensive commenced on 9 April with the British First Army's spectacular storming of Vimy Ridge: one army had clearly absorbed the tactical lessons of the 1916 campaign. A week later French armies struck against the heights north of the river Aisne. Tristani's battalion was waiting in reserve. His men had marched towards the front in high spirits. The enemy were retreating and the *coup de grâce* was about to be delivered by a new commander who had recently triumphed at Verdun. Backed by a numerous and powerful artillery, well armed and, after the Somme, well trained, "Our regiments are superb . . . one large effort is needed. There is everything to hope for." The news of America's declaration of war on the eve of the offensive raised morale and expectations to fever pitch: "now we are sure to win quickly . . . everything looks rosy." "The time has come; courage, confidence; *vive la France!*" Nivelle's order of the day pronounced.[1]

All of these hopes were dashed. The forty-eight-hour breakthrough never materialised. Although the offensive smashed through the German first position in many places, it was blocked by the second. Tristani and his men spent two days milling about in the confusion that gripped the rear areas when the mass of men and *matériel* assembled to exploit the breakthrough could not be moved forward. It rained persistently, dampening the men and their spirits:

Having dreamed of triumphant assaults, of stunning advances, we got demoralising halts in the rain in muddy bivouacs; incessant orders and counter-orders; comings and goings which seemed to have no purpose, and which were never explained, no sleep, defective supplies, in short super-human fatigue. All that was a result of a check experienced by others, because we were not even engaged. We recrossed the Aisne with heavy hearts: we wept for our shattered illusions![2]

Nothing could be done now, Fayolle caustically scribbled, save for what he had done on the Somme: restore order, move the artillery forward and start again. "Nivelle and Mangin's procedures are absurd . . . Especially the strict timetable. It fatally dissociates the fire of the artillery from the movement of the infantry. Why did they wish to deny the Somme and draw abstract principles from the lessons learned there? Pride. Nivelle constantly contrasts the school of Verdun with that of the Somme."[3] In consequence, another long, attritional fight started to set in along the Aisne front.

The British advance had also stalled. At the beginning of April the outpost villages of the Hindenburg Line were seized on Fifth Army's front—in Sixth Army style they were usually flanked and pinched out rather than assaulted directly—with a view to an attack on the main line to support the main offensive. Despite the return of cold weather and two days and nights lying exposed in freezing fields, 1st Australian Division assaulted Hermies with ardour on Easter Monday. "I saw a deed done that earned the VC and another equally deserving," Sergeant Arthur Matthews of the 3rd AIF Battalion remembered, but "the two participants were both killed about 200 yards in front of my section."[4] But Hermies was as far as blind aggression would get them. The new defensive system would never be carried without systematic preparation and a huge weight of artillery fire. The Prussian Guards, however, who counter-attacked the Australians' thin defences "as on parade" at Lagnicourt on 15 April, "received the biggest knock-back they have yet experienced in the war." The initial pre-dawn attack by four German divisions penetrated a mile and a half to the Australians' gun line north of Lagnicourt village, from where they were driven back onto their own wire by I Anzac Corps's own elastic defence and shot to pieces with close-range rifle and machine-gun fire. No quarter was given or expected. The Australians inflicted 2313 casualties, losing 1010 themselves. Twenty-six battalions from four German divisions had been committed to the attack, a diversion from the successful British offensive now under way at Arras, supposedly to improve their morale after the retreat.[5] Instead, they were broken: "It was the first time that I'd seen grown men brought to such a stage of terror and despair that they shrieked like terrified animals, and it was awful and awe-inspiring to watch."[6]

Such were the preliminary moves of the Battle of Bullecourt, Fifth Army's contribution to the Nivelle offensive, Gough's infamous "blood tub," which earned him yet more opprobrium.[7] The counter-attack was proof that the German retreat had stopped, that the war of attrition had resumed, bloodier and more vehement than ever, and that the German army was no longer what it had been: the equivalent of one Australian brigade had repulsed four German divisions, two of them Guards. But blood-lust was ephemeral, and Matthews was in reflective mood when he visited an advanced dressing station:

> One certainly gets very sick of the strong meat of war at times and I felt very much that way that spring morning, as I thought of all these men who sacrificed so much of their lives—and, in many cases, their lives—and live and sleep amongst the filth and wanton destruction of modern warfare, that others may live and sleep how and where they please.[8]

Such growing tension between the justice of the cause and the nature of the fight was to make 1917 a very difficult year for both sides. It marked the onset of the "Grand Disillusion," as Tristani titled the 1917 chapter of his memoir.

NIVELLE'S STRATEGY MADE sense in the context of December 1916, when it was incumbent on France to consider how to deliver the *coup de grâce* to a battered and demoralised enemy.[9] Yet even then there were many who quickly identified the gulf between its tactical principles and its strategic ambition. After his Verdun counter-offensives Nivelle could reassure doubters that "we now have the formula" (whoever's formula it might actually have been),[10] but this was only a formula for capturing enemy defensive positions, not for operating against his armies in manoeuvre warfare. Moreover, what worked on a corps front was not necessarily applicable along the front of several armies. This discrepancy was obvious to Fayolle; and to Haig, who from the start insisted on a contingency plan for an attack against the Belgian coast if Nivelle's offensive failed.[11] As the detailed planning progressed, most other experienced field commanders spotted it too. Pétain, whose Central Army Group was to attack east of Reims to support the main push on the Aisne, certainly favoured concentrating artillery on the first two German positions to guarantee capturing them, rather than dispersing it over four lines of defences to try to push through quickly, and thereby risking failure. Micheler, who commanded the Reserve Army Group charged with exploiting the attack on the Aisne, thought that after the German withdrawal the opportunity which had existed to break a thinned German line had gone.[12] Moreover, by April, when the time came to set the offensive in motion, there had been significant changes in France's political situation and the wider strategic position. As with the Somme the year before, an offensive conceived in one set of circumstances, with one particular purpose, was to be overtaken by events.

Briand's government had not long survived his culling of the high command. Lyautey had not proved a success as War Minister, and his resignation in mid-March, over the persistent issue of the Chamber's right to discuss military affairs, brought down the ministry and plunged French politics into an anxious summer. Briand's successor, the septuagenarian former Finance Minister Alexandre Ribot, a compromise premier who might hold the fractious factions of the *union sacrée* together a little longer, was at least "an excellent patriot, resolute for all-out war."[13] The new War Minister, the mathematician Paul Painlevé, who had been Briand's Minister of Inventions, represented a departure from the previous practice of appointing a soldier to the post. This was further evidence of the shift of strategic control in France to the civilians; and Painlevé was, Haig noted warily, "an extreme socialist" to boot. "Such are the people under whom the British Army has been placed for the forthcoming offensive operations . . . What a crowd of fickle, changeable people are the French Deputies," the British commander concluded. "General Lyautey must have had a most difficult position in a Government of political jugglers with a chamber of semi-lunatics!"[14] The new ministry would certainly exercise close supervision over its own generals, and Nivelle's offensive was the first item on its agenda.

By April, the broader strategic context of the offensive had changed. As in 1916, Anglo-French operations were part of a plan for concerted attacks on all allied fronts. However, the Italians were not ready, while recent events in Russia—the Tsar had just been deposed and the new Provisional Government was getting to grips with the country's serious social and military problems—suggested that no support could be expected on the Eastern Front for some months. America's imminent entry into the war on the allied side also suggested that there were grounds for a thorough strategic rethink. Ironically, in spring 1917 only the British seemed to be playing their part like loyal allies. The preliminary bombardment for Haig's contribution to the offensive, the Battle of Arras, commenced on time on 4 April.

The German withdrawal compromised the details of Nivelle's offensive plan, forcing him to rethink his operation at short notice, and it freed German divisions for reinforcing the Aisne front: owing to woeful French security, the preparations there were no secret. Micheler's and Pétain's doubts now came to the attention of the government. Other generals added their concerns. In early April Messimy (who had been War Minister when war broke out in 1914) informed his former Cabinet colleague Ribot that most of the senior commanders in Micheler's armies agreed with him that the offensive would now produce only heavy losses for no great strategic result.[15]

On 6 April a council of war was called to decide whether the offensive should go ahead. By then, it was perhaps too late: French gunners had already started the artillery preparation and the British were due to attack in three days' time. The atmosphere at the meeting, Painlevé recalled, was "very strained and gloomy," as cold and grey as the wintry weather that had returned at the start of the month.[16] In a confused, acrimonious meeting Nivelle made a strong case that the offensive should not be postponed: to do so would allow the Germans to regain the initiative and attack Italy, Russia or somewhere in the West. His strategic judgement made sense, and it was not for the politicians to question the details of his operational planning: they left that to his subordinates. But his army group commanders, while they did not expect any strategic success, also counselled against cancelling the offensive at this late stage, not least because it would destroy arrangements with the British, which for once seemed to be going smoothly. Nivelle's final trump card was to offer his resignation. If the politicians no longer had faith in him, they should accept it. That would have precipitated yet another political crisis at a time when France needed victories, not vacillation. So the offensive would go ahead; but it would be halted if Nivelle's breakthrough did not occur quickly.[17]

WHEN IT DID not, the collective disappointment precipitated a series of political and social events that were to make the summer of 1917 one of acute crisis, during which France's commitment to seeing the war through to victory came into question. Further government instability, domestic unrest, and above all military mutinies, combined to create a tense, confrontational atmosphere.

At the front it seemed that the generals did not know what they were doing, and were casual with their men's lives. So the soldiers decided to take matters into their own hands. The mutinies in the lower ranks of the French army in late spring and early summer 1917 represented France's most serious wartime crisis. From mid-April there were signs of discontent in the ranks. May saw wide disorder, particularly in the second half of the month. And there was a final, short but even more boisterous outbreak in the first week of June before the trouble subsided. Yet these were not really mutinies in the conventional sense of the term. There was no refusal to fight on, even if calls for peace were heard among the many slogans shouted by disgruntled *poilus,* and the protests were essentially peaceful. There was restraint on the part of the military leadership too. No clashes took place between the mutineers and the cavalry regiments deployed to confront them. Frenchman would not spill the blood of Frenchman. Only after the protests had ended were the mutinies investigated by due judicial process, with "leaders," mainly from the later, more violent outbreaks, being identified and court-martialled. Yet the fact that the collective indiscipline was so widespread prevented real action being taken. The 554 death sentences that were passed symbolised the reinstatement of proper military discipline. Around 50 executions were carried out (no precise figure has been established).[18] It was the Colonials' protest on the Flaucourt plateau in macrocosm: direct action from below in response to failing higher leadership, a potent legacy of 1916.

However, while calls for peace accompanied the protests—the war had gone on too long, and the continued loss of life seemed to be achieving little—they were not for peace at any price. The *poilus* remained determined not to lose the war, and if the enemy had taken advantage of the moment to attack, they would surely have been met with all the fury France's regiments could muster. To that extent, the front-line soldiers had not been infected with the viruses of anti-war propaganda or social revolution as their leaders feared. Those troops who defended the line remained at their posts: the collective indiscipline occurred further back. The facts that the mutinies were concentrated behind the Aisne front and peaked in late May and early June suggest that they were a protest against a return to attrition,[19] as much as a disappointed response to the failure of Nivelle's offensive, for in that sector *grignotage* was to continue throughout the summer. As the French official history later admitted, "this period of the struggle did not deliver the desired results. The attacks by Fifth and Tenth Armies won only insignificant tactical objectives, but with great effort and very heavy casualties."[20]

Tristani's battalion had been sent on to the Craonne plateau in early May to assault a difficult position, the Courtines de Chevreux. The attack was too hastily prepared: "Two days of madness—nothing is ready." It failed and suffered 25 per cent casualties. When the time came to return to the line to try again, the men refused. These were regiments which had passed through the Somme, Verdun, or both, and they had had enough of such costly, indecisive attacks. Tristani and the other officers calmed the protest, which was relatively minor, by assuring their men that they would be the supporting battalion in the next attack.[21]

Such protests, which affected nearly half the divisions in the French army, were a rejection of the method of waging war, not of war itself. They were also an assertion of the fundamental political rights enjoyed by Frenchmen, and a demonstration that the *poilus* were active participants in their war, not its passive victims. France's citizen-soldiers were much more political and politicised than their British or German counterparts. Their democratic, republican traditions dated back to 1789: since then, they had enjoyed rights as men and citizens, alongside duties and obligations to the state and its appointed authorities, in wartime the military command. Yet citizenship gave them rights to dissent and opposition if the state was not fulfilling its obligations to them. Len Smith has suggested that the soldiers' work-to-rule (for, essentially, that is what it was) represented a symbolic political gesture, the reassertion by France's citizen-soldiery of a modicum of control over their own destiny, which for too long had been subject to the whims of an unintelligent and unfeeling high command. At root, the mutineers' purpose was simple, as one soldier of 36 RI expressed it: "rather to attract the attention of the government in making them understand that we are men, and not beasts to be led to the abattoir to be slaughtered."[22] If they were to die for France, they ought to die for a reason, to progress the cause, not merely to lengthen the casualty lists. Once such protest had been expressed and acknowledged, the soldiers could return with satisfaction to their task. If they fought on, it would be through consent rather than coercion. Despite their "blunt objections" to going back up the line, 32 RI's companies acquitted themselves well once there. The renewed attack on the Courtines de Chevreux on 22 May, in which Tristani's men gave valuable support to 66 DI's *chasseurs*, was a success, with many prisoners being taken. On this occasion 32 RI's colonel was held responsible for the failure of the first attack and the indiscipline that followed, and was removed from his command.[23]

The fact that the mutinies were relatively short-lived and peacefully resolved indicates that at base they were merely one of the periodic renegotiations of the relationship between the French state and its citizens which characterise republican politics. Certainly such collective disobedience worried the authorities: but the mutinies were also a manifestation of the French army's commitment to ultimately expelling the invader, of justifying and avenging its heavy losses to date while not adding to them unnecessarily. They indicated a determination that results in future should bear a closer relation to effort.

Under Pétain, the new French commander-in-chief who replaced Nivelle on 15 May, they would. The strike against attrition had worked: a series of G.Q.G. directives reshaped French military policy. From June to August Pétain rested and reoriented his troops, retraining them in the modern *matériel*-intensive tactics he and Fayolle had perfected since 1915. Operations would in future be meticulously prepared, in order to restore the soldiers' confidence in their leaders. Although his method drew heavily on what had gone before, Pétain's plan was new. Usually expressed in Pétain's own convenient but imprecise summary — "We must wait for the Americans and the tanks" — in fact he had not renounced the offensive, as he believed successful offensives were essential to rebuild and sustain the army's morale. French offensives would be limited in strategic objective rather than size. Breakthrough attacks were abandoned. Objectives were to be within artillery range, to minimise infantry casualties. There would be no more *grignotage*; the French army's future bites of the German defences would be large and easily digestible. Moreover, although temporarily on the defensive, the French army was far from idle during the summer. The infrastructure of the whole French front was being improved with a view to launching sudden, surprise assaults in sectors that had been quiet for some time.[24]

THE SUCCESSFUL RESOLUTION of the soldiers' protest through a renegotiation of the parameters of command and duty was one aspect of a broader remobilisation of French public opinion behind the war effort during the difficult six-month period that followed Nivelle's failure. Managing the home front had always involved a careful balance between persuasion and coercion. By such means the war, and the regime that waged it, would win and maintain legitimacy in the public's eyes. Government information services, which managed propaganda to mobilise minds as well as bodies, were a relatively late development. During the first two years of war, the belligerents' respective populations had readily persuaded themselves of the justice of the cause, and the honour and glory of war. However, as the war's depressing reality sank into popular consciousness, from late 1916 persuasion had to increase, and coercion would follow. By the summer of 1917, a growing sense of war-weariness and disillusionment was sweeping through the belligerent states as the conflict entered its fourth year. Many were losing sight of why they had gone to war and what they were fighting for. Peace seemed the real cause now.

The early years of mobilisation had been a boom time for industrial jobs in France, with armaments and munitions factories and other war industries, such as clothing and transport, crying out for workers. The unemployed found jobs, wages rose, and many women and foreign workers entered the industrial economy for the first time. In marked contrast to the usual pattern of France's volatile domestic politics, strikes tailed off in 1914 and 1915. Industrial unrest started to rise again in 1916 as the war dragged on, but these strikes were economic and social in nature, not political: higher wages or better working conditions were the usual demands. In early January 1917 female workers in the Paris clothing industry went on strike for more pay. Next, more worryingly, armaments workers did the same. The unrest spread to the provinces, and in the middle of January the government was forced to raise industrial wages. But at the same time sanctions were imposed on "military workers" (conscripted men who had been redeployed from the army to jobs in armaments factories) who had gone on strike, and compulsory arbitration was enforced. The scene was set for a confrontation between the arbitrary power of government and the collective strength of French labour. Although wages were rising, so were prices, especially for increasingly scarce foodstuffs as the German submarine blockade intensified during 1917. In March building workers went on strike in protest against a slow-down in the construction industry and rising prices. Civil servants, including postal workers and the employees of the Paris Metro, joined them. International Workers' Day, 1 May 1917, saw 250,000 workers on the streets of Paris, and by June munitions workers were out again. In Amiens textile workers, and 380 employees at the British camouflage depot, struck in May and June, demanding better pay and shorter hours.[25] Workers who struck for more money were potential recruits to anti-government or anti-war movements. Recent events in Russia, as well as the growth of a more vociferous pacifist movement in France, threatened to exacerbate socio-economic grievances.[26] Such were the traditional roots of revolution; and revolution was a French tradition. Coinciding with the military mutinies, this wave of industrial unrest produced France's worst domestic crisis for many years.

There was also the spectre of left-wing pacifism to confront. An internationalist peace settlement came on to the wider political agenda during 1917. In April Russia's Provisional Government endorsed the principle, even while affirming Russia's commitment to her allies, but tentative peace overtures to the Central Powers came to nothing.[27] An attempt to organise a meeting of the Socialist International in Stockholm to draw up a socialist peace settlement proved divisive, if ultimately abortive, during the difficult summer. German representatives attended, but Western allied socialist leaders were banned from travelling. Nevertheless, those neutral and far-left delegates who did congregate in September 1917 endorsed the principle of a peace without financial indemnities or territorial annexations. In August, after visiting the Kaiser, the Pope had added his influential voice to the calls for a negotiated peace.[28]

The declared intention of French socialists to attend the Stockholm Conference provoked another stormy secret session in the Chamber at the beginning of June, in which the new government was challenged on France's imperialistic war aims. Ribot's ministry survived a vote of confidence comfortably, but the very fact that the session took place was further proof that the war was now a societal, not merely a governmental, matter: peace was an issue for the whole nation, not one particular party, Ribot argued before the Chamber.[29] Like Britain and Germany, France had extreme left-wing anti-war Deputies who withdrew their support from the government, most notably former premier Joseph Caillaux, who was promoted as a "peacemaking" alternative to Briand and Ribot, as well as pacifist writers and activists on the streets. Notable anti-war publications included *Le Bonnet Rouge* (closed down by the government in July 1917) and *Le Pays,* and left-wing propaganda was distributed to soldiers behind the lines.[30] Tristani blamed the troubles in his regiment on reinforcements from the base depot who had been targeted by such agitators.[31]

Much has been made of this current of hostility to the war, at the time because it smacked of defeatism, and afterwards because it reinforced the paradigm of the horror and pity of war. But all countries experience something similar in wartime, the more so as war goes on and intellectuals start to reflect on it. In France genuine anti-war feeling, as opposed to grumbling about the nature and impact of the conflict, was a trickle against the sustained torrent of patriotic sentiment. French war writing, such as Henri Barbusse's celebrated Goncourt Prize–winning fictionalisation of his trench life, *Le Feu* ("Under Fire," published in November 1916), hid nothing of the horrors of the trenches from the public.[32] Romain Rolland, writing in exile in Switzerland, even won the Nobel Peace Prize in 1916. But these and others like them were merely acceptable voices of dissent in a democratic nation. Many may have sympathised with their sentiments, but few were convinced by their arguments.

"Peace," rather than "victory," increasingly seemed to be the watchword as 1917 progressed.[33] Yet, if France was wearying of the cause, she was not wavering: any peace had to be honourable, reflecting France's efforts, guaranteeing her security and affirming the justice of the Entente cause. For the vast majority, peace at any price was never an option. In a noteworthy hybrid of the old and the new, the Chamber's secret session resolved to continue

> the war imposed on Europe by imperialistic Germany, until the liberation of the invaded provinces, the return of Alsace-Lorraine and just reparations for damages. To reject any idea of conquest or the subjugation of foreign populations . . . to obtain lasting guarantees of peace and independence for peoples great and small, through the organisation, now under way, of a League of Nations.[34]

Even Caillaux pronounced that there could be no peace unless Alsace-Lorraine was returned to France.[35]

There was no peace resolution in France in 1917, as would be passed in the Reichstag in July, and no equivalent of Lord Lansdowne's November 1917 open letter calling for a negotiated peace.* Despite the summer's socio-political crisis, the government was resolved to fight on—as were nearly all of France's soldiers and citizens—and to meet trouble with as much authority as it could muster (although France's war aims became a legitimate subject for political discussion). A more

*See page 450.

vigorous approach to domestic propaganda was pursued through a new umbrella organisation of patriotic associations, the *Union des Grandes Associations contre la Propagande Ennemie* (UGACPE), which held its first meeting on 7 March 1917. Headed by Paul Deschanel, President of the Chamber of Deputies, the new organisation for patriotic education was endorsed by Poincaré and Ribot. It would convey a positive message to the French people over the difficult months of 1917: that the cause was just; that the allies were stronger than Germany and that the enemy could be beaten; that France must and could hold on longer than Germany in order to vanquish autocratic tyranny; that a negotiated peace without victory would be no peace at all. "All of France upstanding for the Victory of Justice" was the inclusive rallying cry behind which the UGACPE remobilised France for a grimmer, tougher war, to reap just reward from the efforts and sacrifices her soldiers and free citizens had already made.[36] Like all effective propaganda organisations, the UGACPE reminded the French of what they already knew, something they had realised at the end of 1916: that they were winning the war with Germany. What they had lost were the false hopes of quick victory that had dominated public perception as Germany was squeezed in the Entente's vise. The defeat of Germany would be slow, but it would be sure; and it would require all that France could give.

Ribot's ministry survived the summer, but in September it lost a vote in the Chamber and Painlevé became premier. By then, the national crisis of confidence was over. Nevertheless, the new ministry lasted barely nine weeks. Remobilised France required her own strong, populist head of government, someone who would fight the war for Frenchmen and Frenchwomen, not just for France. The firebrand nicknamed "the Tiger" could no longer be blocked.

Over a political career that spanned more than half a century, seventy-six-year-old Georges Clemenceau had drifted from the radical left to the patriotic right in France's pliable republican system. From the start, he had been one of the most vocal and trenchant critics of France's dilatory war effort, a scourge on ineffective generals, weak ministers, pacifists and German-sympathisers alike. He had often been talked of as the next premier (he had experience in the post, having held it between 1906 and 1909), but the factional machines that controlled the French Assembly had kept him out of power until there remained no alternative. The Entente's final crisis of 1917, when the Italian front collapsed under a violent German–Austro-Hungarian assault at Caporetto in early November, coincided with the fall of Painlevé's ineffectual and troubled ministry. "Down with Clemenceau! *Vive la République!*" the socialist opposition chanted. But the Tiger was swept to power on a rising tide of nationalist, anti-pacifist agitation that was the inevitable right-wing reaction to a militant left and the troubled summer months.[37]

Clemenceau promised no easy way to victory. "War, nothing but war," was his straightforward, grim promise when he first addressed the Assembly. In 1918 France's war, and by that token the Entente's war, was to assume a new form—*guerre intégrale* (literally "complete war"), as Clemenceau dubbed it—rooted in renewed unity and purpose. Only that way could France eventually put away her "victorious standards, dipped in blood and tears and shredded by shells, the magnificent apparition of our glorious dead" with a clear conscience.[38] Clemenceau's first action was to clear out the old "republican machine" politicians who had been circling in and out of France's wartime governments. His ministry, like Lloyd George's, was filled with younger men of proven managerial ability, many of them socialists; yet his remodelling of the war effort and the war machine was firmly backed by the right-wing press. He kept the job of War Minister for himself.[39] "He was an old man," Jean-Baptiste Duroselle concluded, "but a tremendous one, a man with ferocious energy. He was the ardent patriot and die-hard for whom, after such vacillation, most of the French felt the need."[40] He embodied their will. However, it was not long before he started to be viewed as a dictator. Workers' unrest and left-wing pacifist agitation did not disappear in 1918, and French politics remained volatile, not least in the first half of that year when the Entente suffered its heaviest battlefield defeats since 1914. But under Clemenceau's firm leadership, and with a clearer, more generalised appreciation of the cause and understanding of the all-absorbing nature of modern war, France was better equipped to continue the fight.

In 1917, BRITAIN'S problems were as much political and strategic as social and economic. As Lloyd George and his ministers grew more pessimistic about the war's ultimate outcome, the grudge match between the Prime Minister and Robertson and Haig rumbled on, while German U-boats hacked into Britain's overseas lifelines. As a latecomer to total mobilisation, there remained some resilience in the national spirit; although by the end of the year another attritional campaign, culminating in the muddy hell that borrowed the name Passchendaele, was starting to sap even the Tommies' zest for fighting. This was a different army from that of 1916: the spirit of the enthusiastic volunteers of 1914 and 1915 had been much diluted by the pressed conscripts who had filled the gaps in its ranks during the year. If collectively it became more skilled and was finally taking on the appearance of a veteran force, the British army was less homogeneous and more battered, its fighting men more cautious and cynical than they had been eighteen months earlier.

Nor did Britain escape industrial unrest. May 1917, in particular, was a difficult month on the home front. Some 200,000 workers went on strike in various industries, and 1.5 million working days were lost. Another wave of industrial action followed in November. Over 5.5 million working days were lost to strikes during the year. But since the industrial workforce was now over six million strong, this represented a relatively modest amount of labour unrest compared with pre-war levels: over forty million days had been lost in 1912, for instance. The strikers' grievances were economic rather than political: the price of food was rising under the impact of the German submarine blockade, and increasing government manpower controls and "labour dilution" (the wartime deployment of women and unskilled workers to factories) were undermining hard-won trade union rights. Naturally, the government was worried. Along with concessions on pay and working conditions, it authorised a Commission of Enquiry into Working Class Unrest during the summer. The subsequent report was reassuring: the unrest had not arisen "out of any desire to stop the war . . . One has the impression, in short, of unrest paralysed by patriotism—or, it may be, of patriotism paralysed by unrest."[41] On the whole, compared to France and Germany, the British home front remained reasonably solid.

Inevitably, though, by 1917 more voices were being raised against continuing the war, including some of fighting men. Nevertheless, the protests remained individual rather than collective. One of the most celebrated, not least because it was promoted subsequently by its perpetrator in an influential war memoir, was made by Siegfried Sassoon. While recovering from a wound received in the Battle of Arras, he had taken to reading the anti-war writing of the philosopher Bertrand Russell. His poetry took on a macabre twist: his well-known verse "The General" was penned while convalescing in a London hospital.[42] Sassoon's personal protest involved throwing the ribbon from his Military Cross, which he had won near Fricourt for rescuing the survivors of a

With Our Heroes on the Somme: poster for Germany's January 1917 propaganda film (IWM)

"Comrades in Victory": Punch cartoon marking the Anglo-French capture of Combles on 26 September 1916 (*Mr. Punch's History of the Great War*)

The Australian comforts funds canteen serves food and hot drinks behind the front (Australian War Memorial)

Australian troops practise bayonet drill during the coldest winter in twenty-five years. (Australian War Memorial)

An explosion on the horizon as the German army prepares to retreat
(Australian War Memorial)

Triumphant Australian troops pose on the outskirts of Bapaume
(Australian War Memorial)

German stormtroops advance in March 1918
(© Paris—Musée de l'Armée, Dist. RMN/Pascal Segrette)

Australian and French soldiers man the international post at the junction
of the allied lines near Hangard (Australian War Memorial)

Australian troops and a tank on the Amiens battlefield, 8 August 1918
(Australian War Memorial)

Mont St. Quentin viewed from Australian positions at La Maisonette,
August 1918 (Australian War Memorial)

French refugees in Villers-Bretonneux, 1918 (Corbis)

A French villager raises the tricolour over the ruins of his home on
Armistice Day, 1918 (Australian War Memorial)

The ruins of Péronne, 1919 (Australian War Memorial)

We will remember them: tending makeshift graves behind the French front
(© Paris—Musée de l'Armée, Dist. RMN/Pascal Segrette)

The statue of Foch at
Bouchavesnes on the
Bapaume-Péronne road
(Author)

Nineteen thirties
commemoration ceremony
at the monument to the
missing of the Somme,
Thiepval
(Getty Images)

botched raid in the lead-up to the Somme offensive, into the river Mersey, and circulating his views on the war in a statement to his commanding officer, friends and sympathetic Members of Parliament. It was a one-man mutiny, a refusal of duty in protest at the lack of a statement of peace terms or any governmental inclination to end the war.[43] "I am making this statement as an act of wilful defiance of military authority," he began, "because I believe that the War is being deliberately prolonged by those who have the power to end it." On behalf of the soldiery, he denounced the transition of "a war of defence and liberation" into "a war of aggression and conquest." Notwithstanding "The General," he had no complaint against the military leadership: his protest was against the "political errors and insincerities for which the fighting men are being sacrificed [and the] callous complacency with which the majority at home regard the continuance of the agonies which they do not share, and which they have insufficient imagination to realise."[44]

To what extent this "naïve" protest was a psychological breakdown, one which befitted a self-absorbed Byronic soldier-poet,[45] or an attempt to bring the injustices of the war to a wider audience, a wake-up call to what he characterised as the "British smuggery" that was all too evident once the wounded hero returned home,[46] is difficult to judge. The War Office assumed the former, and a medical board sent Sassoon to Craiglockhart psychiatric hospital to recover from nervous strain. He was back at the front, fighting with his customary verve, in 1918. His encounters at Craiglockhart with the leading shell-shock psychiatrist William Rivers and the talented young poet Wilfred Owen have become the stuff of legend and literature,[47] culturally iconic moments in the transition from one perception of war to another, catalysed by his own fictionalised volumes of memoir, *Memoirs of a Fox-Hunting Man, Memoirs of an Infantry Officer* and *Sherston's Progress.*[48] These books, which contributed to the anti-war Zeitgeist of the early 1930s by charting one media-savvy subaltern's journey from enthusiasm, through disillusion, to redemption, reveal the difficulty of coming to terms with the war and its modernising tendencies. "How could I begin my life all over again," Sassoon concluded his memoir, "when I had no conviction about anything except that the war was a dirty trick played on me and my generation? . . . Yes, my mind was in a muddle."[49] Bertrand Russell at the time recognised the conditionality of Sassoon's protest: "S. S. is not an out-and-out pacifist."[50] Moreover, his actions were highly personal, modest and unremarkable events subsumed amid a war of millions; and not very well received by his peers, before the memory industry got hold of them.[51] The unique refusal of duty and subsequent court-martial of journalist and poet Lieutenant Max Plowman, author of a 1927 memoir, *A Subaltern on the Somme in 1916*,[52] is all but forgotten, but was more an act of anti-war conscience than Sassoon's anticlimactic insubordination. Like Sassoon, Plowman had been a patient of Rivers early in 1917, after being blown up by a shell near Albert. On returning from the front, he protested against the complacent welcome he received in England, a country from which his front-line service had alienated him, publishing an anonymous pamphlet, *The Right to Live*. Subsequently, he became a conscientious objector against war itself, not merely a critic of the present one, and after recovering from shell-shock asked to be relieved of his commission rather than returning to the front. Since he could not be sent by the military authorities for further psychological treatment, he had to be dismissed from the army. This was done in May 1918 and he narrowly avoided imprisonment.[53]

In political circles, too, open criticism of the prolongation of the war appeared. The left wing of the Labour Party, like its European counterparts, started to call for a negotiated internationalist peace. Arthur Henderson, the Labour member of Lloyd George's War Cabinet, resigned in the summer in protest at the government's refusal to allow Labour representatives to attend the Stockholm Conference. Plenty of patriotic Labourites were willing to take his place, but on the far left the Independent Labour Party and the Union of Democratic Control began to lobby, as far as was possible, for an international socialist peace.[54] Their activities kept the question of the nature of the post-war settlement on the political agenda; but, as in France, they did not shake the consensus that the defeat of Germany must be the first item on it. In November the call for a clear statement of allied war aims and a negotiated peace was endorsed in an open letter published in the *Daily Telegraph* by Lord Lansdowne, a pre-war Unionist Foreign Secretary and Minister without Portfolio in Asquith's Cabinet, who had drawn up a draft of British war aims the previous year. However, he did so more from a "hardened reactionary's" fear that British values would not survive the strain of prolonged conflict than from any belief that war was a bad thing.[55] The letter's generally hostile reception suggested that warweariness in Britain had not yet reached crisis point. Nevertheless, there had been a perceptible shift in the nature of British war aims during 1917. The defeat of German militarism remained paramount; but a new world order, and recognition and rewards for those who had brought about the defeat of Germany, supplemented the liberal-imperialist desiderata of the early war years.

For those committed to the war effort and to victory in a liberal society, a more positive message had to answer dissent. Although conscientious objectors, pacifists and defeatists remained relatively few in number, a more dynamic approach had to be taken to sustaining the war effort, not least to ensure that their numbers did not increase. Lloyd George's government endorsed a new multi-party patriotic organisation, the National War Aims Committee (NWAC, a British equivalent of the UGACPE), which chose the third anniversary of the outbreak of hostilities to launch its remobilising message to the public. The country had to be "patient, strong, and above all united," both to win this war and to eliminate war from human enterprise, the Prime Minister pronounced at the NWAC's inaugural meeting.[56] For the rest of the year, the organisation's high-profile speakers toured the country to spread the message about why the nation had to continue the fight. At the same time, the nation was given a new set of social war aims that promised a better, fairer, more prosperous Britain. If the army delivered victory, the conquering soldiers would be rewarded with guaranteed jobs on demobilisation and better living conditions: "homes fit for heroes" in the oft-quoted declaration. The home front's heroes and heroines would also be recognised. In particular, the extension of the franchise to all adult males and women aged over thirty by the 1918 Representation of the People Act would turn post-war Britain into a true democracy for the first time. Nevertheless, the diplomatic settlement remained the central—and most problematic—element of any peace agreement (especially now that Woodrow Wilson's moralistic agenda had to be factored in). Still, it is noteworthy that when Lloyd George finally elaborated Britain's war aims in a speech at Caxton Hall on 5 January 1918, he did so before an audience of trade union leaders.[57]

The inevitable consequence of mass mobilisation was fundamental political restructuring and social change. It brought on the end of liberal Britain as it precipitated a surge on the hitherto steady path to democracy. Recognition of this fact would save Britain from the worst sort of post-war chaos, even if it came at a heavy political cost. Lloyd George took the assassin's knife to both his own political party and the values it held dear in order to keep nation and empire solid, and to win the war.

In similar vein, imperial power was devolved. Those Dominions that had contributed so much to the empire's war effort would in future have more say in its day-to-day conduct. Acknowledging the need for

Dominion representation in the formulation of imperial strategy, in 1917 Lloyd George created an Imperial War Cabinet that included the Prime Ministers of Canada, Australia, New Zealand and South Africa, as well as the Secretary of State for India. South African general Jan Smuts became its permanent representative in the War Cabinet. It would become a valuable makeweight on the Prime Minister's side in his interminable battles with the Western Front–oriented generals.[58]

Throughout 1917 a more focused and direct media effort was needed to sustain civilian engagement with the war. The overarching message remained that the allies were winning, would win and that the rewards of victory were worth the effort. Soldiers themselves were deployed to bolster the home front, slaking civilian curiosity about life in the trenches while mixing in a message of endurance and confidence to counteract signs of defeatism. For example, Jacques Playoust was sent back to Australia for an extended leave in summer 1917 and found himself a minor celebrity. In a country gripped and divided by the war, but physically a long way removed from it, the return of a *poilu* from the other side of the world naturally attracted attention; particularly since he came as a messenger, not as a casualty. Playoust was fêted as a returning hero, "enjoying a well earned leave, and when that is over he will return to the task of hammering the Germans, with all the discomforts the process entails," an interview in the *Sydney Morning Herald* suggested. It continued:

> It is still better to hammer than to be hammered, and as M. Playoust stated, it is the allies now who are doing the hammering. There was a time when the enemy fired 10 shells to the allies' one, but now the allies have the superiority in munitions and guns, and whether in attack or defence the foe's losses are at least as heavy and often heavier than ours . . . For all that no easy or early victory is expected, unless some sudden economic crisis in Germany helps the allies. A military decision will not be achieved quickly. The troops on the Western Front have no delusions on that point, says M. Playoust. There is little of the element of surprise in modern warfare. The continued bombardment always heralds the attack, and the enemy masses his reserves and constructs his new positions accordingly.

Such first-hand testimony was now essential for bringing civilians into contact with the combatants, bridging the gulf between front and rear, and explaining what warfare had now become. Such was Sassoon's and Plowman's purpose too, but they chose to couple it with personal protest. While Playoust went on to describe his grim experiences in the fighting for Combles, he signed off with a realistic but positive evaluation: "M. Playoust states that while there is not much optimism in France as to an early peace there is no doubt as [to] the ultimate result. The Germans cannot win."[59]

For some, the successful prosecution of the war, and a just and inclusive peace, required more than a stronger rallying cry and new socially inclusive war aims. A complete rethink of politics was needed, banishing the class divisions and party factionalism of pre-war Britain, which, three years in, still seemed to be undermining the war effort. For such men, war was an affirmative, purgative, national endeavour and a unifying force. They wished to put forward a positive message to challenge the pacifists and defeatists. After seeing his valiant soldiers cut down at Contalmaison and Pozières, Henry Croft returned from France at the end of 1916 to resume his political career. A young Unionist MP before the war, he had been radicalised by his experience at the front. What had begun as "the Great Adventure" had taken on a grim reality: like so many others, he looked for cause and consequence. In his 1917 memoir of service, dedicated to "my comrades of the battle-field," he appealed to the new men from the Dominions, whom he had fought alongside:

> Do not judge our people by our political rulers, but rather join with the people of the old country to purify a system which is at fault, for whilst our politicians have failed, those also in the younger countries were little more successful in teaching the true path to patriotism—that freedom is of more worth than wealth.
>
> We have learned together how to fight and the blood of our dead is soaking the soil which they together have made immortal; let us now start to design and build a palace on foundations of such depth and thickness that it shall stand for ever, but to secure this end, let us tolerate no half measures in the great memorial to our fallen.[60]

Such "blood and soil" rhetoric was to become commonplace between the wars: Croft's sentiments and words would not have disgraced the most infamous Somme veteran, Adolf Hitler. Croft's solution to the social divisions, national soul-searching and political infighting that he found on his return to Britain was novel. In September 1917 he and half a dozen sympathetic fellow MPs founded a new political force that dubbed itself the National Party, the prototype of the radical right-wing, nationalist, anti-liberal movements that would make European class politics their battleground between the wars. Its aims were avowedly "[a] National as against a class, sectional or sectarian policy. Complete victory in the war and after the war . . . A National social policy based upon the principle that the people shall be reared in such surroundings, in such conditions, and with such opportunities in life, work and play, as will ensure a contented patriotic race." Such an initiative to sweep away the "worn-out party system" and replace it with a unitary state founded on "courage, honesty, thoroughness and industry . . . which the war has taught us to seek for . . . that sense of comradeship which has been won in war" is one example of how war experience changed men's minds; and how that experience, when brought back into civil society, was expected to change their homelands. In late 1917 Croft's was a timely and popular gesture. If parliamentary allies were few (understandably, given the denunciation of the current system in the new party's manifesto), support outside Parliament surged: public meetings were well attended, several "distinguished Admirals and Generals" signed up, and the *Morning Post* and *National Review* rallied behind the new movement.[61] Perhaps as many as a quarter of a million people attended the party's "Win the War" rally in Hyde Park, and more than a million signed its petition calling for the internment of enemy aliens.[62] The consensus politics of the liberal political centre had been discredited by the war; by the killing and the stalemate. The future appeared to lie on the radical political margins, either left or right, which recognised the core unifying values of human society: whether that be just peace or more effective war, both were prepared to campaign for them. While R. H. Tawney gravitated towards the socialist Union of Democratic Control, Croft's National Party was perhaps Sassoon's natural political milieu—the place for someone who "could at least escape from the War by being in it" and felt dislocated when a second wound forced him out of it for good. In the 1930s he admitted:

> It seemed that I had learned but one thing from being a soldier— that if we continue to accept war as a social institution we must also recognize that the Prussian system is the best, and Prussian militarism must be taught to children in schools. They must be

taught to offer their finest instincts for exploitation by the unpitying machinery of scientific war. And they must not be allowed to ask why they are doing it.[63]

AGAINST THIS BACKGROUND of domestic unrest, allied strategy had to be rethought. Nivelle's failure put an end to thoughts of a quick, crushing defeat of the German army. The default position was further attrition, which was readopted as the basis of Anglo-French military strategy in May 1917 because it was working, and was ongoing after the initial, stunning seizure of the Vimy Ridge by the Canadian Corps. The British army's offensive was developing into a three-army attritional battle between Arras and Bullecourt. The French offensive continued on the Aisne front. Haig, Robertson, Nivelle and Pétain, the latter temporarily appointed army chief of staff in the French War Ministry, met in Paris on the morning of 4 May to reformulate strategy. These military principals put their signatures to a joint paper that was presented to an allied political conference in the afternoon. "It is no longer a question of breaking through the enemy's front and aiming at distant objectives," they posited.

> It is now a question of wearing down and exhausting the enemy's resistance, and if and when this is achieved to exploit it to the fullest extent possible. In order to wear him down we are agreed that it is absolutely necessary to fight with all our available forces, with the object of destroying the enemy's divisions . . . We are all of opinion that our object can be obtained by relentlessly attacking with limited objectives, while making the fullest use of our artillery. By this means we hope to gain our ends with the minimum loss possible.[64]

The politicians had no choice but to acquiesce, for the time being at least, in the face of this united military opinion, even if they did not really agree. It marked the restoration of the Somme strategy and tactical method to pre-eminence after the "Verdun school" had failed. "Well, there's the Somme's revenge," Foch had joked with Fayolle the previous month, after Nivelle's disaster.[65] Soon Nivelle was forced from command on the North-eastern Front, with Pétain, the only survivor of the Verdun triumvirate, replacing him. Then the Somme commanders were restored to their proper places at the head of the French army.

Fayolle assumed command of Central Army Group, and responsibility for the attritional battle on the Aisne front (there had been rumours that he was to go to G.Q.G.[66]), while Micheler was demoted to command Fayolle's First Army. Foch replaced Pétain as army chief of staff, with responsibility for overall strategy and the expanding war *matériel* programmes that would provide the French army and its American reinforcements with the guns, tanks and aeroplanes essential for modern offensive warfare. The three wise old professors from the pre-war *École de Guerre,* who had proved their worth in different ways in 1916, were now in place to win the war. For the moment, however, the *poilus'* concerns and Lloyd George's hostility to a strategy of attrition would oblige them to proceed with caution. But these factors would not stop the main event on the Western Front in the second half of 1917, the Third Battle of Ypres, Haig's frequently postponed offensive to secure control of the Belgian coast.

WITH THE DECISION made to focus on that offensive, tranquillity returned temporarily to the shattered land of the Somme. One Sunday on furlough in Bécourt camp, Lieutenant McInnis sat outside his billet writing letters and enjoying the tea and cake brought by his batman: "Surely there is not a war."[67] Over the following weeks he watched as the old Somme battlefield became a training area, explored the sites of the previous summer's great battle, and noted the softening of the harsh scars on the landscape as spring's green shoots and yellow buttercups carpeted the grey-brown earth.[68] The battlefield could now be cleared, and cultivation resumed: every square kilometre of land was needed at a time of intensifying submarine blockade. Some civilians returned to their former homes. The owner of Bécourt château, used as a field hospital and relatively unscathed, turned up, but refused to take his old home back until the trenches and dugouts that criss-crossed its grounds were filled.[69] While labour gangs of prisoners did the filling-in and cleared unexploded munitions, burial and grave registration parties from the recently established Imperial War Graves Commission scoured the empty landscape, interring the many corpses that littered the field, identifying individual graves and improvised cemeteries, recording the details of their occupants if they could, and moving them to the plots that would in time become hallowed ground.

Curious visitors—pilgrims and mourners, photographers and painters—also explored the abandoned battlefield, trying to capture its lost essence, or to conjure up ghosts from the past. Charles Bean began his regular pilgrimages to Pozières, to the shattered stump of its windmill from which he could view the whole of the "ridge more densely sown with Australian sacrifice than any other place on earth."[70] After the war the Australian government would purchase this site, and 1st Australian Division would erect its memorial there. It was hard to believe that less than a year before the countryside had been the scene of such intense fighting: "The country looks like an artist's palette, with the colours changing as the months pass. The white of daisies and the yellow of dandelions and buttercups are now giving way to the red of poppies and the blue of cornflowers and chicory."[71] Indeed, the battlefield was now the place for painters, war artists who set up their easels in order to capture the blasted landscape before nature reclaimed it. William Orpen began his commission as official war artist on the Somme in April 1917, recording the destruction left in the wake of the German retreat. "I shall never forget my first sight of the Somme battlefields," he recollected.

> It was snowing fast, but the ground was not covered and there was this endless waste of mud, holes and water. Nothing but mud, water, crosses and broken Tanks: miles and miles of it, horrible and terrible but with a noble dignity of its own, and, running through it, the great artery, the Albert–Bapaume road, with its endless stream of men, guns, food lorries, mules and cars, all pressing along with apparently unceasing energy towards the front ... towards the hell that awaited them on the far side of Bapaume.[72]

Yet on the old battlefield itself, "All is Peace," Orpen recorded in a poem composed as he mused on the unburied dead.[73] He returned in August to depict it. "I remember an officer saying to me 'Paint the Somme? I could do it from memory—just a flat horizon-line and mudholes and water, with the stumps of a few battered trees,' but one could not paint the smell."[74] Orpen was too late to capture the real Somme: the months of barrages, attacks and counter-attacks; the jarring cacophony of explosions and machine-gun fire; the ubiquitous mud; the everpresent death. Whereas, in the spring Orpen had found "mud, nothing but water, shell-holes and mud—the most gloomy, dreary abomination of desolation the mind could imagine ... now, in the summer of 1917, no words could express the beauty of it." The "dreary dismal mud was baked white and pure" and garlanded in wildflowers. The air was thick with white butterflies and white crosses were everywhere. "Clothes, guns, all that had been left in confusion when the war passed on, had now been baked by the sun into one wonderful combination of colour—white, pale grey and pale gold."[75]

Even Haig availed himself of the opportunity to visit the scene of his first great battle (although he viewed it with a soldier's matter-of-fact eye, rather than through an artist's aesthetic lens). The trip served to reinforce his sense of achievement and confidence in his army. "No one can visit the Somme Battlefield without being impressed with the magnitude of the effort made by the British Army," he reflected.

> For five long months this battle continued. Not one battle, but a series of great battles, were methodically waged by numerous Divisions in succession, so that credit for pluck and resolution has been earned by men from every part of the Empire. And credit must be paid, not only to the private soldiers in the ranks, but also to those splendid young officers who commanded platoons, companies and battalions . . . To many it meant certain death, and all must have known that before they started. Surely it was the knowledge of the great stake at issue, the existence of England as a free nation, that nerved them for such heroic deeds. I have not the time to put down all the thoughts which rush into my mind when I think of all those fine fellows, who either have given their lives for their country, or have been maimed in its service.

Like Croft, Haig had only one complaint: "Later on I hope we may have a Prime Minister and a Government who will do them justice."[76]

OFFICIAL WAR ARTISTS and photographers had quickly found that the narrow viewpoint from trench or observation post, the empty landscape of no man's land, was not the stuff of great paintings or dramatic photographs or newsreels.[77] Images would have to be reconstructed once the battle had moved on, would depict the shattered consequences of fighting or represent heroic acts after the event: factual accuracy vied with artistic interpretation, and in these circumstances the pen might prove mightier than the camera or the paintbrush.

The process by which the memory of the Somme was manufactured began in the summer of 1917. As it had become clear that the war would continue into 1917, it was deemed necessary to explain the wider significance of the great offensive to the public. To that end, John Masefield, poet and author of an officially sanctioned book on the Gallipoli campaign, was commissioned to write an account of the Somme. It proved an impossible task. Despite walking the "vast . . . and very very confusing" battlefield and talking with combatants, Masefield, with no access to official documents, could only complete and publish the preface of what was planned to be a full account of the British offensive. *The Old Front Line*, published towards the end of 1917 and reprinted regularly ever since (most recently in 2006), was a paean to the landscape, "some of it romantic, some of it strange, some unearthly, some savage," which was as much a part of the battle as the men who wrestled back and forth across it.[78] From the fragment that was published, it is clear that Masefield's history would have been more than a simple work of propaganda.[79] He was fashioning an elegy to the men of the New Armies and a panegyric to their achievement with, in Esher's words, "a permanent value in the domain of high literature."[80] It would have been a new sort of democratic military history, representative of the sort of war that was being fought; a story of the men of Britain and her empire, the common soldiery who risked or gave their lives, not the statesmen who had sent them there, or the generals who directed them.

The sustained public desire for information meant that books, articles, films and photographs continued to appear even after the Somme offensive slowed and stopped. British publishing houses especially had a field day with accounts of the battle, or photographic anthologies, in 1917. It had been the British Empire's first great modern battle, a time of shock and awe to the English-speaking world, and it would continue to fascinate for that reason. Masefield's planned account was officially sponsored, while others—such as Buchan's, Perry Robinson's and Bean's collected despatches, as well as Palmer's and Brittain's recollections, published in America—bore the imprint of the censors. Collectively, this contemporary literature and imagery of the Somme juxtaposed death and destruction with purpose, courage and derring-do.

Such eyewitness accounts, bowdlerised though they might be, are representative of the mood of the moment, and contrast starkly with later reconstructions of the battle that shaped a now-dominant image of horror and purposelessness. Battle was a complex experience with both positive and frightening facets. While the war continued, media representation of the Somme brought the reality of industrial battle into public consciousness in a graphic way while maintaining a broader umbrella message of rightful cause, collective effort and worthy sacrifice to sustain civilian morale. In doing so, coverage of the battle was fulfilling a dual purpose, both showing what the war was like and conveying and reinforcing what it meant.[81] To that end, the Somme offensive had been conducted as a public spectacle, and was to leave a rich, vivid archive from which later commentators could fashion the interpretations and myths that have come down to posterity, reinforcing the worst of what it was like but gradually losing touch with what it meant to those who were there. Such negativity undoubtedly had its roots in the difficult year 1917, although then the overriding message was a positive patriotic one, which seemed justified by the relative fortunes of the belligerents at the time.

ALTHOUGH FRANCE WAS plunged into her deepest bout of pessimism and national soul-searching during late spring and summer 1917, this offered little comfort for Germany, which was gripped by a different sort of crisis. While her soldiers stuck stoically to their positions—respect for military discipline and the lack of a tradition of collective citizen action explained that—Germany's home front began to implode. Bethmann Hollweg had recognised before war was even declared that "whichever way the war ends, it will bring about a revolution of everything that exists."[82] Conservative Germany was loath to acknowledge this unpalatable truth, and thereby sealed its own cataclysmic fate.

Germany's home front continued to polarise as 1917 proved even more difficult than earlier years. The Reichstag's peace party grew more vociferous; and the hungry, destitute communities that had already staked everything on victory yet were still being asked for more started to sympathise with the idea of a negotiated peace, even a socialist peace without annexations and indemnities. The crisis point was reached in July. On the 19th the Reichstag, where the conservatives were in a minority, passed its famous "peace resolution." Moderate Deputies from the Social Democratic, liberal Progressive and Catholic Centre parties voted for "a peace of understanding" without "forced acquisitions of territory, and political, economic or financial oppression." The *Burgfrieden* was collapsing. In the same month Bethmann Hollweg, who had always championed political reforms as fair reward for the efforts of the German people, especially the modification of the Prussian franchise to one based on universal suffrage, was forced from office by the high command working hand-in-glove with the conservatives in the Reichstag. He was replaced by a conservative nonentity and military puppet, the Prussian civil servant Georg Michaelis. The latter lasted only three months: in October the Reichstag's growing agitation for political reforms obliged OHL to replace him.[83] The empire's cohesion was also breaking down. In Bavaria, in particular, there was strong sympathy for Bethmann Hollweg's reform programme. The ex-Chancellor had continually warned the conservatives in Prussia that if they did not reform the imperial system, it would be reformed by forces beyond their control, and by late 1917 it was clear that battle lines were being drawn up at home as well as in France.

With "defeatism" gaining the upper hand in the Reichstag, the army decided it had to tighten its grip on power: it would have to win the war before the home front lost it. Its values and methods did not allow for challenge and change. In future, the destiny of the nation would equate with the survival of the military autocracy that had taken Germany to war and increasingly controlled her internal affairs. Its response to the peace resolution was a programme of indoctrination at the front and in the rear. So-called patriotic instruction commenced in the ranks, and the army strengthened its control over the domestic media. But such initiatives were out of touch with the mood of society. Patriotic instruction "was a dismal failure . . . From its title to its schoolmasterly organisation, it totally miscalculated the true mood of the soldier."[84] It did little to discourage the numerous acts of individual desertion and collective indiscipline in German units behind the lines (and a mutiny in the navy) during the summer of 1917.[85]

With the left growing stronger and more willing to challenge the military leadership in Germany, and anti-war sentiment and peace propaganda infecting not only the home front but the army itself, Hindenburg and Ludendorff decided to fight fire with fire. Germans would be reminded of why they were fighting, and rallied to the imperial cause. Old political structures were breaking down, so new ones would have to be erected to replace them. In another chilling anticipation of the militaristic, nationalistic mass parties that would flourish in post-war Europe, in September a new pro-war political party, the Fatherland Party (*Vaterlandspartei*), headed by Admiral von Tirpitz, was founded to promote the values of the regime at home. By 1918, it had over a million members. It was a first, inefficient attempt to create a national community based on universal, right-wing, exclusive values (at exactly the same time as Croft and others were promoting a more radical, socially inclusive, right-wing ideology in the British Empire). Political polarisation was under way in Germany: the Fatherland Party was challenged by an equally numerous, moderate mass movement, the People's League for Freedom and Fatherland (*Volksbund für Freiheit und Vaterland*).[86] In Germany remobilisation, such as it was, would be founded on the conservative, nationalist, militaristic values of the pre-1914 imperial regime. Although workers' labour and middle-class money were vital to the war effort, there was little intimation that these contributions would ever be recognised or rewarded, either in victory or defeat. Authority and subordination, not appreciation and toleration, would be the guiding principles.[87] All this made victory imperative: the fatherland had, willingly or unwillingly, placed its fate in the hands of its generals. The army might just be forgiven its high-handed approach to politics, and the losses at the front might be rationalised, if Germany emerged triumphant. Simply forcing the allies to the peace table would have been considered a triumph by this point. But if that could not be achieved, with the home front polarised between war and peace parties, with the soldiery growing war-weary and more sympathetic to those who called for an end to hostilities, with only discipline and indoctrination keeping the men in the trenches, Germany's last effort would be an all-or-nothing gamble.

FOR WHILE THE domestic consensus in support of the war was fracturing irreversibly, in 1917 the German army enjoyed none of the battlefield success of earlier years. It did hold on in the West, but after breaking the momentum of Nivelle's attack in April it was once again condemned to attrition, on the Aisne and Arras fronts in April and May, and from July to October in Flanders. The casualty lists continued to lengthen, while the tactical position deteriorated markedly. By the end of the 1917 campaign, the German army was in an untenable position in the West. Allied "bite and hold" methods were now effective enough to retake the strategic high ground that the German army had controlled since 1914. The British were fighting hard, and improving all the time. Vimy Ridge fell in April, Messines Ridge in June and the Passchendaele Ridge, dominating the Ypres salient, in October. In November they surprised the enemy in front of Cambrai with a mass tank attack backed by an overwhelming surprise bombardment, a portent of new tactics. The revivified French army added its weight to the fight in late summer. In August, Second Army—now commanded by I CA's General Guillaumat—kicked the enemy off the western Verdun heights, Mort Homme and Côte 304. Then, in the Battle of Malmaison late in October, Sixth Army's massed heavy artillery finally blasted the Germans off the Chemin des Dames, the ridge above the river Aisne on which they had put down deep roots since September 1914: it was "the perfect offensive," in Cyril Falls's judgement.[88] France's military crisis was at an end.

Italy's was just beginning. In late October her Second Army collapsed after German and Austro-Hungarian forces broke the Southern Front in the Battle of Caporetto. British and French divisions were despatched south to shore up the crumbling Italian front. Fayolle was given command of the French contingent, a brief to retrain the Italian army in Western Front methods, and new allies to grumble about in his diary! The allies' political leaders traipsed over the Alps with them, then gathered at the Italian resort of Rapallo to address the latest military crisis. Here a long-overdue, collaborative politico-military organisation was set up with the intention of improving coordination of allied strategy. Unfortunately, the Supreme War Council (SWC), which was to sit at Versailles, proved far from perfect. With its power split between political and military representatives, initially it was merely another forum in which politicians and soldiers could air their disagreements. Lloyd George appointed Henry Wilson as the permanent British military representative on the SWC, and hoped that the new organisation would circumscribe Robertson's influence once and for all. Clemenceau was shrewder. Since before the war, when he had appointed Foch to direct the French army's staff college, he had been impressed by the professor's abilities. If he worked hand-in-glove with the SWC's military members, France's chief of staff could exercise genuine control over allied military strategy. To that end, Weygand was appointed France's permanent military representative in Versailles, while remaining Foch's deputy at the War Ministry. Wilson was an old friend of Foch, while the Italian and American military representatives carried none of the clout of their British and French counterparts. As a result, for the first time, there was an opportunity for the formulation and direction of strategy to be concentrated in the hands of one powerful and respected general—Foch.

It was not to be. The politics of the SWC, and the disputes and compromises that arose in its early meetings, had a familiar ring: disagreements over the length of front to be held by the British and French armies; Italian demands for more support; arguments over peripheral campaigns in the Mediterranean; rows between Lloyd George and Robertson that eventually led to the latter's unwilling resignation in February 1918. Wilson returned to London to become Lloyd George's tame new C.I.G.S. while Rawlinson replaced him at Versailles. Foch at least gained in stature, if not real power. In spring 1918 he was placed at the head of a new body, the SWC's Executive War Board (in practice its military members acting independently from political authority), which was to control an allied General Reserve composed of thirty British, French and Italian divisions. This was a political fudge, war-making by committee, and in reality had no chance of success, since both Haig and Pétain resented the new authority and refused to release any of their divisions to Foch. Instead they made arrangements for mutual support in the event of a German offensive in the West, and Clemenceau quietly let the Executive War Board idea drop.[89] Nevertheless, Foch was clearly in the ascendant. "He is assuming the direction of the war more and more," Fayolle noted in February 1918. Foch did not get on well with Pétain and wanted to replace him, and Fayolle was in no doubt that he was the prime candidate: "I will be his man on the French front, as I was his man in Italy."[90]

ALMOST A YEAR earlier, Nivelle had been caught on the horns of the operational dilemma that had exercised the minds of the high command since the onset of trench warfare. As Robertson perceptively wrote to Haig:

> I cannot help thinking that Nivelle has attached too much importance to what is called "breaking the enemy's front." The best plan seems to me to go back to one of the old principles, that of defeating the enemy's army . . . and that means inflicting heavier losses on him than one suffers oneself. If this old principle is kept in view and the object of breaking the enemy's army is achieved the front will look after itself, and the casualty bill will be less.[91]

The casualty bill was always going to be high—and it had continued to mount up in the unrelenting offensive warfare and attritional battles of 1917—but in Robertson's reasoning there was the nub of a way to beat the German army on the Western Front. This was also Foch's conception, and Nivelle's failure had brought his rehabilitation. It would be many months before he got to put this precept into practice, however. In the meantime, the French army fractured and re-formed, universalising the doctrinal principles that had served it well on the Somme. The British army pressed on regardless, mixing the old and discredited with the innovative and exciting as Haig strove to find his own answer to the Western Front dilemma. The German army strengthened and stuck to its defences, hoping for salvation on the high seas or in the East. All three nations turned inwards, remobilising their populations and resources (in significantly different "national" ways) for the death struggle that the war had become after 1916.

Although momentous events had occurred during the year, as 1917 turned into 1918 an end to the war seemed even further away than it had been twelve months earlier. Domestic difficulties were affecting all the belligerents, civil–military strife rumbled on, and war resources—men, money, *matériel*—were being depleted. However, the Entente was better organised and better armed. France's army had revived and her government was now firm; and, if France was not united, then the anti-war voices were at least quiescent. The British army's methods were growing increasingly modern and effective, and the empire remained squarely behind the motherland. For all its early faults, the SWC, and the accomplished soldier who was its principal personality, promised a great improvement on 1917's game of strategic skittles. And during 1918 America's vast pool of fresh, enthusiastic manpower would start to reinforce the skilled but tiring divisions that controlled the Western Front.

Belatedly, though, Germany had been handed a lifeline. Haig's proviso for winning the war—that Russian effort continued—proved percipient. After her first revolution in February, Russia fought on—albeit for a democratic socialist peace, not for the imperialistic ends of the tsarist regime—honouring her treaties and her obligations to her allies. But Russia's soldiers no longer fought well. Their hearts were not in the war, and their officers' authority was compromised by democratisation of the armed forces and the pervasive influence of Bolshevik anti-war propaganda. Russia's major offensive in the summer of 1917—her last—quickly ground to a halt. There would be no repeat of the General Allied Offensive's tightening stranglehold that had enhanced the Somme's impact on Germany's ability and resolve to make war.

Moreover, in an apparently astute political move that contrasted with his poor strategic sense, Ludendorff had arranged for Lenin, the exiled leader of the radical, left-wing, anti-democratic Bolshevik Party, to return to Russia and agitate for peace. But the First Quartermaster-General could hardly have foreseen that his helping hand would have such swift and dramatic consequences. Just when Germany's military situation was starting to appear desperate, in the autumn of 1917, Lenin mounted a *coup d'état* that crippled Russia's war effort. And this implosion in the East gave Germany one final, unexpected opportunity for victory in the West.

13

Decision in Picardy: 1918

T 04:40 ON 21 MARCH 1918 Picardy's skies exploded into fireworks once again. This time the scale and intensity of the bombardment made the ground tremble as if there were an earthquake and shook the windows in Amiens, forty miles away.[1] The brainchild of the German army's barrage expert, Lieutenant-Colonel Georg Bruchmüller, who had cut his teeth smashing Russian armies on the Eastern Front, this "hurricane" bombardment was designed to be sudden, brief and crushing. The technique was familiar to allied gunners, who had employed it themselves, but it still came as a surprise, since the guns were deployed just before the barrage commenced and went undetected by allied intelligence. Along fifty miles of allied front, from Arras to La Fère, 6473 field and heavy guns and 3532 trench mortars fired for five hours, rising to a five-minute crescendo at the moment of attack. In that brief period 1,160,000 shells (a mixture of high explosive and gas) were launched into the British lines, more than 75 per cent of the number fired by the British artillery in the week before 1 July 1916. The guns were supporting three German armies, fourteen corps with thirty-two divisions in the front line and forty-two in support and reserve,[2] which attacked at 09:40. By then, the bombardment had done its job. The British defences, held by General Sir Julian Byng's Third Army to the north and Gough's thinly stretched Fifth Army to the south, were pulverised and the troops disorientated. The attack was spearheaded by specially trained and equipped assault units, so-called storm-troops, advancing behind a barrage that lifted two hundred metres every four minutes. These elite infantrymen were tasked with pushing through the neutralised British defences and on into their rear, with the aim being to seize the gun line and destroy any parts of the command and communications infrastructure that had been missed by the bombardment. Any resistance that remained would be mopped up by the "battle groups" that followed. Reserve divisions would then push forward to maintain the momentum of the attack.[3] In this way the long-desired break-through might be achieved, and open warfare might resume.

After the October 1917 Bolshevik revolution plunged Russia into anarchy and ended the Russian army's effort on the Eastern Front, Germany had been offered a last chance. The vast swath of Eastern European territory gained by the punitive Treaty of Brest-Litovsk on 3 March 1918 seemed to give substance to the propaganda that preached that Germany could still win the war. It also provided a potential source of food and resources to redress the effects of the allied blockade. Most importantly, victory in the East liberated a reserve of veteran troops who could be redeployed in search of a final victory over the Western allies. However, there were two problems for the German high command: these men could not be fed from the scarce resources left in the West; and many thousands of them did not even reach the Western Front, jumping from the troop trains as they crossed the fatherland.[4] Germany's people had had enough of war, and her soldiers, who knew what awaited them in the West, had little desire to prolong it. There would have to be intense "patriotic instruction" for the troops behind the lines before battle commenced.[5] At best, they were prepared to make one last effort, but as Leutnant Rudolf Hoffman later remembered: "This was the last desperate attempt to bring about a change in our fortunes. Maybe 20 to 30 per cent of our unit were keen because they hoped to find plenty of food and alcohol; they were mostly the young ones. But the rest of us weren't at all enthusiastic; we just wanted to get the war over and get home alive."[6] Gaps remained in the ranks (and the offensive would inevitably produce more), which would be filled by young men of the 1919 class, troops extracted from home garrisons and the rear, and convalescents.[7] It was Germany's final muster.

Events in Russia gave a grim forewarning of what might happen if there was no victory. In Germany, where full mobilisation and sustained military and civil crises had led to a strengthening of military control, authority was increasingly maintained by force rather than consent: when strikes broke out again in 1918, militant workers were sent to the front.

But if the military leadership could keep domestic dissent under control while the army smashed Germany's enemies once and for all, an honourable, victorious peace might yet head off social disorder, and such excesses might be forgiven, if not forgotten, in the euphoria of triumph. At the beginning of 1918 Russia was smashed, Italy cowed, France and Britain apparently weakening. Since the army could no longer hold its defensive positions in France, and had lost the war of siege and attrition, the spring offensive was a last, desperate gamble to save pre-war imperial Germany. The *Kaiserschlacht* ("Kaiser's battle")—as the March 1918 offensive was grandly and, given what was at stake, appropriately titled—was to be the biggest on the Western Front since 1914. It was designed to force the allies to the peace table before the Americans arrived in large numbers.

AT THE BEGINNING of 1918, no one had been willing to predict that the war would be over by year's end, although of course that was the universal hope. If anything, the conflict was intensifying. While Germany had a new reserve of troops, allied war industries were now producing huge amounts of war *matériel* with which to confront them. Churchill, now back in government as Minister of Munitions, was able to replace the guns lost in the March offensive immediately from stockpiles at home, and war workers willingly gave up their Easter holiday to do their bit at a time of acute crisis.[8] But after the intensive attritional warfare of 1916 and 1917, allied ranks were thin and reserves were hard to come by. In 1918 the British army would have to follow the French example and reduce divisions from twelve to nine infantry battalions. Both armies would also have to dissolve some divisions to bring others up to strength. In time the influx of fresh U.S. manpower would restore allied advantage in that area, but in early 1918 they were still waiting for the Americans' arrival.

The timing, scale and speed of the assault caught the Anglo-French armies off-guard. On the eve of the attack the front was so quiet that Fayolle noted: "more and more it seems to be confirmed that the Boche will not attack."[9] However, while Ludendorff achieved something that had always seemed impossible for his allied counterparts—surprise— with millions of men still engaged in the West, and reserves available, this was not enough to force a quick victory. Haig and Pétain had previously arranged to move reserve divisions south or north to support each other's forces in the event of a powerful German attack. On 22 March those arrangements were activated when Pétain started to send divisions northwards. Many more than he had originally expected to redeploy were despatched over the next few days because the crisis

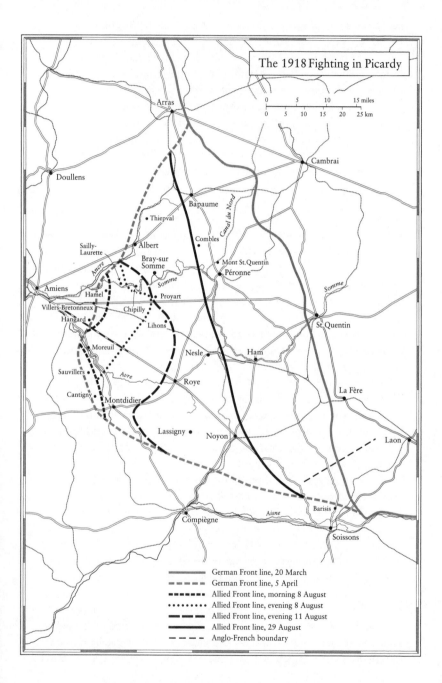

The 1918 Fighting in Picardy

0	5	10	15 miles
0	5	10	15	20	25 km

Arras

Cambrai

Doullens

Bapaume

Thiepval

Combles

Canal du Nord

Sailly-
Laurette

Albert

Bray-sur-
Somme

Mont St.Quentin

Péronne

Ancre

Somme

Amiens

Hamel

Proyart

Somme

Villers-Bretonneux

Chipilly

St.Quentin

Hangard

Lihons

Moreuil

Nesle

Ham

Sauvillers

Avre

La Fère

Cantigny

Roye

Montdidier

Lassigny

Noyon

Laon

Compiègne

Aisne

Barisis

Soissons

	German Front line, 20 March
	German Front line, 5 April
	Allied Front line, morning 8 August
	Allied Front line, evening 8 August
	Allied Front line, evening 11 August
	Allied Front line, 29 August
	Anglo-French boundary

on Fifth Army's front, which was buckling at its southern end under the weight of the German assault, was greater than anything yet experienced. Twenty-one divisions were under orders by 23 March, but it would be some days before they were all concentrated, and Haig felt that they were insufficient for the task.[10] Meanwhile, Gough ordered his battered divisions to conduct a fighting retreat to the line of the river Somme, to gain time for the French reserves to deploy.[11]

Gough was blamed for the disaster—even though he had maintained control of his beleaguered corps and conducted, as far as was possible, a careful withdrawal—and was ordered home in early April. Haig had tried to save Gough, but his offer to take responsibility himself and resign had not been accepted: even Lloyd George was not foolish enough to change his commander-in-chief, whom he did at least consider a fine defensive general, at the height of a crisis.[12] To some extent, Fifth Army's commander was a scapegoat and a victim of circumstance and scarce resources. Over the winter Haig had been obliged to extend the British army's front southwards, to the river Oise, which left Gough's forty-two-mile front, held by twelve infantry and three cavalry divisions, stretched dangerously thin. In comparison, Third Army's twenty-eight-mile line to the north, which was bent backwards between Arras and Albert but did not fracture, was held by fourteen divisions. But Fifth Army's dispositions and defensive tactics were also flawed. In anticipation of a German offensive, it had been building a deep strongpoint-focused defensive system over the winter, but this was unfinished in March. Its units had been trained in defensive tactics (the British army was, in effect, trying to mimic those practised by the German army), but there were too few of them to garrison such a system. The result was that the front positions were defended, but there were inadequate reserves to respond to a major attack.[13] Gough's forces could delay and disrupt, but not check, the German penetration. In Gough's place, Rawlinson returned from his desk job in Versailles to manage the remains of the units retreating across the Santerre plateau, the newly redesignated Fourth Army.

On 22 March the Kaiser decorated Hindenburg with the Iron Cross with Golden Rays, an honour last awarded to Field Marshal Blücher after the overthrow of Napoleon. By then, the German army was several miles into the British defensive positions and Fifth Army's isolated garrisons, bypassed by the leading waves of storm-troops, were being reduced by the supporting divisions. Those outposts that had survived the early bombardment fought tenaciously for as long as they could;

but, cut off from the rear and without orders, one by one they were overwhelmed or surrendered once their ammunition ran out. It initially appeared that the German offensive had finally shattered the front, and that victory was imminent. But this was not the first or the last time on the Western Front that appearances proved deceptive. After two days the remnants of front-line divisions were still holding out in the battle zone, and they had inflicted heavy losses on the attackers. Although the situation on the Picardy front was grave, Fifth Army's front had not collapsed. Its left and centre were withdrawing in good order towards the Somme either side of Péronne. Only on its extreme southern end, where III Corps had been badly beaten and pushed back to the Crozat canal, was a gap opening, but French reinforcements were already deploying to plug it.

FAYOLLE HAD RETURNED from Italy in February to find himself with nothing to do. At the time he suspected that Foch was keeping him in hand to await developments at the front: a senior commander with experience of all three allied armies was just the man to command the General Reserve, should one be established. "It seems probable, almost certain, that I will come back to fight among the English," he believed. The only uncertainty was when.[14] But that was soon decided, for Fayolle was to command the French forces assembled in Picardy to shore up Fifth Army's crumbling front. Third Army's left wing at Arras and the left wing of General Duchêne's French Sixth Army to the south were solid hinges on which the allied armies could close the doors on the advancing masses of German infantry. Fayolle was to be the door-keeper, and Foch the door-maker. On 23 March Fayolle took command of a new Reserve Army Group, composed of General Humbert's Third Army and First Army under his old Somme comrade Debeney, which was to be formed between Duchêne's and Gough's armies, taking over the British front to the Somme. The next day he was appointed to direct the remnants of Fifth Army south of the river.[15] "It was the final battle," Fayolle believed.[16] He was mistaken, because already it was not going according to the German plan, the objective of which had been to break through Third Army's front south of Arras. When Byng's front held firm, Ludendorff opted to switch the main thrust of the attack to his more successful left wing. General von Hutier's Eighteenth Army, supported by Second Army to the north, would drive across the San-terre plain towards Amiens. The enemy was beaten, Ludendorff mistakenly surmised, and all that remained was to separate the British and French armies on either side of the Somme, drive the British into the sea and finally overpower the French.[17] The sort of over-optimistic, under-resourced flight of strategic fantasy that had marred earlier Western Front offensives had reappeared, and it would draw the German army back into the arena of its 1916 torment.

By 24 March the offensive was through the British defensive front and into open country. Fifth Army was in retreat, but it was still a fighting force. Improvisation was rife, as the withdrawal extended Fifth Army's front and stretched its battered divisions even more thinly. But there were many men on a modern army's lines of communication, and Fifth Army's clerks, cooks, drivers, orderlies and instructors could all be sent into the fight. Scratch formations, either organised locally by officers on the ground or pieced together by headquarters, were hastily bundled together from stragglers, rear-area units and battle schools, the cavalry divisions that were the army's immediate reserve, and fresh reinforcements from depots and G.H.Q.'s reserve divisions, and pushed into the cracks in the line to man makeshift trenches or mount local counter-attacks. They would keep the front continuous and in time form nodes around which the defence could consolidate. But Fifth Army's line south of Péronne remained porous, allowing the enemy to penetrate between its isolated outposts, and obliging further withdrawal.[18] To that extent, infiltration tactics were effective and the possibility of holding the line of the Somme was rapidly disappearing. Instead, the high ground in front of Amiens, thirty miles behind Fifth Army's old front line, would have to be organised for defence.

A wave of displaced persons washed before the advancing German armies: refugees from the towns and villages of Picardy; stragglers from the shattered front line; gunners with or without their weapons; transport units and lines-of-communication troops from Fifth Army's rear services and bases. After the German retreat an attempt had been made to put the devastated region back into cultivation. France's relief and reconstruction agency, the *Service de Reconstruction des Régions Libérées,* had repopulated 125 out of 200 evacuated villages and 18,000 civilians had returned to prefabricated homes, herds had been put to pasture and 9000 hectares of land had been re-sown.[19] Now all that effort was undone. The local authorities tried to manage the evacuation of the civilian population, but chaos ensued.[20] M. Mairesse, the secretary of the Somme Industrial Society, was sent by the prefect to organise the evacuation of the villages south of Péronne. Civilian trains arrived late, if at all, although liaison with the British military authorities worked better and army lorries were provided to remove the civilian population. Mairesse's companion, M. Quellien, managed to commandeer eight French military lorries with drivers, who did good work evacuating the populace. They were recommended for the *Croix de Guerre*. Ham, Nesle, Roye and the villages west of the Somme were cleared over the next couple of days, under intensifying German shelling.[21] As the Germans neared Amiens the city authorities gave the order to evacuate the municipal government. Public order was already breaking down under steady aerial bombardment and the barrage of contradictory rumours that always sweeps before an advancing army. Policemen had to be despatched to the municipal air-raid shelters, where the town's pimps and prostitutes were continuing to ply their trade. German prisoner-labourers at least finally had reason to smile.[22] Eighty thousand Amiénois joined the throngs of displaced civilians seeking transport, food and shelter behind the retreating front.[23] Martial law was declared and the assistant provost marshal posted warning notices: "Any man found looting will be shot." The city's treasury and archives were shipped out, along with most of the councillors and their families. Fearing a repeat of the events of 1914, Mayor Duchaussoy fled briefly to Abbeville, but his conscience got the better of him and he returned on 30 March to assist the military authorities as best he could. The Bishop of Amiens had chosen to stay behind: "the *union sacrée* is essential when danger threatens," he assured Duchaussoy.[24] They did not know it, but by then the crisis was already coming to an end.

DESPITE THE RAPID response from G.Q.G. to Haig's call for assistance, and even though more French divisions had been sent north than initially promised, it was soon clear that the arrangements for mutual support were wholly inadequate. Furthermore, mutual mistrust and misunderstanding reappeared. When they met late in the evening of 24 March, Haig felt that Pétain was more concerned with protecting Paris than keeping in touch with the British, while Pétain thought that the British were retiring northwards, away from the French. Haig feared another attack in Flanders, which might break through to his army's communications, and remonstrated that it was Pétain's responsibility to find the troops to cover Amiens.[25] Haig was clearly losing confidence in his opposite number, writing two days later that Pétain "had the appearance of a Commander who was in a funk and had lost his nerve."[26]

However, crisis and panic had always been good for the Anglo-French alliance, and March 1918 was no exception. With the front fracturing, cooperation between Haig and Pétain breaking down, and British and French armies in danger of separation and defeat, a summit was hastily convened at the town hall in Doullens. Only one man could take the lead now: Foch was tasked with coordinating Anglo-French operations on the Western Front. Over the following weeks his responsibilities increased as the crisis continued, and by mid-April he was confirmed as the director of allied military operations on the Western Front, with American, Belgian and Italian troops as well as the British and French forces under his control.[27] The alliance finally had a generalissimo, a single commander without the dual responsibility for a particular national army.

"EVERYONE INTO THE battle," Foch often declaimed. In spring 1918 he got his wish. In March, 192 German divisions, including 50 that had crossed from Russia, lined up against 175 assorted allied divisions: 99 French, 58 British, 12 Belgian, 4 American and 2 Portuguese. More would come from Egypt, Italy, the United States and even Austria-Hungary. There would be no pause in the fighting until one or other of the adversaries had broken. Such was the disorder of the last week of March that Sir James Edmonds admitted defeat when trying to piece events together later: "only a broad outline will, as a rule, be given: the confusion which prevailed must be left to the imagination."[28] Between 24 and 31 March the towns and villages of Picardy fell back into German hands: Moislains, Bouchavesnes, Combles, Péronne, Bapaume, Noyon, Serre, Thiepval, Fricourt, Mametz and Chaulnes; then Maricourt, Albert, Bray, Lihons and Montdidier, which had not been German since 1914. For readers of the daily military communiqués, this mantra of familiar names touched an atavistic nerve. The ground for which so many lives had been sacrificed in 1916 was lost precipitately; to armchair observers, defeat seemed imminent. Yet in the field the old Somme battlefield was vigorously contested the other way in 1918. The South African Brigade sacrificed itself again on 24 March, overwhelmed on the cratered slopes between Bouchavesnes and Combles, just a few miles from the site of its 1916 heroics in Delville Wood.[29]

Yet out of chaos came order. The new generalissimo's priority was to keep the British and French armies together, and to reinforce the battle. "Fine. I will defend Amiens," Foch declared after meeting the operations manager of the *Compagnie du Nord* railway, who had explained the strategic importance of the city's railway junction.[30] For now, the best use had to be made of immediately available reserves. Back to the Somme came the reliable Australians, collected into an Australian Corps of five divisions, three of which were hurried south from Flanders to block the German advance on Amiens. Under Congreve's temporary command, they deployed astride the river Ancre to extend Third Army's right flank to the Somme, where they hoped to link up with Fifth Army's retreating left wing. The French civilians welcomed their slouch-hatted deliverers once more. As the 42nd AIF Battalion passed through Doullens in marching column it presented a welcome contrast to the disorganised procession of civilian refugees and Fifth Army stragglers heading the other way. "*Fini retreat,*" the Australians tried to reassure the distraught civilians in rudimentary French.[31] A white-haired curé raised his hands and blessed the passing column, "a touching and pathetic sight." The estaminet proprietor who handed over all his remaining stock of wine and beer before clearing out was perhaps more appreciated.[32] The empty towns and villages beyond Doullens showed signs of hasty evacuation. Food, drink and other goods were requisitioned by the advancing reserves. Better that the Australians rescue them than they fall into enemy hands: "it was not easy for the authorities to determine, in this as in some previous movements of the main Western Front, where reasonable 'salvage'—or use of abandoned property—ended and reprehensible looting began," Bean later admitted.[33] "We lived like lords in those early days," Lieutenant John Ridley of the 53rd AIF Battalion remembered.[34] The 42nd Battalion took up positions at Sailly-le-Sec in an old line of 1914 French trenches, made themselves comfortable with furnishings and wine from the empty houses in the village, and awaited the arrival of the enemy. Amiens now had a formed, resolved defence.

The countryside, which had not seen fighting since August 1914, was taking on the first flush of spring green, and livestock still grazed in the fields. As then, abandoned cows were a source of daily fresh milk, at least until orders were received to round them up and deliver them to the French authorities in the rear.[35] Briefly, "it gave the place more an aspect of peace than of war." But the enemy had been engaged by cavalry a mile to the Australians' front. As always on the tense eve of battle, the troops were restless. Men could not sleep, and their thoughts were far from their exposed trenches in France: "most of us talked of our chances of seeing Oz again that night. I know I would of [*sic*] given my right arm to have a look at Sydney that night," Private William

Long noted.[36] On 30 March the Australian positions were struck by a hurricane bombardment, behind which German infantry advanced. "The Hun now thought he would resume his victorious advance and capture Amiens, but he didn't succeed as his opponents were Australians, who repeatedly hurled him back as he attacked."[37] The close-quarter battle in front of Amiens was intense and bloody, with grenades and bayonets coming into play, but the Australians held their ground.[38] According to (the Australian) Private Schwinghammer, German tactics lacked the sophistication of 21 March: they "came in mass formation to the top of a hill in front of us and then spread out into single file shoulder to shoulder in a wave, rushing forward towards our trenches. When they came over the rise of the hill they were excellent targets for our machine and Lewis guns, which mowed them down like flies."[39] Three German assaults were shot down during the day: the line was bent back, but the new front held. Although Australian casualties were not light, thousands of German bodies reportedly littered the ground in front of their makeshift trenches.[40] Battle appeared to be resuming the characteristics of 1914.

South of the Somme the situation remained precarious. Here, where Fifth Army's shaky defensive line had had its exposed northern flank turned by German troops advancing north of the river (the Australians had not been able to push forward to cover it in the last days of March), and Rawlinson had not yet had time to reconstitute his Fourth Army, there was a way, if not an open road, to Amiens. A scratch force of cavalry and rear-area troops, including five hundred American engineers, had been concentrated to extend the Australians' line south of the river while 9th AIF Brigade deployed on the Villers-Bretonneux Ridge, the last high ground before Amiens. By the time that the leading German divisions had restored communications and brought up artillery for a large-scale attack against Amiens' last line of defence, a coherent, if thin, protective screen had formed south of the river. The remnants of two Fifth Army divisions were linked by the 35th AIF Battalion, and three more Australian battalions were in close reserve. These connected with the left wing of Debeney's First Army on the Amiens–Roye road. The gap had been plugged.

It had been a tense, trying ten days, but Foch had not panicked. He understood the dynamics of battle. Although the German penetration of the allied front had been deeper and wider than any previously seen in the West, the generalissimo had remained confident that it would be contained. The inevitable inertia of forward movement would slow the

German divisions as they lost men, became tired and disorganised and drew away from their railheads. If the central railway junction at Amiens could be protected, then allied divisions could be moved up more quickly to seal the breach. The front would eventually restabilise of its own accord.

Foch was quite right: by the end of March, the German offensive had spent its energy. Although the German infantry had pushed through into open country, their forward movement was rarely uncontested: the allied retreat was a dynamic rearguard fight, using delaying actions and local counter-attacks, not a rout. Now well beyond the range of their own supporting artillery, the advancing infantry found their progress slowing as allied divisions coalesced into a thickening military mass. The old Somme battlefield, in the path of the main German advance, came to the allies' aid. Previously abandoned trench lines could be hastily manned while new ones were dug further back. The infrastructure was poor, and the going still very muddy. The devastated zone (partly of the Germans' own making during their 1917 retreat) proved very difficult to cross, and the advancing infantry soon outran their supply lines and became short of ammunition and food: "the Somme desert had spoken its last inexorable and mighty word," one German account opined.[41] Furthermore, the Germans grew distracted as soon as they reached the allied rear areas. To soldiers who had been living for months on short rations, black bread and turnips, the cornucopia that awaited them in Fifth Army's positions, camps and supply dumps was irresistible. Indeed, it was proof that their propaganda had been lying to them about the impact of the U-boat blockade on allied resources. Inevitably, the soldiery started eating, then looting, then drinking when rum, beer and wine were discovered. Discipline started to collapse. If not perhaps an acute moral crisis as Edmonds later implied,[42] it was at least an indication that the German troops' mood remained volatile. Although by 26 March it had penetrated twenty-five miles, reaching Bray-sur-Somme and Chaulnes, over the next four days the German army advanced only ten more miles, less every day, until their check by the Australians and Fayolle's fresh divisions. The German tide washing against the allied front was abating, just as Foch had anticipated it would.

Foch's appointment produced a plan, of sorts: to re-form the front covering Amiens and Montdidier by engaging troops straight from their railheads, without artillery and supporting units. Fayolle, on whose world-weary shoulders responsibility would fall, did not like it of course—he wanted to gather a mass of fresh divisions and counter-attack—but it was a start. In fact, the hastily redeployed French divisions did not have the resources and firepower to check the advancing Germans immediately, let alone counter-attack, and they too joined the withdrawal in its later stages, during which Noyon and Roye fell to Eighteenth Army. Gaps in the allied line remained, but it was thickening as ever more units arrived. The enemy broke through to Montdidier on 28 March, causing momentary panic. Fayolle remained "calm, serene, confident."[43] It was Easter, and he had placed his trust in God. Three days later his armies were strong enough to start counter-attacking the German spearheads that had established themselves beyond the river Avre, a tributary of the Somme flowing north-west from Roye to Amiens.[44] For now, the advance had been contained. Foch's planning, Pétain's organisation and Fayolle's direction in the field together saved Amiens, and would go on to win the war for France.[45]

THE MOST INTENSIVE fighting moved north and east from Picardy between April and July, as Ludendorff sought to inflict a decisive defeat on Germany's enemies. But although dented again in Flanders and on the Aisne front, the allied line held. Breakthrough remained a military impossibility when each side had around two hundred divisions in the field, as the first ten days of Operation Michael (as Ludendorff dubbed the first stage of his offensive campaign) had demonstrated. Although his armies had penetrated thirty miles astride the river Somme, the British and French forces had not been separated: their line had been pushed back as far as any operation had managed since 1914, but it had never broken irreparably. The new defensive line had stabilised in front of Amiens: although Bapaume, Péronne and Albert had all been lost, the German advance had been fought to a standstill. The line hugged the heights west of the Ancre and Albert, crossed the Somme at Sailly-Laurette, passing in front of Villers-Bretonneux through Hangard, the new meeting-point of the British and French fronts, and down to the river Avre downstream from Moreuil. From there it meandered south past Sauvillers and Cantigny, curving eastwards round Montdidier, which was now in German hands, past Noyon to Barisis and the old front line. The offensive had created a huge new salient, and stretched the divisions holding it precariously thin on both sides of the line.

The loss of the territory that had cost so much effort and so many lives was inevitably a blow to the morale of the allied troops. Nearly sixty years later, Gunner Hugh Monteith, who had been through Pozières, the Somme winter and the German retreat, had not forgotten the sense of loss occasioned by the German success:

> But over these fields, and down the [Bapaume–Cambrai] road, came the triumphant Germans again in March 1918 . . . It was natural to ruminate on the great price paid on the 1st July 1916 and after . . . and to recall the losses suffered by the Australians in wresting that key-point, Pozières from the stubborn "Hun" . . . In one disconcerting week all our efforts on the Somme seemed to have been wasted. The geography of the disaster left no doubt that all the places with the old familiar names, that had burnt so deeply into the mind, had once more been overrun.[46]

"Despite this bad news we do not despair," Tristani, now in command of his battalion, declared. "Our men are in good form and excellent spirits." Nevertheless, as 32 RI's third battalion deployed to Picardy from an eight-month sojourn on the quiet Lorraine front, they knew that they were returning to real war; albeit a different sort of war. The battalion arrived in open country in front of Montdidier, "in the middle of the hole the Germans have punched between the British and the French," with no artillery support, but no hostile shelling to face either. "[F]ighting will be predominantly with rifle and grenade as in 1914," in isolated fox-holes and strongpoints rather than elaborate entrenched defences. But at least one aspect of the Somme was familiar. In early April the spring rains came: Cresmaux, where the battalion moved on 12 April, was a swamp of mud and filled with a chaotic traffic jam of guns, carts and horses.[47]

By April the allied armies had established a reasonably solid defensive line in front of Amiens, with artillery batteries and a proper logistics network to support it. After a three-day lull while they brought their artillery forward and relieved tired front-line divisions, on 4 April the Germans renewed their assault on the Anglo-French line either side of Hangard, hoping to push into Amiens from the south-east. Seventeen divisions engaged the thickening allied line on a twenty-two-mile front between the Somme and Montdidier, using the same tactics as on 21 March: a three-and-a-half-hour hurricane bombardment followed by an infantry assault led by storm-troops. The Anglo-French line wavered but held. It was raining heavily again and mud greatly hampered the advancing storm-troops. The attack made progress either side of Hangard, with XIX Corps being forced back on to the heights of Villers-Bretonneux plateau, and XXXI CA giving ground slowly on the western bank of the Avre, opposite Moreuil. Further south, the French front held. It was a difficult day, with the Germans gaining more than a mile, but eventually they were fought to a standstill. As their leading waves approached Villers-Bretonneux in late afternoon, an Anglo-Australian counter-attack swept them back to the eastern edge of the plateau and re-established the front. As the troops had feared, "the great offensive battle was threatening to become a battle of attrition on the largest scale" once more.[48] The next day the Germans did not renew their assault in strength, and First Army counter-attacked, but with limited success.[49] Hangard village changed hands twice between 9 and 12 April.[50] On the 16th the leaning Virgin finally fell from the top of the basilica in Albert. British gunners had brought it down for fear that the enemy were using it for observation.[51] For the time being, the battle for Amiens had blown itself out. Casualties on each side were in the region of 250,000 men after just sixteen days' fighting.[52] After committing ninety divisions to the thrust towards Amiens, Prince Rupprecht was forced to admit that "the final result . . . is that our offensive has come to a complete halt, and its continuation without careful preparations promises no success." Ludendorff was obliged to concur.[53]

FAYOLLE JUDGED FROM the recent fighting that military roles had been reversed. The British infantry, reputedly invincible on the defensive, had broken; but they were now much better on the offensive. On the other hand, the French, for all their supposed offensive élan, had defended well but were mediocre when attacking. The Germans had fought more to type: they were still the best organised and trained.[54] The time would soon come to test this theory, for the enemy were still within striking distance of the railhead, and Foch had issued a directive at the start of April that, as far as was possible, local offensive actions should take place to push the enemy away from the vital railway junctions south-east of Amiens.[55] It was a cardinal principle of Foch's defensive strategy that the allied armies should remain as active as possible, pinning down the enemy and improving allied positions with a view to a large-scale counter-offensive, which he always envisaged would be the proper response to the German attacks once these had been contained.

To that end, on 18 April First Army's XXXI CA attacked from the heights west of the river Avre, to clear the western slopes of the valley where the German army had established itself with its 4 April attack. Tristani's battalion was in the centre of 18 DI's attack front, with the difficult task of advancing between two thick woods, Bois Sénécat to the north and Bois du Gros Hêtre to the south. It was, he feared, almost a suicide mission. Yet the French army's methods had developed significantly since 1916. The infantry were to go forward in three waves behind a rolling barrage, while tanks would clear the edges of the woods. The men, well rested after their Lorraine vacation, and with strong artillery and tank support, appeared confident, even if their commander was uneasy: "perhaps for the first time they went to the fight without apprehension, sure of success."[56]

Although, as was only to be expected, the attack did not go like clockwork, the French army's new tactics combining infantry, artillery and tanks, if not yet second nature, proved effective.[57] Thirty-two RI's 3rd Battalion went over the top at 04:50 on 18 April. For an hour beforehand the artillery had swept the glacis of the German position to eliminate any machine-gun posts deployed forward to break up the attack. However, the tanks were late. Advancing blindly into heavy morning mist, the infantry came under machine-gun fire from Bois Sénécat, which forced the left-flank 10th Company to take cover. Then a tank finally arrived; but it started firing on the French infantry! The attack seemed to be collapsing into chaos until the company commander, Captain Velte, grabbed an entrenching tool and hammered on the side of the errant tank to attract its commander's attention. Once it turned its fire on the Germans in the wood the stalled attack gained momentum again, with more tanks clearing the enemy from the battalion's flanks. After beating off a counter-attack from Bois du Gros Hêtre, the right-hand 9th Company was forced to ground short of its objective. One by one the tanks were put out of action or broke down, while the rolling barrage moved off ahead. When further forward progress appeared unlikely, Tristani ordered his men to dig in, stuck forward on a spur in a salient between the two woods, and prepare to fight off the inevitable counter-attacks. By mid-afternoon the battalion on their left had finally secured Bois Sénécat, making their precarious position more secure, but Bois du Gros Hêtre remained in German hands. The afternoon's counter-attacks were repelled with relative ease. Despite the initial confusion and failure to secure the final objectives, the attack, typical of the small-scale local actions of these weeks, should be judged a success. Tristani's battalion had advanced 1200 metres and taken 100 prisoners and 15 machine guns, and inflicted around 200 casualties for the loss of 22 killed and 40 wounded.[58]

At this point the allies scored an important moral victory. Germany's most famous air ace, Baron von Richthofen, was brought down and killed near Amiens on 21 April; whether by a Canadian pilot, Captain Arthur Brown, or Australian ground fire has never been accurately established.[59] At a trying time, such an event had a significant impact on morale, both in the field and on the home front.

The Germans made their final drive against Amiens on 24 April, striking once again at the junction of the allied armies. Again a hurricane bombardment and thick fog facilitated early success. Supported by fourteen new A7V tanks, the assault quickly penetrated 8th Division's front north of Hangard; 131 DI were also driven out of that village, although attacks on the French front around Bois Sénécat were repulsed. The battle was notable for the first meeting of tanks on a battlefield. Inevitably, confusion reigned and accounts of the engagement vary. The British official historian records that three British Mark IV tanks engaged three German A7Vs. The British lost two machines before the German crews abandoned their vehicles.[60] The leading German elements passed through Villers-Bretonneux and into the woods on the ridge behind, from where they caught their first glimpse of Amiens since 1914. It was to be a brief mirage of victory. In the evening Rawlinson committed his reserves, two brigades of 5th Australian Division and one from 18th Division, to a counter-attack on Villers-Bretonneux and Hangard Wood to the south. The Australians attacked from the ridge north of Villers-Bretonneux, where their Western Front memorial now stands proud and assertive above the Amiens plain. It was another bloody close-quarter fight—the Australians encountered Prussian Guards again—but the town was back in allied hands by the next morning, although Hangard Wood could not be retaken.[61] Later the Australians recovered a ditched A7V, *Mephisto,* which survives on display in the Queensland Museum in Brisbane, a unique relic. The battle for Picardy, the "First 1918 Battles of the Somme" as the War Office christened it (for France, it was the "Second Battle of Picardy"), was over, and the Anglo-French armies had won.

However, although the Anglo-French forces had saved Amiens, they had not yet secured it. While out of range of most German artillery pieces, the city was still vulnerable to bombing and long-range heavy artillery. Like the famous "Paris gun," which bombarded the French capital during the summer, there was an "Amiens gun," a 280mm railway-mounted naval gun that shelled the city from a siding near Harbonnières, twenty-five kilometres away. The remaining civilians were evacuated towards the end of April,[62] and the ghost-city belatedly joined the long list of France's "martyred towns." Some seven hundred buildings were totally destroyed and another three thousand damaged. Notwithstanding the threat of the death penalty, many of the surviving buildings were looted too.[63]

The front protecting Amiens remained active throughout the spring and summer, even if there were not yet resources for a major offensive. Australian battalions kept up a policy of constant large-scale raiding of the German trenches, which represented the ultimate refinement of the tactics that had been maturing since 1916. This so-called peaceful penetration was anything but, as AIF battalions and brigades swallowed up the German defences in front of them sector by sector, strongpoint by strongpoint. These were artillery-intensive, small-scale offensives—in one at Morlancourt in June, seven thousand gunners supported an attack by two thousand infantry[64]—which culminated in the celebrated combined-arms *coup de main* at Hamel on 4 July. To mark America's national day, ten companies of U.S. troops made their debut on the Somme battlefield attached to the Australian assault battalions.

Fayolle's armies were not idle, either. First Army mounted similar operations on 3 and 9 May, easing the enemy a little further from Amiens. An American regiment, supported by French artillery, seized the village of Cantigny on 28 May. Meanwhile, Fayolle bided his time, planning a large offensive that would drive the enemy away from Amiens and liberate Montdidier and "give the public confidence . . . without using up too many troops."[65] On 23 July First Army retook the villages of Aubvillers, Sauvillers and Mailly-Raineval on the western Avre heights, along with sixteen hundred prisoners, in a successful attack.[66] At the point when the tide of war was turning in the allies' favour, the scale of Fayolle's operations increased.

THE WAR HAD entered its most intensive, climactic phase. The tactics developed in previous years changed the pace of military operations; but although the stalemate seemed to be broken, that impression was deceptive. The line would never be ruptured, but would flex backwards and forward across northern France and Belgium in a succession of diminishing waves, in Foch's favoured analogy.[67] *Matériel*-intensive offensive tactics could smash in a sector of the front—neither side had any difficulty doing that—but as Ludendorff's succession of attacks in the spring and early summer demonstrated, without a guiding operational rationale such blows would not produce a strategic victory. Ludendorff brashly claimed that "strategic victory follows tactical success"; it was enough to "make a hole, and the rest would take care of itself."[68] He was wrong. Especially as he was now opposed by an operational genius. Foch's theory of industrial battle had been maturing over four years, shaped by his experiences in 1915 and 1916 and codified during his involuntary sabbatical in early 1917. His *bataille générale* combined four key operational principles: tempo, attrition, manoeuvre and *matériel* superiority. Lloyd George recollected watching Foch rain punches and kicks on the British Foreign Secretary, Arthur Balfour, in the gardens at Versailles one summer afternoon in 1918.[69] He was not witnessing the latest outbreak of Anglo-French discord, but a typically energetic demonstration of how the recently appointed allied generalissimo intended to beat the German army. A "series of hammer strokes designed to smash up the German Army" would knock the enemy off-balance and pummel him into submission by relentless, rapid, intensive combat.[70] The allies' ability to destroy German divisions in the field faster than they could be reconstituted, to combine their moral exhaustion with their *matériel* degradation, would bring them victory.

Foch had formulated these methods after the Battle of the Somme, and it was to be in Picardy that the British and French armies—with Rawlinson and Fayolle side-by-side once more—would begin the process of destroying the German army in the victorious "Hundred Days" that crowned the allies' successful attritional campaign on the Western Front. In this final phase the intervals that had allowed Below and Gallwitz to reinforce and repair their shaky defensive front in 1916 would not be allowed. To this end, battle was to be on as large a scale as possible, conducted along the whole Western Front as allied armies engaged in succession. "Although we wished to attack at the outset at one point only," Foch later explained,

> our successive enterprises ought to be prepared as a part of one
> and the same series, so that each of them could, without delay,
> take advantage of the moral ascendancy gained by the previous

one and the disorder brought about in the enemy's dispositions. Also the direction to be given to each must be fixed in such a manner as to arrive finally at one single end.[71]

Thus when the progress of one army was slowed by strengthening German resistance, the epicentre of battle would quickly shift to its flank. While individual offensives would aim to make quick deep advances, no army would push too far beyond the range of its own artillery cover or lines of supply and risk being counter-attacked and driven back. Instead, it would pause, consolidate its gains and move its guns and communications forward while other armies came up on its flanks. Only then did the process resume. Foch had abandoned the idea of breakthrough since he had discovered the principle of "push-back": on the small scale in 1916 on the Somme, and on a front-long scale in 1918. Exploitation of success would take place laterally, not forward, avoiding pockets in which allied armies could be contained and counter-attacked. Foch would soon demonstrate the vulnerability of such pockets to his adversaries.

Away from the Somme the battle remained active throughout the summer, as Hindenburg's armies struck further blows against the front of the French army in May and June. Although individually these were powerful, dangerous attacks that were hard to contain, the enemy was playing into Foch's hands. Foch knew that these were the thrashing blows of a wounded animal: a strong beast, but one that was rapidly losing blood and tiring. Sustained offensive placed the German army in an increasingly difficult situation by wearing out its already stretched divisions and extending its line to such an extent that it could no longer be held in sufficient strength.[72] Furthermore, the German army was effectively abandoning strong prepared defences and moving into a number of vulnerable salients. This was just what Foch needed to put his plan into practice. It had always been his intention to counter-attack when the time was right, "since it was only by offensive action that they could bring the War to a victorious conclusion."[73] That time came in July, after the succession of German offensives had spent themselves along the Western Front and the balance of divisions was swinging back in the allies' favour, with 206 German divisions of very mixed ability facing 195 allied, including 18 fresh, double-sized American divisions.[74]

At the end of May another German push had made a large bulge in the allied front beyond the Aisne, recrossing the Chemin des Dames and reaching the river Marne once more, before French and American reserves halted their progress at Château-Thierry. From here Ludendorff intended to thrust forward yet again: Foch planned to meet this renewed offensive by striking at the Germans' railway communication through Soissons.[75] On 15 July, after another hurricane bombardment, three German armies struck the French Sixth, Fifth and Fourth armies on a ninety-kilometre front from the apex of the Marne salient at Château-Thierry to Champagne. This attack, Ludendorff proclaimed prematurely to the anxious people of Germany, was the final "Peace Offensive."[76] Fourth Army held east of Reims, but Fifth Army's front around Dormans was broken, and the Germans established a bridgehead across the Marne. However, as more German divisions pushed south-east towards Épernay, Foch judged that it was finally time to launch his carefully prepared counter-attack. Timid Pétain demurred, but Foch overruled him.[77] Reserve French and American divisions massed around the German salient, heavily reinforced with guns, tanks and aeroplanes, struck back on 18 July. Fayolle's army group struck from the west, with Charles Magnin (whom Clemenceau had recently rehabilitated following his key role in Nivelle's failed offensive) commanding the main assault by Tenth Army that smashed into the right flank of the German salient. By the end of the day the railway lines through Soissons were in range of the French artillery. The other armies around the salient struck in turn and tightened the noose. The Germans were forced back across the Marne, and by the beginning of August their divisions were in disorderly retreat back to the river Vesle. The turning point of the battle had arrived. At the same time, the recapture of the Marne "pocket" by "French" armies that included British, Italian and American divisions betokened a much more cooperative sort of warfare. With this counter-stroke on the Marne, Foch regained the initiative for the allies. He intended to exploit it without pause, to effect the "definite destruction" of the enemy.[78]

The first counter-attack was losing its momentum by the beginning of August, but Foch had anticipated this. It was vital that the enemy had no time to recover his breath after each battle, so preparations were already under way for a second operation, the "Battle of Picardy and the Santerre,"[79] which would transfer the focus of the fighting elsewhere. French military intelligence had established that after its exertions in the spring the German army was in the grip of a manpower crisis, and Ludendorff would have growing difficulty keeping his already battered units up to strength. The Second Bureau calculated that the enemy still had 67 divisions in reserve on the Western Front at the beginning of August, 39 of them fresh or re-formed after previous battles (the allies had 77, 41 of them fresh): the Marne counter-offensive had engaged 75 enemy divisions (with 59 allied ones), and 13 fresh German reserve divisions had been used up.[80] The allies had regained numerical superiority—and American troops were coming across the Atlantic at the rate of 250,000 per month—but the remaining German reserve divisions would still have to be brought to battle and destroyed if the war were to end. As well as their *matériel* advantage, the allies had regained moral ascendancy, which Foch was determined to exploit to the full.[81] After their successes of the spring and early summer the German soldiery had been gripped by a renewed sense of hope and purpose. After the Marne that disappeared. Meanwhile, allied troops, backed by their masses of guns, aeroplanes, tanks and other war *matériel*, advanced with restored confidence.[82] Victory, peace, finally seemed to be within reach.

The German morale crisis had spread to the home front. While the high command could fill the pages of the press with victories there remained some hope, even if workers and pacifists were becoming more vociferous. But it was impossible to disguise the reversal of fortune at the front after the middle of July, and the brash promises that accompanied the opening of Ludendorff's Peace Offensive evaporated. Allied intelligence monitored the deteriorating home-front situation in Germany closely. Prisoners taken on the Marne were war-weary and unhappy with domestic political developments: a recent bill for the introduction of universal suffrage had been rejected by the conservative-controlled Reichstag. The impact of the blockade on the food supply and home-front living standards was being exacerbated by an influenza epidemic that swept across Europe in the summer of 1918, and even the army was being required to tighten its belt because of food and raw-material shortages.[83] To try to stop the rot, on 4 August Ludendorff issued a typically confident, but clearly specious, message to his troops: "we can await any enemy attack with the greatest confidence. As I have already shown, we should welcome the enemy's offensive, which will only speed the attrition of his forces."[84] He would soon have to swallow his bombastic but empty martial rhetoric. His order of the day was an acknowledgement that the Peace Offensive had failed, and his army was once more on the defensive. The troops knew very well what that entailed. The whirlwind allied counter-attack that ensued, however, went beyond the German command's wildest expectations. On 8 August the morale crisis finally reached the top. The

Amiens–Montdidier offensive would go down in history, in Ludendorff's oft-quoted words, as "the black day of the German army." The First Quartermaster-General was plunged into a state of nervous collapse. Even he had to acknowledge that Germany was now losing the war.

On 30 July Weygand had dined with Fayolle, and discussed the coming offensive in Picardy: "It is an excellent project, which will certainly produce good results."[85] The plan for the Battle of Amiens–Montdidier had been gestating over the summer. If not entirely new in its conception—it owed a lot to the Battle of Cambrai, the Marne counter-offensive and even the tactics employed by the German army in its spring offensive—it was certainly bold. One cheeky Australian soldier at least was struck by its audacity (and the chance for loot): "I wondered, and found the others were of the same mind, who had stuck a pin in some of the Higher Command and taught them the proper way to run a war," Private Marshall noted. "At last we were to have our long held wish. One could not help feeling jubilant at the prospect of at last getting Fritz on the run, and the land of souvenirs likely to be obtained."[86]

Foch circulated a memorandum to the national army commanders, which set out his ideas for the offensive battle. The "turning point of the road" had been reached, he announced confidently, even if it was not yet time to think of "seeking a decision" in 1918. The next series of attacks would for now improve the allies' position on the ground and sustain their newly won initiative, through successive surprise blows limited in extent but rapidly repeated "so as to embarrass the enemy in the utilisation of his reserves and not allow him sufficient time to fill up his units."[87] The specific objective of the Picardy offensive, an operation Haig had mooted on 17 July, was finally to disengage Amiens and the Amiens–Paris railway line, which was within range of German artillery.[88] Fourth Army was to attack astride the Somme, between the rivers Ancre and Luce, with Fayolle's GAN extending the battle southwards, engaging its armies in succession. Rawlinson, perhaps with his 1916 experience in mind, had strongly objected to a joint offensive. However, after his success on the Marne, Foch insisted that the operation be an Anglo-French one, with deep objectives, rather than a limited Fourth Army battle to disengage Amiens.[89] The allied armies mustered forty-three infantry and six cavalry divisions against thirty weakened German divisions.[90]

The main attack would be launched by the Australian and Canadian corps (the latter were secretly railed southwards to surprise the enemy). The northern flank would be covered by Major-General Butler's III Corps, which would seize the high ground of the Chipilly Ridge north of the Somme. Rawlinson had seventeen infantry and three cavalry divisions at his disposal: eight would make the first assault, five (including the American 33rd Division) and the cavalry would follow up, and four were in reserve. Rawlinson's plan was for a combined-arms offensive, with over 2000 guns, 534 heavy, medium and supply tanks and 800 aeroplanes available to support the infantry advance. The objectives were ambitious. But with a sudden hurricane bombardment to break the crust of the German front, two waves of infantry divisions leapfrogging through each other, and mixed columns of cavalry, cyclists and field artillery supported by fast "Whippet" medium tanks and ground-attack aircraft to exploit into the German rear areas, the plan was well within the capacity of Rawlinson's skilled, well-equipped command.[91] Such tactics were more a prevision of the armoured warfare of 1940 than a throwback to the plodding, methodical offensive style of 1916. The British army had come a long way in two years.

Debeney's First Army had four corps in line, with ten divisions in the first wave, three in support, and two infantry and three cavalry divisions in reserve. It was faced by eleven divisions of the German Second and Eighteenth armies. Debeney's men were to engage successively on 8 and 9 August, extending the battle southwards beyond Montdidier. XXXI CA, on Rawlinson's right, which included 42, 153 and 66 DIs, three "shock" divisions that had fought on the Somme in 1916, was to attack at the same time as the British, and in a sensible departure from earlier practice it was put under British command for the first stage: Foch was slowly creating an "allied" army. On 10 August the right wing of Humbert's Third Army, or if the attack was going well his whole army, would join in, extending the front of attack as far as Compiègne. The French attack was to be supported by 1624 light and heavy guns, over 1000 aircraft, and in XXXI CA's sector a regiment of self-propelled artillery and 90 of the new Renault FT17 fast, light tanks, which would swarm into the enemy's rear areas as they had done on the Marne.[92]

By then, Fayolle had lost his loyal chief of staff, Duval, who had been put in charge of France's new bombing force. This promotion was testimony to his understanding of the principles of modern *matériel*-intensive warfare: "Duval was one of the best intellects of the war, because he understood its new character . . . He knew how to industri-

alise war, to replace men with machines."[93] The Battle of Amiens–
Montdidier stands out as the prime example of this new sort of warfare
on the Western Front. The operation was not perfect. Anglo-French
liaison was better than it had been in the past, but the pre-battle meeting
between Major-General Lipsitt of 3rd Canadian Division and General
Deville of 42 DI, where neither had a word of the other's language and
they were reduced to pointing at a map and miming, was farcical. Lipsitt said he had no tanks to spare to assist the French, but at least, it was
later suggested, the meeting was conducted in a spirit of friendship with
a desire to resolve their shared problems. (Four tanks were later found
to help the French mop up enemy machine-gun nests.)[94]

On 7 August 1918 Foch was made a marshal of France. The next day
the systematic destruction of the German army began. It had been a
long, restless night, as all nights were before a big attack. It was also
uncannily quiet: the guns, which usually thundered before a big attack,
were silent. The men in the assault trenches were forbidden from talking or smoking as they carried out their final pre-battle routines. The
only sounds that could be heard were the officers' whispers as they
passed along the line checking that all were ready and in position, and
the snoring of those few apparently nerveless souls who managed to
snatch some sleep. On 42 DI's front at 04:00 a barrage of flares shot up,
and a machine gun raked no man's land, raising momentary fears that
the enemy were waiting for them.[95] Then the unnerving silence, so different from the cacophony of 1 July 1916, fell again.

The quiet was shattered at 04:20 as Fourth and First armies' massed
guns opened up in unison, flashing like sheet-lightning along the Germans' misty horizon. A torrential metal rain saturated the German
defences. Fourth Army's assault troops went forward immediately. "It
was a wonderful sight," Lieutenant Ridley, waiting in reserve, noted:
"some guns roaring away, others being rushed forward, horses and
wagons going up, aeroplanes flying over us and men streaming on."[96]
Forty-two DI's assault battalions—8 BCP and one battalion of the "Bar
le Duc Guards," 94 RI—seized Marburg Trench, the enemy's front line,
at the same moment, from where they would jump off when the time
came: "Perhaps such audacity was unprecedented, but it was logical and
appropriate given the division's morale and the enemy's timidity."[97] It
was also well within the capacity of France's well-led and experienced
infantry. They marched steadily behind the rolling barrage in the now-
customary stance of infantrymen going into the metal whirlwind: heads
bowed, shoulders hunched, teeth gritted. Light machine-gunners led

the way, ready to engage the enemy should the barrage miss any defensive position. They covered the six hundred metres to Marburg Trench—in reality a line of shattered fox-holes—in eighteen minutes, meeting only a few shell-shocked stragglers on the way. The only casualties were the result of soldiers following the French barrage too closely. Behind them followed the mass of 42 DI, six battalions ranged three-deep, with 153 DI and the tanks behind them.[98] At 05:05 XXXI CA's right assault division, 37 DI, left its jumping-off trenches, and 42 DI resumed its advance eighteen minutes later, right on time, when the barrage moved off again. Together they were to smash in the northern flank of the Moreuil salient. Any resistance encountered—and it was invariably weak and isolated—was quickly subdued. But most Germans in the forward defences were surprised, shell-shocked and only too eager to be taken prisoner. As one astonished German captain asked of his captors in fluent French: "Are there many of you?" "There are two divisions behind us." "*Alors, nous sommes foutus!*"[99] The defence had failed completely. The *chasseurs* paused to drink the hot coffee that they found in abandoned German field kitchens. The exuberant men sang and danced with joy until called to order by their officers.[100] Their only worry was that the Canadians did not seem to be advancing as fast on their exposed left flank.[101] The advance continued and expanded as the hot, sunny, happy day rolled on. At 05:05 66 DI had assaulted Moreuil, which fell within the hour. At 08:20 IX CA on XXXI CA's right attacked on the Avre sector; by 11:30 it was across the river.[102] During the afternoon, 42 DI's reserve battalions and 153 DI pushed through with the Renault FT17s across the Santerre plateau. The French army was on the move again.

As so often before it was the German Second Army, now commanded by General von der Marwitz, which was to receive Fourth Army's assault. Any survivors from 1 July 1916 would have been shocked by the ease with which the Australians and Canadians swept through their lines, quickly and with deadly effect. Partly it was the element of surprise, partly their modern tactics and partly their overwhelming *matériel* superiority that gave the British forces such a clear victory. The German defences were not the deep, wired positions of 1916, and British infiltration tactics were now quite up to isolating and overwhelming any points of resistance that had been missed by the artillery or the advancing tanks.

At the same moment as the British bombardment struck the German front position astride the Somme, the infantry and tanks moved off, two hundred yards behind the advancing barrage, covered once more by that helpful morning mist that filled the river valleys. The leading waves swept inexorably on to their first objective, while supporting waves mopped up any pockets of resistance encountered in the enemy's shattered lines (more often than not shell-shocked men emerging from dugouts with their hands up). The first objective was taken within three hours, and the supporting divisions then passed through to push on to the second, which was reached in the Australian sector in the centre by 10:30. Enemy resistance seemed to have melted away beyond the front position, and there was hardly any retaliatory artillery fire. Isolated machine guns were dealt with by tanks or the now second-nature tactic of pinning them with Lewis-gun fire before attacking them from a flank with grenades. Cavalry and armoured cars then pushed forward to the final objective, a line of old French trenches known as the "outer Amiens defence line." Again they met little resistance, and were able to engage many targets of opportunity in the panic-stricken German rear areas. It had taken a long time, but cavalry exploitation on the Western Front was finally proving possible.[103]

The Australians had the easiest and most rewarding time: Private Marshall's mate "Twig" found a box of new Iron Crosses in one abandoned German headquarters. But the victory at Amiens offered far greater reward: "Conversation with various prisoners, the success of the attack and the result convinced me that the war was indeed on the last lap," wrote Marshall.[104] Even the humblest foot soldier could see that a new vigour had been injected into the war: "With the French preference of making the enemy fight rather than living to fight him another day, a victory was assured."[105]

The Canadian Corps on the right met stronger opposition, but it still secured its second objective by 12:30. On the extreme right, alongside XXXI CA, the attack was initially held up by defences in Rifle Wood, but progress accelerated in the late morning and afternoon. By nightfall on 8 August, the Canadians had pushed forward nearly eight miles south of the Somme, the largest single-day advance of the war to date. XXXI CA had progressed five miles on their right flank. The Australians had made six miles. In all, nearly sixteen thousand prisoners and three hundred guns were captured. Nine German divisions were "wiped out," suffering over twenty-seven thousand casualties compared with the British eight thousand.[106] But despite the congratulations that were passed around, the sense of achievement that animated both headquarters and the troops on the ground, the battle had not been a complete success. On the northern flank III Corps's attack on the Chipilly Ridge had been badly mauled by its defenders, and German positions on the spur had inflicted heavy casualties on the Australians' exposed flank south of the river. Only the deep penetration by the Australians to the south allowed III Corps, reinforced by the American 131st Regiment, to complete the occupation of the ridge on 9 August.[107] On that day Private Long was ordered to go forward with other riflemen in one of the tanks—in effect a primitive form of armoured personnel carrier—so that these would have close infantry support as they advanced. He thought it was a suicide mission, but after they jumped from the tank and cleared a German machine-gun post he realised the point of the exercise. The machine-gun lines that had previously delayed progress could now be assaulted effectively, and the advance became a stroll after the enemy's gun line was overwhelmed: "We had no trouble and very little fighting. Now and again we had to scrap with a few but as soon as we worked around them they quickly gave themselves up. We had our rifles slung on our shoulders half the time, and smoking as we walked along."[108] But like all forward moves, it soon slowed—three miles on 9 August, two on the 10th and barely one on the eleventh—as the advancing troops tired and outpaced their supply lines, and the enemy fell back on theirs. The Canadians reached Lihons and Chilly, the old French front line of 1916, but by the eleventh they were encountering firmer resistance as reserve German divisions deployed. Rawlinson knew it was time to stop the battle, rest his divisions and bring forward his heavy artillery for the next set-piece operation.

Within the parameters of the operational method of 1918, Fourth and First armies' offensive was merely the beginning. Now that the "general battle" had been engaged, more armies must attack—the British Third Army to the north, the French Third and Tenth armies to the south—extending the battle laterally and rapidly: the adversary had been seized and would not be released before he was beaten into submission. Foch urged on his generals: "go quickly, straight ahead, manoeuvre, push from behind with all you have until you obtain a decision" was the typically animated instruction to Debeney.[109] Fayolle, "the artisan of [Foch and Pétain's] victories," as usual felt that he would not receive any credit for the initial success. "Patience, patience," he counselled as Foch urged him on into Montdidier and Roye.[110] Ever calm, methodical, thoughtful, Fayolle knew very well how to conduct a battle, how to execute his superior's frenetic commands in the field. As

British progress slowed to the north, First and Third armies expanded the offensive southwards. XXXV CA attacked on 9 August, outflanking Montdidier to the south, and Third Army went forward at dawn the next day, with a swarm of aircraft bombing in front of it, disrupting German communications and breaking up concentrations of reserve troops.[111] Montdidier fell on the morning of 10 August—the town was deserted and in ruins—as Fayolle's armies progressed another eight kilometres. By the time the battle ended, they were up with the British and moving on Roye. The German army had been pushed back twelve miles on a forty-mile front, with British and French troops following up closely and aggressively. But it chose to make a stand on familiar ground, the old Somme front line. A new German army group, commanded by General von Boehn, was created astride the Somme to hold the vital centre section of the German line.

Amiens–Montdidier was the Anglo-French offensive *par excellence*. "Foch wanted an Anglo-French battle because it would be good for him," Fayolle griped, "while I remain in the shadows once again . . . Debeney is under Haig's orders, but it was me who organised the attack east of Montdidier."[112] Although his armies had taken ten thousand prisoners and three hundred guns, even in victory Fayolle's innate tendency to grouse got the better of him. Yet Foch deserved great credit for driving his armies to victory, the most momentous since 1914. In four days his Anglo-French forces had retaken more ground than they had managed on the Somme in four and a half months of 1916, and even the most hostile politicians in London and Paris could not fail to be impressed, if never satisfied. When it came to writing his memoirs, although acknowledging that "the effect of the victory was moral and not territorial," Lloyd George nevertheless, with customary rancour and lack of military sense, could not resist criticising Haig for not throwing his reserves "into the gap" and driving towards a territorial objective, the Somme—exactly the sort of manoeuvre he castigated him for contemplating in 1916![113] The changed nature of warfare seems to have escaped him. The "Battle of the Santerre," as Foch dubbed it, was a "push-back," not the fabled breakthrough. Even if Ludendorff later suggested that the Somme was momentarily open for the taking on 8 August,[114] and if Haig and Foch had started bandying about the Somme crossings, Péronne and Bapaume as ultimate objectives after the stunning initial success[115]—two years later they were still imprinted in their minds—headlong forward rushes were not part of the method of 1918. Moral victory, as ever, was all important, and that could not have been won without the impact of 1916's fighting, as Edmonds acknowl-edged: "After 1916, the year of Verdun and the Somme, the military value of the German officers and men had fallen faster than that of the French and British, whose natural fighting abilities now turned the scale."[116] As well as this moral ascendancy, the 1916 battle had given Foch ideas and experience, as well as the means to operate on the indus-trial battlefield and experienced subordinates who were able to do so: Haig was now "fulfilling a role for which he was admirably adapted: that of second in command to a strategist of unchallenged genius," Lloyd George later admitted, a position in which he "earned high credit."[117] This was the lineage of the Battle of Amiens–Montdidier, and of final victory in the field. XXXI CA's triumphal stele, erected on the ridge above Moreuil, identifies it as the point where the enemy's line was breached and its drive to Haudroy on the Belgian border (where German peace plenipotentiaries crossed through its lines on 7 Novem-ber) began. Now the method was perfected, reconquest of territory would follow inexorably and need not be hurried. All that remained was to complete that victory.

NOW THE WAR assumed a different aspect. "No more topographical objectives, just directions. No more crossing lines according to a timetable, but reinforcements only when needed or even leapfrogging when on the march. It is time to shake off the mud of the trenches," Mangin summarised Fayolle's more detailed instructions as his army entered the battle.[118] French intelligence assessments indicated that the German army was weakening. Although it still nominally had seventy-two divisions in reserve, fifty-seven of them fresh or re-formed, each infantry battalion was being reduced by a company, and assault battal-ions were being broken up to find reserves for the "storm-troop" divi-sions. By August, the German army had engaged the equivalent of 485 divisions in the battle, some up to five or six times. The Second Bureau was also now assessing the morale of the German reserves: very good (18 Divisions); good (26 Divisions); fair (24 Divisions); average (4 Divisions).[119] Over the ensuing weeks and months, G.Q.G.'s accountants would tick off the enemy's forces as they were used up.

"A proposed general advance, by Foch's orders, from the Channel to Switzerland" was about to get under way.[120] This meant re-crossing the old Somme battlefield. Between 21 August, when Fourth and Third armies began the Battle of Albert, and 3 September, when the Battle of

Bapaume finished, all the old Somme sector, north and south of the river, was to be cleared. Fayolle's army group was already on the move. Mangin's Tenth Army on Humbert's right had struck between the Oise and Aisne rivers on 18 August, aiming for Noyon; his Third and First armies resumed their advance sequentially the following week. In the British sector the Battle of Albert began on 21 August, when Third Army attacked the old front-line positions north of Thiepval. As was becoming customary, corps and armies engaged successively, extending the battle along the line as section after defensive section fell in, over-whelmed or outflanked. Fourth Army's left wing attacked south of Thiepval on 22 August, re-taking Albert and Meaulté and reaching Bray-sur-Somme the next day, after which its right attacked south of the river, aiming for Péronne. The French First and Third armies, and General Horne's First Army further north, would soon join in. Six British and French armies were now on the march from the Scarpe to the Vesle.[121] By 27 August, the enemy was retiring back to the line of the Somme. The French First Army reached the river, and Third Army liberated Noyon, two days later. Once their defensive front astride the Somme had been breached the enemy chose to conduct a fighting retreat to their next line of defence—dubbed the "Winter Line" because Ludendorff hoped to dig in and hold here until 1919—extending from Bapaume down to Péronne and then along the formidable natural obstacle presented by the marshy river valley. German resistance, along familiar ridges and trench lines, was violent, and the going was difficult: "We are on the old 1916 battlefield and it is impossible to find solid ground. Everywhere is broken by shell fire," Private Long noted.[122] The vital breakthrough came opposite Bapaume on 25 August. Warlen-court, Contalmaison, Mametz and other points of bitter memory were swept up. Whether their earlier shared trial on the same battlefield came to mind at that moment is not recorded, but on that date Foch sent Haig a warm message of congratulations on his vigorous and successful offensive.[123] It was much easier to be magnanimous when operations were going well.

But when the Australians reached the river line at Péronne, backed by the formidable defences of Mont St. Quentin behind, it seemed that the battle would halt again: there would have to be a pause before the Winter Line could be assaulted in strength. This was not the Aus-tralians' way, and the methods of 1918 were far more fluid than those of 1916. Second and 5th Australian divisions were advancing into the Somme bend, and on 29 August they reached the same blocked position on the Flaucourt plateau where 2 DIC had found itself in July 1916. Lieutenant-General John Monash, commanding the Australian Corps, proposed a flanking manoeuvre to cross the river and take Péronne from the north. Rawlinson, after some hesitation, agreed. It was the very manoeuvre across the Somme that Thierry had suggested in July 1916 and Fayolle had rejected. On 30 and 31 August, Australian brigades crossed the surviving or rebuilt bridges at Feuillères and Ommiécourt. On the thirtieth, the 5th Australian Brigade cleared the enemy from Cléry and established itself within assaulting distance of Mont St. Quentin. The Australians' sudden attack early the next morning took the Germans on the heavily fortified slopes of the hill completely by surprise, and the Australians were able to establish a foothold that was held against German counter-attacks until Mont St. Quentin village was successfully assaulted by 6th Australian Brigade on 1 September. Other brigades converged on Péronne the same day. By the evening, Mont St. Quentin and Péronne were in Australian hands.[124] What Péronne's citizens who returned to reclaim their broken homes thought of the re-named "Roo de Kanga" that now ran through their town is not recorded.[125]

The capture of Mont St. Quentin was one of the most audacious martial feats in four years of warfare on the Western Front, a swift flanking manoeuvre very different from the slow, slogging attrition of popular memory. It was carried out by well-led and skilled veteran troops with very high *esprit de corps* against the best troops—including 2 Guard ID, ordered to hold Mont St. Quentin at all costs[126]—that the German army was able to muster at this late stage of the war. Ultimately, the outcome was determined by *sang-froid* and cold steel: Australians won seven Victoria Crosses on 1 September.[127] "Fifty-fourth v. Fritz. Mob rules" was Marshall's laconic summary of his battalion's storming of Péronne's ramparts.[128] But it was another costly victory. Only fifteen men from Lieutenant McInnis's company were unwounded at the end of the fight.[129] Yet, although the Australians were heavily outnumbered, it was a one-sided contest. "Our brigade met the Prussian guards yesterday and beat them bad," Long noted proudly. "One big Prussian Guard . . . told [me] that their regiment had volunteered to stop the Australian advance and two hours after they arrived on the field they were all killed or captured staff and all . . . they seem too frightened. They think the war is about finished now."[130] Perhaps they remembered their previous encounter at Lagnicourt; more likely, by now, even their best regiments had been eviscerated by attrition.

Signs that the German army was crumbling in the face of unrelenting allied pressure were multiplying. McInnis returned from hospital in England to rejoin his unit at Bray through a landscape marked by the recent battle. What struck him most was the huge amount of equipment and munitions abandoned by the Germans in their hasty flight. In a copse near Chuignes he inspected the gun used by the Germans to bombard Amiens, which some wit had chalked with the injunction, "fragile—handle with care."[131] It was the biggest souvenir that the Australians recovered from the Somme battlefield: it now stands outside the Australian War Memorial in Canberra. The Second Bureau's count of fresh German reserve divisions was down to forty-four. The fighting between 8 and 29 August had engaged 67 German divisions against 61 allied, and there were still 71 allied divisions in reserve.[132] The balance of attrition was working in the allies' favour, as Haig recognised in the revised instruction he issued to his army commanders on 22 August:

> The methods which we have followed, hitherto, in our battles with limited objectives when the enemy was strong, are no longer suited to his present condition.
>
> The enemy has not the means to deliver counter-attacks on an extended scale, nor has he the numbers to hold a position against the very extended advance which is now being directed upon him.

The long-awaited exploitation phase of battle had arrived.[133] Haig's methods, and the sacrifice of the last three years, finally seemed vindicated.

BY SEPTEMBER 1918, THOUGH, both sides were punch-drunk and exhausted, with infantry units well below strength. The Australian brigades that had assaulted Mont St. Quentin were down to around one-third of their established strength.[134] The German divisions that faced them had been reinforced by old men and eighteen-year-olds. In such circumstances morale was crucial. There were rumours that the 54th AIF Battalion, the stormers of Péronne, was to be broken up. If it did not amount to a genuine mutiny, its soldiers' collective protest was enough to cause the corps commander to reverse his decision.[135] Under-strength units with strong *esprit de corps* were much more valuable than big battalions with no collective identity now that the decisive moment of the war was approaching. After a "very impressive memorial service at Quinconce Cemetery, with the unveiling of a beautiful cross in memory of our men who fell there," the 54th Battalion prepared for its final fight.[136] Conversely, the German army was gripped by a morale crisis, with desertion rates rising, shirkers avoiding the front, and huge numbers of men surrendering to the advancing allied forces.[137] It was not surprising. With its Winter Line breached at its vital point—while the Australian Corps had been turning the river line, III Corps had been seizing the defences beyond the Bapaume–Péronne road between Bouchavesnes and Bois St. Pierre Vaast—there remained no other option but to conduct another withdrawal to the Hindenburg Line, from which the offensive had started back in March. All the German soldiers now knew that their efforts and sacrifices of the last six months had amounted to nothing.

Nevertheless, the allied advance through Picardy was not a particularly happy procession. The retiring German army, as was its wont, left destruction in its wake, where allied artillery had not already done its job for it. More villages were burned and larger houses and churches were mined with delayed-action explosives. But France was finally being liberated, and more rapidly than anyone would have dared to predict in the spring. Roye was retaken on 27 August—by the *chasseurs* of 47 DI, directed by Brigadier-General Mangin, who had led 79 RI against Bois Y on 1 July 1916[138]—Combles, Barleux and Noyon two days later, and Bapaume once more on 29 August.

The tide of armed humanity sweeping north-eastwards washed up against the defences of the Hindenburg Line in mid-September. There it might have been expected to stop for another winter, but the armies pressing against it were not what they had been two years earlier, and nor were the men ordered to hold it. Attrition had rendered the German army of late 1918 a hollow force. It still had many divisions (although ten had already been broken up) but their battalions and companies were greatly under-strength. The Australian Private Schwinghammer recorded the pursuit:

> The Germans were now well on the run. They left machine gunners, in strong positions, to give us trouble as their infantry retires. At daylight we started to advance without any barrage from our artillery and the German machine gunners gave us a hot time but there were not many troops in front of us to impede

our progress. I never remember hearing so many machine gun bullets whizzing about before. Of course not having any noise from our artillery seemed to magnify the noise from the machine guns.

German prisoners were starving, dressed in ragged uniforms with holes in their boots. Unsurprisingly, they were "very dejected and down-hearted. They knew that they were losing the war. What a contrast to our troops. We were well clothed, well fed and full of optimism."[139]

Foch determined to press home his advantage, to seek decisive victory before the end of the year. The Belgian army to the north and the Americans on the eastern end of the front would now add their weight to the intensifying pressure on the enemy's flagging forces, drawing in and using up yet more of the diminishing reserve of divisions.[140] After a pause to regroup and bring up munitions, the penultimate phase of the counter-offensive began on 26 September, a sequence of blows that pulverised the enemy's defensive system. Rawlinson and Debeney's armies assaulted the central section of the Hindenburg Line three days later. Fourth Army, with the Australians in the van, stormed into the defences beyond the St. Quentin canal. Debeney's army seized its southern extension over the following days, occupying St. Quentin on 3 October.[141] A defensive system that had been designed to be impregnable in 1916 was in allied hands within a week. Tactical and operational methods had developed enormously in two years, the weight of *matériel* that the allies could bring to bear was unprecedented, and after four years of cumulative attrition there were simply not enough men left in the German army to hold such static positions. The nature of war itself had changed since the Somme of 1916.

With her final defensive bastion overwhelmed, on 4 October, against a background of domestic upheaval, Germany requested an armistice. All that remained was for the allied armies to pursue the German rearguards across northern France and Belgium, maintain the pressure and prevent a new defensive line solidifying. Although there were still 185 German divisions on the Western Front (down from a maximum of 208),[142] they were skeletons of their former selves. According to Second Bureau's calculations, there had been 68 German divisions in reserve in the West before the attack on the Hindenburg Line commenced, with 20 of them fresh; by the time hostilities ceased, there were only 17, with just 2 of them fresh.[143] Most of the allied divisions were tired too, but their sense of purpose, their pride, their knowledge that they were marching to victory carried them over the last miles. Since 8 August the allied armies had taken 385,500 prisoners and captured 6615 guns.[144] Lloyd George's subsequent assertion, after he had read Ludendorff's self-exculpatory memoirs, that his generals did not know their enemy was beaten in autumn 1918 is a final calumny to add to a lengthy list.[145] However violent the resistance of the German army's fanatical rearguards, Foch knew full well that there was very little behind them (although Haig's "unduly restrained" summary of the military situation in mid-October, indicating to Lloyd George that he had worn out his own armies during the counter-offensive, helps to explain the Prime Minister's misrepresentation[146]). Foch was destroying the German army's reserves, the principal objective of a strategy of attrition, and the heavy fighting of October and early November completed that process. A combination of modern combined-arms tactics, a coherent operational method conducted at a rapid tempo, *matériel* superiority and the oft-maligned grinding down of the enemy's manpower and morale—"wearing him out," as Haig was still reminding Churchill at the start of October[147]—finally delivered victory on the industrial battlefield.

GERMAN ARMISTICE NEGOTIATORS arrived in France on 8 November. By then, all Germany's allies had already agreed terms. On 9 November the Kaiser abdicated, and the next day Germany's Social Democrats declared the new republic from the Reichstag balcony. At 05:00 on 11 November, Foch signed the armistice in a railway carriage in the Compiègne Forest. Early that morning British troops re-entered Mons, where they had first engaged the enemy in August 1914. At the eleventh hour of the eleventh day of the eleventh month, in the fifth year of the war, the guns finally fell silent on the Western Front. God had granted the allies victory, Fayolle noted.[148] Church bells rang throughout France. In Amiens one final salvo was heard as the town's anti-aircraft guns fired in celebration. The next day Australian and American military bands entertained the rapturous crowds.[149] The citizens of Péronne marked their final deliverance by burning an effigy of the Kaiser in the town's marketplace.[150] Elsewhere there were more joyous, triumphant celebrations, even if none could put aside the huge efforts and loss of life that had finally brought this long-desired moment. Tristani's men were all very emotional. "We've won . . . Our immense task is finally over. We can think of tomorrow. Death will no longer take us under his wing at any moment . . . We are happy to have done our job till the end, without failing. We've beaten the proud Germany of the Kaiser and Bismarck."[151]

The vaunted German army of 1914 was defeated by a combination of the professionalism of the French army and the dogged determination of the British Empire's forces. This had proved an effective, if sometimes tense, alliance. Haig, self-centred and petty-minded to the end, still resented the fact that his allies were winning much of the glory, or at least that he was not being given enough credit for the success by politicians and the British press, which was "cracking up the French and running down the British military methods and Generals!"[152] One final time he felt obliged to tell the generalissimo "a few 'home truths' for when all is said and done, *the British Army has defeated the Germans this year,* and I alone am responsible to the British government for the handling of the British troops, *not* Foch."[153] Haig was adamant when the Prime Minister questioned him on armistice terms that the French army was "worn out and has not really been fighting latterly . . . The British alone might bring the enemy to his knees."[154] Even in their shared moment of victory he could not resist a final sour rant at those who had frustrated and annoyed him throughout his tenure of command:

> For the past three years I have effaced myself, because I felt that, to win the war, it was essential that the British and French Armies should get on well together. And in consequence I have patiently submitted to Lloyd George's conceit and swagger, combined with much boasting as to "what *he* had achieved, thanks to his *foresight* in appointing Foch C. in C. of the Allied Forces . . . " The real truth, which history will show, is that the British Army has won the war in France in spite of L. G. and I have no intention of taking part in any triumphal ride with Foch, or with any pack of foreigners, through the streets of London mainly to add to L. G.'s importance and to help him in his election campaign.[155]

It had been a rancorous war, a prolonged strain on nerves and a test of character. Neither Haig nor Lloyd George emerged from it unblemished. Now, with hostilities at an end, the battle of words, the fight for reputations, was starting. Subsequent perceptions of the predominant role played by British arms in the final "advance to victory" have their roots in Haig's long-nurtured discontent with his French allies: an impression which was to be enshrined in the memoir of his command that his private secretary later published at his behest.[156] As the end approached, Haig was happy to record, Foch was "suffering from a swollen head, and thinks himself another Napoleon!"[157] At that moment, though, Foch's pride was surely justified. The Western Front had certainly been a coalition campaign, but one in which France and Frenchmen took the leading role,[158] a fact that Haig and other Englishmen frequently resented. Although Haig and Foch might never have been close friends, they had been comrades-in-arms for four years, which had fostered a certain degree of mutual respect (Haig's repeated petulant outbursts notwithstanding) and common experience that held them to the joint task of defeating Germany. They and the other successful commanders in the Battle of the Somme ultimately led the allied armies to victory. Foch was at their head and, despite the vicissitudes of civil–military relations, Haig was still the unchallengeable commander of the British army in France. Fayolle directed the other army group that swept the Germans out of Picardy and northern France: "In 1918 he seized [the enemy] by the throat and played a preponderant role in the victory," his citation for the *Médaille Militaire* acknowledged.[159] Alongside him, Rawlinson led Britain's most successful army, the Fourth, which protected Amiens, drove the enemy back across the old Somme battleground and smashed through the Hindenburg Line, with the Australians in the van. He and his troops had learned much about industrial war since July 1916. Others who learned their military trade on the Somme would play their parts in the 1918 victory. Corps commanders Horne, Birdwood, Debeney, Berthelot and Guillaumat were all commanding armies by then.

Although Joffre partnered Foch in the victory parade down the Champs-Élysées in 1919, he was by then a forgotten man: many spectators did not even recognise him. Gough and Micheler both rose after the Somme, only to be brought down by subsequent military disasters: the latter was sacked by Clemenceau after his army was overrun on the Aisne in May 1918. But all three deserve to be remembered among those who contributed to the General Allied Offensive of 1916, who took on the German army breast-to-breast, who held it and beat it, laying the foundations for military victory.

Above all, however, the hundreds of thousands of middle-ranking and junior officers, non-commissioned officers and men who passed through the tempering fires of the Somme and made the British and French armies the skilled modern forces that were capable of overcoming the proficient German military machine deserve acknowledgement. Some of them, such as Weygand, Debeney and Gamelin, Montgomery and Brooke, would go on to make greater military reputations in peacetime or the next war. Others, such as Commandant Tristani and Sergeant James Erswell Philpott MM—who served the guns in 168 Brigade Royal Field Artillery near Albert on 1 July 1916, on the Ancre in October 1916 and during the retreat across the Somme and the advance from Amiens in 1918—put off their uniforms and returned to quiet civilian lives. The names of thousands more live for evermore in the cemeteries and on the memorials scattered throughout the now-somnolent battlefield.

ON 2 NOVEMBER FAYOLLE attended a mass for the dead.[160] It was important that the fighting had not been in vain, that the sacrifice had not been made for nothing. "We do hope our people will not make peace until the enemy is properly beaten, and our work complete," Lieutenant McInnis reflected as the end drew near. "We all have the greatest confidence in Foch, and expect he will do it properly, as he has done everything since he took command."[161] Even with both sides exhausted after eight months of the most intensive combat of the war, the need to secure a decisive allied victory was recognised, even by the humblest soldier. "We were disappointed with the [armistice] terms and stopping just when the Huns wanted to," Private Long wrote. "I am sick of the game, but I would be quite satisfied to carry on through to Germany and give them a little of what they gave the poor French people."[162] Perhaps Foch had not done it properly. His next planned attack, a Franco-American offensive into Lorraine scheduled for 14 November, would have taken the war into German territory for the first time since 1914. It might have shown Germany that she was really beaten; convinced her militarists that they had lost.[163] Tristani, whose regiment waited behind the Lorraine front, remembered that his men were anticipating delivering the *coup de grâce* to the Germans: "Our march into Lorraine would have inflicted the most crushing defeat the world had ever seen."[164] It was not to be: why incur another fifty thousand casualties if by the terms of the armistice the allies could march across the Rhine unopposed, Foch had reasoned. But in hindsight it would have been a small price to pay to avert the second conflict that German hubris was to inflict on Europe two decades later.

Although the Kaiser had gone, like a rat deserting a sinking ship,[165] and there was Bolshevik-inspired violence in Germany's streets, German militarism had not been extirpated. "The country is beaten," Fayolle noted as he led French forces to the Rhine,

> but not crushed. Her armies are retiring in good order and the soldiers' Soviets are nothing but eyewash. If she manages to muster the 12 million Germans of Austria she will still remain a formidable power and continue to threaten us . . . The country does not give the impression of a beaten people. Order, prosperity, wealth are everywhere. Germany is far from exhausted. If she is left to her own devices she will start the war again in ten years, maybe even sooner.[166]

Such prosperity gave a rather different impression to those returning German soldiers. The civilians had clearly not been making the sacrifices that the soldiers had endured for the fatherland. From such convictions emerged one of the foundation myths of Adolf Hitler's National Socialist movement: that Germany's army had been stabbed in the back by defeatists on the home front. Fayolle's fear that war would come again was reasonable.

RICHARD BESSEL HAS concluded that between 1914 and 1918 "the Germans initially were rather successful in mobilising economic, military and psychological resources for what ultimately was an unsuccessful and probably impossible project that never should have been attempted."[167] Fayolle put it more simply: "It was Germany's own fault that she was beaten . . . what would have happened if she had acted loyally and humanely" by not invading Belgium or by refusing to declare unrestricted submarine warfare and thereby bringing the wrath of the liberal world upon herself?[168] The allies felt that they had punished German arrogance, avenged German atrocities and reined in German militarism, for a while at least. But it had been "a terrible apprenticeship in modern war that left all the belligerents floundering."[169] In the race to victory the allies were slow starters, but they were always going to overhaul Germany and the other Central Powers in the end. Although too many British soldiers lost the race to the enemy's parapet on 1 July 1916, their war was a marathon, not a sprint. The Entente endured, improved and triumphed. Germany fought "against considerable odds and with remarkable tenacity,"[170] but, despite her own unprecedented effort, after 1916 she could never win a war of attrition.

14

Aftermath and Memory

I N 1 9 6 3 T H E first British television historian, A. J. P. Taylor, published a short but long-lived and influential history of the First
World War. His account of the Battle of the Somme is grim, dog-
matic, condemnatory. By the time the offensive was launched it "had no
longer any purpose."

> Nothing had been learned from previous failures except how to
> repeat them on a larger scale . . . each day the same tragic story
> was repeated on a diminishing scale . . . on 14 July the British
> infantry saw a sight unique on the Western Front; cavalry riding
> into action through the waving corn with bugles blowing and
> lances glistening. The glorious vision crumbled into slaughter as
> the German machine guns opened fire . . . the fighting dragged
> stubbornly forward to no purpose . . . The surprise of a really
> heavy attack by tanks was lost . . . The front churned into
> mud . . . Then the battle, if such it can be called, came to its dis-
> mal end . . . the battle of the Somme was an unredeemed
> defeat . . . Idealism perished on the Somme . . . The Somme set
> the picture by which future generations saw the First World War:
> brave helpless soldiers; blundering obstinate generals; nothing
> achieved . . . Only disenchantment was the result.[1]

Taylor was, like so many others, imposing himself on the history of
that battle. Ten years old at the time, he had grown up under the influence of a mother and an uncle who objected to the war on grounds of
conscience. That experience determined the strength of his liberal pacifist political convictions. By the time Taylor was writing his history, the
painful aftermath of the Great War had been played out. Rather than

the "war to end all wars," it had proved the precursor to a troubled half-century of economic crisis and social upheaval, an even greater and more barbarous world war and the ideological division of the globe which now lived under the terrible threat of nuclear holocaust. If Taylor could blame the Somme for a parting of the ways between generations, his own factual inaccuracies, half-truths and clichés anthologised all that popular memory came to believe was wrong with the Somme and the war of which it was the central event; seizing on the disenchantment that followed, he himself "set the picture" for later generations. His short book was a one-sided synthesis and distillation of arguments that had gone before, written in punchy style with wittily captioned photographs: not an attempt to find meaning in that conflict, but a rumination on the folly of war in general and on the failure of the First World War to end war in particular.

But his history would sell well, not only because of his public reputation, but also because of the *Zeitgeist* of the early 1960s, a period in which the advent of a new mass medium, television, coincided with another generational shift in British society. If Taylor's opinions tell us more about himself and the mood of 1960s Britain than they do about the Battle of the Somme, nonetheless, they must be acknowledged as part of the process of remembering that has reshaped the history of the battle down the years, one strand in a complex tangle of memory woven through subsequent experience.

THE END OF the Great War had left Europe exhausted, depopulated, partially devastated, nearly bankrupt, socially divided and psychologically scarred. Whoever won eventually, a long war of attrition was always going to be mutually destructive.[2] The greater irony was that it was also indecisive. Gloomy comparisons made at the time with Rome's recurrent Punic Wars with Carthage were well founded.[3] The military defeat of Germany did not presage her political reconciliation with the values of the Western allies, even if the collapse of the imperial regime that coincided with the armistice negotiations seemed to suggest as much. Nor would France prove magnanimous in victory. The festering wound of 1870 had not been healed by the war, but continued to suppurate. Tristani's men, part of the French army of occupation in the Rhineland, found themselves billeted in Ems in the summer of 1919. In this pretty spa town Bismarck had drafted the forged telegram that had ignited the 1870 war. Above it a proud bronze eagle stood atop a granite plinth. Scratched into the stone, as into so many Franco-Prussian War monuments, was the legend "Occupation française 1918–1919, 32 RI."[4]

A peace settlement that primarily addressed the international rivalries of pre-1914 Europe rather than the social conflicts of 1917 and 1918 would not resolve the problems of the new Europe, even if, as Michael Howard has suggested, it gave democracy another chance.[5] But it was really only half a chance after belligerent, disillusioned war veterans returned home to societies ill-prepared to receive them.

By the time Edward, Prince of Wales, then president of the Imperial War Graves Commission, presided at the inauguration of the Thiepval Memorial to the Missing on 1 August 1932, it was becoming apparent that the end of the war had not brought the hoped-for peace between nations, or social harmony and prosperity, that its warriors had fought on for after 1916. Indeed the inauguration ceremony itself, originally scheduled for 16 May, had had to be postponed after President Paul Doumer, who himself had sacrificed four sons for France during the war, had been assassinated by a mentally unstable, anarchist Russian émigré in early May. This was just one of many signs that the world was forgetting about the last war, and domestic and international politics were entering a new confrontational phase. Although wreaths could be laid on behalf of the *anciens combattants* at Albert's war memorial, and Senator Jovelet, president of the departmental council, could make a speech favouring peace, such gestures and words were starting to seem hollow, repetitive platitudes. As the day approached there was concern that communist and socialist protests against Doumer's successor, Albert Lebrun, the man they dubbed *"Président de la reaction,"* might disrupt the ceremony.[6] Early the next year the election of embittered Somme veteran Adolf Hitler in Germany would set Europe inexorably on the path to another reckoning. In May 1940 German troops would be back in Péronne and Amiens, and French and British forces would fight another last-ditch battle along that infamous river. Picardy would be occupied once more and liberated by British forces before the continent finally settled into an uneasy, divided peace.

IT SHOULD HARDLY surprise that, looking back ninety years later, Jay Winter could conclude that First World War memory was founded on ruins: "the ruin of the hope of progress, the ruin of millions of lives, the ruin of the idea that war can be predicted, controlled, imagined."[7]

Between the two world wars the predominant theme in the rituals of remembrance through which the survivors mourned had been sacrifice, not slaughter: this later motif arose out of the bitter, class-determined disputes which followed a second world war. For post–Great War Europe, engulfed in collective grief and pausing to come to terms with its enormous human loss, it was "the weight of the dead on the living" that determined their memory of the war and its battles.[8] Churchill was caught up in this ritual as much as anyone else in Britain in the 1920s, and he ended his account of the Somme with a personal tribute to Britain's citizen soldiers: "Martyrs not less than soldiers, they fulfilled the high purpose of duty with which they were imbued . . . Unconquerable except by death, which they had conquered, they have set up a monument of native virtue which will command the wonder, the reverence and the gratitude of our island people as long as we endure as a nation amongst men."[9] Martyrs should have their monuments and memorials. When James Beck and H. E. Brittain toured the Somme in 1916, they already appreciated that "we were standing on ground which will forever be regarded as holy. Martyrdom hallows, and wherever a man has laid down his life for a country that he loves or a cause in which he believes . . . that spot must be for ever sacred."[10] The landscape of the Somme was to be woven into the history and memory of the battle, to become the physical manifestation of a psychological phenomenon: a site of memory and for mourning across which cemeteries, memorials and monuments proliferated to mark the battle's key events and sacrificial places.[11] As John Masefield later explained, *The Old Front Line* had been published with this as much as a historical record in mind: "Feeling that perhaps some who had lost friends in the battle might care to know something of the landscape in which the battle was fought, I wrote a little study of the position of the lines."[12] The process of hallowing the ground that Masefield, Bean and Buchan began in 1916 would in time impose a patina of eternal memory on the Somme's fields, subsequently restored to the green and pleasant land that the fallen always believed they were fighting towards.

The Thiepval Memorial was the culmination of a process that started as soon as the German army retired in March 1917, and encompassed soldiers and civilians, communities and individuals. In the 1920s in particular the laying out and landscaping of military cemeteries and the erection of unit and national memorials manufactured a sacred site for Britain and her empire: whether commemorative or triumphal was never entirely clear. This eternal occupation was instigated alongside the resurrection of the shattered ground, whose farmers were in future to share their fields and villages with those whom the battle had left behind, "the glorious dead," and the living who came to mourn and remember them.

While hostilities continued, capturing the essence of the battlefield before nature began the process of reclamation was necessarily a hasty, disorganised business. Bean's excursions to the Pozières heights with an official photographer, Masefield's wanderings through the maze of deserted trenches, Orpen's sketches of unburied bodies and Malins's films of the now-empty uplands were all elements of the commemorative and sanctifying process. But no visual record could capture the essence of the battle. Masefield's peregrinations only evoked confusion and despair. "I probably know more of the Somme field than any of the soldiers who fought there," he confided to his wife. "Parts of it do not attract me, parts repel me, some of it is romantic, some strange, some unearthly, some savage."[13] Missing from the deserted battlefield were the men and activity that had made it what it was, as significant a part of memory as the land itself. But the reality of industrial battle could not be captured or told. "I prefer not to write about my surroundings," one young officer wrote home to his mother shortly before his death in autumn 1916. "I stood on a slope yesterday and watched through a telescope . . . the heaviest bombardment and fiercest attack that has taken place in the war. I watched a good many other things, but we'll leave them out. So don't ask for news or realism as I can't give one and won't the other."[14] If the horrific details were inexplicable, the whole could be reduced to a simple, virtuous mirage. Bean's realisation of the Australian soldiers' stoical heroism while sharing their ordeal in a shell-strafed dugout near Pozières; Messimy's grief at the losses among his brave *chasseurs*; and across the line Queri's paeans to the bravery and self-sacrifice of the ordinary German soldier: from such representations the casualties of the First World War—the killed, the maimed, the missing, the survivors who would never speak of the world of noise, fire and blood that they endured—became its heroes. Their patriotic cause may have been just in 1914, but two years later the fight was no longer merely for the values of the living. Nations and armies fought on to avenge the dead, to justify unprecedented human sacrifice and to win through to a better world in which the survivors would live in peace; where the glorious dead could rest in peace; and where, as the inscription on the stone of remembrance that sits in Britain's military cemeteries assured, "their name liveth for evermore."

One way to fill the void in perception and society was to represent the ranks of the absent by the concrete in a sanctified sacrificial landscape: neat rows of white Imperial War Graves Commission headstones, or crosses in French and German cemeteries. As well as its material form in the cemeteries and monuments that would mark for ever the sites of memory where the carnage was at its worst, this commemorative process would have symbolic, metaphysical elements: quasi-religious, but non-denominational, so that soldiers and their families of all creeds and none could be included in the mourning rituals. Sacred places, the hallowed words of participants, society-wide rites of commemoration, the active participation of veterans and war widows would combine to restore structure and meaning to a war that around 1916 had lost its way, had itself been re-formed at the same time as it reshaped its world. What emerged, in the short term at least, Stéphane Audoin-Rouzeau and Annette Becker have suggested, was a "crusade myth" to justify society's effort, emotional investment and sacrifice.[15]

In time such activities took on a life of their own, which Jay Winter has dubbed "historical remembrance," even as the events that were being remembered faded in memory and changed their meaning.[16] Charles Bean's visits to Pozières served his need to acknowledge and mourn his fallen comrades—an essential need that he shared with the rest of humanity—and the practical need to record the contours of the battlefield for posterity for his planned Australian War Memorial. He was absorbing the sites, revisiting key places and events, committing them to his own memory the better to describe and interpret them in the official history of Australia's Great War that he was to write. Curious sightseers followed in his footsteps, such as Lieutenant McInnis, who wanted to see High Wood, the 1 July starting-line and the mine craters, where quasi-mythical great events had taken place.[17] Winston Churchill too took the opportunity to visit the battlefield during a visit to France in September 1917:[18] his impressions are not recorded. Others came with sadder purpose. After his aunt had written from Australia asking him to do so, Private Schwinghammer walked the battlefield searching for the grave of his cousin, killed at Morlancourt on 10 June 1918.

> I walked for four hours looking at every cross I came to and at last came across a large wooden cross on the top of a ridge. It had a map of Australia, worked in tin, on it, and on which were written the names of fourteen Aussies who were all buried in the one

grave. My cousin's name was amongst them. I buried a little bunch of pressed flowers (that his mother had posted over to me) on the grave and got a little tobacco tin of earth from the grave and which I posted back to my aunt in Australia.[19]

That small familial ritual, the laying of flowers on a loved one's grave, was to be repeated countless times down the years.

On 1 July 1917 church services were held in England to mark the first anniversary of the opening of the Somme offensive and to honour the memory of the fallen:[20] that date had immediately assumed a central place in the British Empire's rituals of commemoration. McInnis, Churchill and Schwinghammer were just a few early representatives of that flood of humanity — curious, mournful, questing: often all three — that has driven down the Somme's narrow roads and walked through its fields in search of faded memory, lost progeny or ancestry or a glimpse at history. The opening of the fateful offensive is ritually observed to this day, even if, down the years, the Somme's shifting history and enduring memory have co-existed in uneasy association.

THE FORMATION OF memory is a complex, controversial process. Over time the recollections of individuals coalesce into a collective consciousness; yet this cultural memory, an element of "historical remembrance," remains amorphous, shifting and contested. It should come as no surprise that since its end, writing about the Great War has diversified as the societies that fought in it have changed, and as later generations have taken up the tasks of analysing and understanding. Jay Winter and Antoine Prost have identified three different "configurations" to this history, succeeding and supplementing but never entirely supplanting one another.[21] Immediately after the war its diplomacy and military conduct were the principal themes for historical enquiry. Its origins preoccupied scholars, while in a "battle of the memoirs" generals and politicians presented their competing explanations for how their war had been waged and won.[22] This history, told with the authority of high office and experience, was largely partisan and inaccurate. Meanwhile, official historians worked to filter the complex details of military operations, and to construct a useful narrative for post-war armies. The combatants themselves were sidelined in this first configuration: although soldiers' memoirs proliferated, they were tangential to the formal historical record.

At the time the mighty Somme battle could be inspiring, terrible yet somehow majestic and important. H. E. Brittain's account of his tour of the front could quote the words of Abraham Lincoln's Gettysburg Address with no sense of irony: "The brave men, living and dead, who struggled here have consecrated it far above our poor power to add or detract. The world will little know, nor long remember, what we say here; but it can never forget what they did here."[23] This collective summation was difficult to reconcile with personal experience. At war's end memory was opaque; it had to be captured. "I must go over the ground again," Edmund Blunden, embarking on an academic career in Japan, acknowledged in his memoir *Undertones of War,* begun in 1924: "how thickly and innumerably yet it was spread with the facts or notions of war experience." He had been wrestling with "the image and horror of it" in poetry ever since.[24] In a later volume of more conventional memoirs charting his life after he was invalided home from the Somme with gastric fever, Siegfried Sassoon acknowledged that his wartime journey reached its end only in 1920, after he returned from a lecture tour of America where he was labelled "England's young soldier poet." He would now begin another career.[25] It took Robert Graves, Sassoon's comrade in arms and friend, rather longer to say *Goodbye to All That.* In his flippant yet readable 1929 memoir he was trying to come to terms with his chaotic post-war existence as well as to put his defining war experience behind him; to take meaning from it in a world from which post-war hopes were slowly disappearing. It proved to be the best-seller he had consciously crafted.[26] Its measured satire was a milestone on the road to disillusionment with the war which veterans in all countries were starting to take, marching under the cloud of economic depression which presaged the gathering storm of another war. Although Somme survivors' tales differed in detail, collectively they were starting to establish its image of waste and futility, just around the time when the efforts and achievements of the Somme were losing their lustre and people were starting to ask whether it had all been worth it. Blunden was commissioned to write his recollections of and reflections on the battle for the BBC's new magazine, *The Listener,* in 1929. For him the soldiers waiting to go over the top on 1 July 1916 had become the "intended victims" of war. By the end of that fateful day, it was clear that "neither race had won, or could win, the war. The war had won and would go on winning." But what Blunden called "the experiment of the century" would still go remorselessly on, "a slow, slaughtering process," ending in "a desperate mud-field . . . there was no sign yet of the fabled green country beyond the Somme battle."[27] It was a metaphor for his sorrow-numbed and ideologically riven world a decade after his war had ended.

IN PART, THESE memoirists were trying to reclaim the memory of their war from the politicians, generals and historians whose salvoes of self-justifying political and military memoirs had drowned out the voice of the ordinary soldier. The controversies about the Somme offensive's strategic conduct which sprung up in the 1920s reverberate to this day.

Back in November and December 1916, only a few senior commanders had seen the full picture of the Battle of the Somme. It would be the task of writers and historians—they had already begun—to piece together the battle's details and interpret them for its participants, spectators and future generations. Douglas Haig set the post-mortem enquiry in motion with his official despatch published at the end of the battle. Haig noted that the Somme constituted "a feat to which the history of our nation records no equal. The difficulties and hardships cheerfully overcome, and the endurance, determination and invincible courage shown in meeting them, can hardly be imagined by those who have not had personal experience of the battle."[28] Decades later, imagination still fails: the novelty, enormity and singularity of the campaign defy comprehension. But the utility or futility of the Battle of the Somme remains a live issue: in 2006 the plethora of public events, books and scholarly conferences that marked the ninetieth anniversary of 1 July 1916 attested to that.

In his despatch at the end of his first campaign, with its provocative title, "The Opening of the Wearing-out Battle," Haig was making a case for the continuation of his strategy, one of attrition. It was the practice and consequences of attrition, not its principles, that were problematic, then and subsequently. In 1916, Haig's tactics, not his strategy, had been singled out for criticism, with Churchill and Lloyd George most prominent among the detractors. And ever since, his operational and tactical methods have been revisited, scrutinised and decried. This has had the unfortunate effect of skewing the focus of subsequent evaluations of the Somme offensive, which encompassed much more than a misconceived British battle plan.

More than ninety years later, assessing the 1916 battle—manoeuvring through the opinions that have formed, been refuted and restated, summarising the debates and identifying their errors and omissions, falsehoods and misapprehensions—is difficult but crucial if the Somme's nature and significance are to be grasped. Although British generals also wrote their memoirs,[29] it was the politicians' more eloquent voices which the public heard, and so the "anti-Somme" still tends to hold the field. The parameters of any assessment have not changed greatly from those of early 1917. However, for the sake of clarity, one should emphasise the distinction between the tangible and the intangible. The visible and measurable—ground gained, casualties, manpower and reserves and the *matériel* balance of the war—have all been assessed, then and subsequently. Less quantifiable, more subjective variables—morale, strategic initiative, the international and domestic political situations, command and operational art—also have to be considered. Often they are closely intermingled and the separation of one from another is artificial. Arguments balancing the whole are complex and convoluted, from which there is no real escape, because the battle and the war were complex, multifaceted events.

Any analysis of the Somme has the nature and effectiveness of strategic attrition at its heart. This element of the 1916 strategy has remained the most controversial, and provoked the longest-running arguments, yet even now its centrality to strategic thinking is poorly understood. Carter Malkasian, who has written the most complete study of attrition as a strategic theory to date, persists in the misperception that attrition on the First World War battlefield "mainly occurred as a by-product of failed attempts at manoeuvre warfare."[30] In fact it was the key principle of both Falkenhayn's offensive at Verdun and the allies' counter-offensive on the Somme, even if the focus on Haig's operational planning (which Malkasian perpetuates) suggests otherwise. Of course, attrition, essentially killing on account, was and is a controversial and emotive approach to making war: morally bankrupt in Jack Sheldon's judgement.[31] But loathsome as it may be, the Somme and the First World War cannot be understood without engaging with it. Moreover, it worked.

In the summer of 1916 everyone, soldier or statesman, accepted the attritional nature of the conflict, and the logic behind this reasoning. On 5 July 1916 Castelnau had explained "the object and scope of the great offensive on the Somme" to Repington. Although "the movement aimed at Bapaume, Maubeuge, Liège, and Luxemburg . . . We should not be hypnotised by the German lines, as it was not the trenches or the barbed wire that stopped us but the men." The next day Charteris endorsed this: "our present action was to kill Germans. The strategic objective in this area was a secondary consideration."[32] Without attrition, the capture of ground would never be possible on the Western Front.

Like all the other politicians who conducted the war and afterwards regretted its cost and consequences, Churchill accepted the inevitability of attrition, even if its outcomes were to be comprehensively denounced. In May 1916 he endorsed the Chantilly strategy in one of his anti-Asquith parliamentary interjections. Instead of the limited and ill-coordinated offensives of 1915 (for which he was also reluctant to assume his due share of responsibility), he advocated surrounding the Central Powers with "armies which show a real, substantial preponderance of strength, then the advantage of their interior situation will be swamped and overweighed, and then the idea of decisive victory will be at hand."[33] But while fancying himself a grand strategist, Churchill had also been a common soldier. His sojourn at the front had furnished the final piece of his conception of the war—that it was the "trench population" which suffered and died, while the "non-trench population" behind the lines lived a safe and comfortable existence.[34] Once the offensive got under way he conflated this typical front-line soldier's view with a partial grasp of strategy. The elements from which Churchill's eloquent words were later to construct our enduring perception of the Battle of the Somme were in place. Esher wisely and perceptively forewarned Haig on the eve of a visit by Churchill to G.H.Q. in 1917: "He handles great subjects in rhythmical language, and becomes quickly enslaved by his own phrases. He deceives himself in the belief that he takes broad views, when his mind is fixed upon one comparatively small aspect of the question."[35] Although such faults are obvious in his war history, Churchill's influential 1920s assessment of attrition makes an appropriate starting point for analysis, because his summation determined how the Western Front has been viewed ever since. It should be recognised that in *The World Crisis* Churchill was not condemning an attritional strategy *per se,* merely the generals' conduct of it on the Western Front. Essentially he was making a case, rooted in his July 1916 evaluation, that more ground could have been won at less cost. His "inaccurate conclusions" and "unsound theories" were not accepted by the well informed at the time.[36]

In *The World Crisis* he republished his 1 August 1916 memorandum that the Cabinet had rightly dismissed at the time, and developed from this example a wider critique of attrition, which he dubbed, with his customary emotive turn of phrase, "The Blood Test."[37] In doing so he took a specific concern that Haig was using up too many men in an attempt to break through the enemy's defensive positions during the first month of the Somme offensive—which was a reasonable criticism, given the heavy casualties incurred in Haig's early operations, if divorced from the wider scheme of things—and transmuted it into one of the strongest and most enduring critiques of the high command's conduct of the Western Front campaign. It was Churchill's broader contention that the attritional operations of 1915–17 were always more costly to the allies than to the German army, by a factor of 3:2 on the British front and 1.5–2:1 on the French.[38] To argue that a modern industrial war could have been won without any attrition would certainly be spurious, and Churchill never suggested that. However, he does imply that attrition was effective only if one side lost fewer men than the other. Such an absolute measure denies the significance of manpower reserves for sustaining attritional war. He posits, moreover, that because in 1916 Germany called up more men than she lost, her reserves were potentially inexhaustible.[39] In fact, during 1916 Germany was still moving to full mobilisation. The wearing-out fight, as Haig's despatch posited, was under way in earnest, putting German military manpower under increasing strain and compelling reinforcement. The relative ability of each side's manpower reserves to bear military casualties as well as sustain the wider war effort was the keystone of a strategy of attrition: the *Reichsarchiv*'s post-war assessment of the battle admitted that "[our] grave loss of blood affected Germany very much more heavily than the Entente."[40] Also, the less measurable but vital moral impact of attrition has to be factored into any numerical calculations. Churchill's simplistic answer to this was that the stimulus to the defender "crouched by his machine gun," of "long lines mowed down, wave after wave," counted more than constant losses and continually giving up ground.[41] This is a fine example of what might be termed "1 July 1916 syndrome," a malady that Churchill caught in July 1916 and passed on to subsequent commentators. Above all, one unfortunate month on one section of the Somme front is not a valid yardstick against which to judge the battle, let alone the attritional strategy of the war as a whole.

Indirectly, Churchill is arguing his own contention, as he presented it in October 1917, that "success will only be achieved by the scale and intensity of our offensive effort within a limited period. We are seeking to conquer the enemy's army and not his position."[42] Such a statement would have come as no surprise to the generals: Castelnau and Char-teris have confirmed as much to Repington in July 1916, Kitchener had argued much the same, unsuccessfully, to Joffre at the start of 1915 and Foch had reached the same conclusion a year later. In 1916, as the Somme campaign demonstrated, as yet this was easier said than done. By October 1917, when Churchill was back in the Cabinet as Minster of Munitions, he was strongly arguing the case for intensifying and speeding up attrition.[43] He had become an advocate of Foch's *matériel-intensive* strategy. It was no coincidence that in May 1917 he had toured the front, meeting Haig, Fayolle and Foch: "indeed I think the French soldiers see very clearly the truths of this front," he confided to his wife of his strategic education.[44]

Churchill's post-war alternatives to the Somme strategy were short battles fought on the model of those of 1917, or compelling the enemy to attack, rather than "prolonged offensives on the largest scale in order to wear down the enemy by attrition."[45] But these would never have been possible without the new weaponry and re-skilled armies that were the legacies of 1916: thus Churchill torpedoes the logic in his argument. The arch-amateur strategist was quite emphatic in telling generals how they could have done their jobs better, unaware (or unwilling to acknowledge) that many of them had worked this out for themselves much more quickly. Politicians and policies had delayed delivery, if indeed delivery could have been any quicker than it was. Notwithstanding his subsequent critique, somewhere, at some time, a "blood test" had to take place before the war could be decided.

Churchill's analysis provided the framework for other disputes, none more virulent than that over his treatment of casualty figures. The details of the "casualty controversy" are convoluted and abstruse, and they will be summarised as briefly as possible here. The losses on the Somme are a factor in a wider argument about the physical impact of attrition. Allied casualties on the Somme have been established with reasonable exactitude. Captain Wilfrid Miles's official history volume for 1916 gives a figure of 419,654 British casualties up to 30 November (including 5 per cent "absentees" who subsequently returned).[46] The French official history records 154,446 Sixth Army and 48,131 Tenth Army casualties between 1 July and 20 November 1916, a grand total of 202,577.[47] The German casualties are much more disputed. Estimates for German losses on the Somme range between 400,000 and 680,000 killed, wounded, missing and prisoners, the latter figure being that cited in the British official history.[48] In absolute numbers this represents a heavy toll, but there is also a huge variation between the lowest and the highest estimates.

In his "blood test" statistics Churchill lumped Somme losses into broader figures for Western Front casualties, including trench wastage (in two separate periods, July to October and November to December), so they are not directly comparable with calculations of losses in the offensive. He suggests that the British army suffered 513,279 casualties in the second half of 1916; the French 434,000, including the fighting at Verdun and in the Near East.* For Germany, over the same months, Churchill's aggregate figure is 630,192.[49] Although Sir James Edmonds unofficially advised Churchill on his Somme chapter and queried his figures (which were based on incomplete official British, French and German calculations issued in the early 1920s),[50] Churchill was sufficiently convinced of their accuracy to venture into print. After publication his Somme statistics provoked a strong rebuttal from Sir Charles Oman, the distinguished historian of the Napoleonic Wars who in wartime had been responsible for calculating the German figures from casualty returns published in German newspapers. Oman estimated German casualties on the Somme at 530,000.[51] Edmonds himself later challenged Churchill's figures in the Somme official history volumes. He claimed—but with no clear substantiation—that German casualty returns, which were made only every ten days, did not include lightly wounded men who returned quickly to the front, and were therefore not directly comparable with allied numbers. Adding another 30 per cent to the published German figures to account for this gave a real German loss of 680,000 men, significantly more than the combined allied totals. Edmonds's recalculation has not been accepted by all: in the early 1960s M. J. Williams argued that his estimate was spurious.[52] The argument did not end there, though, and the search for accurate German casualty figures goes on. Most recently, James McRandle and James Quirk's analysis based on statistics in the German official medical history has tried to factor German lightly wounded back into the comparative statistics (at 11 per cent of the total). They suggest 597,000 German casualties on the Western Front from July to October 1916 (compared with Churchill's 538,000), and another 132,000 (as opposed to 93,000) from November to December. Although their figures are consistently higher than Churchill's, McRandle and Quirk's analysis supports his basic contention that the German army inflicted more

---

*The latter is only admitted when Churchill's tables are repeated as an appendix.

casualties than it suffered on the Somme, and during the attritional campaign as a whole. However, they recognise that this ability to inflict disproportionate damage on the allies diminished as the war went on, "as the Allied armies moved up their learning curve, and as the course of the war and its attrition eroded the initial advantages in training and leadership of . . . the German army."[53]

As one astute United States staff officer who had been tasked with calculating the relative losses of the two sides at the end of 1916 observed, the victor was far more likely to publish accurate casualty statistics at the end of a war than the vanquished.[54] The complex, often incomplete and contradictory nature of German statistical returns is not in dispute.[55] An accurate figure for German casualties on the Somme will never be established, but undoubtedly it lies somewhere between the lower and higher estimates given. From the available evidence, however, it can be inferred that they were heavy, and difficult to bear in an army that was increasingly stretched by intensifying campaigns on two fronts. For the period January to October 1916, *Der Weltkrieg*, the official German history, indicates that the German army suffered 1,400,000 irreplaceable casualties, 800,000 of them from July onwards (an indeterminate figure of recoverable lightly wounded men had been excluded from this estimate).[56] These figures included casualties from the Somme, Verdun (330,000 casualties at least, the majority before July, compared with around 378,000 French),[57] the Eastern Front battles and general trench wastage. By inference, the heaviest casualties suffered by the German army in 1916 were on the Somme, probably more than 500,000 irreplaceable losses. The British official history later corrected its figure to 582,919, reportedly based on the figure issued by the German Casualty Enquiry Office.[58]

Whatever the precise number, there is much evidence to suggest that German units fighting on the Somme were decimated. Ernst Jünger, who arrived on the Somme in August, was informed in late September by the nurse who treated his first wound that 30,000 casualties had passed through that casualty clearing station alone in the previous few weeks. He was back in hospital in November. The "procession of corpses" that passed through it daily was testimony enough to the heavy losses inflicted on the German army on the Somme: not surprisingly, it brought on "an attack of the glooms."[59] Franz von Papen's 4 Guards ID, which did three tours of duty on the Somme, suffered 8842 casualties there, 4272 of those killed in their intensive fight on 15 September.[60] Such anecdotal evidence supports the view that the

German army was bled consistently throughout the summer, with casualty rates intensifying through the September battles, in which the balance of attrition tipped in the allies' favour, even if it could not be sustained once the weather worsened. While Churchill's snapshot of July 1916 might be judged a fair estimation of the relative losses on the British front in the battle's early weeks, they are not representative of the battle as a whole.

As Robin Prior, the most thorough analyst of Churchill's casualty statistics, rightly observed, "a 'blood test' is a very crude way of comparing the ability of modern states to wage war." Other key components of mass mobilisation—such as relative manpower reserves, industrial capacity, agricultural productivity and the resilience of the financial resources required to support multi-million-man armies in the field—were all factors in an attritional war effort, making "the question of why the war was won (or lost)... a good deal more complicated than the tables of figures produced by Churchill would indicate."[61] After all, the allies won the war, despite losing this quantitative blood test.

The intangible, immeasurable factors of strategy were perhaps decisive. The exhaustion and declining morale of the German army from 1916 has been outlined. Cumulatively, the effects of attrition combined with repetitive and increasingly frequent battlefield defeats were to bring on its eventual collapse. In October 1918 Ludendorff stated that it was this breakdown of morale, rather than a shortage of troops, which persuaded him to seek an armistice.[62] In his critique of allied military method, Churchill was willing to acknowledge that "the moral effect upon the German Army of seeing position after position, trench after trench, captured and its defenders slaughtered or made prisoners, was undoubtedly deeply depressing ... The effect was lasting ... never again did the mass of German rank and file fight as they fought on the Somme."[63]

LIKE CHURCHILL, LLOYD George was to write a popular yet contentious set of memoirs, which dwelt on his own importance to the British war effort, and contrasted himself with the stolid and unimaginative high command that saw the Western Front and attrition as the be-all and end-all of the war. This dichotomy had its roots in the events of 1916 and, it has been suggested, "a deep sense of guilt at not having stopped the carnage."[64] With an advocate's eloquence and attention to detail, Lloyd George set out the case why he was right, and how he might have shortened the war and saved British lives. Yet his selective memory is conspicuous, and, like Churchill's, his basic premises and particular assumptions are open to question. His attempt to dissociate himself from the attritional strategy and battle that brought him to power and won his war, while understandable, grossly misrepresents his role. Andrew Suttie's study of Lloyd George's memoirs concluded fairly that his "relentless attacks on . . . generals and their strategy and conduct of the war and military operations ultimately rebounds to his own discredit and cannot fail to detract from his own significant and genuine wartime achievements."[65]

Although he was a "wizard" when it came to organising the supply of the army in the field, and was an inspirational leader of a nation in arms, Lloyd George had difficulty with the nature and impact of industrial war, and an uncertain grasp of how to fight and win that war. That was hardly surprising: after all, he was no soldier. Haig, meanwhile, stuck to the policy of attrition that Kitchener had formulated and the method of beating the main enemy in the field that he had haltingly begun on the Somme. By doing so, he earned the perpetual scorn of Lloyd George for his blinkered attitude. Robertson backed the commander-in-chief firmly; hence the divisive civil–military relations that characterised the British war effort. Nevertheless, beneath the swirling dust of post-war controversy, whatever his personal reservations and however often he tried to intervene to change their minds, Lloyd George always backed—or at least never overruled—soldiers when they were united in their opinion. Joffre and Haig in November 1916, Nivelle and Haig in January 1917, Pétain and Haig in May 1917, Robertson and Haig in June 1917, Foch and Haig in 1918: all convinced Lloyd George of the centrality of a Western Front offensive strategy to ultimate victory. No matter how reluctant he was to accept it at the time, or acknowledge it subsequently, the harsh fact was that the majority of British forces were committed to the continent, were supporting the French army and would have to fight on in France. It was always a question of when, where and how, not if, a new Western offensive would take place.

If Lloyd George could naïvely suggest in January 1917 that "to win we must attack a soft front, and we could not find that on the Western Front,"[66] he was denying the reality of industrial war, in which there were no soft fronts. The idea that in a war against an enemy operating with the advantage of interior lines of communication, which allowed the rapid redeployment of strategic reserves to any threatened point, forces could be moved to open space elsewhere was illusory. Even discounting the logistical problems of transport and supply, the trench stalemate which also set in at the Dardanelles, at Salonika and on the Italian front, and the fact that much of the Eastern Front was as entrenched as the Western, bore witness to the fact that industrial battle was one of fixed positions and *matériel*.[67] This was the consequence of an unfavourable force-to-space ratio (too many men in too little space, which allowed the accumulation of strategic reserves behind the fixed fronts), which negated manoeuvre. This was at its worst on the Western Front, where millions of heavily armed men were squeezed into a narrow strip of fortified land for four years. The force would have to be reduced before the space could be fully utilised: by attrition. Strategic diversions, real or imaginary, outside North-west Europe were merely window-dressing for this unpalatable truth. An Eastern strategy was never "a substitute for, but . . . a supplement to the great offensive in the west."[68] Churchill was at least prepared to excuse, if not forgive, Haig for the "tragedies of 1916," which for him inevitably followed from the events of 1915.[69] Lloyd George, who unlike Churchill, bore a huge share of the responsibility himself, could never do so.

Like it or not, British politicians were embroiled in a war of attrition, which had fundamentally changed its nature after 1916. Lloyd George certainly did not like it, and tried everything to sidestep this unpalatable yet fundamental truth, pursuing what Brock Millman has called a "New Eastern" strategy, focused on imperial interests outside Europe.[70] But still the Prime Minister could not escape the grip of the main theatre. Cavil as they did, condemn as they would, Lloyd George, and later Churchill, were responsible for providing the men and the guns with which Haig maintained his strategy of attrition of the German army from 1916.

Their resentment of this fact triggered the most enduring controversy about the conduct of the war on the Western Front, that over military command. The Battle of the Somme should be recognised as one moment—the critical one—during the sequence of fundamental military changes that occurred between 1914 and 1918. Over these four years generals experienced rapid professional change. If command on the Somme was far from perfect, it reflected the military realities of 1916. Individually, all the senior commanders showed both strengths and weaknesses. Haig was determined and meticulous; perhaps too much so, as he interfered with his subordinates and was an awkward and self-important ally. Rawlinson, while understanding the core principles of industrial battle, gave his subordinates too much freedom and did not coordinate operations. Gough, on the other hand, interfered too much in the detailed preparations of his front-line formations. These were the two extremes of an army learning the business of operational command. Fayolle micromanaged the artillery barrage, as befits a professor of artillery, but left his subordinates alone as long as the basic tactical method was sound and they were delivering. He sacked only one corps commander during the offensive. Micheler lacked the confidence, as well as the resources, to make much of his belated offensive, which was methodical rather than dynamic. Joffre had only one eye on the battle, as he was perpetually distracted by the politicians. Perhaps that was no bad thing, though, for there were already quite enough cooks. Foch struggled to coordinate everything, and did a fair job in the circumstances, although he antagonised both allies and subordinates. Gallwitz and Below conducted a careful, methodical defence while those above them fussed.

In the linear siege that was the Western Front, generals were the professional technicians of the industrial–scientific battlefield. To expect acts of Napoleonic genius from them is to place them in the wrong age. The expression "military machine" is appropriate in these circumstances. Generals directed huge, mass armies—"commanded" is not the right word—in ponderous rhythm. That is, if they could be directed at all, for as Foch had identified before the war: "The Armies have outgrown the brains of the people who direct them. I do not believe there is any man living big enough to control these millions. They will stumble about and sit down helplessly in front of each other, thinking only of their means of communication to supply these vast hordes."[71] Operating these slow, cumbersome lethal machines, which combined the communications technology of the nineteenth century with the killing power of the twentieth, was not a simple business. What would today be called the weapons systems at their disposal contained some apparent anachronisms—in particular the oft-maligned cavalry—tested and effective weapons of industrial war, such as the quick-firing gun and the machine gun, and novelties, such as the aircraft and the tank. All had to be combined into an effective military tool, and employed to break the stalemate. On the Somme, this process of professional learning reached an important crossroads. Yet, essentially, the commanders were wrestling with the age-old military dilemma of combining firepower with shock action. Haig prioritised the moral, the value of shock, in his strategy and planning: Bapaume was never a specific operational objective, but the conditional target if the defence collapsed. Fayolle and Rawlinson focused on the material. Foch was striving to integrate the two. Everyone who met him commented on his great intellect: on the Somme he demonstrated that he could get the stalled allied armies moving again, slowly at first, but inexorably towards Germany and victory.

BRIAN GARDNER, WHO published the first popular account of the Battle of the Somme in 1961, felt able to summarise the French contribution to the battle in half a paragraph. Subsequent English-language studies did not bother to develop his superficial impression of "a persistent and bloody, but unspectacular, advance."[72] Perhaps, however, the British commander-in-chief should bear ultimate responsibility for the French army's marginalisation. In October 1916 Haig could still delude himself, and argue to no less a person than the King, that "[t]here is no doubt that the French have not really exerted themselves on the north of the Somme: but then they rather meant to save their troops and avoid casualties in view of their losses at Verdun and previously."[73] Joffre and Foch were hardly mentioned in Haig's authorised account of his command,[74] and in evaluating France's contribution on the Somme, Britons undoubtedly took their cue from their commander-in-chief, never a man to downplay his own role or laud that of his ally. Recognising the French army's important but near-forgotten contribution to the Somme offensive, from both senior commanders and *poilus*, delivers an important corrective to the prevailing, Anglo-centric view of command deficiencies and poor strategy. For France, as Fayolle recognised, 1916 would always be the year of Verdun (which itself was the Somme's conjoined twin, the two forming "one great battle").[75] Thus, over the years, the British have usurped and the French forgotten the latter's major role in the Battle of the Somme: there are no enduring controversies over its conduct in French histories of the war. Churchill, for example, dismissed Joffre's strategy and army with sweeping rhetorical generalisations about faulty methods and losses in the 1914 and 1915 battles that have echoed down the years: a repetitious catalogue of "gross mistakes . . . glaring errors . . . insensate obstinacy and lack of comprehension . . . without any novel mechanical method, without any pretence of surprise or manoeuvre" establishes Joffre's and Foch's culpability relentlessly rather than cogently. Haig and the British army of 1916 could quite easily be substituted for France in 1915 in this emotive diatribe. If Churchill at least acknowledged that in 1916 "French and German losses were much less unequal," this was not on account of the great lessons in tactics and operations that had been absorbed from the earlier battles, but simply because "the brunt of the slaughter was borne by the British."[76] If strictly factually accurate, Churchill's condemnation is unqualified and based on limited knowledge and spurious reasoning.

French tactics on the Somme demonstrated that an alternative already existed to costly British methods, in which general inexperience and insufficient *matériel* were the real causes of the lengthy casualty lists. Lloyd George was right to look to them for an explanation, even if Haig seemed too concerned with scoring points at his ally's expense to learn from their methods in the way his subordinates were prepared to.

More important than the tactics, Foch's operational method promised more productive battles in future, even if these would be grander, require much more *matériel*, and prove even more costly.[77] If at the start of 1916 there was still some residual belief that battlefield attrition would, sooner or later, reverse the imbalance between force and space on the Western Front and restore movement and manoeuvre to warfare—Haig, Joffre and Castelnau all hoped for this even though they feared the opposite—the July–December grapple on the Somme demonstrated the naïvety of this belief (although it never entirely disappeared). Instead a new operational method, based on coordinated, interconnected and mutually supporting deep attacks, emerged from the primordial swamp of Picardy. While each individual operation would be limited in time and space, with careful sequencing "modern" war, in its 1918 incarnation, would be both attritional and disruptive of fixed defences. The Somme confirmed that the war, warfare itself, had changed profoundly. This had not, however, altered the fundamental strategic truth that military victory, if it were achievable, had to be won against the enemy's main army in the principal theatre.

These controversies over strategy and command, and shifting soldiers' memories of war experience, constitute the dual strands of Winter and Prost's first configuration of memory, established between the wars. In Hankey's account of the Somme, written in 1938 after his retirement from the post of Cabinet Secretary, he was clearly trying to balance the battle's two contradictory facets. He strove to establish the offensive's grandeur and achievement, woven in with the grim experience.

> Considerable success was achieved at first ... but there was nothing in the nature of a breakthrough ... the battle of the Somme degenerated into a mere "slogging match," in which we

always had the upper hand, but were never able to obtain a big strategical victory. Division after division took its turn, each adding to its escutcheon the name of some captured town or village, hitherto unheard of and already unrecognizable, or maybe of some scraggy eyesore of a wood, High Wood, Delville Wood, Mametz Wood, etc.—otherwise unknown to history—where imperishable deeds had been performed. Moreover, the moral effect must not be lost sight of . . . our new, hastily improvised volunteer army . . . had shown themselves a match, and perhaps a little more than a match for the most highly trained and organized army that the world has ever seen.

He accepted the mistakes in execution as inevitable with a hastily improvised army. Nevertheless, the Somme revealed to the enemy that "his army was not only not invincible, but that he had no monopoly of the offensive. More especially must this realization have come home to him now that the failure of his attacks at Verdun had revealed the unwelcome fact that the war had become one of attrition." But despite this impression of hard-fought attritional victory and heroic effort (which was magnanimous, given Hankey's hostility to the battle), he finished his recollections of the Somme with a quote from Dante's *Inferno:* no doubt thoughts of his dead brother were in his mind.[78] By the time Hankey's memoir was eventually published in 1961, it would be the hellish experience and ultimate failure, not the "considerable success" and moral victory, that were stressed by Taylor and his successors.

IF THE INTER-WAR generation had struggled to come to terms with the reasons for and meaning of their sacrifice, the events of 1939–45 robbed the Great War of what flimsy meaning it had retained, so a new one had to be found. The second historical configuration, which brought the ordinary soldier and civilian and their experiences to the fore, while denigrating their leaders, was more contentious. While it drew on the dual strands of memory of the pre–Second World War years, in the radically different Cold War political climate it warped them into a parable of class confrontation: one of uncaring, high-command butchers and helpless soldier victims. In the 1960s and 1970s such patterns of social division seemed appropriate for explaining Europe's great tragedy. In this configuration the Great War shifted from a meaningful sacrifice into an irredeemable tragedy: although the memorials left in Picardy's fields between the wars were concrete, memory proved abstract and mutable. Once the "war to end all wars" proved the precursor to an even more extensive, more vicious contest, with its own four horsemen—attrition again, aerial bombardment, occupation and above all genocide—in which the settling of old scores between Germany, France and Britain merely set the stage for a truly global conflict, the futility of the earlier conflict seemed self-evident. In contrast, the new war was prosecuted with vigour and purpose. Its failures are generally heroic, its victories just, and this "good war"— Taylor's expression, in conscious contrast with the bad one of his formative years—presents none of the supposed senselessness of its predecessor.

Even its battles seemed to live up to expectations. When writing of the "hinge of fate" of the Second World War, Eighth Army's desert victory at El Alamein, Churchill could not help but contrast its 13,500 casualties with the notorious losses of 1 July 1916. "It may almost be said, 'Before Alamein we never had a victory. After Alamein we never had a defeat,'" he pronounced. The Somme had been the equivalent combat in the previous war, after which victories were to outnumber defeats. But Churchill was never able to acknowledge the 1916 turning point: his comparators remained the 1917 Battle of Cambrai and the British army's 1918 victories, on account of the "forward inrush of the tanks." Yet Churchill could not avoid identifying Alamein with the set-piece offensives of the First World War. Montgomery, Eighth Army's commander, had learned the central component of such a battle— "artillery in its heaviest concentration, the 'drum-fire barrage'"—on the Somme. Alamein was the elusive "breakthrough battle" of 1916 finally perfected, but it could no more win its war than greater operational success on the Somme would have won the First World War.[79]

IN THE POPULAR coverage of the fiftieth anniversary of the First World War the new mass medium of television was instrumental in bringing the history of the trenches to a wider audience. *The Great War,* a BBC television series first broadcast in 1964, was groundbreaking both as a documentary and as a history of the war. It stirred memories in an older generation, and interest in a younger one. In particular, for the first time veterans could give their testimony to camera—which made the war more vivid, even though their recollections were fifty years old—while the words of now-dead war leaders,

Lloyd George and Haig prominent among them, could be intoned solemnly to the viewing public. The series set the tone for engaging with the war for the next forty years. The testimony of veterans, on the one hand, the rivalries of soldier and statesman, on the other, represented the bifurcated lines of memory of the conflict.

After the trauma of another war the veterans' children and grandchildren belatedly enjoyed a better world. Europe's prolonged postwar boom, one of opportunity and promise, of social democracy, full employment, consumerism and liberalisation, took off in the 1950s and lasted throughout the 1960s (although the shibboleth of class conflict, whether domestic or international, remained). As a consequence, rediscovery and repopularisation of the war were intertwined with its reevaluation in the light of fifty years of subsequent history. Against this background of belated progress and prosperity the bitter, costly, hellish, disillusioning stalemate war firmly muscled aside the popular, patriotic, meaningful conflict. Joan Littlewood's 1963 Theatre Workshop play, *Oh! What a Lovely War,* and the film version that followed six years later, in particular slewed younger generations' perceptions of the conflict in which their fathers and grandfathers had fought, moulding it into the class-political parody that has for a long time dominated public perception. Yet, as Theatre Workshop's historical consultant Raymond Fletcher, the military correspondent of the left-wing journal *Tribune* (and later revealed as a Soviet spy), was happy to admit, his advice had been "one part me, one part Liddell Hart, the rest Lenin!"[80]

It should hardly be surprising that old history was bundled out along with the values of previous generations. Historians colluded in this paradigm shift. Between the wars historians had taken a firm hold of the Somme and the wider war. Drawing on the competing accounts of the principals, the earliest influential popular and supposedly well-informed histories had appeared, Liddell Hart's *The Real War* prominent among them, which was revised and reprinted as *The First World War* after the Second. Such works were the roots of A. J. P. Taylor's influential work. His widely read *The First World War: An Illustrated History* (tellingly dedicated to Joan Littlewood) cemented this modern view of the Western Front in the British public's eyes. Littlewood's cast, who absorbed Taylor's book avidly, were reportedly "delighted that a serious scholar—well, more or less serious—confirmed the version of the war they were putting on stage." In turn, Littlewood's play inspired Taylor's valedictory lecture at Oxford University, which confirmed "how far historical research endorsed Joan's version."[81] Needless to say, there was little historical research in Taylor's throwaway account. Generals with antediluvian methods were cited as the guilty parties, but craven politicians shared the blame: "The civilian rulers were pushed aside in every country, and were often glad to shelter behind the military leaders ... Many ministers had doubts whether the war could be won by going on in the old way, but they were at a loss what else to do."[82]

The seeds of our folk memories of the Great War were sown by Littlewood and Taylor, if hardly reflecting how the war generation saw their fight, or the military, social and political importance of the struggle. In the meantime the ordinary soldiers of the Great War—and their shock troops, the modern poets—had become the heroes of the Western Front. Generals' reputations were at an all-time low, as such pseudo-historical books as Norman Dixon's *On the Psychology of Military Incompetence* sought to tarnish Haig's character and reputation further with an unfavourable comparison with the then paradigm of the "victor of Alamein," General Montgomery. Dixon (citing Montgomery's own skewed memoir of the earlier war) proposed that the only thing "Monty" had learned from Haig's twentieth-century battle school was to despise generals who had a complete disregard for human life.[83] The lions led by those donkeys of increasingly ill-repute found their own historians in Martin Middlebrook, Lyn Macdonald and Bill Gammage, among others, and their tales continue to be retold by the likes of Malcolm Brown and Peter Hart.

It should come as no surprise that the veterans themselves were sucked in by this post-facto generalisation of the nature of their war. As their own memories faded they looked to the slew of books that reminded them of their war and purported to explain the bigger picture that they had glimpsed only from the bottom of a trench. Churchill and Liddell Hart, then Taylor and Middlebrook, offered clear, convincing interpretations intertwined with memories of fallen comrades and details of harsh post-war realities. As a result, these books often convinced the combatants of the overarching futility and tragedy of their youthful fight.[84] But such a collective memory has its pitfalls. Take the example of Sassoon, who has been numbered among the leading pacifists on account of his 1917 protest. He certainly had a problematic relationship with his war as it was going on, which translated into support for the prevailing pacifist mood in the 1920s and 1930s. But it is rarely acknowledged that later events, and greater maturity, caused him to re-evaluate his wartime dissent. The rise of a new German menace led him to reject his liberal-pacifist stance and resume a more considered point of view that in certain circumstances war was just, even if it was always terrible. In a later, post–Second World War memoir he acknowledged the youthful folly and conflicted personal motives in his wartime protest: "in the light of subsequent events it is difficult to believe that a Peace negotiated in 1917 would have been permanent. I share the general opinion that nothing on earth would have prevented a recurrence of Teutonic aggressiveness."[85]

YET THE TROOPS on the Western Front were not the victims that twentieth-century history has made them, but complicit actors in their own sacrificial epiphany.[86] By the 1980s and 1990s, as Europe shifted out of its Marxist mind-set, in a third historical configuration the war was assessed as a cultural phenomenon as much as a social conflict. War experience, memory and representation became themes for investigation and the war became a living, breathing thing once more as the grandchildren and great-grandchildren of that fading generation picked up on their experiences in a "memory boom" that tried to comprehend the unattainable and unknowable while there were still some survivors left to retell their tales.

By that time the voice from the trenches had become strong, even strident. It has not, however, drowned out the voice from the châteaux and conference rooms entirely. Memory and history are obviously separate things, and Jay Winter rightly recognises that they can never be conflated:[87] memory cascades down the generations; history is a product of its time. Yet ultimately historians and their histories shape collective memory. Some mould and reinforce it. After Taylor came Sir John Keegan and Sir Martin Gilbert, eminent authorities (and scions of the post-1945 world) who added the weight of their reputations to the superstructure that supports the view of the First World War as an overriding human tragedy.[88] On the other hand, the worker-bees of the historical profession make their reputations by engaging with and challenging such dominant, quasi-mythical perspectives. Historians have engaged with the Somme since the 1960s, with greater or lesser success. The arguments about British strategy and performance on the battlefield, the parameters of which were set by the likes of Churchill, Liddell Hart and Taylor, ultimately became stale and repetitive. The perspective needed to be widened, beyond that of generals and soldiers to that of armies and nations, methods and myths.

The eightieth anniversary of the conflict shifted the process of memory and remembrance once more, following another socio-political earthquake in Europe. The Great War was a growing "presence" in the cultural landscape of *fin de siècle* Europe that had entered a new post–Cold War phase of its history, a phase that shockingly saw a return of war and genocide to Europe in the Balkans, with atavistic 1914 connotations. The final few, long-lived veterans, the last tangible links with that increasingly distant and unknowable conflict, became icons of a lost world.[89] The process had begun in the 1960s, with the foundation of the transnational *Ceux de la Somme* association, based in Péronne, to identify and bring together veterans from both sides of the line. France and Germany had set aside their seventy-five-year feud and were leaders of a new Europe taking its first tentative steps on the road to union. The sharing of their common memory of adversity was one symbol of the future. The Western Front Association was founded in Britain in 1980, its purpose "remembering."[90] In time it branched out into France, Germany and the United States. Part veterans' association, part history society, part genealogical organisation, part pressure group, the wide reach and broad range of activity of the Western Front Association ensured that the memory of the war would live on, even if its participants were inexorably disappearing.

Societies were never going to forget their war, even if they would remember it in different ways now that Western European states were no longer fighting among themselves. Representative of this new approach was the foundation of the *Historial de la Grande Guerre* in Péronne, an international museum and research centre opened in 1992. This was not to be a static, judgemental museum, but a space for historical remembrance, for engaging with and exploring modern society's changing relationship with its foundation myth. It was to be an international "experiment in collaborative history," a joint endeavour of French, British and German historians and museologists, "former allies and former enemies," funded by the regional government, recognising and representing their shared heritage while acknowledging its mutable, disputed nature.[91] The authority of the Somme's timeless victory monuments would in future be challenged by a vibrant intellectual memorial, "a midpoint between . . . cold, dispassionate, precise history and warm, evocative, messy memory."[92]

Down the years the war had become a contested cultural phenomenon, as well as a contentious historical event. The evolving history of the Somme must be understood in this framework. Recently authors have endeavoured to explain why the so-called literary memory of the war, grounded in the poetry and memoirs of the inter-war years, has eclipsed its historical reality. Studies of popular culture, propaganda, art and film, memorials and remembrance have spread engagement with the nature and legacy of the war far and wide. But they have muddied as much as clarified the picture. Similarly, a concerted scholarly counter-offensive to explain the nature and impact of industrial war has been mounted against the prevailing perception of the Western Front as an arena for futile slaughter.[93] This approach may finally be showing some signs of success; but reconciling divergent perceptions of the war and resolving the tensions between cultural memory and historical accuracy will be long, hard and perhaps impossible tasks. Futility can be refuted, but the slaughter, "the weight of the dead on the living," can never be removed.

UNDERSTANDING THE GREAT War at this distance may be an impossible aspiration. There is "a breach in understanding . . . The sense of obligation, of unquestioned sacrifice, which held most people in its tenacious, cruel clutches for so long and so profoundly, and without which the war could never have lasted as long as it did, is no longer acceptable" in contemporary society. Without it, however present the First World War may be in modern Europe, paradoxically "it is as though we wished to understand the Great War more than ever before without being sure of ever having the means to do so."[94]

For example, when the newly founded *Historial de la Grande Guerre* sought to represent the Battle of the Somme, its local and seminal campaign, it was at a loss. Winter (one of its instigators) explained that "without even a notice," it would offer

a metaphoric representation of something not there—the Battle of the Somme itself. In this museum dedicated to the history of the war in general, and to the war on the Western Front in particular, it may seem odd that there is no visual representation of this battle. Nor can there be. By leaving a blank wall on your right between rooms 2 and 3 . . . we confront you with the radical impossibility of representing the Battle of the Somme. This absence, or silence, or void, if you will, is an anti-monument, a challenge to our comprehension of the anticipation, the terror and exhaustion, and pain, and anguish, and ugliness which constituted the battle . . . We bring the visitor, through a set of spatial metaphors, to the limits of representation itself.[95]

Some may dismiss this as postmodern pretension (or a cop-out), and the original idea of having the Somme experience represented by a mannequin of a *poilu* enduring endless rain reinforces such an impression.[96] However, for the battle's ninetieth anniversary the *Historial* was able to come up with a clear representation of the Somme as a "world arena."[97] But the point that the more it recedes into history, the more a world event like the Somme becomes unknowable and incomprehensible is a fair one. Since the cultural responses to the Great War have been so diverse and prolonged, they have been interpreted, interrogated and misappropriated in equally varied and ultimately nugatory ways. With such a world event its meaning was too complex, too fragmented to allow a simple, unitary explanation. Yet while its history may evolve, arguments may continue and new books be published, the essential facts of the Battle of the Somme persist, reaffirmed to the point where they have become mythical.

REMEMBERING THE PAST is a contentious and flawed process, especially when war is the subject. The history of war, they say, is written by the victors. But the history written by the victors of 1914–18, and the Somme's place in it, is curious. Although the Battle of the Somme looks unlikely to be forgotten, it is, Geoff Dyer has suggested, "deeply buried in its own aftermath."[98] When it was still ongoing the writers, cameramen and memoirists were already harvesting its image for future generations. If we can still go there today thanks to Malins's camerawork, it is to a sepia-tinted pseudo-past that has been shaped and reshaped by later events and muddled recollections. Throughout that process the Somme, or its 1916 incarnation at least, has held the foreground in British national memory, commemorated annually, its story retold regularly, reinforced and embellished repeatedly as new memoirs of the horror of it all surface year-on-year from a publishing industry keen to satisfy an apparently insatiable need to revisit the trenches. If a fuller picture of the Somme is belatedly presented here, it can never be complete, and it is certainly squeezed between the national memories and histories of the participants. Myth, of course, is always simple; an easily digestible, factually anaemic, unsophisticated parody of the truth. While history is complex; more so as it recedes and the contemporary voices are lost, and a babble of later opinions competes for attention. As an event, the mythical Somme has lost most of its context, the majority of its French and German dimensions, its deeper roots and medium- and long-term consequences, and its central place in the war efforts of all three belligerents. What has persisted is a caricature, a snapshot of a British imperial army fighting an unfamiliar and vicious war, of tactical disasters and civil–military tensions magnified into strategic stalemate, which in combination nourish impressions of futility. These myths are rooted in the experiences and recollections of contemporary observers, and have grown in their telling and retelling, but they were inevitably distilled and relocated over time. They will continue: they have a life of their own now. The "first day of the Somme" has become shorthand for military disaster and command incompetence; 15 September and the deployment of the tanks represents a wasted opportunity; and the mud, cold and wet of the winter is evidence of uncaring, detached authority and innocent soldier victims. These are all British myths: an Anglo-Saxon Battle of the Somme, not the complete three-empire encounter. Germany and France manufactured their own myths: of heroic defence against the odds and sacrifice, respectively, which deserve equal recognition in the history and memory of the world-changing encounter in Picardy.

Reflection

O N 1 JULY 2006, CHARLES, Prince of Wales, heir to the British throne, led the mourners at a memorial service before the imposing Thiepval Memorial to the Missing of the Somme. "The magnitude of the allied losses on 1 July 1916 are unimaginable . . . these days . . . but even ninety years ago they caused a profound shock to our nations and left scars that remain with us today," Prince Charles told the five-thousand-strong crowd.[1] Ninety years on, the anniversary of the opening of the Battle of the Somme remained very alive in Britain's collective memory, the centrepiece of several months of commemoration of the British Empire's most iconic battle. Laurence Binyon's emotive stanza from his 1914 poem "For the Fallen" was recited once more:

> They shall not grow old as we that are left grow old;
> Age shall not weary them, nor the years condemn.
> At the going down of the sun, and in the morning,
> We will remember them.

A bugler from the Royal Irish Regiment played "The Last Post" to honour the memory of the 73,357 British and South African fallen from the 1916 and 1917 Somme battles who have no known grave, whose names line up in serried ranks on the columns of Britain's largest war memorial. The flags of old comrades' associations—the British Legion, the former empire's veterans' associations, France's *anciens combattants*—were inclined in homage. Wreaths of poppies were laid. With only a single superannuated veteran, 110-year-old Henry Allingham, now in attendance to remember his youthful pals, a party drawn from Great War living history groups marched along the Western Front kitted out like 1916 infantrymen, re-enacting the appointment with destiny of their grandsires. If the Tommies of the Somme were finally passing from living memory into history, there were modern ways of commemoration, new forms of pageantry and theatre that would keep the memory alive after those few who could still remember had finally rejoined their forever-young comrades.

The experience of the trenches has been anthologised and reiterated to the point where familiarity breeds acceptance and reassurance (in this case it could never breed contempt). Unfortunately, though, we have increasingly remembered the men of the Great War, their efforts and sacrifice, while forgetting their mind-set and motives, and misunderstanding their milieu and methods. Gunner Hugh Monteith, for example, always wrestled with the juxtaposition of the two in his personal remembrance: " 'Extolling the virtues of war' is sometimes flung at Anzac Day. It was hardly that from my experience. Rather it is a day to nudge aside the mind from the daily dollar, and think about the digger-spirit and the price paid. 'Paid for what?'—some may ask. 'The price paid for going to the help of others, thousands of miles away?' "[2]

THE REST OF us remain caught in that veteran's essential conundrum. As Monteith himself identified sixty years on after revisiting his old front line, "The once seemingly clear-cut issues of the Great War have become blurred in the minds of many." Germany was now one of the Western allies. But, he reasserted, he had joined up to contest the spread of German militarism and remained certain "that grim would have been the lives of many, if Germany and her allies had gained final victory in the 1914–18 war."[3]

Those who fought the Battle of the Somme judged it a success, despite its copious human sacrifice: a necessary evil subsumed within a greater just cause. Subsequently, however, its contemporary critics spoke up, drowning out those voices that attested its importance, although these never quite fell silent. It is perhaps surprising that an event that changed so much has come down to posterity as an indecisive, futile encounter. At the end of the Somme offensive everything may have seemed the same, including the lines on the map, more or less, but the world was profoundly different. By the 1960s, the Somme had become the root of all the evil that had come after: "Who can say how far the troubles of the world today did not stem, via the Second World War, from the Battle of the Somme," the novelist John Harris could assert in his derivative 1966 summary.[4]

If, over the ensuing years, the distorting lenses of memory and history dulled the perception of victory, the purpose of the Somme, if it was not to defeat the German army, was never adequately redefined. In his post-battle despatch Haig listed its rationales, as if they were its achievements: to relieve Verdun; to help other allies; to wear down the enemy; to blood and train Britain's New Armies. These were indeed objectives of the Somme offensive, but neither singly nor collectively could they explain its course. Its purpose was to show the enemy that her army would be beaten; to show the French and British, both soldiers and civilians, that the German army could be beaten, if not quickly and easily, then eventually and conclusively. By instilling this belief in the allied armies, and by gaining the initiative and advantage in the land war that up to that point had lain with Germany, the Somme was the decisive victory of the attritional war of which it was the centrepiece: a moral victory based on growing *matériel* predominance and improving tactical and operational ability.

Of course, the Battle of the Somme did not live up to Joffre's grandiose December 1915 conception in scale, method or results. But much more than the Anglo-French offensive plan changed during the course of 1916. As the offensive was scaled down over the first half of 1916, there was a growing realisation that it would not be the decisive battle; not on its own, at least, but if combined with other allied offensives it might help crush the fighting spirit of the Central Powers and turn the war to the allies' advantage. By September 1916, when the Somme was at its height, the General Allied Offensive had Germany in a death grip. A better British or more numerous French army might have pulled off a real victory. That the Somme offensive could not fulfil more grandiose aims was due to three factors: an admirable defence by a professional army at the height of its powers, which demonstrated tenacity and bravery in the face of overwhelming odds; the chaos and confusion in the combat zone, which ensured that when battle was engaged actions and reactions would be slow and opportunities would be missed; and the inability of the logistical chain to sustain such an intense fight over prolonged time and through congested space. However, viewed as an Anglo-French operation, the offensive was conducted according to a coherent strategic and operational scheme (for which Foch should take much of the credit), with tactics that were appropriate and responsive to the conditions (at least as far as any could be) and came closer to breaking German resistance than is generally supposed.

· · ·

NOW THAT THE echoing, plaintive voices of the Great War's veterans have fallen silent, the Somme and its war can finally become a part of history, "a terrible and tragic event certainly, but one that must be understood in its context, and explained rather than condemned." The great battle can be examined in its entirety and complexity, and even rehabilitated as the "quest for blame" of the post-war decades subsides.[5] Critics of the Somme fixed upon the heavy losses and meagre gains in the British sector in its first month, and this distorting prism has obscured the real Somme ever since. Anglophone historians have been returning to the battle regularly for more than ninety years, to refight old skirmishes and reopen old wounds. An agenda set by the antagonists in Britain and her empire's contemporary political discourse and subsequent civil turmoil has been mulled over, distilled and regurgitated, sometimes as scholarship, sometimes as polemic. Certain themes recur: command, the soldiers' experience, tactical and operational method and the possibilities of the 1916 battlefield, for cavalry and tanks in particular. This study has given its own commentary and perspective on these hoary controversies. But it has also suggested that what Haig and Rawlinson contemplated in 1916 was not impossible, even if it was improbable, given the lack of training and inexperience of Kitchener's Army. Comparison with the better-equipped and more experienced French army demonstrates that battlefield method was sophisticated enough to break into the German defensive system, and that it was not poor tactics, but undynamic operational methods that stymied offensive battle in 1916. It has also established that the ensuing long, attritional battle was both anticipated and possessed structure and purpose. It was neither an afterthought nor a self-justification by a beleaguered British commander-in-chief.

The longevity of Churchill's criticisms of July 1916, valid as they may have been at that moment, does not suggest subsequent reflection on the nature of attrition, which requires both *matériel* superiority and battlefield skill to be effective. The British army lacked both of these in July and August 1916; although by September its deficiencies were being addressed. Unfortunately, ever since July 1916, the narrative of the Somme has been determined by what went wrong, rather than what went right. Haig's phlegmatic, realistic appreciation of the nature of the war was more apposite, but unpalatable. "Our French friends are already talking of the Rhine and how soon they can be there!" he noted as his armies went forward on the first day of the Somme. "We have many hard fights yet before that objective is reached."[6] The fact that it was reached some two years later was largely due to events on the Somme, and the hard fights that followed.

There were two battles on the Somme in 1916: the sudden shock offensive of early July, the tragic last hurrah of an old style of warfare; and the sustained, multi-army advance of September, the first manifestation of modern operational warfare. Two gruelling attritional phases between these dates and after—the July–August one dictated by *matériel* and moral factors; and the October–November one dictated by logistics and weather—gave the battle its enduring character. Thus revolutionary changes in warfare have been swamped by mud and blood. The fifth and sixth segments of the Somme offensive—the persistent forward-sapping of winter 1917, which rendered the German front untenable; and the pursuit of the retreating German army in the spring—are rarely even identified as phases of the same battle.

In the actual circumstances of battle and "total" war in 1916, any quest for breakthrough and decisive battlefield victory was chimerical. The breaches which Sixth Army made on the Somme, products of momentary *matériel* superiority and modern offensive methods executed by highly skilled troops, represented the height of tactical finesse on the industrial battlefield. But they were too small and too easily sealed by an adversary with adequate reserves and the advantage of logistics ever to amount to much. Even if they had been exploited, the defence would merely have consolidated on the next natural obstacle or the next man-made defensive position. Such was the case in March 1918 when Ludendorff's vaunted storm-troops seemed to have smashed the enemy's front. Of greater military significance were the developments in the operational art of war that Foch conceived and applied on the battlefield scale in September 1916, and on the battlefront scale in summer and autumn 1918. These shattered a static front, pushing it back without breaking through it, and brought the Anglo-French armies victory. For a brief interlude, from late 1914 until early 1918, tactics, logistics and communications fell out of synchronisation with operational ambition. But as the former improved and the latter diminished, in 1918 battle once again became an instrument of strategy.

The way to defeat Germany had been grasped early on: an attritional strategy, coordinated high-tempo operations and *matériel*-intensive tactics. It was political imperatives more than military errors that hurried its implementation and made it so fraught. But victory would take time and cost lives and money whatever happened and whoever was ultimately responsible. That truth was inherent to the nature of industrial war.

AFTER MORE THAN ninety years the commemoration of the Somme seems to be growing rather than shrinking, in spite of the passing of the battle's veterans. These days what the authorities in Picardy have dubbed *"tourisme de mémoire"* is bringing visitors of all ages in increasing numbers to the Somme's monuments and museums.[7] It suits many of them to remember the battle as a tragedy and to mourn its victims. Yet, despite its horrors and hardships, many others view it as a triumph, or certainly as a worthwhile sacrifice. If there had been no purpose to it, why would the generation of 1914–18 have sent their sons to fight the menace to civilisation that revived in extreme form twenty years later?

That is the essential, barely comprehensible paradox of war, especially so for a campaign like that fought on the Western Front. Placing the events of 1916 in context shows how they were a product of what had gone before, and would determine what came after; not just on the battlefield, but in the corridors of power and the towns and villages of the belligerents. It would be wrong to attribute all that came after to the Somme alone, because the wider war, and the unsatisfactory peace that concluded it, played greater roles. Nevertheless, the First World War was a life-or-death struggle that reached beyond the battlefield, an attritional fight for survival between Europe's industrial empires. Placing the Somme, its attritional fulcrum, right at its centre—where it should be, not just in time, but in the motion of societies from one epoch to the next—emphasises its seminal significance.

It is easy to say that the Battle of the Somme should not have been fought; that other strategies offered more direct or less costly paths to victory. But those who were there did believe that it had to be fought, because they believed in what they were fighting for. Even its contemporary critics decried only the blunt method and its consequences, not the need or the purpose. It was the military turning-point of the war—Ludendorff confirmed as much after the conflict[8]—although two further years of hard fighting were necessary to assert Anglo-French superiority on the battlefield and deliver the *coup de grâce*. After that point, however, whatever strategic expedient she tried, Germany could win only if allied resolve broke. In the event it wavered, but ultimately it held.

No one would dispute that the Battle of Stalingrad was the turning-point of the Second World War, horrific attritional campaign though it was. It is celebrated as a victory, despite its appalling human cost (which was on a par with that of Great War battles) and the fact that years of hard fighting still lay ahead. So why is the Battle of the Somme viewed in a different light? The answer lies in the enduring ambiguities of that battle, represented by its shifting military fortunes, its combatants' varied experiences, its disputed outcome and its profound consequences. Commandant Courtès grasped this essential paradox when reflecting on his own small part in the huge battle: "Pursued successfully owing to our incontestable superiority on this part of the front, it was nothing but a hecatomb for us, who were thrown in at a moment when the Boche had recovered himself. I cannot assess its consequences, but they were considerable."[9] A tactical stalemate still meant a strategic victory: a pyrrhic victory that nevertheless reversed the fortunes of war; and a genuine moral victory in adversity, over circumstances, the elements and the enemy. Even Charles Bean, who denounced the "dull, determined strategy" of 1916, reconciled it with "the devotion of an inexperienced army."[10] A battle could deliver a salutary lesson to the British army and a vindication of sound principles to the French, as well as a shock to the system of the German: the latter eventually died of wounds received on the Somme battlefield. "To this extent the battle marked a definite step towards the winning of the war," Bean conceded. "But the cost was dangerously high."[11] This cost has overshadowed the consequences ever since.

Only when studied in the round does the Somme's significance emerge. It could mark the death-knell for old ways of warfare and the cradle of modern combat. It could be both a necessary evil and a useless slaughter: a meaningful struggle and a futile sacrifice. It could uplift and motivate nations and empires, while at the same time cleaving them to the core. It could be both a victory and a defeat, for both sides at the same time, or over time. Battle on such a scale would be deterministic, if not immediately decisive. From 1914 to 1918 the young men of France, the British Empire and Germany fought for their past, present and future in Picardy. In the manufactured landscape of remembrance that the Somme battlefields have become we will continue to honour their efforts, at the same time endeavouring to understand them better.

Nowadays a human tragedy cannot be a military triumph. It was different in the early twentieth century, before the experience of industrial war fractured the societies that fought it irreversibly. That 1916 on the Western Front was, in Churchill's emotive summation, "a welter of slaughter," will never be forgotten.[12] Exactly why it was so, and what consequences it had, deserves to be remembered.

Appendix

A Note on Military Organisation, 1914-1918

Divisions

The division was the basic tactical formation for battlefield operations in the First World War, composed of infantry and artillery regiments with supporting cavalry, engineer, medical and administrative units. In 1914 French and German infantry divisions both had two infantry brigades, each of two regiments of three battalions (around 1000 men each). Battalions would be further subdivided into companies and platoons. French regiments for example had twelve companies, numbered 1 to 12. The British in contrast divided divisional infantry into three brigades of four battalions, drawn from different regiments, with a thirteenth pioneer battalion attached. German divisions had greater firepower than their French equivalents. Each German infantry regiment in 1914 had a separate machine-gun company of six Maxim guns, while the French attached a two-gun machine-gun section to each battalion. Each French division had an artillery regiment with thirty-six 75mm field guns, in nine batteries of four guns. German divisional artillery regiments had fifty-four guns, organised into three groups of three batteries with six guns each; one group was equipped with heavier 105mm field howitzers, the other two with 77mm field guns. British divisional artillery comprised three brigades, each with three six-gun batteries of 18-pounder field guns and one six-gun battery of 4.5-inch howitzers, and four 60-pounder medium guns (until 1915) — seventy-two guns in total in 1916. On mobilisation reserve formations generally had fewer guns and older equipment.

Cavalry divisions had a similar composition, comprising cavalry regiments, horse artillery and attached light infantry, sometimes mounted on bicycles for greater mobility. The French divisions had nine cavalry regiments, and one brigade of artillery (eight 75mm field guns), plus a single *chasseur* cycle company; the German three cavalry brigades, twelve 77mm field guns and one *Jäger* battalion, with a cycle company; the British four brigades each of three regiments, twenty-four 13-pounder field guns in two brigades of two batteries each, but no infantry component, the troopers being trained to fight as mounted infantry.

By 1916, although the basic shape of the division had not changed, its components had been reorganised around the new weapons of industrial war. The number of machine guns had increased and new machine-gun units had been added to the division. Heavy artillery was strengthened and lighter trench artillery became integral to divisional organisation. The infantry were equipped with new weapons of trench warfare—light and sub-machine guns, grenades and grenade-launchers, and light trench cannons—and organised into sections of specialists.

By 1918 the division had taken on a more modern form as its infantry component shrank and its artillery and supporting weapons elements expanded in response to the industrial battlefield. The standard lowest level infantry formation was now the platoon of mixed specialists, divided into sections. In late 1916 the French reorganised their divisions into two brigades, one of infantry (three regiments of three battalions) and one of artillery, which now included four 155mm field howitzers. The Germans had adopted this structure from the end of 1914. In early 1918 the British reduced their infantry brigades from four to three battalions. (Although colonial divisions retained the four-battalion structure almost till the end of the war, shortage of replacements meant that these battalions were always considerably below strength.)

Army Corps

An army corps was a grouping of divisions (either infantry or cavalry), usually two in number in 1914, but later in the war generally between three and five. The corps was an administrative and strategic formation, which would control non-divisional units such as heavy artillery and aviation, and later tank units. The number of corps in the field at any one time varied. Corps were customarily designated by Roman numerals (XV Corps), and this convention has been followed in this book to distinguish them from divisions designated by Arabic numerals (7th Division), and armies, spelled out in full (Fourth Army).

Armies and Army Groups

Each nation grouped its army corps into armies. France began the war with five armies on the Western Front, and formed three more by the end of 1914 (numbered Sixth, Ninth and Tenth). Germany deployed seven armies to the Western Front in August 1914, which she had supplemented with a number of "army detachments" by July 1916. These armies would later be grouped into larger army groups, intermediate formations between the commander-in-chief and the armies. France created three army groups from late 1914, the *Groupes des Armées du Nord, du Centre* and *de l'Est*. The *Groupe des Armées du Nord* was responsible for the Somme front. German army groups, organised from 1916, were named after their commanders. For example army group Crown Prince Rupprecht was responsible for the Somme front. The six-division British Expeditionary Force was equivalent to a small army in August 1914, and expanded to five armies by 1916, by which point it was equivalent in size to an over-large army group, although it retained the status of an independent allied army throughout the war.

FOR FURTHER INFORMATION, see John Ellis and Michael Cox, *The World War I Databook: The Essential Facts and Figures for All the Combatants* (London: Aurum Press, 1993).

Notes

Abbreviations

AAT   *Archives de l'armée de terre,* Vincennes

ADS   *Archives départmentales de la Somme,* Amiens

AFGG   French official history (Service historique de l'armée, *Les armies françaises dans la grande guerre*)

AIF   Australian Imperial Force

AK   *Armeekorps* (German army corps)

AWM   Australian War Memorial, Canberra, ACT

BCA   *Brigade/bataillon des chasseurs alpins* (French mountain infantry brigade/battalion)

BCP   *Brigade/bataillon des chasseurs à pied* (French light infantry brigade/battalion)

BEF   British Expeditionary Force

BI   *Brigade d'infanterie* (French infantry brigade)

BIC   *Brigade d'infanterie coloniale* (French Colonial army infantry brigade)

CA   *Corps d'armée* (French army corps)

CAC   *Corps d'armée colonial* (French Colonial army corps)

CC   *Corps de cavalerie* (French cavalry corps)

C.I.G.S.   Chief of the Imperial General Staff

DI   *Division d'infanterie* (French infantry division)

DIC   *Division d'infanterie coloniale* (French Colonial army division)

FD   Émile Fayolle's diary (from *Cahiers secrets de la grande guerre,* ed. Henry Contamine)

GAN   *Groupe des Armées du Nord* (Northern Army Group)

G.H.Q.   British General Headquarters

G.Q.G.   *Grand Quartier Général* (French General Headquarters)

HD   Sir Douglas Haig's Diary

ID   *Infanteriedivision* (German infantry division)

IR   *Infanterieregiment* (German infantry regiment)

JJ   Joseph Joffre's journal (from *Journal de marche de Joffre, 1916–19,* ed. Guy Pedroncini)

JMO   *Journal de marche et d'opérations*

LHCMA   Liddell Hart Centre for Military Archives, King's College London

NCO   Non-commissioned officer

NWAC   National War Aims Committee

OH   British Official History (Sir James Edmonds et al., *Military Operations, France and Belgium, 1914–1918*)

OHL   *Oberste Heeresleitung* (German Supreme Army Headquarters)

RAK   *Reservearmeekorps* (German reserve army corps)

RI   *Régiment d'infanterie* (French infantry regiment)

RIC   *Régiment d'infanterie colonial* (French Colonial army infantry regiment)

RID   *Reserveinfanteriedivision* (German reserve infantry division)

RIR   *Reserveinfanterieregiment* (German reserve infantry regiment)

SPD   German Social Democratic Party

SWC   Supreme War Council

TNA   The National Archives, Kew

UGACPE *Union des grandes associations contre la propagande ennemie*

Engagement

1. Walter Page to Woodrow Wilson, 21 July 1916, in Burton J. Hendrick, *The Life and Letters of Walter H. Page* (London: William Heinemann, three vols., 1925), *iii*, p. 307.

2. *La Bataille de la Somme: Un Espace mondial* (Paris: Somogy éditions d'art, 2006), pp. 7 and 12.

3. Ernst Jünger, *Storm of Steel*, trans. Basil Creighton (London: Chatto & Windus, 1930), p. 110.

4. Winston S. C. Churchill, *The World Crisis* (London: Thornton Butterworth, six vols., 1923–31). This work used a later revised edition, *The World Crisis, 1911–1918* (London: Odhams Press, two vols., 1938).

5. Quoted in Ian Beckett, "Frocks and Brasshats," in *The First World War and British Military History*, ed. Brian J. Bond (Oxford: Clarendon Press, 1991), p. 97.

6. Churchill, op. cit., *ii*, pp. 1083–89.

7. Robin Prior, *Churchill's* World Crisis *as History* (Beckenham: Croom Helm, 1983), p. iii.

8. Lord Sydenham et al., *The World Crisis by Winston Churchill: A Criticism* (London: Hutchinson, n.d.), p. 6.

9. Prior, op. cit., p. 90.

10. Alan J. P. Taylor, *The First World War: An Illustrated History* (London: Hamish Hamilton, 1963).

11. Gary Sheffield, *Forgotten Victory. The First World War: Myths and Realities* (London: Review, 2002), p. 177.

12. Sydenham, op. cit., p. 23.

13. Viscount Esher, *The Tragedy of Lord Kitchener* (London: John Murray, 1921), p. 207.

14. John Masefield, *The Old Front Line* (Barnsley: Pen & Sword, 2003 [originally published 1917]), pp. 75–76.

15. Ibid., p. 11.

16. Basil H. Liddell Hart, *The Real War, 1914–1918* (London: Faber & Faber, 1930). Subsequently republished with revisions and additional material under the titles *A History of the World War* and *History of the First World War*.

17. Brian J. Bond, *The Unquiet Western Front: The First World War in Literature and History* (Cambridge: Cambridge University Press, 2002). A selection of the contemporaneous writing on the 1916 battle is anthologised in *The Fierce Light: The Battle of the Somme, July–November 1916*, ed. Anne Powell (Ceredigion: Palladour Books, 1996).

18. Robin Prior and Trevor Wilson, *The Somme* (New Haven, Conn.: Yale University Press, 2005), p. 4.

19. Basil H. Liddell Hart, *History of the First World War* (London: Pan Books, 1972), p. 231.

20. Peter Hart, *The Somme* (London: Weidenfeld & Nicolson, 2005), p. 11.

21. John Terraine, *The Smoke and the Fire: Myths and Anti-Myths of War, 1861–1945* (London: Leo Cooper, 1992), p. 45.

22. Churchill, op. cit., *ii*, p. 1077.

23. Hew Strachan, *The First World War: A New Illustrated History* (London: Simon & Schuster, 2003), p. 161.

24. Pierre Miquel, *Les Oubliés de la Somme: juillet–novembre 1916* (Paris: Éditions Tallandier, 2001), p. 8.

25. F. Scott Fitzgerald, *Tender Is the Night* (London: Penguin Books, 1983 edn. of 1934 original), p. 67.

26. Sheffield, op. cit., pp. 170–71.

27. Captain von Hentig, quoted in Gary Sheffield, *The Somme* (London: Cassell, 2003), p. 155.

28. Ernst Jünger, *Storm of Steel*, trans. M. Hoffmann (London: Allen Lane, 2003), p. 69.

29. General Ludendorff, *My War Memories, 1914–1918* (London: Hutchinson, two vols., 2nd edn., n.d.), *i*, p. 304.

30. Quoted in Churchill, op. cit., pp. 915–16.

Chapter 1: Three Armies

1. Henri Douchet's journal, 28 August 1914, quoted in Helen McPhail, *The Long Silence: Civilian Life Under the German Occupation of Northern France* (London: I. B. Tauris, 2001), p. 18.

2. Ibid.; Henri Malo, *Villes de Picardie* (Paris: Librairie d'art et d'Histoire, 1920), pp. 49–50.

3. *La Bataille en Picardie: Combattre de l'antiquité au XX siècle*, ed. Philippe Nivet (Amiens: Encrage, 2000).

4. Malo, op. cit., p. 41.

5. Richard Holmes, *Fatal Avenue* (London: Jonathan Cape, 1992), p. 3.

6. Malo, op. cit., pp. 48–49.

7. Michael Howard, *The Franco-Prussian War* (London: Collins, 1967), pp. 396–97 and 403–6.

8. Comte de Caix de Saint-Aymour, *La Marche sur Paris de l'aile droite allemande: ses derniers combats, 26 août–4 septembre 1914* (Paris: Henri Charles-Lavauzelle, 1916), p. 9.

9. Brig.-Gen. Sir James Edmonds et al., *Military Operations, France and Belgium, 1914* (London: Macmillan, 1937) (hereafter, *OH*), *i*, pp. 20–21 and 493–59.

10. Général Barthélemy E. Palat, *La Grande Guerre sur le front occidental* (Paris: Berger-Levrault, fourteen vols., 1917–29) (hereafter Palat), *v*, p. 140.

11. "307 RI, bataille de Moislains (Somme) du 28 août 1914," *Archives de l'armée de terre*, Vincennes (hereafter AAT), 1Kt169.

12. Palat, *v*, pp. 141–43.

13. Douglas Porch, *The March to the Marne: The French Army, 1871–1914* (Cambridge: Cambridge University Press, 1981), pp. 213–31.

14. Michel Goya, *La Chair et l'acier: l'armée français et l'invention de la guerre moderne, 1914–1918* (Paris: Tallandier, 2004), pp. 69–112.

15. Grandmaison (1908) quoted in ibid., p. 99.

16. Goya, op. cit., pp. 95–96 and 162.

17. Ibid., pp. 82–85 and 136–41.

18. Marshal Joffre, *The Memoirs of Marshal Joffre*, trans. Col. T. Bentley Mott (London: Geoffrey Bles, two vols., 1932), *ii*, p. 20.

19. Général Desmazes, *Joffre: La Victoire du caractère* (Paris: Nouvelles Éditions Latines, 1955), pp. 2–61.

20. Basil H. Liddell Hart, *Reputations: Ten Years After* (London: John Murray, 1925), p. 7.

21. Note by de Langler, in Fonds Duchaussoy, *Archives départementales de la Somme*, Amiens (hereafter ADS), 14J48.

22. Untitled note in ibid.

23. Bruno Barbier, *La Grande guerre à Amiens* (Amiens: Encrage, 1992), p. 34.

24. Ibid., pp. 47–48; note by de Langler, op. cit.

25. Liddell Hart, *History of the First World War*, p. 78.

26. "La Bataille Circulaire," 25 August 1914, in Gabriel Hanotaux, *Pendant la grande guerre (aôut–décembre 1914)* (Paris: Librairie Plon, 1916), p. 75. This volume is a collection of his articles in *Le Figaro* and *La Review hebdomadaire*.

27. "Carnet du Brigadier Réquin," 5th dragoons, Historial de la Grande Guerre, Péronne, 29952.

28. Oberstleutnant Randebrock, quoted in Jack Sheldon, *The German Army on the Somme, 1914–1916* (Barnsley: Pen & Sword, 2005) (hereafter Sheldon), p. 12.

29. Service historique de l'armée, *Les Armées françaises dans la grande guerre* (Paris, Imprimerie nationale, eleven tomes in one hundred and three vols., 1922–1937) (hereafter AFGG), tome I, volume 4, pp. 148–65.

30. Eugène Pic, *Dans la tranchée, des Vosges en Picardie* (Paris: Perrin et Cie, 1917), pp. 30–31.

31. Sheldon, pp. 12–16.

32. Palat, *vii*, p. 178.

33. Quoted in Sheldon, p. 34.

34. A. Joubaire, *Pour la France: Carnet de route d'un fantassin* (Paris: Perrin et Cie, 1917), pp. 105–7.

35. Basil H. Liddell Hart, *Foch: The Man of Orleans* (London: Eyre & Spottiswoode, 1931), p. 68.

36. Palat, *vii*, pp. 182–83.

37. Cited in Liddell Hart, *Reputations*, p. 29.

38. Palat, *vii*, pp. 198–99.

39. XX CA to II Armée, 29 September 1914, "Compte rendu du fin de journée," AFGG, I/4, ax. 1662; Palat, *vii*, p. 211.

40. Général H. Colin, *Les Gars du 26e: Souvenirs du commandement du 26e RI de la division de fer, 1914–1915* (Paris: Payot, 1932), pp. 132–45.

41. Ibid., pp. 153–63.

42. "Psychologie de cette guerre," 26 August 1914, Hanotaux, op. cit., p. 80.

43. Fayolle diary, 30 August 1914, in Maréchal Fayolle, *Cahiers secrets de la grande guerre*, ed. Henry Contamine (Paris: Plon, 1964) (hereafter FD).

44. Palat, *vii*, pp. 172 and 203.

45. G.Q.G., "Note pour les commandants d'armée," 15 October 1914, AFGG, I/4, ax. 2864.

46. Strachan, *First World War*, p. 169.

47. "Psychologie de cette guerre" and "L'usure," 31 August 1914, Hanotaux, op. cit., pp. 81 and 100.

48. Joubaire, op. cit., pp. 110–29.

49. Ibid., pp. 122–23.

50. Palat, *vii,* pp. 277–78.

51. Palat, *viii,* pp. 173–77, 188–89 and 262–63; 4 November 1914, Raymond Poincaré, *Au Service de la France, vol. V: L'invasion, 1914* (Paris: Librairie Plon, 1928), p. 423.

52. Maurice Barrès, *L'âme française et la guerre: Les Saints de la France,* "Au milieu des saints de la France," 23 November 1914 (Paris: Emile-Paul Frères, 1915), p. 121; Joubaire, op. cit., p. 138.

53. Barrès, op. cit., p. 122.

54. 23 October 1914, Pic, op. cit., pp. 37–38.

55. Anon., *Souvenirs de guerre d'un sous-officier allemand* (Paris: Payot & Cie, 1918), pp. 114–15.

56. 24 September 1914, Commandant Bréant, *De l'Alsace à la Somme* (Paris, Librairie Hachette, 1917), p. 75.

57. 20 September, 21, 26, 29 October and 1 November 1914, 11 and 21 January 1915; Commandant Henri Bénard, *De la mort, de la boue, du sang: lettres de guerre d'un fantassin de 14–18,* ed. Françoise Lautier (Paris: Jacques Grancher, 1999), pp. 33, 40–43, 49, 54, 88 and 91.

58. Palat, *viii,* p. 9.

59. Joubaire, op. cit., pp. 150–51.

60. Pic, op. cit., p. 45.

61. 19 and 21 December 1914, Bénard, op. cit., pp. 73–74.

62. Palat, *ix,* pp. 43–44.

63. Pic, op. cit., pp. 46–47.

64. Quoted in McPhail, *The Long Silence,* p. 24.

65. Helen McPhail and Philip Guest, *St. Quentin, 1914–1918* (Barnsley: Leo Cooper, 2000), pp. 31–43; *Souvenirs de guerre d'un sous-officier allemand,* pp. 115–16.

66. "La Guerre par ceux qui l'ont faite: historique du 205e régiment d'infanterie de 1914 à 1918," AAT, IKt43.

67. Pic, op. cit., p. 49; Joubaire, op. cit., p. 151.

68. Sheldon, p. 54.

69. Pic, op. cit., p. 50.

70. *Souvenirs de guerre d'un sous-officier allemand,* p. 122.

71. Sheldon, pp. 49–50.

72. Ibid., p. 56; 21 January 1915, Bénard, op. cit., p. 91.

73. 14 and 16 November and 15 and 25 December 1914, Bénard, op. cit., pp. 57, 72 and 76.

74. 4 November 1914, Poincaré, *Au Service, v,* p. 423.

75. See Sheldon, pp. 62–64.

76. 17 and 19 March 1915, Bénard, op. cit., pp. 116–18.

77. AFGG, III/2, pp. 113–19; Sheldon, pp. 68–70.

78. Palat, *ix,* pp. 366–67.

79. Quoted in W. Michael Ryan, *Lieutenant-Colonel Charles à Court Repington: A Study in the Interaction of Personality, the Press and Power* (London: Garland Publishing, 1987), p. 149.

80. *The Times,* 5 August 1914.

81. Ibid.

82. Quoted in Ryan, op. cit., pp. 150–51.

83. *The Times,* 15 August 1914.

84. 25 August 1914, *Parliamentary Debates: Lords,* 5th Series, XVII (1914), 501–4.

85. Sir George Arthur, *Life of Lord Kitchener* (London: Macmillan, three vols., 1920), *iii*, pp. 2–3.

86. Ibid., p. 7.

87. "Rapport de mission en Angleterre," by Commandant Stirn, 23 October 1911, AAT, 7N1243.

88. Quoted in Edward Spiers, "The Late Victorian Army, 1868–1914," in *The Oxford History of the British Army*, ed. David Chandler and Ian Beckett (Oxford: Oxford University Press, 1996), p. 210.

89. Maj.-Gen. Sir Charles Callwell, *Experiences of a Dug-out, 1914–1918* (London: Constable, 1920), p. 4.

90. Edward Spiers, *Haldane: An Army Reformer* (Edinburgh: Edinburgh University Press, 1980).

91. Peter Simkins, "The Four Armies, 1914–1918," in *Oxford History of the British Army*, p. 235. See also Peter Simkins, *Kitchener's Army: The Raising of the New Armies, 1914–1916* (Manchester: Manchester University Press, 1988) and *A Nation in Arms: A Social Study of the British Army in the First World War*, ed. Ian Beckett and Keith Simpson (Manchester: Manchester University Press, 1985).

92. Simkins, "Four Armies," p. 239.

93. William Turner, *The "Accrington Pals': A Pictorial History* (Preston: The Lancashire Library, 1986), pp. 2–3.

94. Simkins, "Four Armies," p. 239.

95. Ibid., p. 240.

96. Nigel Hamilton, *Monty: The Making of a General, 1887–1942* (London: Hamish Hamilton, 1981), pp. 69–79.

97. David Fraser, *Alanbrooke* (London: Collins, 1982), pp. 60–63.

98. Basil Liddell Hart, *The Memoirs of Captain Liddell Hart* (London: Cassell, two vols., 1965), *i*, pp. 10–13.

99. Helen McPhail and Philip Guest, *On the Trail of the Poets of the Great War: Robert Graves and Siegfried Sassoon* (Barnsley: Pen and Sword, 2001), pp. 23–32.

100. John Joliffe, *Raymond Asquith: Life and Letters* (London: Collins, 1980), p. 188.

101. Roy Jenkins, *Asquith* (London: Collins, 1964), p. 378.

102. John Grigg, *Lloyd George: From Peace to War, 1912–1916* (London: HarperCollins, 1997), pp. 169–71.

103. "One of the fifty nine thousand," by J. R. Bamford, Australian War Memorial, Canberra (hereafter AWM), MSS0844.

104. *Lieutenant Owen William Steele of the Newfoundland Regiment: Diary and Letters*, ed. David Facey-Crowther (Montreal: McGill-Queen's University Press, 2002), pp. xiii–xix and 7–8.

105. H. C. O'Neill, *The Royal Fusiliers in the Great War* (London: William Heinemann, 1922), pp. vii and 2–5.

106. Roy Jenkins, *Churchill* (London: Macmillan, 2001), p. 300.

107. Sheldon, pp. 74–78.

108. Liddell Hart, *Memoirs, i,* p. 13.

109. A. F. Becke, *History of the Great War: Order of Battle, Part 4: The Army Council, G.H.Q.s, Armies and Corps* (London: HMSO, 1945), p. 92.

110. Capt. D. Sutherland, *War Diary of the Fifth Seaforth Highlanders, 51st Highland Division* (London: John Lane, The Bodley Head, 1920), pp. 32–35.

111. Maj. F. W. Bewsher, *The History of the 51st (Highland) Division, 1914–1918* (London: William Blackwood and Sons, 1921), pp. 33–34.

112. Sutherland, op. cit., pp. 42–43.

113. Bewsher, op. cit., pp. 39–42.

114. Ibid., p. 45.

115. General von der Marwitz, Commander of VI AK, quoted in Sheldon, p. 97.

116. Annika Mombauer, *Helmuth von Moltke and the Origins of the First World War* (Cambridge: Cambridge University Press, 2001), pp. 16–17.

117. F. J. Stephens and Graham Maddocks, *The Imperial German Army, 1900–1918* (London: Altmark Publishing, 1975), pp. 7–10.

118. Mombauer, op. cit., pp. 34–41.

119. Martin Samuels, *Doctrine and Dogma: German and British Infantry Tactics in the First World War* (Westport, Conn.: Greenwood Press, 1992), pp. 101–7.

120. Callwell, *Dug-out*, pp. 4–5.

121. Hanotaux, *Pendant la grande guerre*, p. 149.

122. Brig.-Gen. Sir James E. Edmonds, *Military Operations, France and Belgium 1916, vol. I: Sir Douglas Haig's Command to the 1st July: Battle of the Somme* (London: Macmillan, 1932), pp. vii–viii.

Chapter 2: The Strategic Labyrinth

1. William Philpott, "Britain, France and the Belgian Army," in *Look to Your Front: Studies in the First World War,* ed. Brian Bond (Staplehurst: Spellmount, 1999), pp. 121–35.

2. Joffre, *Memoirs, ii,* pp. 409–12.

3. William Philpott, "Squaring the Circle: The Higher Coordination of the Entente in the Winter of 1915–16," *English Historical Review,* 114 (1996), pp. 875–98.

4. Ibid., pp. 890–92.

5. "Written Statement of the Conference Held at Chantilly, December 6th 1915," The National Archives, Kew (hereafter TNA), WO 106/1454.

6. "Plan of Action Proposed by France to the Coalition," ibid., app. 2.

7. Haig diary, 12 March 1916, Field Marshal Earl Haig of Bemersyde papers, National Library of Scotland, Edinburgh, NLS accession no. 3155 (hereafter HD). Two editions of Haig's diary have also been published: *The Private Papers of Douglas Haig, 1914–1919,* ed. Robert Blake (London: Eyre & Spottiswoode, 1952), and *Douglas Haig: War Diaries and Letters, 1914–1918,* ed. Gary Sheffield and John Bourne (London: Weidenfeld & Nicholson, 2005). This work draws on all three sources.

8. Joffre, *Memoirs, ii,* pp. 413–14.

9. Ibid.

10. "Conclusions come to at the Conference," WO 106/1454.

11. "Note for the Conference of 6th December," ibid., app. 1.

12. Maj.-Gen. Charles E. Callwell, *Field Marshal Sir Henry Wilson: His Life and Diaries* (London: Cassell, two vols., 1922), *i,* p. 80.

13. Jean de Pierrefeu, *French Headquarters,* trans. C. J. C. Street (London: Geoffrey Bles, 1924), pp. 22–23.

14. Liddell Hart, *Foch,* p. 109.

15. Callwell, op. cit., *i,* pp. 78–79.

16. Charteris diary, 20 February 1916, Brig.-Gen. John Charteris, *At G.H.Q.* (London: Cassell, 1931), p. 193.

17. John Terraine, *Douglas Haig: The Educated Soldier* (London: Hutchinson, 1963).

18. Ibid., pp. 4–5.

19. Gerard J. De Groot, *Douglas Haig, 1861–1928* (London: Unwin Hyman, 1988), pp. 27–30.

20. Ibid., pp. 13–14.

21. George S. Duncan, *Douglas Haig as I Knew Him* (London: Allen & Unwin, 1966), p. 43, quoted in "The Reverend George Duncan at G.H.Q.," ed. Gerard J. De Groot, in *Military Miscellany I* (Stroud: Sutton Publishing Limited for the Army Records Society, 1996), p. 270.

22. Quoted in Terraine, op. cit., p. 10.

23. Ibid., p. 12.

24. De Groot, *Haig,* pp. 106–8.

25. Quoted in Terraine, op. cit., p. 35.

26. "Notices sur personnages," 1911, AAT, 7N1241.

27. Quoted in Terraine, op. cit., p. 173.

28. "Instructions for General Sir Douglas Haig," 28 December 1915, in *OH: 1916, i,* app. 5.

29. HD, 3 December 1916.

30. HD, 28 March and 4 April 1916.

31. William Philpott, "Haig and Britain's European Allies," in *Sir Douglas Haig: Seventy Years On,* ed. Brian J. Bond and Nigel Cave (Barnsley: Leo Cooper, 1999), pp. 128–44.

32. HD, 26 May 1916.

33. See for example HD, 8 March 1916.

34. HD, 17 December 1915.

35. Wilson diary, 12 May 1916, Field Marshal Sir Henry Wilson papers, Imperial War Museum, London, collection 73/1.

36. Jean des Vallières, *Au Soleil de la cavalerie; avec le général des Vallières* (Paris: André Bonne, 1965), p. 140.

37. Roy A. Prete, "Joffre and the Question of Allied Supreme Command," *Proceedings of the Annual Meeting of the Western Society for French History* (1989), pp. 329–38.

38. HD, 1 January 1916; Des Vallières, op. cit., pp. 139–40; Elizabeth Greenhalgh, *Victory Through Coalition: Britain and France During the First World War* (Cambridge: Cambridge University Press, 2005), pp. 89–90.

39. Joffre, *Memoirs, ii,* p. 417.

40. HD, 23 December 1915.

41. "Notes for CGS on taking over more French line," 21 December 1915, Haig papers.

42. HD, 26 December 1915.

43. HD, 29 December 1915.

44. Joffre to Haig, 25 December 1915, TNA, WO 158/14/74.

45. William Philpott, *Anglo-French Relations and Strategy on the Western Front* (London: Macmillan, 1996), pp. 113–16.

46. Des Vallières to Joffre, 19 January 1916, Marshal Joseph Joffre papers, AAT, 1K268: 1K268/3/50.

47. "Précis of an interview between Commandant Gemmeau and Haig," 18 January 1916, AAT, 1K268/3/48.

48. "Paper by the General Staff on the Future Conduct of the War," 16 December 1915, *OH: 1916, i,* app. 2.

49. Asquith to Venetia Stanley, 6 August 1914, in *H. H. Asquith: Letters to Venetia Stanley,* ed. Michael and Eleanor Brock (Oxford: Oxford University Press, 1982), p. 158.

50. Panouse to *ministre de la guerre,* 19 August 1914, AAT, 7N1228.

51. Robertson to Haig, 5 January 1916, Blake, op. cit., p. 122.

52. William Philpott, "Kitchener and the 29th Division: A Study in Anglo-French Strategic Relations, 1914–1915," *The Journal of Strategic Studies*, 16 (1993), pp. 375–407.

53. Lord Hankey, *The Supreme Command, 1914–1918* (London: Allen & Unwin, two vols., 1961), *i*, pp. 348–49.

54. Dardanelles Committee minutes, 20 August 1915, TNA, CAB 42/3/16.

55. Philpott, "Kitchener and the 29th Division"; Rhodri Williams, "Lord Kitchener and the Battle of Loos: French Politics and British Strategy in the Summer of 1915," in *War, Strategy and International Politics*, ed. Lawrence Freedman, Paul Hayes and Robert O'Neill (Oxford: Oxford University Press, 1992), pp. 117–32.

56. Kitchener to Haig, 22 December 1915, Blake, op. cit., p. 120.

57. Hankey, op. cit., *ii*, pp. 467–69.

58. War Committee minutes, 28 December 1915 and 13 January 1916, TNA, CAB 42/6/14 and 42/7/5.

59. Philpott, *Anglo-French Relations*, p. 85.

60. Robin Prior and Trevor Wilson, *Command on the Western Front: The Military Career of Sir Henry Rawlinson, 1914–18* (Oxford: Blackwell, 1992), p. 3.

61. Robertson to Haig, 5 January 1916, Blake, op. cit., pp. 122–23.

62. Robertson to Haig, 13 January 1916, ibid., p. 124.

63. HD, 30 January 1916.

64. Lloyd George to Haig, 8 February 1916, Blake, op. cit., p. 128.

65. War Committee minutes, 13 January 1916.

66. Robertson to Haig, 13 January 1916, Blake, op. cit., p. 126.

67. HD, 11 January 1916.

68. Des Vallières to Joffre, 10 and 16 January AAT, 1K268/3/39 and 46.

69. Précis of interview between Haig and Gemmeau, op. cit.

70. Note by 3rd Bureau (Operations), G.Q.G., 19 January 1916, AAT, 1K268/3/51.

71. Joffre to Haig, 23 January 1916, TNA, WO 158/14/81.

72. HD, 20 January 1916.

73. Haig to Asquith, 26 January 1916, Blake, op. cit., p. 126.

74. Des Vallières, *Au Soleil*, p. 142; FD, 30 April and 21 May 1916.

75. HD, 28 January 1916.

76. Robertson to Haig, 28 January 1916, Haig papers.

77. HD, 31 January 1916.

78. Haig to Joffre, 1 February 1916, TNA, WO 158/14/82.

79. Des Vallières to Joffre, 5 February 1916, AAT, 1K268/3/59.

80. HD, 11 and 12 February 1916.

81. "Note sur la conduite des opérations en 1916 sur le front occidental," by G.Q.G., 10 February 1916, TNA, WO 158/14/85.

82. Des Vallières, *Au Soleil*, p. 142.

83. Ibid., p. 143.

84. HD, 7 February 1916; King Albert's diary, 7 February 1916, *Albert Ier: Carnets et correspondance de guerre, 1914–1918*, ed. Marie-Rose Thielemans (Louvain-la-Neuve: Duculot, 1991), pp. 248–49.

85. G.Q.G. note, op. cit.; HD, 11 February 1916.

86. HD, 14 February 1916; Joffre, *Memoirs, ii*, pp. 417–19; Des Vallières, *Au Soleil*, pp. 143–45.

87. Terraine, *Haig*, p. 49.

88. HD, 11 February 1916; Haig to Robertson, 19 February 1916, Blake, op. cit., p. 130.

89. Des Vallières, op. cit., pp. 146–48.

90. HD, 29 March 1916.

91. Philpott, *Anglo-French Relations*, pp. 124–25.

92. HD, 29 March 1916.

93. Robertson to Haig, 6 March 1916, in *The Military Correspondence of Field Marshal Sir William Robertson, December 1915–February 1918,* ed. David Woodward (London: The Bodley Head for the Army Records Society, 1989), p. 40.

94. Hankey, *Supreme Command, ii,* p. 486.

95. Hankey diary, 7 April 1916, Hankey papers, Churchill College, Cambridge, HNKY 1/1; War Committee minutes, 7 April 1916, TNA, CAB 42/12/5.

96. Hankey diary, 2 May 1916, HNKY 1/1.

97. War Committee minutes, 30 May 1916, TNA, CAB 42/14/12.

98. "Note sur la situation militaire et les projets militaires de la Coalition," by G.Q.G., 26 March 1916, AFGG, IV/1, ax. 1539.

99. HD, 9 June 1916.

100. Philpott, op. cit., pp. 121–26.

101. Charteris diary, 1 May 1916, *At G.H.Q.,* p. 143; HD, 29 May and 5 June 1916; Esher journal, 28 May 1916, Viscount Esher papers, Churchill College, Cambridge, ESHR 2/16.

102. Robertson to Haig, 6 March 1916, quoted in HD, 8 March 1916.

103. "Operations on the Western Front, 1916," Haig papers, 213a.

104. 17 May 1916, Raymond Poincaré, *Au Service de la France, vol. VIII: Verdun* (Paris: Plon, 1931), pp. 223–25.

105. Esher's notes of conversations with Briand and Poincaré, 23 May 1916, David Lloyd George papers, Parliamentary Archives, London, D25/1/1.

106. HD, 26 May 1916.

107. 24 May 1916, Poincaré, op. cit., *viii,* p. 235.

108. HD, 31 May 1916.

109. Ibid.

110. 31 May 1916, Poincaré, op. cit., *viii,* p. 251.

111. Haig to Bertie, 5 June 1916, quoted in Sheffield and Bourne, *Douglas Haig: War Diaries,* p. 189.

112. 31 May 1916, Poincaré, op. cit., *viii,* p. 251.

Chapter 3: Planning the Attritional Battle

1. Marc Stéphane, *Verdun: Souvenirs d'un chasseur de Driant* (Paris: Librairie René Liot, 1929), pp. 68–78.

2. Robert T. Foley, *German Strategy and the Path to Verdun: Erich von Falkenhayn and the Development of Attrition, 1870–1916* (Cambridge: Cambridge University Press, 2005); Arden Bucholz, *Hans Delbrück and the German Military Establishment: War Images in Conflict* (Iowa City: University of Iowa Press, 1985).

3. Quoted in "Marx and Engels on Revolution and War," in W. B. Gallie, *Philosophers of Peace and War* (Cambridge: Cambridge University Press, 1978), pp. 92–93.

4. Mombauer, *Helmuth von Moltke,* pp. 14–41.

5. Moltke to his wife, 29 January 1905, quoted in Foley, op. cit., p. 73.

6. Quoted in David Stevenson, *The First World War and International Politics* (Oxford: Clarendon Press, 1988), p. 89.

7. Foley, op. cit., pp. 106–8.

8. Mombauer, op. cit., pp. 266–69.

9. Ibid., pp. 198–205.

10. Foley, op. cit., pp. 109–26.

11. Quoted in Ian Passingham, *All the Kaiser's Men: The Life and Death of the German Army on the Western Front, 1914–1918* (Stroud: Sutton Publishing, 2003), p. 92.

12. Lt. Raymond Jubet, quoted in Strachan, *First World War*, p. 185.

13. Joffre, *Memoirs, ii*, p. 450.

14. Ibid., p. 452.

15. For example in Hubert Essame, *The Battle for Europe 1918* (London: B. T. Batsford, 1972), p. 86.

16. Ferdinand Foch, *Des Principes de la guerre* (Paris: Berger-Levrault, 1903) and *De la Conduite de la guerre* (Paris: Berger-Levrault, 1904).

17. Michael Neiberg, *Foch: Supreme Allied Commander in the Great War* (London: Brassey's, 2003), pp. 9–12.

18. FD, 28 November 1915 and 20 March 1916.

19. Ferdinand Foch, *The Memoirs of Marshal Foch*, trans. Col. T. Bentley Mott (London: William Heinemann, 1931), p. 75.

20. See Goya, *La Chair et l'acier.*

21. Untitled note, 12 December 1915, AAT, 1K129/1.

22. Foch, "Enseignements à tirer de nos dernières attaques," 6 December 1915, AAT, 1K129/1.

23. 31 May 1916, Poincaré, *Au Service, viii*, p. 251.

24. Foch, op. cit., p. 217.

25. Note by Madame Foch, 6 May 1916, AAT, 1K129/9.

26. See for example FD, 12, 20 and 23 March 1916.

27. "Guerre de position," n.d. AAT, 1K130/14, and note by Foch for Painlevé (summer 1917), AAT, 1K129/1.

28. "Enseignements à tirer de nos dernières attaques," op. cit., and untitled note, 12 December 1915, AAT, 1K129/1.

29. "Le GAN, année 1916: La bataille de la Somme," AAT, 1K130/3; Foch's war diary, AAT, 1K129/2.

30. Joffre, op. cit., *ii*, p. 419.

31. Ibid., *ii*, pp. 449 and 465.

32. G.H.Q., "The Opening of the Wearing-out Battle," 23 December 1916, in *Sir Douglas Haig's Despatches*, ed. Lt.-.Col. John H. Boraston (London: Dent, 1919), p. 21.

33. GAN, "Projet d'attaque dans la région entre Somme et Oise," 16 March 1916, AFGG, IV/1, ax. 1282.

34. Joffre to Foch, 22 March 1916, AFGG, IV/1, ax. 1428.

35. Joffre to Foch, 26 April 1916, AFGG, IV/1, ax. 2223.

36. "L'offensive de la Somme," n.d., AAT, 1K129/1.

37. G.Q.G. directive, 21 June 1916, quoted in ibid.

38. Prior and Wilson, *Command*, p. 10.

39. Ibid., pp. 70–72.

40. Ibid., pp. 83–91.

41. Ibid., pp. 132–33.

42. The plan has been analysed most thoroughly in ibid., pp. 137–70.

43. Rawlinson diary, 14 March 1916, Field Marshal Lord Rawlinson papers, Churchill College, Cambridge, RWLN 1/5.

44. Rawlinson to Wigram, 27 February 1916, op. cit.

45. Rawlinson diary, 27 February 1916, RWLN 1/5.

46. Rawlinson diary, 10 March 1915, RWLN 1/1.

47. Rawlinson to Wigram, 27 February 1916, quoted in Prior and Wilson, op. cit., p. 139.

48. Prior and Wilson, *Command,* p. 70.

49. Ibid., p. 25.

50. "Plan for Offensive by Fourth Army," 3 April 1916, TNA, WO 158/233/7.

51. Rawlinson to Wigram, 25 March 1915, quoted in Prior and Wilson, *Command,* p. 78.

52. Rawlinson to Kitchener, 15 March 1915, quoted in ibid., p. 77.

53. HD, 5 April 1916.

54. Sheffield and Bourne, *Douglas Haig: War Diaries,* p. 184, n.1.

55. Haig's marginalia on "Plan for Offensive by Fourth Army," op. cit.

56. Haig's note for CGS, 5 April 1916, on ibid.

57. Haig's marginalia, op. cit.

58. "Projet d'attaque dans la région entre Somme et Oise," op. cit.

59. Rawlinson diary, 2, 4 and 5 March 1916, RWLN 1/5.

60. Joffre to Haig, 27 March 1916, TNA, WO 158/14/104.

61. Greenhalgh, *Victory Through Coalition,* pp. 55–63.

62. Des Vallières to Joffre, 22 March 1916, AAT, 18N147/1.

63. Des Vallières to Joffre, 5 April 1916, AAT, 18N147/1; Joffre to Haig, 14 April 1916, TNA, WO 158/14/106.

64. Des Vallières to Joffre, 22 March 1916.

65. Joffre to Haig, 14 April 1916, op. cit.

66. Des Vallières to Joffre, 5 April 1916, op. cit.; Haig to Joffre, 10 April 1916, TNA, WO 158/14/105; Joffre to Haig, 14 April 1916, op. cit.

67. Joffre to Haig, 14 April 1916, op. cit.

68. Note by Foch, 13 April 1916, AFGG, IV/1, ax. 2015.

69. Brig.-Gen. Edward L. Spears, *Prelude to Victory* (London: Jonathan Cape, 1939), p. 127.

70. Joffre, *Memoirs, ii,* p. 470.

71. FD, 3–24 August 1914; Henry Bordeaux, *Le Maréchal Fayolle* (Paris: G. Cres et Cie, 1921), p. 26.

72. Ibid., pp. 122–24.

73. Ibid., p. 15.

74. FD, 30 August 1914.

75. FD, 30 January and 5 February 1916.

76. FD, 25 April 1916.

77. FD, 12 April 1916.

78. Spears, op. cit., pp. 126–27.

79. FD, 4 February 1915.

80. Bordeaux, op. cit., p. 28.

81. FD, 29 and 30 November 1914, and letter of 1 January 1915, in *Cahiers secrets,* p. 72.

82. FD, 28 November and 31 December 1915 and 21 January 1916.

83. FD, 21 May 1916.

84. FD, 23 March 1916.

85. FD, 22 February 1916.

86. FD, 21 January 1916.

87. FD, 14 March 1916.

88. FD, 23 March 1916.

89. FD, 27 March 1916.

90. Foch, "Instruction générale, personnelle et secrète pour Messieurs les généraux commandant les VIe et Xe armées," 14 April 1916, AFGG, IV/1, ax. 2030.

91. Foch, "Instruction particulière personnelle et secrète pour Monsieur le général commandant la VIe armée," 14 April 1916, AFGG, IV/1, ax. 2031.

92. FD, 6 May and 13 June 1916.

93. Foch, "Instruction personnelle et secrète pour M. le général commandant la VIe armée," 25 May 1916, AFGG, IV/2, ax. 581.

94. Fayolle, "Instruction personnelle et secrète pour M. M. les généraux commandant les 20e CA, 1er CAC et 35e CA," 31 May 1916, AFGG, IV/2, ax. 738.

95. Fayolle, "Note relative à la préparation et à l'exécution des attaques," 8 June 1916, AFGG, IV/2, ax. 1019.

96. Fayolle to Foch, 5 June 1916, AFGG, IV/2, ax. 944.

97. Untitled note, 12 December 1915, AAT, 1K129/1.

98. Kiggell to Rawlinson, 13 April 1916, TNA, WO 158/233/12.

99. Joffre to Foch, 21 June 1916, AFGG, IV/2, ax. 1385.

100. Kiggell to Rawlinson, 12 and 13 April 1916, TNA, WO 158/233/11 and 12.

101. Rawlinson to G.H.Q., 19 April 1916, TNA, WO 158/11/233/12a.

102. Haig's marginalia on ibid.

103. "Plan for Offensive by Fourth Army," TNA, WO 158/233/7.

104. The debate on and nature of the bombardment is analysed thoroughly in Prior and Wilson, *Command*, pp. 141–53 and 167–70.

105. Terraine, *Haig*, pp. 33–34; de Groot, *Haig*, pp. 96–104; Richard Holmes, *The Little Field Marshal: Sir John French* (London: Weidenfeld & Nicolson, 2004), pp. 152–63.

106. Memorandum by Joffre, 20 May 1916, AFGG, IV/2, ax. 395.

107. Haig's marginalia on Rawlinson to G.H.Q., 19 April 1916, TNA, WO 158/11/233/12a.

108. Kiggell to Rawlinson, 13 and 28 April 1916, and Kiggell to Gough, 28 April 1916, TNA, WO 158/233/12 and 23.

109. Prior and Wilson, *The Somme*, p. 45.

110. Haig to Joffre, 4 June 1916, TNA, WO 158/14/121.

111. Des Vallières to Joffre, 3 June 1916, AFGG, IV/2, ax. 870, Des Vallières's italics.

112. Joffre to Haig, 6 June 1916, TNA, WO 158/14/121a: the sentence in italics was underlined on Haig's copy.

113. Tim Travers, *The Killing Ground: The British Army, the Western Front, and the Emergence of Modern Warfare, 1900–1918* (London: Unwin Hyman, 1990), p. 86. See also HD, 18 January 1916.

114. HD, 9 June 1916.

115. A critique started by Lloyd George. See David Lloyd George, *War Memoirs* (London: Odhams Press, two vols., 1938 edn.), *ii*, pp. 1315–18.

116. James M. Beach, "British Intelligence and the German Army, 1914–1918," Ph.D. thesis, University of London, 2004.

117. Ibid., pp. 157–64.

118. Haig to Rawlinson, 16 and 21 June 1916, *OH: 1916, i*, apps. 13 and 15.

119. Haig to Rawlinson, 20 June 1916, ibid., *i*, app. 14.

120. Rawlinson diary, 17 and 21 June 1916, RWLN 1/5.

121. Haig to Rawlinson, 16 June 1916, *OH: 1916, i*, app. 13; Philpott, *Anglo-French Relations*, pp. 122–28.

122. Rawlinson diary, 4 April 1916, RWLN 1/5.

123. "Report of the Army Commander's Remarks at the Conference Held at Fourth Army Headquarters, 22 June 1916," quoted in Prior and Wilson, *The Somme*, pp. 52–53.

124. HD, 23 January 1916.

125. HD, 28 June 1916.

126. Note by Foch, 21 November 1916, AAT, IK129/6.

127. HD, 7 June 1916.

128. HD, 9 June 1916.

129. Rawlinson diary, 19 March 1916, RWLN 1/5.

130. Rawlinson diary, 30 March 1916, RWLN 1/5.

131. HD, 27 June 1916.

132. Sheffield, *The Somme*, p. 19.

133. 31 May 1916, Poincaré, *Au Service, viii*, p. 251.

134. Masefield, *The Old Front Line*, p. 76.

135. Shelford Bidwell in Norman Dixon, *On the Psychology of Military Incompetence* (London: Pimlico, 1994), p. 12.

136. 30 June 1916, Charteris, *At G.H.Q.*, p. 151.

137. Haig to Lady Haig, 30 June 1916, in *Military Miscellany II*, p. 294.

138. 31 May 1916, Poincaré, op. cit., *viii*, p. 251.

139. FD, 22 June 1916.

140. Sheldon, p. 398.

141. FD, 21 January 1916.

142. Ibid.

Chapter 4: Preparing the Big Push

1. Capt. Percy Chapman diary, 30 June 1916, National Library of Australia, Canberra, MS 7509.

2. Kennedy diary, 15 April 1916, Major-General Sir John Kennedy papers, Liddell Hart Centre for Military Archives, King's College London (hereafter LHCMA), 1/3.

3. Gunner Keith Buchanan diary, 24 August 1916, AWM, PR 83/192.

4. Pvt. Henry William Freeman diary, 20 June 1916, AWM, PR 87/104.

5. Harry P. Robinson, *The Turning Point: The Battle of the Somme* (London: Heinemann, 1917), p. 1.

6. Sutherland, *5th Seaforth Highlanders*, pp. 44–47.

7. Quoted in Malcolm Brown, *The Imperial War Museum Book of the Somme* (London: Pan Books, 2002), pp. 40–41.

8. "Projet d'attaque dans la région entre Somme et Oise," op. cit.

9. Rémy Porte, *La Mobilisation industrielle: "Premier front" de la grande guerre* (Paris: 14–18 Éditions, 2005).

10. Quoted in Philip Magnus, *Kitchener: Portrait of an Imperialist* (London: John Murray, 1958), p. 284.

11. Robinson, op. cit., p. 1.

12. Andrew Suttie, *Rewriting the First World War: Lloyd George, Politics and Strategy, 1914–1918* (Basingstoke: Palgrave Macmillan, 2005), pp. 60–77.

13. Quoted in *The Times,* 20 September 1914.

14. *OH: 1916, i*, pp. 119–20.

15. HD, 14 January 1916; Esher journal, 1 June 1916, ESHR, 2/15.

16. Robertson to Haig, 13 January 1916, in Blake, *Private Papers,* pp. 124–25.

17. HD, 30 January 1916.

18. "The Final Despatch," 21 March 1919, Boraston, p. 321.

19. *OH: 1916, i*, pp. 121–24.

20. Lloyd George, *War Memoirs, i*, pp. 61–70.

21. Ian Ousby, *The Road to Verdun* (London: Pimlico, 2003), pp. 56–60.

22. Jere C. King, *Generals and Politicians: Conflict Between France's High Com-*

mand, Parliament and Government, 1914–1918 (Berkeley: University of California Press, 1951), pp. 108–25.

23. Ibid., p. 64.

24. HD, 11 September 1916.

25. Alain Hennebicque, "Albert Thomas and the War Industries," in *The French Home Front, 1914–1918,* ed. Patrick Friedenson (Oxford: Berg, 1992), pp. 89–132.

26. "Fourth Army: Tactical Notes," May 1916, *OH: 1916, i,* app. 18.

27. "Enseignements à tirer de nos dernières attaques," 6 December 1915, AAT, 1K129/1.

28. Jonathan B. A. Bailey, "The First World War and the Birth of Modern Warfare," in *The Dynamics of Military Revolution, 1300–2050,* ed. Macgregor Knox and Williamson Murray (Cambridge: Cambridge University Press, 2001), pp. 132–53.

29. Beach, "British Intelligence," p. 21.

30. Panoramas and maps have been reproduced in Peter Barton, *The Somme: A New Panoramic Perspective* (London: Constable, 2006).

31. Rawlinson to Haig, 3 April 1916, TNA, WO 158/233/7.

32. Général Émile Mangin, *Un Régiment lorrain; le 7–9; Verdun–le Somme* (Payot: Paris, 1934), pp. 189–91.

33. Général Amédée Thierry, "Souvenirs de la Somme, 1916," AAT, 1Kt698, p. 11.

34. "The Opening of the Wearing-out Battle," Boraston, p. 22.

35. Sheldon, p. 123.

36. Sixth Army, "Moyens prévus en aviation," n.d., AAT, 22N2420.

37. Rawlinson diary, 25 June 1916, RWLN 1/5; Foch to Haig and Butler to Foch, 23 June 1916, TNA, WO 158/27/132 and 133.

38. Beach, op. cit., pp, 82–83.

39. Ibid., pp. 80–81.

40. Ibid., pp. 126–29.

41. 79 RI "Compte Rendu," 14, 17 and 19 June 1916, AAT, 24N219.

42. Beach, op. cit., pp. 45–63.

43. Joffre Journal, 29 June 1916, in *Journal de Marche de Joffre, 1916–19,* ed. Guy Pedroncini (Vincennes: Service Historique de l'Armée de Terre, 1990) (hereafter, JJ); G.Q.G. Second Bureau, "Les renforts allemands dans les zones d'attaque," 28 June 1916, AFGG, IV/2, ax. 1714.

44. Beach, op. cit., pp. 161–62.

45. GAN, "La bataille offensive," 20 April 1916, AAT, 18N148.

46. Thierry, "Souvenirs," p. 6.

47. "La bataille offensive," op. cit.

48. Ibid.

49 "Fourth Army: Tactical Notes," op. cit.

50. Thierry, op. cit., p. 35.

51. Ibid., pp. 7 and 37.

52. 11 DI, *Journal de marche et d'opérations* (hereafter, JMO), 1 May–31 October 1916, AAT, 26N289.

53. Prior and Wilson, *The Somme,* p. 64.

54. Fraser, *Alanbrooke,* pp. 72–73.

55. "Fourth Army: Tactical Notes," op. cit.

56. "La bataille offensive," op. cit.

57. Ibid.; "Note on the use of gas shells," I CAC, 1 June 1916, SHAT, 22N2420/5.

58. *OH: 1916, i,* p. 286.

59. Ibid., p. 295.

60. "La bataille offensive," op. cit.

61. *OH: 1916, i,* p. 296.

62. "Fourth Army: Tactical Notes," op. cit.

63. G.Q.G., "Note relative aux enseignements à tirer des affaires de Verdun," 30 March 1916, AFGG, IV/1, ax. 1603.

64. Goya, *La Chair et l'acier,* pp. 262–64 and 272–80.

65. Général Berdoulat, "Note sur les enseignements de la bataille de la Somme," AAT, 22N2420.

66. "Fourth Army: Tactical Notes," op. cit.

67. Quoted in Lyn Macdonald, *Somme* (London: Michael Joseph, 1983), p. 83.

68. Prior and Wilson, *The Somme,* pp. 112–15.

69. Travers, *The Killing Ground,* p. 144.

70. "Fourth Army: Tactical Notes," op. cit.

71. Prior and Wilson, op. cit., p. 60.

72. Churchill, *The World Crisis,* pp. 1074–77, quotes from a German account of 8th Division's attack.

73. Prior and Wilson, op. cit., p. 73.

74. "Attaque d'une division sur le front Canal de la Somme-route Cappy à Dompierre," n.d., and "Plan d'engagement du 2 DIC," 15 June 1916, AAT, 22N2420/1a and 2.

75. Note by Berdoulat, op. cit.

76. HD, 4 May 1916.

77. Charles E. W. Bean, *The Official History of Australia in the War of 1914–1918, vol. III: The Australian Imperial Force in France, 1916* (Sydney: Angus & Robertson, 1941), p. 87.

78. Bernard Ayre to his mother, 21 April 1916, Janet Ayre Miller Collection (Coll-158), Archives and Special Collections, Queen Elizabeth II Library, Memorial University, St. John's, Newfoundland, 4/01/003.

79. *OH: 1916, i,* p. 262, n. 1.

80. Lt. William Colyer, quoted in Hart, *The Somme,* p. 55.

81. Capt. George McGowan, quoted in Brown, op. cit., p. 35.

82. HD, 30 June 1916.

83. Quoted in Simkins, *Kitchener's Army,* p. 306.

84. John Baynes, *Far from a Donkey: The Life of General Sir Ivor Maxse* (London: Brassey's, 1995), pp. 135–43.

85. Fraser, *Alanbrooke,* pp. 68–73.

86. "La bataille offensive," op. cit.

87. HD, 10 May 1916.

88. Maxse, quoted in Baynes, op. cit., p. 135.

89. "La bataille offensive," op. cit.

90. Martin van Creveld, "World War I and the Revolution in Logistics," in *Great War, Total War: Combat and Mobilization on the Western Front, 1914–1918,* ed. Roger Chickering and Stig Förster (Cambridge: Cambridge University Press, 2000), pp. 57–72.

91. 37 RI's report on the Vaux sector, June 1916, AAT, 24N219.

92. *OH: 1916, i,* pp. 272–74.

93. Ibid., p. 274.

94. Ibid., pp. 266–68.

95. Ousby, op. cit., pp. 3–4.

96. 5 August 1916, Duchaussoy, "Une municipalité du Union Sacrée," 15 May 1916–20 March 1918, ADS, 14J97.

97. Brig.-Gen. Henry Page Croft, *Twenty-two Months Under Fire* (London: John Murray, 1917), p. 201.

98. Gerald Brenan, *A Life of One's Own: Childhood and Youth* (London: Jonathan Cape, 1975), p. 198.

99. *OH: 1916, i,* pp. 112–13.

100. Ibid., p. 283.

101. Lt. F. P. Bethune, quoted in Bean, op. cit., p. 70.

102. Quoted in Brown, op. cit., p. 42.

103. Laurence V. Moyer, *Victory Must Be Ours* (London: Leo Cooper, 1995), pp. 134–37.

104. Hew Strachan, *The First World War: Vol. 1: To Arms* (Oxford: Oxford University Press, 2001), pp. 1014–40.

105. Moyer, op. cit., pp. 148–49.

106. Leutnant M. Gerster, quoted in Sheldon, p. 114.

107. Foley, *German Strategy,* pp. 237–50.

108. General Erich von Falkenhayn, *General Headquarters, 1914–1916 and Its Critical Decisions* (London: Hutchinson, n.d. [1919]), p. 239.

109. Field Marshal Erich von Manstein, *Aus Einem Soldatenleben, 1887–1939* (Bonn: Athenäun-Verlag, 1958), p. 45.

110. Robert B. Asprey, *The German High Command at War: Hindenburg and Ludendorff and the First World War* (London: Little, Brown, 1991), p. 244. Asprey surprisingly confuses Fritz von Below with his cousin Otto.

111. Foley, op. cit., pp. 246–49. See also *OH: 1916, i,* pp. 316–19.

112. Sheldon, pp. 70–72, 80–83 and 110–13.

113. Foley, op. cit., p. 248; SS491: "The Construction of Field Defences," 20 June 1916, reprinted by G.H.Q., 16 October 1916, Field Marshal Sir Archibald Montgomery-Massingberd papers, LHCMA, 7/5.

114. General Freiherr von Soden, quoted in Sheldon, pp. 111–12.

115. "The Opening of the Wearing-out Battle," Boraston, p. 23.

116. Quoted in Sheldon, p. 80.

117. Samuels, *Doctrine and Dogma,* pp. 59–72.

118. Sheldon, pp. 116–19.

119. Mangin, op. cit., pp. 189 and 191.

120. JJ, 29 June 1916.

121. Thierry, op. cit., p. 39.

122. Churchill, op. cit., *ii,* p. 1072.

123. Precise allied figures are difficult to pin down. Herwig, *The First World War: Germany and Austria-Hungary, 1914–1918* (London: Arnold, 1997), p. 199, gives 2960 allied artillery pieces, compared with 844 German. Alain Denizot, *La Bataille de la Somme* (Paris: Perrin, 2002), p. 54, gives 1550 French and 1335 British and 844 German. *Der Weltkrieg, 1914–1918: vol. 10: Die Operationen des Jahres 1916* (Berlin: E. S. Mittler & Sohn, 1936), pp. 340 and 349, gives 1599 British and 1404 French guns, against 844 German. *OH: 1916, i,* p. 486, gives a figure of 1,627,824 shells fired before the British assault.

124. Thierry, op. cit., p. 47.

125. Quoted in Hart, op. cit., p. 90.

126. AFGG, IV/2, pp. 229–30; "Rapport sommaire sur les opérations du 1er Corps d'Armée Colonial pendant la période du 24 juin au 5 juillet 1916," 20 July 1916, AAT, 19N1089/1a; 6 BI, "Rapport sommaire sur les opérations du 1er au 4 juillet 1916," AAT, 22N2421/3.

127. Prior and Wilson, *Command,* pp. 167–69.

128. *OH: 1916, i,* p. 300, gives a figure of 427 British heavy guns, plus 20 lent by the French. Robert Doughty, *Pyrrhic Victory: French Strategy and Operations in the Great*

War (Cambridge, Mass.: The Belknap Press, 2005), p. 292, gives a figure of 552 heavy French guns in addition to those integral to the corps artillery. Denizot's figures, op. cit., p. 54, are 854 French heavy guns, and 467 British.

129. Fraser, op. cit., pp. 71–72.

130. *OH: 1916, i,* pp. 299–307 and 485–88; Prior and Wilson, *Command,* pp. 171–76.

131. Prior and Wilson, op. cit., pp. 175–76.

132. 28 June 1916, Charteris, *At G.H.Q,* p. 150.

133. Rawlinson diary, 29 June 1916, RWLN 1/5.

134. "La bataille offensive," op. cit.

135. Ibid.

136. Rawlinson diary, 24 June 1916, RWLN 1/5.

137. Quoted in Brown, op. cit., pp. 40–41.

Chapter 5: "Varying Fortune": 1 July 1916

1. Quoted in Martin Middlebrook, *The First Day of the Somme* (London: Penguin Classics, 2001), p. 116.

2. Lance-Cpl. J. Quinn, quoted in Graham Maddocks, *Liverpool Pals: 17th, 18th, 19th and 20th Battalions The King's (Liverpool Regiment)* (London: Leo Cooper, 1991), p. 89.

3. "The Attack," *Westminster Gazette,* August 1916, reprinted in R. H. Tawney, *The Attack and Other Papers* (Nottingham: Spokesman, 1981), pp. 11–20: p. 12.

4. Quoted in Malcolm Brown, *The Imperial War Museum Book of the Western Front* (London: Pan Books, 2001), p. 117.

5. Quoted in Prior and Wilson, *The Somme,* pp. 73–74.

6. Quoted in Brown, *Somme,* p. 62.

7. Quoted in Sheldon, p. 135.

8. Quoted in Brown, *Western Front,* pp. 118–19.

9. Quoted in Brown, *Somme,* p. 91.

10. Basil Liddell Hart, *Through the Fog of War* (London: Faber & Faber, 1938), p. 252.

11. *OH: 1916, i,* p. 324; Sheffield, *The Somme,* p. 185, n. 10.

12. Baurès, "La prophétie du tante Anna" (typescript, 1973), AAT, 1Kt326.

13. *OH: 1916, i,* p. 344.

14. Colin, *La Division de fer, 1914–1918* (Paris: Payot, 1930), pp. 48–53.

15. Mangin, *Un régiment lorrain,* p. 189.

16. Ibid., pp. 189–91 and 200.

17. Colin, op. cit., p. 116.

18. Mangin, op. cit., pp. 201–3 and 213–15; Palat, *ix,* pp. 41–43.

19. Carnet de Charles Barberon, Historial de la Grande Guerre, Péronne, pp. 117–18.

20. AFGG, IV/2, p. 231; Colin, op. cit., pp. 116–17.

21. Lt.-Col. N. Fraser-Tytler, quoted in Maddocks, op. cit., pp. 82–83.

22. Lance-Cpl. J. Quinn, quoted in ibid., p. 89.

23. Ibid.

24. *OH: 1916, i,* p. 328, n. 3.

25. Ibid., p. 327.

26. Ibid., p. 322.

27. Quoted in Brown, *Western Front,* p. 117.

28. Maddocks, op. cit., p. 86.

29. Pvt. W. Gregory, quoted in ibid., pp. 87–88.

30. Quoted in Brown, *Somme*, pp. 87–88.

31. Sheldon, pp. 162–64.

32. *OH: 1916, i,* p. 335.

33. Ibid., p. 337; Leibrock quoted in Sheldon, pp. 164–66.

34. Gerald Gliddon, *When the Barrage Lifts: A Topographical History of the Battle of the Somme* (Stroud: Allan Sutton Publishing, 1987), p. 344.

35. Capt. George McGowan, quoted in Brown, *Somme,* p. 90.

36. *OH: 1916, i,* p. 336.

37. FD, 1 July 1916.

38. Palat, *ix,* p. 40.

39. Thierry, "Souvenirs," p. 51.

40. *OH: 1916, i,* p. 347.

41. HD, 30 June 1916.

42. Quoted in Brown, *Somme*, pp. 64–65; *OH: 1916, i,* p. 347.

43. *OH: 1916, i,* pp. 349–50.

44. Brown, *Somme,* p. 77.

45. Sheldon, p. 162.

46. Prior and Wilson, *The Somme*, p. 104.

47. Middlebrook, op. cit., pp. 103–4.

48. Lt. Will Mulholland, quoted in Brown, *Western Front,* p. 119.

49. *OH: 1916, i,* p. 355.

50. Sassoon diary, 1 July 1916, *Siegfried Sassoon Diaries, 1915–18,* ed. Rupert Hart-Davis (London: Faber & Faber, 1983), p. 82.

51. *OH: 1916, i,* pp. 357–59.

52. Liddell Hart, *Memoirs, i,* pp. 21–22.

53. *OH: 1916, i,* pp. 360–61.

54. Quoted in Brown, *Somme*, p. 67.

55. *OH: 1916, i,* p. 379.

56. John Harris, *Covenant with Death* (London, Pan Books, 1963), p. 364.

57. Gerster, *Die Schwaben an der Ancre,* quoted in Churchill, *The World Crisis, ii,* p. 1076.

58. Liddell Hart, *First World War,* pp. 240 and 242.

59. Lloyd George, *War Memoirs, ii,* p. 1247.

60. Themes explored in Dan Todman, *The Great War: Myth and Memory* (London: Hambledon & London, 2005).

61. Middlebrook, op. cit.; Macdonald, *Somme*; Brown, *Somme.*

62. Sheldon; Passingham, *All the Kaiser's Men.*

63. Battalion history, quoted in Prior and Wilson, *The Somme,* p. 75.

64. Ibid., pp. 76 and 88–89.

65. Ibid., p. 99.

66. *OH: 1916, i,* p. 427.

67. Prior and Wilson, *The Somme,* pp. 73–74.

68. Musketier Karl Blenk, cited in Middlebrook, op. cit., p. 157.

69. Otto Lais, quoted in Sheldon, pp. 142–43.

70. Quoted in Middlebrook, op. cit., pp. 150–51.

71. Percy Croney, *Soldier's Luck: Memoirs of a Soldier of the Great War* (Ilfracombe: Arthur H. Stockwell, 1965), pp. 119–34.

72. 8 Bavarian RIR report, in Sheldon, pp. 225–28.

73. *OH: 1916, i,* pp. 379, 403 and 443–44.

74. William Turner, *Pals: The 11th (Service) Battalion (Accrington) East Lancashire Regiment* (Barnsley: Wharnecliffe Publishing, n.d.), p. 179.

75. Gliddon, op. cit., p. 259.

76. *OH: 1916, i*, p. 402.

77. Sheldon, p. 225.

78. Michael Steadman, *Battleground Europe: Thiepval* (London: Leo Cooper, 1995), p. 65.

79. Ibid., pp. 66–68; *OH: 1916, i*, pp. 404–7.

80. Quoted in Gliddon, op. cit., p. 45.

81. *OH: 1916, i*, p. 436.

82. Lt.-Cpl. F. Riggs, "A tribute to our noble officers, Capt. E. S. Ayre, Lieuts. G. & W. Ayre," 26 August 1916, Janet Ayre Murray papers, 6.02.

83. Quoted in Middlebrook, op. cit., pp. 188–89.

84. *OH: 1916, i*, p. 436.

85. HD, 1 July 1916.

86. Ibid.

87. FD, 1 July 1916.

88. JJ, 1 and 2 July 1916.

89. *OH: 1916, i*, p. 344; *Der Weltkrieg, x*, pp. 351–53.

90. Middlebrook, op. cit., pp. 212–13 and 286–89.

91. Thierry, op. cit., p. 49.

92. Sheldon, pp. 174–75.

93. Ibid., p. 172.

94. AFGG, IV/2, pp. 231–32.

95. Sassoon diary, 1 July 1916, op. cit., p. 84.

96. *OH: 1916, i*, p. 416.

97. Ibid., pp. 410–11.

98. Quoted in Middlebrook, op. cit., p. 210.

99. *OH: 1916, i*, pp. 413–15.

100. Quoted in Middlebrook, op. cit., p. 209.

101. Sheldon, pp. 148–55 and 225–28; Steadman, op. cit., pp. 79–80.

102. *OH: 1916, i*, pp. 416–23.

103. Quoted in Sheldon, p. 156.

104. Ibid.

105. A. Stuart Dolden, *Cannon Fodder: An Infantryman's Life on the Western Front, 1914–18* (London: Blandford Press, 1980), pp. 71–72.

106. Kennedy diary, 1 July 1916, LHCMA, 1/3.

107. Tawney, "The Attack," pp. 19–21.

108. Quoted in Sheldon, p. 176.

109. Quoted in Brown, *Western Front*, p. 119.

110. Quoted in Brown, *Somme*, p. 92.

111. Quoted in Sheldon, p. 176.

112. AFGG, IV/2, pp. 229–33.

113. Reichsarchiv, *Somme-Nord: Die Brennpunkte der Schalct im Juli 1916* (Berlin: Oldenburg ID, two vols., 1927), *i*, pp. 22–84.

114. Joffre, *Memoirs, ii*, pp. 470–71.

115. JJ, 2 July 1916.

116. Quoted in Brown, *Somme*, p. 68.

117. Nugent to his wife, 2 July 1916, in *Major General Oliver Nugent and the Ulster Division, 1915–1918*, ed. Nicholas Perry (Stroud: Sutton Publishing for the Army Record Society, 2007), p. 80.

118. Liddell Hart, *First World War,* p. 242.

119. Quoted in Brown, *Somme,* p. 70.

120. Rt. Hon. J. E. B. Seely, *Adventure* (London: William Heinemann, 1930), p. 251.

121. Tawney, "The Attack," p. 20.

122. Barberon, op. cit., p. 119.

123. Fritz von Below, "Experience of the German 1st Army in the Somme Battle," 30 January 1917 (trans. General Staff [Intelligence] G.H.Q., 3 May 1917), p. 3.

Chapter 6: Exploitation

1. Charles E. W. Bean, *Letters from France* (London: Cassell, 1917), p. 77.

2. Kennedy diary, 4 July 1916, LHCMA 1/3.

3. Sassoon diary, 2 July 1916 op. cit., p. 85.

4. *OH: 1916, i,* p. ix.

5. "The Opening of the Wearing-out Battle," Boraston, p. 31.

6. Rawlinson diary, 2 July 1916, RWLN 1/5.

7. Cosby report, Paris, 7 July 1916, National Archives at College Park, College Park, Md., Record Group 165: War Department General and Specific Staffs, microfilm publication M1024A: M1024A/8690–441.

8. Gen. Sir Anthony Farrar-Hockley, *The Somme* (London: Pan Books, 1983), pp. 159–60; Peter Hart, *Somme Success: The Royal Flying Corps and the Battle of the Somme* (Barnsley: Leo Cooper, 2001), p. 95.

9. Palat, *ix,* p. 49.

10. Thierry, "Souvenirs," p. 52.

11. Fritz von Lossberg, *Meine Tätigkeit im Weltkrieg, 1914–1918* (Berlin: E. S. Mittler & Sohn, 1939), p. 219.

12. "Rapport du Général Maziller, commandant du 2e DIC sur les opérations du 1er au 4 juillet inclus," 11 July 1916, AAT, 24N2956/1.

13. Général Maurice Abadie, *Flaucourt* (Paris: 1933), pp. 57–58.

14. "Rapport sommaire sur les opérations du 1er Corps d'Armée Colonial pendant la période du 24 juin au 5 juillet 1916," AAT, 19N1089/1a.

15. Abadie, op. cit., p. *i.*

16. 11 DI daily situation report, 3 July 1916, AAT, 24N219/1.

17. Capt. Wilfrid Miles, *Military Operations, France and Belgium, 1916, vol. 2: 2nd July 1916 to the End of the Battles of the Somme* (London: Macmillan, 1938), p. 7.

18. HD, 2 July 1916.

19. HD, 2 July 1916; Rawlinson diary, 2 July 1916, RWLN 1/5; *OH: 1916, ii,* pp. 8–10.

20. JJ, 2 July 1916.

21. JJ, 3 July 1916.

22. Wilson diary, 5 July 1916; Callwell, *Wilson, i,* p. 287.

23. HD, 3 July 1916.

24. JJ, 3 July 1916.

25. *OH: 1916, ii,* p. 19.

26. JJ, 4 July 1916.

27. Note of 4 July meeting at Dury and des Vallières, "Résumé des intentions du commandement brittanique," 6 July 1916, AFGG, IV/2, axs. 1987 and 2071.

28. JJ, 4 July 1916; Pierrefeu, *French Headquarters,* p. 76.

29. Philpott, *Anglo-French Relations,* p. 127.

30. JJ, 4 July 1916.

31. Madame Foch's diary, 31 July 1916, AAT, IK129/9.

32. *OH: 1916, i,* p. 319.

33. G.Q.G. Second Bureau, "Les renforts allemands dans les zones d'attaque," 28 June 1916, AFGG, IV/2, ax. 1714.

34. JJ, 3 July 1916.

35. Liddell Hart, *Foch,* p. 278.

36. "L'offensive de la Somme," Foch papers, AAT, 1K129/1.

37. Liddell Hart, *First World War,* pp. 231–53.

38. JJ, 3 July 1916.

39. JJ, 4 and 6 July 1916.

40. Ibid. The French verb *entraîner* can mean to lead, pull or train.

41. Des Vallières, "Memorandum pour le chef du 3e bureau," 5 August 1916, AFGG, IV/2, ax. 2746.

42. "Note sommaire sur les projets et intentions de manoeuvre des généraux Foch et Fayolle," 6 July 1916, AFGG, IV/2, ax. 2070.

43. Thierry, op. cit., p. 55.

44. Abadie, op. cit., pp. 40–41.

45. See *The Great War: The Standard History of the All-Europe Conflict, vol. VII,* ed. H. W. Wilson and J. A. Hammerton (London: The Amalgamated Press, 1916), pp. 536–37.

46. "Rapport sommaire . . . du I CAC," op. cit.

47. Wilson diary, 5 July 1916; Callwell, *Wilson, i,* p. 287.

48. 6 BIC, "Rapport sommaire sur les opérations du 1er au 4 Juillet 1916," 8 July 1916, AAT, 22N2421/3; Thierry, op. cit., p. 52.

49. Carnet de Charles Barberon, pp. 119–20.

50. Thierry, op. cit., pp. 55–56.

51. FD, 3 July 1916.

52. AFGG, IV/2, pp. 237–39.

53. "Rapport sommaire . . . du I CAC"; Palat, *ix,* pp. 52–53.

54. AFGG, IV/2, pp. 140–42.

55. Pierrefeu, op. cit., pp. 76–77.

56. Abadie, op. cit., p. 47.

57. FD, 8 July 1916.

58. AFGG, IV/2, pp. 248–49.

59. "Rapport sommaire du Général de division Bonnier sur la participation de la 16e DIC à la Bataille de la Somme (juillet 1916)," 23 July 1916, AAT, 22N1715.

60. Ibid.; Palat, *ix,* pp. 60–63.

61. FD, 11 July 1916.

62. "Rapport sommaire du Général Bonnier," op. cit.

63. Ibid.

64. Ibid.

65. Thierry, op. cit., p. 66.

66. Rawlinson diary, 2 and 3 July 1916, RWLN 1/5; *OH: 1916, ii,* p. 9.

67. HD, 4 July 1916; Haig to Lady Haig, 8 and 10 July 1916, in Sheffield and Bourne, *War Diaries,* pp. 199–202.

68. *OH: 1916, ii,* p. 24.

69. G.H.Q. intelligence summary, 10 July 1916, quoted in Beach, "British Intelligence," pp. 166–67.

70. Wilson diary, 5 July 1916, Callwell, *Wilson, i,* p. 287.

71. Quoted in Passingham, *All the Kaiser's Men,* p. 114.

72. Quoted in ibid.

73. Quoted in ibid.

74. Sheldon, pp. 180–87.

75. Rupprecht diary, 17 July 1916, quoted in ibid., p. 202.

76. Abadie, op. cit., pp. 56–59.

77. Rupprecht diary, op. cit.

78. Lt.-Gen. Sir Tom Bridges, *Alarms and Excursions* (London: Longmans Green, 1938), pp. 156–59; *OH: 1916, ii*, pp. 12–18.

79. Kuhn report, Berlin, 30 August 1916: "The Battle of the Somme in July," M1024A/8690–464.

80. Mangin, *Un régiment lorrain*, pp. 220–23; 11 DI JMO, 5 July 1916; 79 RI JMO, 5 July 1916, AAT, 9M366; "Historique des opérations," op. cit.

81. Mangin, op. cit., pp. 224–25; 11 DI JMO, 6 July 1916; 79 RI JMO, 6 July 1916.

82. Prior and Wilson, *The Somme*, pp. 126–27.

83. Mrs. C. Maxwell, *Frank Maxwell VC: A Memoir and Some Letters* (London: John Murray, 1921), pp. 153–61, quoted in Baynes, *Far from a Donkey*, pp. 145–46.

84. *OH: 1916, ii*, p. 28.

85. Quoted in Fraser, *Alanbrooke*, p. 73.

86. *OH: 1916, ii*, pp. 29–31.

87. "Fourth Army: Tactical Notes," May 1916, *OH: 1916, i*, app. 18.

88. Croft, *Twenty-two Months*, pp. 207–11.

89. Colin Hughes, *Mametz: Lloyd George's Welsh Army and the Battle of the Somme* (Norwich: Gliddon Books, 1990), pp. 85–94.

90. Llewelyn Wyn Griffith, *Up to Mametz* (London: Faber & Faber, 1931), p. 201, quoted in Hughes, op. cit., p. 90.

91. Griffith, op. cit., pp. 205–6, quoted in Hughes, op. cit., pp. 90–91.

92. HD, 9 July 1916.

93. Hughes, op. cit., pp. 29–30.

94. Maj. G. P. L. Drake-Brockman, quoted in Hughes, op. cit., p. 137.

95. Hughes, op. cit., p. 97.

96. Sgt. Tom Price, quoted in Hart, *Somme*, p. 256.

97. Hughes, op. cit., pp. 103–16.

98. *OH: 1916, ii*, p. 54, n. 1.

99. Griffith, op. cit., pp. 208–13, quoted in Hughes, op. cit., pp. 118–19.

100. David Kelly, *39 Months with the "Tigers," 1915–1918* (London: Ernest Benn, 1930), quoted in Hughes, op. cit., p. 142.

101. Prior and Wilson, *The Somme*, p. 127.

102. *OH: 1916, ii*, p. 54, n. 1.

103. FD, 12 July 1916.

104. Quoted in Farrar-Hockley, op. cit., p. 180.

105. Liddell Hart, *Memoirs, i*, p. 24.

106. Prior and Wilson, *Command*, pp. 190–95.

107. Col. E. Craig-Brown, quoted in Brown, *Somme*, p. 130.

108. *OH: 1916, ii*, p. 83.

109. Farrar-Hockley, op. cit., p. 193.

110. *OH: 1916, ii*, pp. 83–87.

111. Prior and Wilson, *The Somme*, pp. 133–36.

112. Ibid., p. 134.

113. Prior and Wilson, *Command*, pp. 192–93.

114. Prior and Wilson, op. cit., p. 201.

115. 2nd Lt. F.W. Beadle, quoted in Macdonald, *Somme*, pp. 137–38.

116. Richard Holmes, *Tommy: The British Soldier on the Western Front, 1914–1918* (London: HarperCollins, 2004), pp. 440–41.

117. Ibid.; *OH: 1916, ii,* pp. 86–87; Farrar-Hockley, op. cit., pp. 192–96; Prior and Wilson, *The Somme,* p. 139.

118. JJ, 14 July 1916.

119. HD, 14 July 1916.

120. JJ, 11 July 1916.

121. JJ, 14 July 1916.

122. 5 July 1916, in Lt.-Col. Charles à Court Repington, *The First World War* (London: Constable & Company, 2 vols., 1920), *i,* p. 257.

123. JJ, 11 July 1916.

124. Elizabeth Greenhalgh, " 'Parade Ground Soldiers': French Army Assessments of the British on the Somme in 1916," *Journal of Military History,* 63 (1999), pp. 283–312.

125. *OH: 1916, ii,* pp. 87–88.

126. Mangin, op. cit., p. 231.

127. FD, 16 July 1916.

Chapter 7: The Battle of Attrition: August 1916

1. Pvt. Henry William Freeman diary, 12 July 1916, AWM, PR87/104.

2. Ibid., 15 August 1916.

3. Bridges, *Alarms and Excursions,* p. 160.

4. Rawlinson diary, 9 October 1916, RWLN, 1/7.

5. Gary Sheffield, "Hubert Gough as an Army Commander on the Somme," in *Command and Control on the Western Front: The British Army's Experience, 1914–1918,* ed. Gary Sheffield and Dan Todman (Staplehurst: Spellmount, 2004), pp. 71–95.

6. Croft, *Twenty-two Months,* p. 221.

7. Prior and Wilson, *The Somme,* p. 176.

8. "Cadmus" (Lance-Cpl. D. Horton), *Posiers,* State Library of New South Wales, Sydney, ML MSS 1991, p. 4.

9. Ibid.

10. Ibid., p. 6.

11. *OH: 1916, ii,* pp. 141–46; Prior and Wilson, op. cit., p. 177.

12. Cadmus, op. cit., p. 7.

13. Prior and Wilson, op. cit., p. 177; Reichsarchiv, *Somme-Nord, ii,* p. 145.

14. Gunner Hugh Monteith, "Observations of No. 5886," AWM, MSS0810.

15. Sapper J. Julin letter to George W. Watson (December 1916), AWM, PR03081.

16. Pvt. William G. Holford diary, AWM, PR01746.

17. Pvt. Athol Dunlop, 28 July 1916, AWM, PR00676.

18. Bean, *AIF in France, 1916,* p. 862.

19. Charles Bean diary, 19 September 1916, AWM, 3DRL 606/59.

20. Athol Dunlop to Florence Dunlop, 31 July 1916, AWM, PR00676.

21. Craig Melrose, " 'A Praise That Never Ages': The Australian War Memorial and the 'National' Interpretation of the First World War," Ph.D. thesis, University of Queensland, 2005, p. 186.

22. Ibid., p. 197; Ken Inglis, *C. E. W. Bean: Australian Historian* (St. Lucia: University of Queensland Press, 1970), p. 23.

23. "Pozières Heights, 1916," Australian War Memorial, Canberra, ACT.

24. Melrose, op. cit., pp. 197–98.

25. Bean, *AIF in France, 1916*, p. 716.

26. Melrose, op. cit., p. 172.

27. Pvt. Raymond E. Membrey memoirs (1986), AWM, PR02022.

28. Melrose, op. cit., pp. 18–19.

29. Athol Dunlop to Florence Dunlop, op. cit.

30. Inglis, op. cit., p. 21.

31. Bean, *AIF in France, 1916*, pp. 871–77; Peter Charlton, *Pozières 1916: Australians on the Somme* (North Ryde: Methuen Haynes, 1986); John Laffin, *British Butchers and Bunglers of World War One* (Stroud: Alan Sutton, 1988); Denis Winter, *Haig's Command: A Reassessment* (London: Viking, 1991); Prior and Wilson, *Command and The Somme*.

32. Christopher Pugsley, *The ANZAC Experience: New Zealand, Australia and Empire in the First World War* (Auckland: Reed, 2004); Bill Rawling, *Surviving Trench Warfare: Technology and the Canadian Corps, 1914–18* (Toronto: University of Toronto Press, 1992); Simon Robbins, *British Generalship on the Western Front, 1914–18: Defeat into Victory* (London: Frank Cass, 2005); Andy Simpson, *Directing Operations: British Corps Command on the Western Front, 1914–18* (Stroud: Spellmount, 2006); Sheffield and Todman, op. cit.

33. Bean diary, 4 February 1917, AWM, 3DRL 606/69.

34. Ibid., 19 September 1916, AWM 3DRL 606/59.

35. *OH: 1916, ii*, p. 91.

36. Ian Uys, *Delville Wood* (Rensburg: Uys Publishers, 1983).

37. Terry Norman, *The Hell They Called High Wood: The Somme, 1916* (London: Kimber, 1984).

38. Pierrefeu, *French Headquarters*, pp. 77–78.

39. Prior and Wilson, *The Somme*, p. 159.

40. J. Rickard, *Max von Gallwitz, German General, 1852–1937*, <http://www.historyofwar.org/ articles/people_gallwitz_max.html> (accessed 3 December 2007).

41. Quoted in Sheldon, p. 222.

42. Ibid., p. 210.

43. Hew Strachan, "Attrition," in *The Oxford Companion to Military History*, ed. Richard Holmes (Oxford: Oxford University Press, 2001), p. 106.

44. Kennedy diary, 15 July 1916, LHCMA 1/3.

45. Rawlinson diary, 15 October 1916, RWLN 1/7.

46. Palat, *ix*, pp. 77–78.

47. Hughes, *Mametz*, p. 93.

48. Palat, *ix*, p. 77.

49. Sheldon, p. 210.

50. *OH: 1916, ii*, p. 102, n. 1.

51. Palat, *ix*, pp. 72–79.

52. "Rapport sommaire de la 16 DIC," AAT, 22N1715.

53. AGFF, IV/2, pp. 267–68.

54. AGFF, IV/2, pp. 262–64 and 268.

55. Berdoulat to Fayolle, 8 August 1916, AAT, 22N2420/11.

56. Thierry, "Souvenirs," pp. 68–79; Berdoulat to Commander Third Army, 23 August 1916, AAT, 22N2420/11; JJ, 14 and 15 August 1916.

57. Sixth Army, "Note pour les commandants des CA," 15 August 1916, AAT, 19N1050.

58. Quoted in Sheldon, pp. 239–41.

59. "Note pour les commandants des CA," op. cit.

60. R. G. Nobécourt, "Gamelin, 14–18," unpublished typescript, pp. 195–209.

61. AGFF, IV/2, pp. 266–67.

62. "Historique des opérations de la 11 Division" and "Rapport concernant les opérations auxquelles a pris part la 39e division pendant la période du 1 juillet au 10 août 1916," AAT, 22N1346/2.

63. FD, 4 and 6 August 1916.

64. AFGG, IV/2, p. 280.

65. JJ, 28 July 1916.

66. FD, 28 July 1916; AFGG, IV/2, pp. 269–70.

67. 79 RI JMO, 7–9 August 1916, AAT, 9M366; "Historique des opérations de la 11 Division," op. cit.

68. AFGG, IV/2, p. 285.

69. Nobécourt, op. cit., pp. 210–12; AFGG, IV/2, pp. 287–92.

70. Greenhalgh, " 'Parade Ground Soldiers,' " p. 297.

71. Congreve diary, 6 July 1916, quoted in Greenhalgh, *Victory Through Coalition*, p. 64; Rawlinson diary, 17 July 1916, RWLN 1/5.

72. AFGG, IV/2, pp. 271–72.

73. FD, 4 August 1916.

74. Logan, "Politico-Military Notes," Paris, 7 September 1916, M1024A/8690–476.

75. See AFGG IV/2, pp. 256–60, 276–87 and 306; Denizot, *La Somme*, pp. 106–13.

76. JJ, 21 July and 11 August 1916.

77. JJ, 25 July 1916.

78. Haig to von Donop, 18 July 1916, in Sheffield and Bourne, *War Diaries*, p. 207.

79. HD, 30 July and 11 August 1916.

80. JJ, 5, 11 and 12 August 1916.

81. HD, 22 August 1916.

82. Haig to Rawlinson and Gough, 2 August 1916, *OH: 1916, ii*, app. 13.

83. Prior and Wilson, *The Somme*, pp. 146–51.

84. Ibid., pp. 164–67.

85. *OH: 1916, ii*, p. 204.

86. Ibid., *ii*, p. 148.

87. "IX Corps notes on information collected from various sources, including troops who have been engaged in the recent fighting," 31 July 1916, 19th Division, "Notes on the recent operations," 19 July and September 1916, and X Corps, "Questions relating to an initial attack after lengthy preparation," 16 August 1916, in "Notes on the Somme Fighting," Montgomery-Massingberd papers, LHCMA, 7/3.

88. Shea to Montgomery, 2 August 1916, Montgomery-Massingberd papers, LHCMA, 7/3.

89. 30th Division conference notes, 25 July 1916, ibid.

90. Rawlinson diary, 8 August 1916, RWLN 1/5.

91. Note by Haig, 19 July 1916 (Haig's emphasis) on "Telephone message received from Fourth Army, 7:50 a.m.," 19 July 1916, TNA, WO 158/234.

92. Rawlinson diary, 4 July 1916, RWLN 1/5.

93. Rawlinson diary, 16 September 1916, RWLN 1/5.

94. "Rapport sommaire de la 16 DIC," AAT, 22N1715.

95. JJ, 23 July 1916.

96. 79 RI JMO, 19 July 1916, AAT, 9M366; Berdoulat, "Note sur les enseignements de la bataille de la Somme," AAT, 22N2420.

97. Dr. J. Forderer, quoted in Sheldon, p. 247.

98. JJ, 23 July, 16 and 28 August, and 8 September 1916.

99. Note by Duval, 10 July 1916, AFGG, IV/2, ax. 2217.

100. The full series of papers is in "Sixth Army: Somme Offensive Correspondence," AAT, 19N1050.

101. See, for example, note by Berdoulat, op. cit.

102. Quoted in Sheldon, p. 240.

103. Aaron Norman, *The Great Air War* (New York: Macmillan, 1968), pp. 96–106 and 146–47; Hart, *Somme Success*, pp. 51–53; Chaz Bowyer, *Albert Ball, VC* (London: William Kimber, 1977), pp. 44–45.

104. Hart, op. cit., pp. 70–71; John H. Morrow Jr., *German Air Power in World War I* (Lincoln: University of Nebraska Press, 1982), p. 61.

105. HD, 31 July 1916.

106. Bowyer, op. cit., pp. 61 and 88; Hart, op. cit., p. 194.

107. Norman, op. cit., pp. 218–21.

108. Bowyer, op. cit., pp. 85–86 and 97.

109. Norman, op. cit., pp. 222–23.

110. Quoted in Sheldon, p. 209.

111. Ibid., pp. 238–40.

112. Hart, op. cit., p. 128.

113. G.Q.G., "Note sur la liaison d'infanterie par avion," 6 July 1916; Balfourier to Fayolle, 14 July 1916, enclosing reports by Capitaine Lalanne and Capitaine Cayatte on "la liaison avec l'infanterie par avion et par ballon," 10 and 11 July 1916, AAT, 19N1050.

114. JJ, 3 August 1916.

115. Sheldon, p. 249.

116. Prior and Wilson, *The Somme,* pp. 155–66.

117. 46 DI, "Ordre général d'opérations no. 13," 25 August 1916, AAT, 24N1121.

118. 46 DI, "Ordre général d'opérations no. 12," 23 August 1916, AAT, 24N1121.

119. Frederick Robinson, quoted in Brown, *Somme,* p. 265.

120. Nicholas Reeve, "Through the Eyes of the Camera: Contemporary Cinema Audiences and Their 'Experience' of War in the Film *Battle of the Somme,*" in *Facing Armageddon: The First World War Experienced,* ed. Hugh Cecil and Peter H. Liddle (London: Leo Cooper, 1996), pp. 780–98.

121. Baurès, "La prophétie du tante Anna," AAT, 1Kt326.

122. *The Wartime Diaries and Letters of Jacques Playoust: Or "Seven Aussie Poilus" 1914–1918,* ed. Jacqueline Dwyer (Glebe: Fast Books, 1995), State Library of New South Wales, ML MSS 6586/1, p. 51.

123. Robert Graves, *Goodbye to All That* (London: Penguin, 2nd edn., 1960), pp. 180–83.

124. 27 July 1916, Playoust, op. cit., p. 57.

125. Quoted in Palat, *ix,* p. 122.

126. Quoted in Ball, *The Guardsmen,* p. 57.

127. Gaston Lefebvre, quoted in Denizot, op. cit., p. 119.

128. Leutnant Freiherr von Salmuth, quoted in Sheldon, pp. 252–53.

129. Kennedy diary, 4 July 1916, LHCMA, 1/3.

130. Brooke diary, 21 September 1916, Field Marshal Lord Alanbrooke papers, LHCMA, 1/1/9.

131. Capt. Oscar Viney memoir, provided by Charlie Viney.

132. Quoted in Palat, *ix,* p. 187.

133. Jünger, *Storm of Steel.*

134. Quoted in Sheldon, pp. 205–6.

135. Ibid., pp. 204–5.

136. For example, "Souvenirs de guerre du capitaine Gartner (29e régt d'artillerie de campagne), 1914–1918," AAT, 1Kt156.

137. Viney, op. cit.

138. Quoted in Denizot, op. cit., p. 120.

139. 7–9 July and 23 August 1916, Playout, op. cit., p. 59.

140. 27 July 1916, ibid., pp. 56–57.

141. Rawlinson diary, 14 and 15 July 1916, RWLN 1/5.

142. Fourth Army Order, 10 August 1916, TNA, WO 95/431.

143. Soldier in 3rd Ersatz Regiment, n.d., in "Extracts from letters found on Germans during the Somme battle" (April 1917), AWM, 27/310/72.

144. Jünger, op. cit., p. 90.

145. Second Army Supplementary Order, 17 July 1916, in Sheldon, p. 207.

146. Jünger, op. cit., p. 69.

147. Ibid., p. 119.

148. Quoted in Denizot, op. cit., p. 120.

149. Quoted in Uys, *Delville Wood*, pp. 54–55.

150. HD, 28 July 1916.

151. Freeman diary, 21 August 1916, op. cit.

152. Rawlinson diary, 2 August 1916, RWLN 1/5; Palat, *ix*, p. 111.

153. Fourth Army Order, 20 and 25 August 1916, TNA, WO 95/431.

154. Beach, "British Intelligence," pp. 171–75.

155. Palat, *ix*, p. 125.

156. Rawlinson diary, 1 August 1916, RWLN 1/5.

157. *OH: 1916, ii*, p. 234, n. 3.

158. Undated letter in "Extracts from letters," op. cit.

159. Falkenhayn, *General Headquarters*, pp. 266–67.

160. JJ, 8 and 12 August 1916.

161. JJ, 18 August 1916.

162. FD, 17–20 August 1916.

163. Général Hély d'Oissel, "Journal de guerre," 5 August 1916, AAT, 1Kt444.

164. Note handed to Sir Douglas Haig, 19 July 1916, AFGG, IV/2, ax. 2491.

165. JJ, 15 August 1916.

166. FD, 25 August 1916.

167. FD, 28 August 1916.

Chapter 8: Behind the Lines

1. Micheler to Antonin Dubost, 9 June 1916, in Colonel E. Herbillon, *Le Général Alfred Micheler, 1914–1918* (Paris: Librairie Plon, 1933), p. 76.

2. Quoted in King, *Generals and Politicians*, p. 117.

3. Ibid., pp. 116–23; JJ, 9 July 1916.

4. Esher to Asquith, 23 June 1916, in *Journals and Letters of Reginald, Viscount Esher*, ed. Maurice and Oliver Brett (London: Nicholson & Watson, four vols., 1934–38), *iv*, pp. 35–36.

5. Esher journal, 26 August 1916, ESHR 2/16.

6. 2 July 1916, Duchaussoy, "Une municipalité du Union Sacrée," op. cit.

7. Pierrefeu, *French Headquarters*, pp. 75–76.

8. Cosby, "General Estimate of the Forces and Resources of the Enemy," Paris, 21 August 1916, M1024A/8690–462.

9. Parker, "Report on Trip to the Front, 19–24 August 1916," Paris, 28 September 1916, M1024A/8690–489.

10. HD, 29 March 1916.

11. Esher journal, 22 July 1916, in *Journals and Letters, iv*, pp. 38–39.

12. Robertson to Gwynne, 18 June 1916, in Woodward, *Military Correspondence*, p. 56.

13. Robertson to Kiggell, 5 July 1916, in ibid., pp. 64–66.

14. Robertson to Rawlinson, 26 July 1916, in ibid., pp. 72–73; Paul Harris and Sanders Marble, "The 'Step-by-Step' Approach: British Military Thought and Operational Method on the Western Front, 1915–1917," *War in History*, 15 (2008), pp. 17–42: 30–33.

15. Robertson to Haig, 5 July 1916, in Woodward, op. cit., pp. 66–67.

16. Esher journal, 22 July 1916, in *Journals and Letters, iv*, pp. 38–39.

17. Esher to Hankey, 3 August 1916, in ibid., *iv*, pp. 43–44.

18. Churchill memorandum, 1 August 1916, in Martin Gilbert, *Winston S. Churchill, vol. III: Companion* (London: William Heinemann, two vols., 1972), *ii*, pp. 1534–39.

19. Esher to Hankey, op. cit.

20. Note by Smith, 1 August 1916, in Gilbert, op. cit., *ii*, p. 1534.

21. 20 July and 1 August 1916, Repington, *First World War, i*, pp. 287 and 294.

22. George Cassar, *Asquith as War Leader* (London: The Hambledon Press, 1994), p. 192.

23. Churchill, *World Crisis, ii*, pp. 1084–89. He removed the memorandum from the later abridged version of *The World Crisis* (London: Thornton Butterworth, 1931), p. 653.

24. Prior and Wilson, *The Somme,* pp. 195–96.

25. Martin Gilbert, *Winston S. Churchill, vol. III: 1914–1916* (London: William Heinemann, 1971), p. 697.

26. Winston Churchill to John Churchill, 15 July 1916, in Gilbert, *Churchill: Companion, ii,* pp. 1530–31.

27. Brig.-Gen. the Lord Croft, *My Life of Strife* (London: Hutchinson, 1948), p. 110.

28. Churchill, House of Commons, 31 May 1916, quoted in Gilbert, *Churchill,* p. 775.

29. Churchill, House of Commons, 23 May 1916, quoted in ibid., p. 774.

30. Robertson to Haig, 29 July 1916, and Haig note thereon, Haig papers.

31. Robertson to Haig, 1 August 1916, in Woodward, op. cit., pp. 76–77.

32. Robertson to Haig, 5 August 1916, in ibid., p. 78.

33. HD, 3 August 1916.

34. HD, 9 August 1916.

35. Robertson to Haig, 7 and 8 August 1916, in Woodward, op. cit., pp. 79–80.

36. "Summary of the Military Situation in the Various Theatres of War for the Seven Days Ending 3rd August, with Comments by the General Staff," 3 August 1916, TNA, CAB 42/17/2.

37. Robertson to Murray, 16 October 1916, in Woodward, op. cit., pp. 96–97.

38. Brown, *Somme,* pp. 261–62.

39. *OH: 1916, i,* pp. 145–47.

40. Logan, "Peace Activities of Radical Socialists in French Parliament," Paris, 26 September 1916, M1024A/8690–484.

41. Haig to Robertson, 3 June 1916, in Woodward, op. cit., pp. 55–56.

42. HD, 23 July and 2 August 1916.

43. *OH: 1916, i,* p. 148.

44. 8 July 1916, Charteris, *At G.H.Q.,* p. 153.

45. 11–13 July, Repington, *First World War, i,* p. 274.

46. HD, 10 July 1916.

47. 31 October 1916, Charteris, op. cit., p. 176.

48. HD, 3 August 1916.

49. HD, 25 October 1916.

50. Kennedy diary, 4 October 1916, LHCMA, 1/3.

51. Robertson to Haig, 1 August 1916, in Woodward, op. cit., pp. 76–77.

52. *The King Visits His Armies in the Great Advance,* Imperial War Museum Film and Video Archive, IWM 192/01&02 P1 A35.

53. Pieter Geyl, "A Visit to Fricourt," Parts 1 to 3, *De Nieuwe Rotterdamsche Courant,* 31 August and 4 September 1916, W. Geyl papers, AWM, MSS1092.

54. Malins, *How I Filmed the War,* pp. 208–13.

55. Ibid., p. 207.

56. Esher to Haig, 22 July 1916, in Blake, *Private Papers,* pp. 161–62.

57. HD, 10 August 1916.

58. HD, 17 September 1916.

59. HD, 8 and 12 August 1916.

60. Message from George V, 15 August 1916, Haig papers.

61. Asquith to Sylvia Henley, 26 August 1916, quoted in Cassar, op. cit., p. 194.

62. Hankey, *Supreme Command, ii,* pp. 512–13.

63. Margot Asquith diary, 17 September 1916, Margot Asquith papers, Bodleian Library, Oxford, MS Eng d.3215.

64. Cassar, op. cit., pp. 194–96.

65. War Committee minutes, 12 September 1916, TNA, CAB 42/19/9.

66. Asquith to Gwynne, 11 October 1916, RWLN 1/8.

67. Robertson to Haig, 7 August 1916, Woodward, op. cit., p. 79.

68. 12 August 1916, Charteris, op. cit., p. 164.

69. Esher to Haig, 17 August 1916, in *Journals and Letters, iv,* p. 52.

70. Lloyd George, *War Memoirs, i,* p. 1247.

71. Haig to Lloyd George, 23 September 1916, Lloyd George papers, E/1/6/6.

72. Haig to Lady Haig, 13 September 1916, in Blake, *Private Papers,* p. 166.

73. Undated note by Eugene Rochard (Foch's cousin), AAT, 1K129/2.

74. HD, 17 September 1916.

75. Lloyd George, op. cit., *ii,* p. 1752.

76. Ibid., p. 2036.

77. Hankey diary, 18 October and 1 November 1916, HNKY 1/1.

78. Quoted in Stephen Roskill, *Hankey, Man of Secrets* (London: Collins, three vols., 1970–74), *i,* p. 286.

79. Rawlinson diary, 10 December 1916, RWLN 1/7.

80. 22 July 1916, Charteris, op. cit., p. 158.

81. Brown, op. cit, p. 264.

82. Ibid., pp. 271 and 278.

83. Bean to Buchan, 15 August 1916, Bean correspondence, January 1915–February 1917, AWM, 3DRL 6673/271.

84. Quoted in Henry Wickham Steed, *Through Thirty Years, 1892–1922: A Personal Narrative* (London: William Heinemann, two vols., 1924), *ii,* p. 121.

85. Quoted in Gilbert, *Churchill,* p. 791.

86. Reeve, "Through the Eyes of the Camera," pp. 792–93.

87. "On Leave to a New England," in Bean, *Letters from France,* pp. 175–80; Bean diary, 19 September 1916, AWM, 9DRL 606/59.

88. Page to Wilson, 21 July 1916, in *Life and Letters, iii,* pp. 307–8.

89. Bean diary, op. cit.

90. Tawney, "The Attack," pp. 11–20.

91. Tawney, "Some Reflections of a Soldier," *Nation,* October 1916, reprinted in *The Attack,* pp. 21–22.

92. Ibid.

93. Martin Kitchen, *The Silent Dictatorship: The Politics of the German High Com-

mand Under Hindenburg and Ludendorff, 1916–1918 (London: Croom Helm, 1976), pp. 21–23.

94. Herwig, *First World War*, p. 134.

95. John Wheeler-Bennett, *Hindenburg: The Wooden Titan* (London: Macmillan, 1967), pp. 79–80.

96. Kitchen, op. cit., pp. 17–18.

97. Moyer, *Victory Must Be Ours*, pp. 179–80.

98. Herwig, op. cit., pp. 229–30; Kitchen, op. cit., pp. 19–20.

99. Jürgen Kocka, *Facing Total War: German Society, 1914–1918* (Oxford: Berg, 1984), pp. 42–44.

100. 5 August 1916, Charteris, op. cit., p. 161.

101. Kocka, op. cit., p. 61.

102. Ibid., pp. 77–113, passim.

103. Kitchen, op. cit., pp. 26–29.

104. "Killed in Action," *Evening Telegram*, 7 July 1916.

105. Bean diary, 1 September 1916, AWM, 3DRL 606/60.

106. 1 June 1916, Charteris, op. cit., p. 145.

107. Bean diary, 1 June 1916, AWM, 3DRL 606/44.

108. Quoted in Donald Horne, *Billy Hughes: Prime Minister of Australia, 1915–1923* (Melbourne: Black Inc., 2000), p. 109.

109. Bean diary, op. cit.

110. Membrey memoirs, op. cit.

111. Horne, op. cit., p. 120.

112. Bean diary, 9 September 1916, AWM, 3DRL 606/59.

113. Bill Gammage, *The Broken Years: Australian Soldiers in the Great War* (Harmondsworth: Penguin Books, 1975), p. 19.

114. Horne, op. cit., pp. 11 and 121.

115. HD, 16 October 1916; Bean diary, 22 October 1916, AWM, 3DRL 606/62.

116. Leslie Robson, *The First A.I.F.* (Melbourne: Melbourne University Press, 1970), p. 116.

117. Horne, op. cit., pp. 112 and 116–19.

118. JJ, 12 July 1916.

119. JJ, 13 and 15 July 1916; King, op. cit., pp. 126–29.

120. JJ, 17 July 1916.

121. King, op. cit., pp. 131–34.

122. Logan, "Politico-Military Notes," Paris, 7 September 1916, M1024A/8690/476.

123. Joffre, *Memoirs, ii*, pp. 496–500.

124. HD, 27 August 1916.

125. Handwritten note on "travail Pagezy," n.d., Weygand papers, AAT, 1K130/9/3.

126. Glenn Torrey, "The Entente and the Romanian Campaign of 1916," in Glenn Torrey, *Romania and World War I: A Collection of Studies* (Oxford: The Centre for Romanian Studies, 1998), pp. 154–72: 154.

127. Glenn Torrey, "Romania's Entry into the First World War: The Problem of Strategy," in ibid., pp. 137–53: 140.

128. King Albert's diary, 2 July and 11 September 1916, in *Albert Ier: Carnets et correspondance de guerre*, pp. 273 and 281.

129. "The Opening of the Wearing-out Battle," Boraston, p. 20.

130. Kitchen, op. cit., p. 37.

131. 24 November 1916, Charteris, op. cit., p. 179.

132. Briggs, Vienna, 7 August 1916 and 2 September 1916, M1024A/8690–451 and 466.

133. *Morning Post*, 20 July 1916, RWLN 1/6.

134. Bean diary, 3 October 1916, AWM, 3DRL 606/60.

135. Georg Queri, *Banderbuch vom Blutigen Westen* (Weimar: Alexander Duncker Verlag, 1917).

136. Queri, "In the Offensive Section of the Western Front," 19 July 1916, in Kuhn report, Berlin, 24 July 1916, M1024A/8690–450.

137. Queri, "The Artillery Battle at the Somme," 3 September 1916, in Kuhn report, Berlin, 5 September 1916, M1024A/8690–469.

138. Ibid.

139. Kocka, op. cit., pp. 40–42.

140. Moyer, op. cit., pp. 156–70.

141. Ibid., p. 377, n. 47.

142. Rainer Rother, "The Experience of the First World War and the German Cinema," in *The First World War and Popular Cinema: 1914 to the Present,* ed. Michael Paris (Edinburgh: Edinburgh University Press, 1999), pp. 217–46: 220–22.

143. "Note on letter received from C.I.G.S. dated 29th July," Haig papers.

144. Poillon, "Dutch Military Authority's Comment on the Present Offensive of the Allies," The Hague, 22 July 1916, M1024A/8690–448.

145. Geyl, "A visit to Fricourt, part 3."

146. James Beck, "Foreword," in H. E. Brittain, *To Verdun from the Somme: An Anglo-American Glimpse of the Great Advance* (London: John Lane, The Bodley Head, 1917), p. viii.

147. Brittain, op. cit., p. 64.

148. HD, 31 July 1916; JJ, 7 August 1916.

149. Beck, op. cit., p. xi.

150. Logan, "Politico-Military Notes," Paris, 7 September 1916, M1024A/8690/467.

151. Joyce Grigsby Williams, *Colonel House and Sir Edward Grey: A Study in Anglo-American Diplomacy* (Lanham, Md.: University Press of America, 1984), pp. 104–5.

152. Quoted in ibid., pp. 99–100.

153. House diary, 28 January 1916, quoted in ibid., p. 94, n. 53.

154. Quoted in ibid., pp. 83–84.

155. House–Grey memorandum, 17 February 1916, quoted in ibid., pp. 86–87.

156. Ibid., p. 106.

157. Ibid.

158. Ibid., pp. 100–101 and 108–10.

159. Logan, "Politico-Military Notes," op. cit.

160. Quoted in Williams, op. cit., p. 110.

161. Falkenhayn, *General Headquarters,* pp. 268–69.

162. Kuhn, "German Opinion of Ammunition Deliveries by Neutral Countries," Berlin, 29 September 1916, M1024A/8690–485; Queri, "The Artillery Battle at the Somme," op. cit.

163. Kuhn, "America's Neutrality," Berlin, 30 August 1916, M1024A/8690–463.

164. 2 August 1916, Charteris, op. cit., p. 161.

165. Williams, op. cit., pp. 113–14.

166. Micheler to Dubost, 15 October 1916, quoted in Herbillon, op. cit., p. 97.

167. Rawlinson diary, 22 September 1916, RWLN 1/5.

168. Haig to Robertson, 1 August 1916, Haig papers.

169. Cavan to H. P. Crawley, 11 October 1916, Margot Asquith papers, MS Eng d.3319.

170. Haig to Lady Haig, 11 September 1916, in Blake, op. cit., p. 166.

171. Lloyd George, op. cit., p. 323.

Chapter 9: The Tipping Point: September 1916

1. Palat, *ix*, pp. 189–91; 6 BCA JMO, 12 September 1916, AAT, 26N557.

2. Esher journal, 10 September 1916, ESHR, 2/16.

3. F. F., "Les opérations de guerre: la bataille de la Somme," *Journal de Genève*, 25 March 1917, AAT, 1K129/1.

4. GAN, "Instruction pour M. le Général commandant le IX CA," 23 October 1916, AAT, 22N581.

5. HD, 6 August 1916.

6. JJ, 7 September 1916.

7. AFGG, IV/3, p. 30.

8. Queri, "The Battle on the Somme: The English Soldiers," 15 September 1916, M1024A/8690–475.

9. "Historique résumé des opérations auxquelles les corps de la 2e division ont pris part dans la bataille de la Somme," n.d., AAT, 22N22.

10. FD, 3 September 1916.

11. Farrar-Hockley, *The Somme*, p. 223.

12. Sixth Army, "Ordre général d'opérations no. 1309" and "Instruction personnelle et secrète no. 1313," 31 August and 2 September 1916, AFGG, IV/3, axs. 13 and 30; I CA, "Compte rendu," 3 September 1916, AFGG, IV/3, ax. 59.

13. 66 DI, "Ordre général d'opérations no. 88," AFGG, IV/3, ax. 25.

14. I CA "Compte rendu," op. cit.

15. AFGG, IV/3., pp. 82–85; Palat, *ix*, pp. 170–72.

16. FD, 4 September 1916.

17. *OH: 1916, ii*, pp. 254–57.

18. History of Fusilier Regiment No. 35, quoted in *OH: 1916, ii*, pp. 204–6.

19. Ibid., *ii*, pp. 282–84.

20. Edmund Blunden, *Undertones of War* (London: Penguin Books, 2nd edn., 1936), pp. 76 and 98–102.

21. *OH: 1916, ii*, pp. 278–82.

22. Sheffield, *The Somme*, pp. 28 and 108. Farrar-Hockley and Prior and Wilson do not mention Tenth Army's offensive at all.

23. Pierrefeu, *French Headquarters*, pp. 78–79.

24. II CA, "Instruction personnelle et secrète no. 2," 2 September 1916, AFGG, IV/3, ax. 34.

25. AFGG, IV/3, p. 94.

26. XXXV CA, "Ordre général d'opérations no. 906," 28 August 1916, AFGG, IV/3, ax. 10.

27. AFGG, IV/3, p. 34.

28. Micheler to his CA commanders, 14, 16, 21 August and 2 September 1916, AAT, 22N660/1, 24N1062 and 24N2406, and AFGG, IV/3, ax. 32.

29. "Rapport sur l'attaque de Vermandovillers par la 132e division du 4 au 8 septembre 1916," 26 September 1916, AAT, 24N2460/2.

30. AFGG, IV/3, p. 81.

31. Tenth Army, "Bulletin de renseignements," 4 September 1916, AFGG, IV/3, ax. 86.

32. "Rapport sur l'attaque de Vermandovillers," op. cit.

33. AFGG, IV/3, pp. 95–97.

34. JJ, 4 September 1916.

35. AFGG, IV/3, pp. 99–105.

36. Weygand to Micheler, 5 September 1916, AFGG, IV/3, ax. 113.

37. Foch to Micheler, 8 September 1916, AFGG, IV/3, ax. 212.

38. Palat, *ix*, p. 175, citing Humbert, pp. 129–30.

39. Note by Micheler, 5 September 1916, AFGG, IV/3, ax. 129.

40. 6 BCA JMO, 4 September 1916, AAT, 26N557.

41. Palat, *ix*, pp. 176–79.

42. AFGG, IV/3, pp. 86–93.

43. Sixth Army, "Ordre général d'opérations no. 1323," 5 September 1916, AFGG, IV/3, ax. 124.

44. AFGG, IV/3, pp. 106–7; note by Foch to Fayolle, 5 September 1916, AFGG, IV/3, ax. 115.

45. Palat, *ix*, pp. 180–81.

46. Note by Guillaumat, 6 September 1916, AFGG, IV/3, ax. 164.

47. 6 BCA JMO, 5–11 September 1916.

48. FD, 3, 4, 6, 7 and 9 September 1916; JJ, 7 September 1916.

49. Sixth Army, "Instructions personnelle et secrète nos. 1326, 1327 & 1332" and "Ordre particulier no. 1334," 6 and 8 September 1916, AFGG, IV/3, axs. 156, 157, 221 and 222.

50. Greenhalgh, *Victory Through Coalition*, p. 65.

51. Haig to Joffre, 11 September 1916, TNA, WO 158/15/137; AFGG, IV/3, pp. 155–57.

52. HD, 8 September 1916.

53. FD, 12 September 1916.

54. "Aux corps d'armée et divisions," 11 September 1916, AFGG, IV/3, ax. 301.

55. FD, 12 September 1916.

56. Pierrefeu, op. cit., p. 78.

57. 2 DI historique, 12–14 September 1916, AAT, 22N22.

58. "Compte rendu des événements survenus sur le front du 1er CA, du 11 septembre (18h) au 12 septembre (18h)," 12 September 1916, AFGG, IV/3, ax. 332.

59. XXXIII CA, "Compte rendu de fin de journée," 12 Septembre 1916, AFGG, IV/3, ax. 339.

60. Pierrefeu, op. cit., p. 78.

61. Sixth Army, "Ordre général d'opérations no. 1342," 12 September 1916, AFGG, IV/3, ax. 327.

62. Sixth Army, "Ordre particulier no. 1343," 13 September 1916, AFGG, IV/3, ax. 353.

63. Nobécourt, "Gamelin," pp. 221–22 and 225–27.

64. Note by Foch for Nudant, 13 September 1916, AFGG, IV/3, ax. 348.

65. Sixth Army, "Ordre général d'opérations no. 1342," op. cit.

66. Sixth Army, "Ordre général d'opérations no. 1346," 13 September 1916, AFGG, IV/3, ax. 354.

67. Sixth Army, "Bulletin de renseignements no. 738," 17 September 1916, AFGG, IV/3, ax. 469.

68. De Bazelaire to Fayolle, 14 September 1916, AFGG, IV/3, ax. 392.

69. FD, 13 and 14 September.

70. Des Vallières to Joffre, 8 September 1916, AFGG, IV/3, ax. 211.

71. HD, 14 September 1916.

72. HD, 4 and 6 September 1916.

73. HD, 13 and 14 September 1916 and Haig's pencilled note, 13 September 1916, on Joffre to Haig, 12 September 1916 (Haig's italics), Haig papers.

74. Haig to Joffre, 11 September 1916, TNA, WO 158/15/137.

75. Robertson to Haig, 29 August 1916, in Woodward, *Military Correspondence,* pp. 84–85.

76. Haig to Robertson, 1 September 1916, Haig papers.

77. Sixth Army, "Ordre général d'opérations no. 1349," 14 September 1916, AFGG, IV/3, ax. 385.

78. Rawlinson to G.H.Q., 31 August 1916, TNA, WO 158/235.

79. Prior and Wilson, *Command,* pp. 227–32; Prior and Wilson, *The Somme,* pp. 216–20.

80. Haig to Joffre, 11 September 1916, TNA, WO 158/15/137.

81. Beach, "British Intelligence," pp. 171–75.

82. Kiggell to Rawlinson, 31 August 1916, TNA, WO 158/235.

83. Note by Haig for CGS (Kiggell), 29 August 1916, TNA, WO 158/235.

84. HD, 9 September 1916.

85. Rawlinson to G.H.Q., 31 August 1916, op. cit.

86. Prior and Wilson, *Command,* pp. 230–32.

87. *OH: 1916, ii,* pp. 299–302.

88. Kiggell to Gough and Rawlinson, 13 September 1916, Fourth Army Operational Order, 13 September 1916, and notes of conference between Haig and Rawlinson, 14 September 1916, TNA, WO 158/236.

89. G.H.Q. scheme of operations, 13 September 1916, TNA, WO 158/236.

90. Prior and Wilson, *The Somme,* pp. 221–22.

91. Fourth Army, "Cavalry Objectives," 11 September 1916, TNA, WO 158/236.

92. Prior and Wilson, *The Somme,* p. 222.

93. Fourth Army Operational Summary, 11 September 1916, TNA, WO 158/236.

94. HD, 13 September 1916.

95. Kiggell to Rawlinson, 31 August 1916, op. cit.

96. HD, 9 September 1916.

97. Rawlinson to G.H.Q., 31 August 1916, op. cit.

98. *OH: 1916, ii,* pp. 291–92.

99. Churchill, *World Crisis, ii,* pp. 508–12.

100. Ibid., *ii,* pp. 1082–83.

101. Hankey, *Supreme Command, ii,* pp. 513–14.

102. *OH: 1916, ii,* pp. 232–35.

103. Ibid., *ii,* p. 292.

104. Ibid., *ii,* p. 293, n. 1, and p. 294, ns. 1–3.

105. Prior and Wilson, *Command,* p. 233.

106. Ibid., pp. 235–36.

107. Reserve Leutnant Herman Kohl, quoted in Sheldon, p. 292.

108. *OH: 1916, ii,* pp. 321–23.

109. Patrick Wright, *Tank: The Progress of a Monstrous War Machine* (London: Faber & Faber, 2000), p. 38.

110. Sheldon, op. cit., pp. 294–95.

111. Prior and Wilson, *The Somme,* p. 231.

112. Ibid., p. 233.

113. Viscount Chandos (Oliver Lyttelton), *The Memoirs of Lord Chandos* (London: The Bodley Head, 1962), pp. 62–64.

114. Ball, *The Guardsmen,* pp. 58–64.

115. Prior and Wilson, *The Somme,* pp. 235–36; Prior and Wilson, *Command,* p. 241; *OH: 1916, ii,* pp. 331–36.

116. <www.firstworldwar.bham.ac.uk/donkey/barter.htm> (accessed 28 January 2008).

117. David Campbell, "A Forgotten Victory: Courcelette, 15 September 1916," and Anon., "Story of the 22nd Battalion, September 15th, 1916," *Canadian Military History,* 16/2 (Spring 2007), pp. 27–48 and 49–58; Rawling, *Surviving Trench Warfare.*

118. Franz von Papen, *Memoirs,* trans. Brian Connell (London: André Deutsch, 1952), pp. 66–67.

119. *Der Weltkrieg, vol. 11: Die Kriegführung in Herbst 1916 und im Winter 1916–17* (Berlin: E. S. Mittler & Sohn, 1938), p. 70.

120. Fourth Army, "Extracts from Reports on the Use of Tanks," n.d., TNA, WO 158/236.

121. Trevor Pidgeon, *The Tanks at Flers* (Cobham: Fairmile Books, 1995), pp. 198–200.

122. G.H.Q., "Note on Use of Tanks," 5 October 1916, TNA, WO 158/236.

123. Pidgeon, op. cit., p. 205.

124. Quoted in ibid., p. 190.

125. Sheldon, pp. 295–96.

126. Wright, op. cit., pp. 38–53.

127. Fourth Army Order, 20:00, 15 September 1916, TNA, WO 158/236.

128. Prior and Wilson, *Command,* pp. 242–43.

129. Joffre to Haig, 17 September 1916, Haig to Joffre, 19 September 1916 and "Special Order of the Day," 16 September 1916, Haig papers.

130. HD, 16 and 17 September 1916.

131. FD, 17 September 1916.

132. Quoted in Passingham, *All the Kaiser's Men,* pp. 122–23.

133. AFGG, IV/3, pp. 138–39.

134. Ibid., pp. 145–49.

135. *OH: 1916, ii,* p. 356; Prior and Wilson, *The Somme,* pp. 242–43.

136. *OH: 1916, ii,* pp. 370–72.

137. AFGG, IV/3, pp. 154–55; Fayolle to CA commanders, 13 September 1916, and "Répartition d'AL du GAN," 17 September 1916, AFGG, IV/3, axs. 352 and 459.

138. Commandant Courtès, "Souvenirs de guerre du Commandant Courtès, 1914–1918," AAT, 1Kt15, pp. 305–12.

139. AFGG, IV/3, p. 164; Sixth Army, "Instruction pour l'attaque," 17 September 1916, AFGG, IV/3, ax. 471; XXXII CA, "Mesure prise en vue de l'exploitation éventuelle de succès," 24 September 1916, AAT, 24N992.

140. "Instruction personnelle et secrète pour M. Le général commandant le CC ayant traversé la ligne ennemie," 16 September 1916, AFGG, IV /3, ax. 444.

141. "Note pour les généraux commandant de CA," 10 September 1916, AFGG, IV/3, ax. 283.

142. *OH: 1916, ii,* p. 289, n. 2.

143. JJ, 25 September 1916.

144. JJ, 23 September 1916.

145. "Carnets de guerre d'André l'Huillier," 25 September 1916, AAT, 1Kt185.

146. Commandant Bouchacourt, *L'infanterie dans la bataille: étude sur l'attaque étude sur la défense* (Paris: Charles-Lavauzelle, 1931), p. 182; 42 DI, "Compte-rendu des 24 heures, 25–26 septembre 1916," 26 September 1916, AAT, 24N992.

147. 42 DI, "Compte-rendu des événements du 26–27 septembre 1916," 27 September 1916, AAT, 24N992.

148. 83 BI, "Rapport sur les événements du 18 au 28 septembre," 2 October 1916, AAT, 24N992.

149. Foch to Joffre, 17 September 1916, AFGG, IV/3, ax. 468.

150. Palat, *ix,* pp. 210–14.

151. *OH: 1916, ii*, pp. 384–85.

152. Prior and Wilson, *The Somme*, pp. 241–47.

153. Nobécourt, op. cit., p. 230.

154. Courtès, op. cit., pp. 314–33.

155. AFGG, IV/3, pp. 174–78; 10 DI, "Compte-rendu des événements du 26 au 27 septembre," 27 September 1916, AAT, 24N992; "Rapport d'ensemble sur les opérations de la 10e Division d'infanterie sur la Somme," 10 December 1916, AAT, 24N201.

156. *OH: 1916, ii*, pp. 399–401.

157. Ivor Maxse, "The 18th Division and the Battle of the Ancre," December 1916, quoted in Baynes, *Far from a Donkey*, p. 149.

158. Maxse to Montgomery, 31 July 1916, Montgomery-Massingberd papers, LHCMA, 7/3.

159. Maxse to Mary Maxse, 7 October 1916, quoted in Baynes, op. cit., p. 163.

160. Ibid., pp. 149–63; *OH: 1916, ii*, pp. 403–7 and 416–17.

161. Oberstleutnant Alfred Bischler, 180 RI, quoted in Passingham, op. cit., p. 123.

162. Palat, *ix*, pp. 216–17.

163. Passingham, op. cit., p. 118.

164. Palat, *ix*, pp. 208–9; AFGG, IV/3, p. 152.

165. Passingham, op. cit., p. 121.

166. Second Army, "Usure allemand sur la Somme," 20 September 1916, AFGG, ax. 537.

167. Gelibrand to Walter Gelibrand, 26 September 1916, Gelibrand papers, AWM, 3DRL 6541/2.

168. Ludendorff, *War Memories, i*, pp. 276–78.

169. Field Marshal Paul von Hindenburg, *Out of My Life*, trans. F. A. Holt (London: Cassell, 1920), p. 217.

170. Ludendorff, op. cit., *i*, pp. 265–74; Asprey, German High Command, pp. 268–69.

171. Ludendorff, op. cit., *i*, p. 268.

172. 10 DI, "Compte-rendu des événements du 19 septembre au 20 septembre," 20 September 1916, AAT, 24N992.

173. Sheldon, p. 308.

174. V CA, "Instruction particulière," 19 September 1916, AFGG, IV/3, ax. 521.

175. Hart, *Somme Success*, pp. 157–59 and 174–79.

176. Norman, *Great Air War*, pp. 153–54.

177. Morrow, *German Air Power*, p. 62.

178. Ibid., pp. 204–8.

179. Ibid., pp. 215–16; Norman, op. cit., p. 172.

180. Haig to War Office, 30 September 1916, in H. A. Jones, *The War in the Air*, vol. II (Oxford: The Clarendon Press, 1928), pp. 296–97.

181. AFGG, IV/3, pp. 62–63.

182. Norman, op. cit., pp. 218 and 223–24; John Morrow, *The Great War in the Air* (Washington, D.C.: Smithsonian Institution Press, 1993), pp. 135–37.

183. 83 BI, "Rapport sur les événements du 18 au 28 septembre," 2 October 1916, AAT, 24N992.

184. Max Hoffmann, quoted in Asprey, op. cit., p. 257.

185. On the cover of Sheffield, *Somme*, for example.

186. Quoted in Denizot, op. cit., p. 145.

187. "The Opening of the Wearing-out Battle," Boraston, p. 57.

188. AFGG, IV/3, pp. 150 and 155–62.

189. HD, 27 August 1916.

190. 42 DI, "Compte-rendu de contrôle de la correspondance," 28 August 1916, AAT, 24N992.

191. Quoted in Général Yves Gras, *Castelnau: ou l'art de commander, 1851–1944* (Paris: Éditions Denoël, 1990), p. 319.

192. King Albert's diary, 21 September 1916, *Carnets*, pp. 284–85.

193. 42 DI, "Contrôle de correspondance," op. cit.

Chapter 10: Muddy Stalemate: October–December 1916

1. Commandant Tristani, '3e bataillon du 32 RI pendant la Grande Guerre, 1914–1918," typescript, January 1922, AAT, 1Kt49, pp. 164–65.

2. Rawlinson diary, 19 September 1916, RWLN 1/5.

3. Haig to the War Committee, 7 October 1916, Haig papers.

4. HD, 17 October 1916.

5. AFGG, IV/3, pp. 60–63; Joffre to Foch, 16 October 1916, AFGG, IV/3, ax. 1052; *OH: 1916, ii*, p. 459.

6. Bean diary, 1 November 1916, AWM, 3DRL 606/63.

7. Col. Lorieux, *Le Service des Routes Militaires*, cited in Col. Astouin and Chef d'Escadron Izard, *Le Train des équipages et le service automobile pendant la grande guerre, 1914–18: aperçu historique* (Paris: Association Nationale des Anciens Combattants du Train, 1934), p. 202.

8. Bean diary, 29 October 1916, AWM, 3DRL 606/63.

9. Tristani, op. cit., pp. 167–68.

10. AFGG, XI, pp. 363 and 377; Astouin and Izard, op. cit., pp. 210–11.

11. Fayolle to Foch, two letters, 19 September 1916, and Commandant Rozet (G.Q.G. liaison officer with Sixth Army), "Compte rendu," 21 September 1916, AFGG, IV/3, axs. 515, 516 and 558.

12. AFGG, XI, pp. 377–80.

13. HD, 5 October 1916, and Haig to Joffre, 11 October 1916, Haig papers.

14. Geddes, "Roadstone Traffic" and "Light Railway Development on the British Front, France," both 11 September 1916, Haig papers.

15. HD, 12 September 1916.

16. Ian Malcolm Brown, *British Logistics on the Western Front, 1914–1919* (Westport, Conn.: Praeger, 1998), pp. 109–34.

17. Ibid., pp. 139–51.

18. Quoted in Hart, *Somme Success*, p. 192.

19. John F. Williams, *Corporal Hitler and the Great War, 1914–1918: The List Regiment* (London: Frank Cass, 2005), p. 11.

20. Capt. Fritz Wiedemann, quoted in ibid, pp. 147–60.

21. HD, 29 September 1916.

22. HD, 30 September and 2 October 1916.

23. FD, 2 and 4 October 1916.

24. Spears, *Prelude to Victory*, pp. 127–28.

25. Prior and Wilson, *The Somme*, p. 275.

26. Leutnant Wulf, quoted in Sheldon, pp. 311–12.

27. *OH: 1916, ii*, pp. 432–37 and 466–74.

28. Ibid., p. 441.

29. Ibid., pp. 430–31 and 437–38.

30. Note by Fayolle, 18 September 1916, AFGG, IV/3. ax. 491.

31. 32 CA, "Destruction du village de Sailly-Saillisel," 3 October 1916, AAT, 18N149.

32. FD, 18, 21 and 24 October 1916.

33. Général Pentel, "Note pour les 18e et 152e Divisions," 22 September 1916, AAT, 22N581.

34. 18 DI, "Opérations sur la Somme du 10 au 21 octobre et du 4 au 21 novembre 1916," November 1916, AAT, 24N373.

35. Tristani, op. cit., pp. 177–79.

36. FD, 1 November 1916.

37. Denizot, *La Somme*, p. 158.

38. Palat, *ix*, pp. 236–39.

39. Capitaine Terrasse, *Avant l'oubli: L'histoire vécue du 355e régiment d'infanterie: Grande Guerre, 1914–1918* (Nice: Imprimerie Don-Bosco, 1964), p. 128.

40. 101 RI historique, AAT, 1Kt353; Palat, *ix*, pp. 240–41; AFGG, IV/3, p. 193.

41. 101 RI historique, op. cit.

42. AFGG, IV/3, pp. 198–208; Sheldon, pp. 339–49.

43. "Opérations du 22e bataillon de Chasseurs dans la tranchée de Reuss du 1er au 8 novembre 1916," AAT, 26N557.

44. FD, 4 November 1916.

45. Bouchacourt, *L'infanterie dans la bataille*, pp. 181–200.

46. AFGG, IV/3, pp. 239–46, 254–59 and 265.

47. FD, 3 and 4 November 1916.

48. 6 BCA JMO, 5 November 1916, AAT, 26N557.

49. AFGG, IV/3, pp. 226 and 237; XXXIII CA, "Ordre général d'opérations," 26 October 1916, AAT, 22N1833.

50. Palat, *ix*, p. 252.

51. XXXIII CA, "Situation de la 77e DI le 30 octobre," 30 October 1916, and General de Fonclare to Fayolle, 8 November 1916, AAT, 22N1833; Palat, *ix*, pp. 254–55.

52. Palat, *ix*, p. 243.

53. 25 RI JMO, 13 October 1916, AAT, 9M305; Denizot, op. cit., pp. 159–60.

54. Ibid., pp. 252–53.

55. Prior and Wilson, *The Somme*, p. 274.

56. Bean, *AIF in France, 1916*, p. 926.

57. Terrasse, op. cit., p. 124.

58. Sheldon, pp. 311–22.

59. "Carnets de guerre d'André l'Huillier," 15 and 27 October 1916, AAT, 1Kt185.

60. Rawlinson diary, 24 September 1916, RWLN, 1/5.

61. Simon Wessely, "The Life and Death of Private Harry Farr," *Journal of the Royal Society of Medicine*, 99 (2006), pp. 440–43.

62. Rawlinson diary, 25 September 1916, RWLN 1/5.

63. Unteroffizier Feuge, quoted in Sheldon, pp. 344–45.

64. Carnet de Charles Barberon, p. 147.

65. Williams, op. cit., pp. 154–57.

66. Lt. Philip Chattaway diary, 6 October 1916, Cheshire Regiment Archive, Chester, 0017.01.41.

67. Captured diary from 66 IR, 52 ID, in "Extracts from letters found on Germans during the Somme battle," April 1917, AWM 27/310/72.

68. "Héröique épopée du 86 RI: glorieux régiment de Velay, sur le champs de bataille de la Somme," AAT, 1Kt70.

69. Tristani, op. cit., pp. 184–87.

70. Pierre Loti, "Il pleut sur l'enfer de la Somme," 25 November 1916, *L'Illustration*, no. 3847.

71. *Historique du régiment de marche de la Légion Étrangère* (Paris: Berger-Levrault, n.d.), p. 60.

72. Courtès, "Souvenirs," p. 337.

73. Quoted in Sheldon, p. 350.

74. Sixth Army to corps and divisional commanders, "Note du service," 4 October 1916, SHAT, 24N373.

75. Sheldon, op. cit.

76. "Carnets de guerre d'André l'Huillier," 28 October 1916.

77. Tristani, op. cit., pp. 179–81.

78. FD, 21 October and 5 November 1916.

79. Rawlinson diary, 14 October and 1 and 16 November 1916, RWLN 1/7.

80. Quoted in Farrar-Hockley, *The Somme*, p. 244.

81. Haig to the King, 5 October 1916 (Haig's emphasis), in Sheffield and Bourne, *War Diaries*, p. 237.

82. Sheffield, *The Somme*, p. 143.

83. Cavan to Fourth Army, 3 November 1916, TNA, WO 158/236; *OH: 1916, ii*, p. 472; Prior and Wilson, *The Somme*, pp. 275–77.

84. Rawlinson to Kiggell, 3 November 1916 and unsigned and undated G.H.Q. memorandum OAD 305, TNA, WO 158/236.

85. Untitled report by Gort, 3 November 1916, TNA, WO 158/236.

86. Rawlinson to G.H.Q., 7 November 1916, TNA, WO 158/236.

87. "Summary of conference held by Fourth Army Commander at Heilly," 6 November 1916, TNA, WO 158/236.

88. "Fourth Army: Note of C.-in-C." s Instructions," 8 November 1916, TNA, WO 158/236.

89. Petit journal, 31 October and 1 November 1916, quoted in Denizot, op. cit., p. 158.

90. Quoted in Palat, *ix*, p. 256.

91. Chattaway diary, 12 October 1916.

92. Buchanan diary, 24 January 1917, AWM, PR83/19224.

93. Bean, *AIF in France, 1916*, p. 901.

94. Carnet de Charles Barberon, pp. 133–34 and 138–46.

95. Tawney, "Some Reflections," p. 25.

96. 16 September 1916, Charteris, op. cit., p. 166.

97. Cosby report, Paris, 7 July 1916, M1024A/8690-441.

98. Note by Foch, 2 November 1916, AAT, 1K130/9.

99. Knox to War Office, 2 December 1916, quoted in Torrey, "The Entente and the Romanian Campaign," p. 170.

100. Horne, *Verdun*, pp. 267–68.

101. Ousby, *Road to Verdun*, pp. 255–56.

102. Joffre, *Memoirs, ii*, pp. 488–95.

103. Horne, op. cit., p. 309.

104. Quoted in Robin Neilland, *Attrition: The Great War on the Western Front, 1916* (London: Robson Books, 2001), p. 289; Doughty, *Pyrrhic Victory*, pp. 304–8.

105. Ousby, op. cit., p. 210.

106. FD, 29 October and 14, 16 and 17 November 1916.

107. Palat, *ix*, p. 254, n.1.

108. *OH: 1916, ii*, pp. 456–60.

109. Ibid., *ii*, pp. 498–501.

110. Ibid., *ii*, pp. 494–97.

111. Ibid., *ii*, p. 493, n. 2.

112. Ibid., *ii*, p. 524, n. 1, and p. 527; Prior and Wilson, *The Somme*, pp. 294–99.

113. Blunden, *Undertones of War*, p. 138.

114. Lt. Arthur Waterhouse to his mother, 23 August 1917, AWM, PRO 3288.

115. Prior and Wilson, *The Somme*, p. 288.

116. Quoted in David Woodward, *Lloyd George and the Generals* (Newark, N.J.: University of Delaware Press, 1983), p. 113.

117. 19 July 1916, Repington, *First World War*, pp. 283–85.

118. Hankey, *Supreme Command, ii*, p. 556.

119. Cassar, *Asquith as War Leader*, pp. 198–200.

120. Ibid., pp. 310–11; King, *Generals and Politicians*, pp. 135–36.

121. Lloyd George to Robertson, 11 October 1916, in Woodward, *Military Correspondence*, pp. 93–96.

122. Woodward, *Lloyd George*, pp. 110–11.

123. Robertson to Northcliffe, 11 October 1916, and Robertson to Gwynne, 11 October 1916, in Woodward, *Military Correspondence*, pp. 91–92.

124. HD, 22 October and 2 November 1916.

125. Haig to Lady Haig, 21 October 1916, in Sheffield and Bourne, op. cit., p. 244.

126. Lloyd George to Haig, quoted in HD, 11 November 1916.

127. HD, 7, 10 and 16 October 1916; Robertson to Haig, 9 October 1916, Haig papers.

128. Gwynne to Rawlinson, 11 October 1916, RWLN, 1/8.

129. Haig to the King, 18 October 1916, in Sheffield and Bourne, op. cit., pp. 243–44.

130. Prior and Wilson, *The Somme*, p. 314.

131. War Committee minutes, 3 November 1916, TNA, CAB 42/23/4.

132. Robertson to Kiggell, 29 September 1916, in Woodward, *Military Correspondence*, pp. 87–88.

133. 13 November and 6 December 1916, Charteris, *At G.H.Q.*, pp. 177–80.

134. Prior and Wilson, *The Somme*, pp. 310–11.

135. Quoted in Woodward, *Lloyd George*, pp. 119–20.

136. HD, 15 and 16 November 1916; Cassar, op. cit., pp. 202–5.

137. Cassar, op. cit., p. 203.

138. HD, 16 November 1916.

139. Rawlinson diary, 21 November 1916, RWLN 1/7.

140. Cassar, op. cit., pp. 202–10.

141. T. A. White, *The Fighting Thirteenth: The History of the 13th Battalion AIF* (Sydney: Tyrells, 1924), p. 80.

142. Buchanan diary, 21 January 1917.

143. Bean, *AIF in France, 1916*, pp. 894 and 950.

144. Bean diary, 12 October 1916, AWM, 3DRL 606/60.

145. Bean, *AIF in France, 1916*, p. 900.

146. Bean diary, 5 November 1916, AWM, 3DRL 606/63.

147. Sergeant Arthur E. Matthews, "The Campaign on the Somme—in the Winter of 1916–17," AWM, PR91/119, p. 6.

148. Bean, *AIF in France, 1916*, pp. 902–15.

149. Bean diary, 17 November 1916, AWM, 3DRL 606/66.

150. Ibid., 18 November 1916.

151. Matthews, op. cit., p. 8.

152. Buchanan diary, 24 January 1917.

153. Tristani, op. cit., p. 184.

154. Buchanan diary, 26 January 1917.

155. Matthews, op. cit., p. 12.

156. Bean diary, 18 November 1916, AWM, 3DRL 606/66; Chapman diary, 7 January 1917.

157. Tristani, op. cit., pp. 188–90.

158. Matthews, op. cit., pp. 9–11.

159. Address on Lt. Norman Heathcote letter to his parents, 19 November 1916, AWM, PR02015.

160. Gunner Cecil Giffin diary, State Library of New South Wales, ML MSS 1025.

161. Matthews, op. cit., pp. 2–3 and 15.

162. Chapman diary, 7 January 1917.

163. Buchanan diary, 21 January 1917.

164. Matthews, op. cit., p. 16.

165. Heathcote to his parents, 1 January 1917.

166. Lt. Ronald A. McInnis diary, 13 and 23 February 1917, AWM, PR00917.

167. Ibid., 5 November 1916.

168. Bean diary, 18 February 1917, AWM, 3DRL 606/70.

169. Matthews, op. cit., p. 19.

170. Heathcote to his parents, 10 February 1917.

171. Lt. Ulric Walsh to Austin Walsh, 25 December 1916, AWM, PR01801.

172. Bean diary, 24 and 25 December 1916, 3DRL 606/68.

173. Pvt. E. G. King diary, 25 December 1916, AWM, PR83/018.

174. Leutnant Mandl diary, 24 December 1916, AWM, 2DRL/1167.

175. Ibid., 12 December 1916.

176. Quoted in Bean diary, 16 December 1916, AWM, 3DRL 606/68.

177. Mandl diary, 12 December 1916.

Chapter 11: "Victory Inclining to Us"

1. FD, 5 January 1917.

2. FD, 8 January 1917.

3. Joffre, *Memoirs, ii,* p. 500.

4. FD, 8 February 1917.

5. King, *Generals and Politicians,* pp. 106 and 137; Doughty, *Pyrrhic Victory,* pp. 319–22.

6. FD, 9 January 1917.

7. Doughty, op. cit., p. 320.

8. FD, 8, 9, 10 and 28 January 1917.

9. Cassar, *Asquith as War Leader,* pp. 206–10

10. 19 December 1916, *Parliamentary Debates: Commons,* lxxxviii, 1339–40.

11. Note, 19 April 1917, Foch papers, AAT 1K130/9/6.

12. 10 and 20 December 1916, Charteris, *At G.H.Q.,* pp. 180–81.

13. HD, 13 January 1917.

14. HD, 7 February 1917.

15. Hankey diary, 15 February 1917, HNKY 1/1.

16. HD, 20 and 28 December 1916.

17. Philpott, *Anglo-French Relations,* pp. 100–102.

18. FD, 8 January 1917.

19. Wilson diary, 31 December 1916, Callwell, *Wilson, i,* p. 306.

20. Robinson, *The Turning Point.*

21. Waterhouse to his mother, 23 August 1917, AWM, PR03288.

22. Masefield, *The Old Front Line,* pp. 75–76.

23. "The Opening of the Wearing-out Battle," Boraston, p. 58.

24. Edwin Montagu, "Memorandum on the Fall of the Coalition," 9 December 1916, Bonham Carter papers, Bodleian Library, Oxford, 668.

25. Esher journal, 6 August 1916, in *Journals and Letters, iv,* p. 45.

26. Military attaché, Buenos Aires, to Washington, 15 November 1916, enclosing letter from Germany, September 1916, M1024A/8690–514.

27. Ibid.

28. Esher journal, 29 July 1916, in *Journals and Letters, iv,* p. 40.

29. Observer, "The War Week by Week," *Evening Telegram,* 3 July 1916.

30. "The Opening of the Wearing-out Battle," Boraston, p. 51.

31. FD, 8 January 1917.

32. Esher journal, 30 July 1916, in *Journals and Letters, iv,* pp. 40–43.

33. Rawlinson diary, 30 December 1916, RWLN 1/7.

34. Colonel Treadwell, U.S. Marine Corps, "Visit to British War Zone in France," 28 December 1916, M1024A/8690–533.

35. HD, 5 November 1916.

36. Major Moses, "Present Strength of Armies," 1 December 1916, and Kennedy, "Fighting Numbers engaged in European War with losses to date," 11 December 1916, M1024A/8690–510 and 516.

37. Montagu, "Memorandum," op. cit.

38. Charteris, "Fluctuations in German Strength during 1916," 23 October 1916, RWLN, 1/8; Herwig, *First World War,* pp. 247–48.

39. Herwig, op. cit., p. 264.

40. Passingham, *All the Kaiser's Men,* p. 121.

41. "Comparative Strengths of German Forces in 1916 and 1917 on Western Front," RWLN, 1/8.

42. G.H.Q. intelligence summary, "Reduction of Battalion Strengths," 30 April 1917, RWLN, 1/8.

43. Herwig, op. cit.

44. *OH: 1916, ii,* p. 578, n. 3.

45. Robertson to Haig, 25 August 1916, in Woodward, *Military Correspondence,* p. 83.

46. Woodward, *Lloyd George,* pp. 174–75.

47. G.H.Q. intelligence summary, "German Divisions Engaged on the Somme," 4 October 1916, RWLN, 1/8.

48. *Ordres de bataille des grandes unités* (1924), AFGG, X.

49. Maj. A. F. Becke, *History of the Great War: Order of Battle* (London: HMSO, five vols., 1935–45).

50. "Captures in the Somme Battle," RWLN, 1/8.

51. *OH: 1916, ii,* p. 555.

52. Joffre to *Ministre de la Guerre,* 18 July 1916, in Pédroncini, *Journal de Marche,* p. 58.

53. Joffre, *Memoirs, ii,* pp. 496–99.

54. Ibid., pp. 497–98.

55. "The Opening of the Wearing-out Battle," Boraston, p. 59.

56. Rawlinson diary, 6 November 1916, RWLN 1/7.

57. Paddy Griffith, *Battle Tactics of the Western Front: The British Army's Art of Attack, 1916–1918* (New Haven, Conn.: Yale University Press, 1994), pp. 76–79.

58. See the comparison of the two in Spears, *Prelude to Victory,* app. 1, pp. 517–26.

59. *OH: 1916, ii,* pp. 567–68.

60. F. F., "Les Opérations de guerre: la bataille de la Somme," *Journal de Genève,* 25 March 1917, AAT, 1K129/1.

61. Repington, *First World War, i*, pp. 256–58.

62. Debeney, Speech on the unveiling of the statue of General Foch, 1926, AAT, 1K129/3.

63. Untitled and undated memorandum (ca. January 1917), in "Après la Somme: études," AAT, 1K130/3/J.

64. Falkenhayn, *General Headquarters*, p. 266.

65. Sheldon, *passim*.

66. Gerhard Ritter to Hermann Witte, 16 May 1917, quoted in Herwig, op. cit., p. 204.

67. Rawlinson to G.H.Q., 7 November 1916, TNA, WO 158/236; *OH: 1916, ii*, p. 555.

68. Charteris to Macdonough, 26 October 1916, quoted in Beach, "British Intelligence," p. 180.

69. Mandl diary, 26 February and 3 March 1917, AWM, 2DRL/1167.

70. "The Opening of the Wearing-out Battle," Boraston, p. 53.

71. Fourth Army, "Tactical Notes," May 1916, *OH: 1916, i*, ax. 18.

72. Quoted in Passingham, op. cit., p. 125.

73. 2 January 1917, J. C. Dunn, *The War the Infantry Knew, 1914–1919* (London: Abacus, 1994), p. 288.

74. Below, "Experience of the German 1st Army," p. 6.

75. Waterhouse, op. cit.

76. Treadwell, "Visit to British War Zone in France," op. cit.

77. F. F., "Les Opérations de guerre," op. cit.

78. HD, 15 January 1917.

79. Queri, "The Artillery Battle at the Somme," op. cit.

80. "Ordre général no. 55," 17 November 1916, 6 BCP JMO, AAT, 26N557.

81. Untitled and undated memorandum (ca. January 1917), op. cit.

82. Waterhouse, op. cit.

83. "The Opening of the Wearing-out Battle," Boraston, p. 57.

84. Thierry, "Souvenirs," p. 75.

85. Playout to his "brothers and sisters," 10 December 1916, in *Seven Aussie Poilus*, p. 66.

86. Logan, "Politico-Military Notes," Paris, 29 November 1916, M1024A/8690–518.

87. Ibid.

88. *OH: 1916, ii*, p. 554, n. 2.

89. Logan, "Politico-Military Notes," op. cit., reporting Albert Favre's 25 October 1916 speech in the Chamber.

90. Military attaché, Buenos Aires, to Washington, op. cit.

91. Logan, "Politico-Military Notes," op. cit.

92. Dunn, 9 December 1916, op. cit., p. 284.

93. Ebba Dahlin, *French and German Pubic Opinion on Declared War Aims, 1914–1918* (New York: Ams Press, 1971).

94. Robertson to Lansdowne, 1 December 1916, Woodward, *Military Correspondent*, p. 119.

95. Chapman diary, 7 January 1917.

96. Logan, "Politico-Military Notes," op. cit.

97. Colonel Lassiter, "Notes on the Progress of the War as Affecting England," London, 29 December 1916, M1024A/8690–532.

98. Bean diary, 12 January 1917, AWM, 3DRL 606/68.

99. Ibid., 13 January 1917.

100. 19 December 1916, *Parliamentary Debates: Commons,* lxxxviii, 1335.

101. David Stevenson, *The First World War and International Politics* (Oxford: Clarendon Press, 1988), pp. 136–38.

102. Charles Bean, *The Official History of Australia in the War of 1914–1918, vol. IV: The Australian Imperial Force in France, 1917* (Sydney: Angus & Robertson, 1933), pp. 50–51; quoted in Horne, *Verdun,* p. 331.

103. Wilhelm Deist, "The German Army, the Nation State and Total War," in *State, Society and Mobilization in Europe During the First World War,* ed. John Horne (Cambridge: Cambridge University Press, 1997), pp. 160–72: 166.

104. Military attaché, Buenos Aires, to Washington, op. cit.

105. McInnis diary, 6 February 1917, AWM, PR00917.

106. Mandl diary, 6 and 15 January and 28 March 1917.

107. Moyer, *Victory Must Be Ours,* pp. 170–71.

108. Herwig, op. cit., pp. 260–61.

109. Taylor, War College Division, "Fighting Numbers Engaged in European War with Losses to Date," Washington (ca. 11 December 1916), M1024A/8690–510.

110. Falkenhayn, op. cit., pp. 289–91.

111. Jünger, *Storm of Steel,* pp. 121–24.

112. G.Q.G., "Ligne Hindenburg," 26 March 1917, AFGG, V/1, ax. 1026; Passingham, op. cit., pp. 139–44.

113. Cyril Falls, *Military Operations: France and Belgium,1917, vol. I: The German Retreat to the Hindenburg Line and the Battle of Arras* (London: Macmillan, 1940), p. 110.

114. Ibid., pp. viii–ix.

115. Jünger, op. cit., pp. 127–28.

116. Buchanan diary, 1 May 1916.

117. Matthews, op. cit., p. 21.

118. AFGG, V/1, pp. 371–75.

119. G.Q.G., "Indices de préparation d'un repli du front allemand sur la ligne Lille, Cambrai, St.-Quentin, massif de la Fôret de St.-Gobain," 25 February 1917, AFGG, V/1, ax. 731.

120. AFGG, V/1, pp. 375–76.

121. Pierrefeu, *French Headquarters,* pp. 139–40.

122. Ibid.

123. AFGG, V/1, pp. 388–91; Doughty, op. cit., pp. 335–36.

124. "Note pour les groupes d'armées," AFGG V/1, ax. 972; FD, 15–31 March 1917.

125. AFGG, V/1, p. 385.

126. *OH: 1917, i,* pp. 61–63.

127. Pierrefeu, op. cit., pp. 138–39.

128. Cecil Giffin diary, op. cit.

129. AFGG, V/1, pp. 378–79.

130. AFGG, V/1, p. 378.

131. McInnis diary, 9–19 March 1917; Bean, *AIF in France, 1917,* pp. 116–18.

132. Bean, op. cit., p. 120.

133. Ibid., pp. 125–26.

134. Ibid., p. 144.

135. Bamford, "One of the fifty-nine thousand."

136. Bean diary, 18 March 1917, AWM, 3DRL 606/73.

137. Bean, *AIF in France, 1917,* pp. 205–6.

138. Malins, *How I Filmed the War,* p. 287.

139. Monteith, "Observations of No. 5886," *i*, p. 76.

140. Buchanan diary, 16 April 1917.

141. Seely, *Adventure*, pp. 256–58.

142. Waterhouse, op. cit.

143. Ibid.

144. Giffin diary, op cit.

145. White, *The Fighting Thirteenth*, p. 91.

146. Bean diary, 19 March 1917, AWM, 3DRL 606/73.

147. Waterhouse, op. cit.

148. Malins, op. cit., p. 250.

149. 17 and 18 March 1917, Commandant Louis Girard, *Sur le front occidental avec la 53e division d'infanterie, vol. III: Fin-août 1916 à octobre 1917* (Paris: Brodard et Taupin, n.d.), pp. 70–73.

150. Ibid., p. 72; notes by M. Mairesse, 28 December 1921, ADS, Fonds Duchaussoy, 14J51.

151. Third Army, "Bulletin de renseignements no. 121," 18 March 1917, AFGG, V/1, ax. 925.

152. Malins, op. cit., pp. 279–80.

153. "Ravitaillement de la population civile du département de la Somme, 1914–21," ADS, 6M1996.

154. Prefectorial report to Minister of the Interior, October 1915, ADS, 6M1856.

155. Seely, op. cit., p. 255; Bean, *AIF in France, 1917*, p. 141.

156. McInnis diary, 3 April 1917.

157. Pierrefeu, op. cit., pp. 140–42.

158. Malins, op. cit., pp. 254–70.

159. Reeve, "Through the Eyes of the Camera," p. 782.

160. Pierrefeu, op. cit., pp. 140–42.

161. Passingham, op. cit., p. 138.

162. AFGG, V/1, p. 375.

163. McInnis diary, 18 March 1917.

164. Third Army, "Bulletin de renseignements no. 121," op. cit.

165. First Army, "Bulletin de renseignements no. 135," 17 March 1917, AFGG, V/1, ax. 903.

166. Third Army, "Bulletin de renseignements no. 121," op. cit.

167. Seely, op. cit., pp. 255–56.

168. *OH: 1917, i*, pp. 113–14.

169. Wilson diary, 31 December 1916, Wilson papers.

170. Wilson diary, 26 October and [14] November 1916, in Callwell, *Wilson, i*, p. 296.

171. Brock Millman, *Pessimism and British War Policy, 1916–1918* (London: Frank Cass, 2001), pp. 26–27.

172. Quoted in Sheldon, p. 396.

173. Quoted in Passingham, op. cit., p. 124.

174. Quoted in *OH: 1916, ii*, p. 555.

175. Herwig, op. cit., p. 230.

176. 21 December 1916, *Parliamentary Debates: Commons*, lxxxviii, 1719.

177. "Post-war Recollections of Haig as Commander-in-Chief," n.d., ESHR 19/5.

178. Le Roy Lewis to Lloyd George, 23 November 1916, Lloyd George papers, E/3/14/30.

179. Notes by Foch, 2 and 3 November and 15 December 1916, AAT, 1K130/9/6.

Chapter 12: Remobilisation

1. Tristani, "3e bataillon du 32 RI," pp. 192, 203 and 209.

2. Ibid., pp. 210–14.

3. FD, 16 April 1917.

4. Matthews, "The Campaign on the Somme," pp. 24–27.

5. Bean, *AIF in France, 1917,* pp. 393–400.

6. Matthews, op. cit., pp. 28–31.

7. Jonathan Walker, *The Blood Tub: General Gough and the Battle of Bullecourt, 1917* (Staplehurst: Spellmount, 1998).

8. Matthews, op. cit., p. 38.

9. HD, 20 December 1916.

10. Doughty, *Pyrrhic Victory,* p. 324.

11. Haig to Nivelle, 6 January 1916, TNA, WO 158/37/8.

12. Doughty, op. cit., pp. 338–39.

13. Briand, quoted in ibid., p. 337.

14. HD, 16, 18 and 24 March 1917.

15. Doughty, op. cit., pp. 341–42; Spears, *Prelude to Victory,* pp. 356–58.

16. Spears, op. cit., p. 364.

17. Ibid., pp. 364–77; Doughty, op. cit., pp. 342–44.

18. Doughty, op. cit., pp. 361–64; Leonard V. Smith, "Remobilizing the Citizen-Soldier Through the French Army Mutinies of 1917," in Horne, *State, Society and Mobilization,* pp. 144–59.

19. Tristani, op. cit., p. 217.

20. AFGG, V/1, p. 765.

21. Tristani, op. cit., pp. 221–33.

22. Quoted in Smith, op. cit., p. 153.

23. Tristani, op. cit., pp. 234–35.

24. AFGG, V/2, *passim.*

25. Barbier, *Grande guerre à Amiens,* pp. 133–38 and 178–80.

26. Jean-Baptiste Duroselle, *La Grande guerre des français, 1914–1918* (Paris: Perrin, 1994), pp. 199–202.

27. Stevenson, *International Politics,* pp. 148–52.

28. Ibid., pp. 156–62.

29. Duroselle, op. cit., pp. 296–98.

30. Ibid., pp. 294–95.

31. Tristani, op. cit., p. 233.

32. Duroselle, op. cit., p. 291.

33. Ibid., p. 290.

34. Quoted in ibid., pp. 297–98.

35. Ibid., p. 296.

36. John Horne, "Remobilizing for 'Total War': France and Britain, 1917–1918," in Horne, *State, Society and Mobilization,* pp. 195–211: 198–99.

37. Duroselle, op. cit., pp. 303–7.

38. Quoted in ibid., p. 322.

39. Ibid., pp. 316–17.

40. Ibid., p. 306.

41. Gerard DeGroot, *Blighty: British Society in the Era of the Great War* (London: Longman, 1996), pp. 110–21; Arthur Marwick, *The Deluge: British Society and the First World War* (Basingstoke: Palgrave Macmillan, 2006), pp. 243–50.

42. Max Egremont, *Siegfried Sassoon* (London: Picador, 2005), pp. 133–35.

43. Ibid., p. 151.

44. Quoted in ibid., pp. 143–44.

45. Ibid., pp. 145–46.

46. Quoted in ibid., p. 135.

47. Pat Barker, *Regeneration* (London: Viking, 1991).

48. Siegfried Sassoon, *The Complete Memoirs of George Sherston* (London: Faber & Faber, 1937 [originally published separately in 1928, 1930 and 1936]).

49. Ibid., p. 803.

50. Quoted in Egremont, op. cit., p. 153.

51. Ibid., pp. 151–53 and 158.

52. "Mark VII" (Max Plowman), *A Subaltern on the Somme in 1916* (London: Dent, 1927).

53. Malcolm Pittock, "Max Plowman and the Literature of the First World War," *Cambridge Quarterly*, 33 (2004), pp. 217–43; "Protesters Against the War: The Contrasting Cases of Siegfried Sassoon and Max Plowman," in Brian Bond, *Survivors of a Kind: Memoirs of the Western Front* (London: Continuum, 2008), pp. 93–112.

54. Marwick, op. cit., pp. 255–56.

55. Ibid., p. 255; DeGroot, op. cit., p. 290.

56. Horne, "Remobilizing for 'Total War,'" pp. 199–200.

57. Stevenson, op. cit., pp. 191–93.

58. Millman, *Pessimism and British War Policy*, pp. 135–40.

59. *Sydney Morning Herald*, 31 July 1917, in Dyer, *Seven Aussie Poilus*, pp. 80–81.

60. Croft, *Twenty-two Months*, pp. xiii–xv.

61. Croft, *My Life of Strife*, pp. 129–33.

62. Ibid., p. 139.

63. Sassoon, *Complete Memoirs*, pp. 802–3.

64. Statement by Robertson, 4 May 1917, quoted in Lloyd George, *War Memoirs, i*, pp. 925–26.

65. FD, 21 April 1917.

66. FD, 26 April 1917.

67. McInnis diary, 6 May 1917.

68. Ibid., 26 May and 16 June 1916.

69. Ibid., 21 June 1917.

70. Quoted in John Coates, *An Atlas of Australia's Wars* (Melbourne: Oxford University Press, 2001), p. 56.

71. McInnis diary, 21 June 1917.

72. Sir William Orpen, *An Onlooker in France, 1917–19* (London: Williams & Norgate, 1921), p. 18.

73. "Beaumont-Hamel: A Memory of the Somme," in ibid., p. 23.

74 Ibid., p. 20.

75. Ibid., p. 36.

76. HD, 31 March 1917.

77. Todman, *Myth and Memory*, p. 16.

78. "Introduction," in John Masefield's *Great War: Collected Works*, ed. Philip W. Errington (Barnsley: Pen & Sword, 2007), pp. 16–22. The quotations are from his contemporary letters to his wife.

79. John Masefield, "The Battle of the Somme," in ibid., pp. 263–305 (originally published 1919).

80. 19 October 1916, Charteris, *At G.H.Q.*, p. 174.

81. Todman, op. cit., p. 22.

82. Quoted in Deist, "The Nation-State and Total War," p. 166.

83. Roger Chickering, *Imperial Germany and the Great War, 1914–1918* (Cambridge: Cambridge University Press, 2nd edn., 2004), pp. 161–63; Kocka, *Facing Total War,* pp. 129–30.

84. Strachan, "The Morale of the German Army, 1917–18," in Cecil and Liddle, *Facing Armageddon,* pp. 383–99: 386–87.

85. Ibid., pp. 387–88.

86. Chickering, op. cit., pp. 162–63.

87. Deist, op. cit., pp. 166–72.

88. Cyril Falls, *The First World War* (London: Longman, 1960), p. 385.

89. Philpott, *Anglo-French Relations,* pp. 150–54.

90. FD, 17 February 1918.

91. Robertson to Haig, 20 April 1917, in Woodward, *Military Correspondence,* pp. 178–80.

Chapter 13: Decision in Picardy: 1918

1. 21 March 1918, Duchaussoy, "Une municipalité du Union Sacrée," part 2, ADS, Fonds Duchaussoy, 14J98.

2. Brig.-Gen. Sir James Edmonds, *Military Operations, France and Belgium, 1918, vol. I: The German March Offensive and Its Preliminaries* (London: Macmillan, 1935), pp. 152–53.

3. Ibid., pp. 155–60; Martin Middlebrook, *The Kaiser's Battle, 21 March 1918: The First Day of the German Spring Offensive* (London: Allen Lane, 1978), pp. 51–55.

4. Strachan, "Morale of the German Army," pp. 393–94.

5. Middlebrook, op. cit., p. 62.

6. Quoted in ibid., p. 64.

7. Ibid., pp. 58–59.

8. Churchill, *World Crisis, ii,* pp. 1295–96.

9. FD, 20 March 1918.

10. Elizabeth Greenhalgh, "Myth and Memory: Sir Douglas Haig and the Imposition of Allied Unified Command in March 1918," *The Journal of Military History,* 68 (2004), pp. 771–820: 786–89.

11. *OH: 1918, i,* pp. 265–66.

12. Woodward, *Lloyd George,* p. 292.

13. Middlebrook, op. cit., pp. 74–82.

14. FD, 19 March 1918.

15. *OH: 1918, i,* pp. 392–93.

16. FD, 22 March 1918.

17. *OH: 1918, i,* pp. 395–97.

18. Ibid., p. 401.

19. 21 March 1918, Duchaussoy, "Une Municipalité," part 2.

20. Philippe Nivet, "Les Civils dans la bataille de Picardie (1918)," in Nivet, *La Bataille en Picardie,* pp. 171–93.

21. M. Mairesse, notes on 21–24 March 1918, 28 December 1921, ADS, Fonds Duchaussoy, 14J51.

22. 25 March 1918, Duchaussoy, op. cit.

23. Barbier, *Grande guerre à Amiens,* pp. 148–49.

24. 26–30 March 1918, Duchaussoy, op. cit.

25. *OH: 1918, i,* pp. 448–50; Greenhalgh, "Myth and Memory," pp. 789–96.

26. HD, 26 March 1918.

27. Greenhalgh, "Myth and Memory," pp. 806–7; Philpott, *Anglo-French Relations*, pp. 155–56.

28. *OH: 1918, i,* p. 401.

29. Ibid., pp. 416–17.

30. 28 March 1918, Duchaussoy, op. cit.

31. Charles Bean, *The Official History of Australia in the War of 1914–1918, vol. V: The Australian Imperial Force in France, During the Main German Offensive, 1918* (Sydney: Angus & Robertson, 1941), pp. 174–77.

32. Pvt. Verdi G. Schwinghammer diary, 1916–1919, AWM, 2DRL/0234.

33. Bean, *AIF in France, 1918, v,* p. 188.

34. Lt. John G. Ridley diary, AWM, 2DRL/0775.

35. Bean, *AIF in France, 1918, v,* pp. 191–92; Nivet, "Les civils," p. 175.

36. Pvt. William B. Long diary, 29 March 1918, AWM, PR01406.

37. Schwinghammer diary.

38. Long diary, 30 March 1918.

39. Schwinghammer diary.

40. Ibid.; Bean, *AIF in France, 1918, v,* pp. 230–35.

41. Kapitän G. Goes, *Der Tag X: Die grosse Schlacht in Frankreich 21 Marz–5 April 1918* (Berlin: Kolk, 1933), quoted in *OH: 1918, i,* p. 533.

42. *OH: 1918, i,* p. 533.

43. Henry Bordeaux quoted in Contamine, *Cahiers secrets,* p. 266.

44. AFGG, VI/1, pp. 380–90.

45. FD, 28–31 March 1918.

46. Monteith, "Observations of No. 5886," *i,* p. 88, and *ii,* pp. 9–10.

47. Tristani, 32 RI," pp. 258–65.

48. Goes quoted in Brig.-Gen. Sir James Edmonds, *Military Operations, France and Belgium, 1918, vol. II: March–April: Continuation of the German Offensive* (London: Macmillan, 1937), p. 120.

49. AFGG VI/1, pp. 412–8; *OH: 1918, ii,* pp. 121–29.

50. AFGG VI/1, pp. 462–63.

51. Brown, *Somme,* p. 317.

52. Ibid., p. 316.

53. Quoted in *OH: 1918, ii,* p. 137.

54. FD, 13 April 1918.

55. Foch, "Directive générale no. 2," 3 April 1918, AFGG, VI/1, ax. 1374.

56. Tristani, op. cit., pp. 266–70.

57. IX CA, "Instruction relative au combat de rupture," 1 April 1918, AAT, 24N378.

58. Tristani, op. cit., pp. 272–80; 32 RI JMO, AAT, 3M312.

59. Dale M. Titler, *The Day the Red Baron Died* (London: Ian Allan, 1973).

60. *OH: 1918, ii,* p. 392. See also Heinz Guderian, *Achtung Panzer* (London: Arms & Armour Press, 1992), pp. 184–86, who suggests the German tanks fared rather better in the engagement.

61. *OH: 1918, ii,* pp. 396–405.

62. Nivet, "Les civils," p. 176.

63. Barbier, op. cit., p. 152.

64. Coates, *Atlas of Australia's Wars,* p. 1.

65. FD, 19 May 1918.

66. FD, 23 July 1918.

67. Maj.-Gen. Sir George Aston, *The Biography of the Late Marshal Foch* (London: Hutchinson, 1929), p. 226.

68. Quoted in Asprey, *German High Command,* pp. 365–68.

69. Lloyd George, *War Memoirs, ii,* p. 1844.

70. Ibid., p. 1868.

71. Foch, *Memoirs,* p. 393.

72. Strachan, "The Morale of the German Army," p. 390.

73. Foch, op. cit., p. 390.

74. Contamine, *Cahiers secrets,* p. 286.

75. Foch, op. cit., pp. 390–92.

76. Barrie Pitt, *1918: The Last Act* (London: Cassell, 1962), p. 178.

77. FD, 16 July 1918.

78. Foch, op. cit., pp. 404–24.

79. Ibid., p. 423.

80. AFGG, VII/1, pp. 161–62.

81. Foch memorandum, 24 July 1918, in Foch, op. cit., pp. 425–29.

82. AFGG, VII/1, p. 163; Strachan, "The Morale of the German Army," pp. 390–96.

83. G.Q.G. Second Bureau, "Bulletin d'information (extrait)," 7 August 1918, AFGG, VII/1, ax. 525.

84. G.Q.G. Second Bureau, "Compte rendu de renseignements no. 1525 (extrait)," 21 August 1918, AFGG VII/1, ax. 775.

85. FD, 30 July 1918.

86. Pvt. J. I. Marshall, "My Story of the Big War," State Library of New South Wales, ML MSS 1164.

87. Foch memorandum, 24 July 1918, op. cit.

88. Foch, op. cit., pp. 433–34.

89. Paul Harris with Niall Barr, *Amiens to the Armistice* (London: Brassey's, 1998), p. 65.

90. AFGG, VII/1, pp. 172–73.

91. Prior and Wilson, *Command,* pp. 301–8; Harris, op. cit., pp. 59–86.

92. AFGG, VII/1, pp. 167–71.

93. Henry Bordeaux quoted in Contamine, *Cahiers secrets,* p. 274.

94. Lt.-Col. A. Grasset, *Montdidier: Le 8 août 1918 à la 42e division* (Paris: Éditions Berger-Levrault, 1930), pp. 35, 44 and 69–70.

95. Ibid., pp. 80–81.

96. Ridley diary, op. cit.

97. Grasset, op. cit., pp. 25–26.

98. Ibid., pp. 83–87.

99. Ibid., pp. 99–104.

100. Ibid., p. 118.

101. Ibid., pp. 130–31.

102. AFGG, VII/1, pp. 175–76.

103. Harris, op. cit., pp. 89–92.

104. Marshall, op. cit.

105. Ibid.

106. Brig.-Gen. Sir James Edmonds, *Military Operations, France and Belgium 1918, vol. IV: 8th August–26th September: The Franco-British Offensive* (London: Macmillan, 1947), pp. 84–86.

107. Ibid., *iv,* pp. 106–14.

108. Long diary, 9 August 1918.

109. Foch to Debeney, 9 August 1918, AFGG, VII/1, ax. 571.

110. FD, 8 August 1918.

111. AFGG, VII/1, pp. 184–85.

112. FD, 9 and 10 August 1918.

113. Lloyd George, op. cit., *ii*, pp. 1868–70.

114. Quoted in ibid., p. 1869.

115. AFGG, VII/1, p. 190.

116. *OH: 1918, iv*, p. 15.

117. Lloyd George, op. cit., *ii*, p. 1876.

118. Fayolle, "Note pour les armées," 9 August 1918, and Mangin, "Note pour les armées," 10 August 1918, AFGG, VII/1, axs. 579 and 609.

119. AFGG, VII/1, pp. 226–27; G.Q.G. Second Bureau, "Répartition des divisions allemands sur le front occidental à la date du 20/8/18," 21 August 1918, AFGG, VII/1, ax. 774.

120. Marshall, op. cit.

121. AFGG, VII/1, pp. 230–34.

122. Long diary, 30 August 1918.

123. Foch to Haig, 25 August 1918, AFGG, VII/1, ax. 835.

124. *OH: 1918, iv*, pp. 356–61 and 366–75; Harris, op. cit., pp. 157–62.

125. Marshall, op. cit.

126. Harris, op. cit., p. 159.

127. *OH: 1918, iv*, p. 374, n. 1, and p. 375, n. 1.

128. Marshall, op. cit.

129. McInnis diary, 12 September 1918.

130. Long diary, 29 and 31 August 1918.

131. McInnis diary, 9 September 1918.

132. AFGG, VII/1, pp. 257–58; G.Q.G. Second Bureau, "Répartition des divisions allemands sur le front occidental à la date du 31 août 1918," 1 September 1918, AFGG, VII/1, ax. 922.

133. Haig to army commanders, 22 August 1918, TNA, WO 158/241.

134. *OH: 1918, iv*, p. 367.

135. McInnis diary, 24 and 26 September 1918.

136. Ibid., 21 September 1918.

137. Strachan, "Morale of the German Army," pp. 394–95.

138. Général E. Mangin, *Les Chasseurs dans la bataille de France: 47e Division (juillet–novembre 1918)* (Paris: Payot, 1935), pp. 116–18.

139. Schwinghammer diary.

140. Foch, op. cit., pp. 467–72.

141. Ibid., pp. 482–83.

142. Brig.-Gen. Sir James Edmonds, *Military Operations, France and Belgium 1918, vol. V: 26th September–11th November: The Advance to Victory Offensive* (London: Macmillan, 1947), p. 569.

143. AFGG, VII/2, app. 2, pp. 412–21.

144. *OH: 1918, v*, p. 557.

145. Lloyd George, op. cit., *ii*, pp. 1882–84.

146. Ibid., pp. 1968–70.

147. Haig to Churchill, 3 October 1918, in Blake, *Private Papers*, p. 329.

148. FD, 11 November 1918.

149. Barbier, op. cit., p. 155.

150. Arthur Streeton to Tom Roberts, 28 November 1918, in *Letters from Smike: The Letters of Arthur Streeton, 1890–1943*, ed. Ann Galbally and Anne Gray (Melbourne: Oxford University Press, 1989).

151. Tristani, op. cit., p. 358.

152. Haig to Lady Haig, 11 October 1918, in Blake, op. cit., pp. 331–32; HD, 17 October 1918.

153. HD, 24 October 1918.

154. HD, 19 October 1918.

155. HD, 30 November 1918.

156. George Dewar and Lt.-Col. John Boraston, *Sir Douglas Haig's Command, 1915–1918* (London: Constable, two vols., 1922).

157. HD, 27 October 1918.

158. Greenhalgh, *Victory Through Coalition, passim.*

159. FD, 27 October 1919.

160. FD, 2 November 1918.

161. McInnis diary, 18 October 1918.

162. Long diary, 11 November 1918.

163. Contamine, *Cahiers secrets,* p. 313.

164. Tristani, op. cit., p. 358.

165. McInnis diary, 10 November 1918.

166. FD, 27 November and 26 December 1918.

167. Bessel, "Mobilizing German Society for War," in Chickering and Förster, *Great War, Total War,* pp. 437–51: 451.

168. FD, 27 November 1918.

169. Stevenson, "French Strategy on the Western Front, 1914–1918," in Chickering and Förster, op. cit., pp. 297–326: 324.

170. Ibid.

Chapter 14: Aftermath and Memory

1. Taylor, *The First World War,* pp. 99–105.

2. Michael Howard, "The First World War Reconsidered," in *The Great War and the Twentieth Century,* ed. Jay Winter, Geoffrey Parker and Mary R. Habeck (New Haven, Conn.: Yale University Press, 2000), pp. 13–29: 27.

3. Letter from Germany, September 1916, enclosed with military attaché, Buenos Aires, to Washington, 15 November 1916, M1024A/8690–514.

4. Tristani, "3rd Battalion of 32 RI," pp. 375–76.

5. Howard, op. cit., p. 28.

6. Correspondence and papers in "Thiepval Memorial," ADS, KZ248.

7. Jay Winter, *Remembering War: The Great War Between Memory and History in the Twentieth Century* (New Haven, Conn.: Yale University Press, 2006), p. 237.

8. Stéphane Audoin-Rouzeau and Annette Becker, *1914–1918: Understanding the Great War* (London: Profile Books, 2002), p. 1.

9. Churchill, *World Crisis, ii,* pp. 1091–92.

10. James Beck, "Foreword," in Brittain, *To Verdun from the Somme,* p. xv.

11. Jay Winter, *Sites of Memory, Sites of Mourning: The Great War in European Cultural History* (Cambridge: Cambridge University Press, 1995).

12. John Masefield, "Battle of the Somme," in *John Masefield's Great War,* p. 265.

13. Masefield to Constance Masefield, 23 May 1917, quoted in ibid., p. 21.

14. Chattaway to his mother, 29 September 1916, Cheshire Regiment Archive, Chester, 0017.01/45.

15. Audoin-Rouzeau and Becker, op. cit., pp. 9 and 92–93.

16. Winter, *Remembering War,* pp. 1–13.

17. McInnis diary, 20–22 April 1917.

18. Martin Gilbert, *Winston S. Churchill, vol. IV: 1916–1922* (London: William Heinemann, 1975), p. 47.

19. Schwinghammer diary.

20. Brown, *Somme,* p. 297.

21. Jay Winter and Antoine Prost, *The Great War in History: Debates and Controversies, 1914 to the Present* (Cambridge: Cambridge University Press, 2004), pp. 6–33.

22. Bond, *The First World War and British Military History,* passim.

23. Beck, "Foreword," in Brittain, *To Verdun from the Somme,* p. xvi.

24. "Preliminary," in Blunden, *Undertones of War,* p. viii.

25. Siegfried Sassoon, *Siegfried's Journey, 1916–1920* (London: Faber & Faber, 1982 [originally published 1945]), p. 224.

26. "Robert Graves and Goodbye to All That," in Bond, *Survivors of a Kind,* pp. 1–11.

27. Edmund Blunden, "The Somme Still Flows," 10 July 1929, reprinted in *The Listener,* 19 January 1989.

28. "The Opening of the Wearing-out Battle," Boraston, p. 53.

29. Most significantly Haig's authorised memoir, Dewar and Boraston, *Sir Douglas Haig's Command,* and Field Marshal Sir William Robertson, *Soldiers and Statesmen, 1914–1918* (London: Cassell, two vols., 1926).

30. Carter Malkasian, *A History of Modern Wars of Attrition* (Westport, Conn.: Praeger, 2002), p. 36.

31. Sheldon, p. 398.

32. Repington, *First World War, i,* pp. 256–58.

33. Churchill, House of Commons, 23 May 1916, quoted in Gilbert, *Churchill, iii,* p. 774.

34. Gilbert, op. cit., *iii,* p. 771.

35. Esher to Haig, 30 May 1917, quoted in Gilbert, op. cit., *iv,* pp. 64–65.

36. Sydenham, *The World Crisis: A Criticism,* p. 5.

37. Churchill, *The World Crisis, ii,* p. 950.

38. Ibid., pp. 965–66.

39. Ibid., p. 968.

40. Quoted in Passingham, *All the Kaiser's Men,* p. 124.

41. Churchill, op. cit., ii, pp. 966–67.

42. Ibid., p. 955.

43. Memorandum to the War Cabinet, 21 October 1917, in ibid., pp. 1178–84.

44. Gilbert, op. cit., *iv,* pp. 64–65.

45. Churchill, op. cit., *ii,* pp. 970–71.

46. *OH: 1916, ii,* p. xvi.

47. AFGG, IV/3, pp. 522–23.

48. *OH: 1916, ii,* p. 553.

49. Churchill, op. cit., *ii,* pp. 963–64.

50. Prior, *World Crisis as History,* pp. 221–23.

51. Sir Charles Oman, "The German Losses on the Somme, July–December 1916," in Sydenham, op. cit., pp. 40–65: 52–53.

52. M. J. Williams, "Thirty Per Cent: A Study in Casualty Statistics" and "The Treatment of German Losses on the Somme in the British Official History," *Journal of the Royal United Service Institution,* CIX (1964), pp. 51–55, and CXI (1966), pp. 69–74.

53. James McRandle and James Quirk, "The Blood Test Revisited: A New Look at German Casualty Counts in World War I," *The Journal of Military History,* 70 (2006), pp. 667–702: 692.

54. Kennedy, "Fighting Numbers Engaged in European War with Losses to Date," 11 December 1916, M1024A/8690–510.

55. *OH: 1916, ii*, p. xv, n. 4.

56. Cited in Williams, "Treatment of German Losses," p. 70.

57. Ousby, *Road to Verdun*, p. 5.

58. *OH:1916, ii*, p. xv and "addenda and corrigenda"; Williams, "Treatment of German Losses," p. 73, suggests this is actually Churchill's *Reichsarchiv* figure which applied to the whole Western Front.

59. Jünger, *Storm of Steel*, pp. 105 and 116–17.

60. Papen, *Memoirs*, p. 67.

61. Prior, op. cit., p. 229.

62. Cited in Alexander Watson, *Enduring the Great War: Combat, Morale and Collapse in the German and British Armies, 1914–1918* (Cambridge: Cambridge University Press, 2008), p. 184.

63. Churchill, op. cit., *ii*, p. 1081.

64. Suttie, *Rewriting the First World War*, p. 6.

65. Ibid., p. 4.

66. HD, 15 January 1917.

67. *OH: 1916, ii*, pp. 551–52.

68. Hankey to Robertson, 9 November 1916, HNKY 4/8.

69. Churchill, op. cit., *ii*, p. 1090.

70. Millman, *Pessimism and British War Policy, passim*.

71. Quoted in Seely, *Adventure*, pp. 150–51.

72. Brian Gardner, *The Big Push* (London: Cassell, 1961), p. 148.

73. Haig to the King, 18 October 1916, in Sheffield and Bourne, *War Diaries*, p. 243.

74. Dewar and Boraston, op. cit.

75. FD, 26 April 1917.

76. Churchill, op. cit., *ii*, p. 996.

77. Note by Foch (ca. end of November 1916), AAT, IK129.

78. Hankey, *Supreme Command, ii*, pp. 511–16.

79. Winston S. Churchill, *The Second World War, vol. IV: The Hinge of Fate* (London: Cassell, 1951), pp. 539–41.

80. Quoted in Bond, *The Unquiet Western Front*, p. 59.

81. Alan J. P. Taylor, *A Personal History* (London: Hamish Hamilton, 1983), p. 243.

82. Taylor, *The First World War*, pp. 84–85.

83. Dixon, *On the Psychology of Military Incompetence* [originally published 1976], pp. 355–69.

84. Audoin-Rouzeau and Becker, *Understanding the Great War*, p. 2.

85. Sassoon, *Siegfried's Journey*, pp. 55–57; Egremont, *Siegfried Sassoon*, pp. 144–45.

86. Leonard Smith, *The Embattled Self: French Soldiers' Testimony of the Great War* (London: Cornell University Press, 2007), *passim*.

87. Winter, *Remembering War*, p. 222.

88. John Keegan, *The First World War* (London: Hutchinson, 1998); Martin Gilbert, *The Somme: The Heroism and Horror of War* (London: John Murray, 2006).

89. Audoin-Rouzeau and Becker, op. cit., pp. 3–8.

90. Western Front Association, <http://www.westernfrontassociation.com/>.

91. Audoin-Rouzeau and Becker, op. cit., p. 11; Winter, *Remembering War*, pp. 222–37.

92. Winter, op. cit., p. 224.

93. Sheffield, *Forgotten Victory*.

94. Audoin-Rouzeau and Becker, op. cit., pp. 10–11.

95. Winter, op. cit., p. 232.

96. Ibid., p. 224.

97. Historial de la Grande Guerre, Péronne, *The Battle of the Somme: A World Arena* (La Bataille de la Somme: Un Espace mondial) (Paris: Somogy éditions d'art, 2006).

98. Geoff Dyer, *The Missing of the Somme* (London: Phoenix Press, 2001 [originally published 1994]), p. 6.

Reflection

1. <http://news.bbc.co.uk/1/hi/uk/5135878.stm> (accessed 8 August 2008).

2. Monteith, "Observations of No. 5886," *ii*, p. 24.

3. Ibid., *i*, p. 21, and *ii*, p. 22.

4. John Harris, *The Somme: Death of a Generation* (London: Zenith Books, 1966), p. 127.

5. Michael Howard, "Introduction," in *A Part of History: Aspects of the British Experience in the First World War* (London: Continuum, 2008), p. xiv.

6. Haig to Esher, 1 July 1916, ESHR 4/6.

7. Bob Bushaway, " 'There Is Still the River': The Somme Ninety Years On," *Journal of the Centre for First World War Studies*, 3 (2007), pp. 97–101, <http://www.jsww1.bham.ac.uk/fetch.asp?article=issue6_bushaway2.pdf> (accessed 28 January 2008).

8. Cited in "Post-war Recollections of Haig as Commander-in-Chief," n.d., ESHR 19/5.

9. Courtès, "Souvenirs," p. 338.

10. Bean, *AIF in France, 1916*, p. 943.

11. Ibid., p. 946.

12. Churchill, *World Crisis, ii*, p. 1091.

Notes on Sources and Further Reading

The literature, published and unpublished, on the Battle of the Somme and the war of which it was a part, is vast. Eighty years later Fred Van Hartesveldt's *The Battle of the Somme, 1916: Historiography and Annotated Bibliography* (London: Greenwood Press, 1996) listed 704 works, predominantly in the English language. By the time of the ninetieth anniversary commemorations in 2006 hundreds more volumes on the Somme or other aspects of the First World War had appeared. This book cannot make a claim to have drawn on even a fraction of the sources which might be used for the history of the 1916 Somme offensive, or the operations that preceded and followed it, but its range of sources and international perspective is wider than that of any work on the battle to date. Reference to the endnotes suggests the broad range of material used and these notes merely outline key works on the Somme.

British and Australian Sources

Extensive archive sources on military operations and political decisionmaking exist in the British National Archives at Kew. These have been used extensively before in English-language accounts, and for this study only limited use has been made of G.H.Q. papers and Cabinet records. A key source for military operations remains the multi-volume British official history, *Military Operations, France and Belgium* (London: HMSO, fourteen vols., 1925–48). The two volumes on the 1916 Somme battle, authored by Brigadier-General Sir James Edmonds (1932) and Captain Wilfrid Miles (1938), should be a first resource for all interested in the development of field operations. These detailed chronologies underpin everything that has been written afterwards.

Both Sir Douglas Haig and Sir Henry Rawlinson kept diaries of their commands. The former, with extensive interpolated documents, is deposited in the National Library of Scotland, Edinburgh. Extracts have been published in two editions, Robert Blake's *The Private Papers of Douglas Haig, 1914–1919* (London: Eyre & Spottiswoode, 1952) and Gary Sheffield and John Bourne's *Douglas Haig: War Diaries and Letters, 1914–1918* (London: Weidenfeld & Nicolson, 2005). Rawlinson's unpublished diary and accompanying papers are deposited in the Churchill Archives Centre, Cambridge University. Further contemporary sources of note are *The Military Correspondence of Field Marshal Sir William Robertson*, edited by David Woodward for the Army Records Society (London: The Bodley Head, 1989), and intelligence chief John

Charteris's published "diaries" (in fact based on letters written to his wife at the time), *At G.H.Q.* (London: Cassell & Co., 1931).

Secondary studies of the Somme offensive are numerous and take two forms. First, there are studies of the military operations. The most important, chronologically, are Anthony Farrar-Hockley's *The Somme* (London: Pan Books, 1983 [originally published in 1964]), Gary Sheffield's *The Somme* (London: Cassell, 2003) and Robin Prior and Trevor Wilson's *The Somme* (New Haven, Conn.: Yale University Press, 2005). Prior and Wilson's study of Rawlinson's professional development, *Command on the Western Front: The Military Career of Sir Henry Rawlinson, 1914–18* (Oxford: Blackwell, 1992), should receive an honourable mention alongside such studies. Second, there are studies of the experience of battle. Martin Middlebrook's *The First Day of the Somme* (London: Penguin Classics, 2001) remains a classic of military history. Lyn Macdonald's *Somme* (London: Michael Joseph, 1983) covers the whole battle. Malcolm Brown's *The Imperial War Museum Book of the Somme* (London: Pan Books, 2002), and *The Imperial War Museum Book of the Western Front* (London: Pan Books, 2001), both draw extensively on that museum's collection of unpublished first-hand accounts, as does Peter Hart's more recent *The Somme* (London: Weidenfeld & Nicolson, 2005).

Bill Gammage has done the same for Australian soldiers on the Western Front in *The Broken Years: Australian Soldiers in the Great War* (Harmondsworth: Penguin Books, 1975), supplementing Charles Bean's official history of Australia in the war, volumes three to six of which, *The Australian Imperial Force in France* (Sydney: Angus & Robertson, 1929–42), cover the fighting on the Western Front. Extensive unpublished personal accounts and official papers are deposited at the Australian War Memorial, Canberra, ACT.

French Sources

Official and private documents on the battle exist in the French army's *Archives de l'armée de terre,* in Vincennes, on the outskirts of Paris. This work has drawn on only a fraction of the huge amount of material held there. It supplements the *Service historique de l'armée*'s unwieldy official history of the First World War, *Les Armées françaises dans la grande guerre* (Paris: Imprimerie nationale, eleven tomes in one hundred and three vols., 1922–37). The more concise history written between the wars by the official historian Général Bathélemy Palat, *La Grande guerre sur le front occidental* (Paris: Berger-Levrault, fourteen vols., 1917–29), gives a clear narrative. Volume nine (1925) covers the 1916 Somme offensive. Two studies of the offensive have appeared in French in recent years, Pierre Miquel's *Les Oubliés de la Somme: juillet–novembre 1916* (Paris: Éditions Tallandier, 2001) and Alain Denizot's *La Bataille de la Somme, juillet–novembre 1916* (Paris: Perrin, 2002), but both draw heavily on English-language sources and present a limited French narrative. Robert Doughty's *Pyrrhic Victory: French Strategy and Operations in the Great War* (Cambridge, Mass.: Belknap Press, 2005) provides a thorough survey of French high politics and strategy between 1914 and 1918.

German Sources

The destruction of the Prussian army's First World War archives in an allied Second World War bombing raid limits the existing German primary source material. The best surviving collection is that in the *Bayerisches Kriegsarchiv,* the Bavarian army's archive in Munich. This was drawn on extensively in the key English-language text for following the German side of the 1916 battle, Jack Sheldon's *The German Army on the Somme, 1914–1916* (Barnsley, U.K.: Pen & Sword, 2005). The *Reichsarchiv* published two official histories between the wars: the fourteen-volume *Der Weltkrieg, 1914–1918,* of which volume 10, *Die Operationen des Jahres 1916* (Berlin: E. S. Mittler & Sohn, 1936), covers the Somme battle; and the unfinished multi-volume series of battle accounts, two of which, *Somme Nord* (parts 1 and 2, Berlin: Oldenburg, 1927), cover the July operations north of the river in dense, tactical detail. Information in these histories can be supplemented from the many detailed German regimental histories published between the wars. Enough material exists in translation to form a clear picture of the German high command's conduct of the Somme battles, and their soldiers' experience. Robert Foley's *German Strategy and the Path to Verdun: Erich von Falkenhayn and the Development of Attrition, 1870–1916* (Cambridge: Cambridge University Press, 2005) is the most thorough study of the battle's preliminaries from the German side, while Holger Herwig's *The First World War: Germany and Austria-Hungary, 1914–1918* (London: Arnold, 1997) relates military, political and social developments in the Central Powers' war effort. Like the allied principals, most of the German senior commanders published war memoirs, some of which are available in English translation: General Erich von Falkenhayn, *General Headquarters, 1914–1916 and Its Critical Decisions* (London: Hutchinson, n.d.); General Erich Ludendorff, *My War Memories, 1914–1918* (London: Hutchinson, two vols., n.d.); Field Marshal Paul von Hindenburg, *Out of My Life* (London: Cassell, 1920).

For the Battlefield Visitor

Numerous guides to the battlefields of the Western Front have been published in recent years. Now in its twelfth edition, Rose Coombs's *Before Endeavours Fade* (Old Harlow: Battle of Britain International, 2006) was the first and is still valuable. For the Somme battlefields, Martin and Mary Middlebrook's *The Middlebrook Guide to the Somme Battlefields* (Barnsley, U.K.: Pen & Sword, 2007) covers battles from Crécy to the Second World War, and Tonie and Valmai Holt's *Major and Mrs. Holt's Battlefield Guide to the Somme* (London: Leo Cooper, 1996) focuses on the 1916 battle. Gerald Gliddon's gazetteer, *Somme, 1916: A Battlefield Companion* (Stroud: Sutton Publishing, 2006) gives historical details for all the key locations in the British sector of the 1916 battlefield. The volumes in Pen & Sword's *Battleground Europe* series (general editor Nigel Cave) are fuller historical guides to locations on the Somme. These are essentially Anglo-centric volumes. Jean-Pascal Soudagne, *Guide de la Somme, 1914–1918: lieux de combats et de mémoire* (St. Cloud: 14–18 Éditions [2006]), also includes some French sites, as does the Circuit of Remembrance self-guided tour established in 2006 by the *Comité du Tourisme de la Somme.* Visits to the Thiepval Visitor Centre and the Historial de la Grande Guerre in Péronne will orientate battlefield visitors to the 1916 battle and the First World War context.

Index